T0253635

Ueberhuber · Numerical Compu

Springer

Berlin
Heidelberg
New York
Barcelona
Budapest
Hong Kong
London
Milan
Paris
Santa Clara
Singapore
Tokyo

Christoph W. Ueberhuber

NUMERICAL COMPUTATION 1

Methods, Software, and Analysis

With 157 Figures

 Springer

Christoph W. Ueberhuber

Technical University of Vienna
Wiedner Hauptstrasse 8-10/115
A-1040 Vienna
Austria

e-mail: cwu@uranus.tuwien.ac.at

Title of the German original edition: *Computer-Numerik 1.* Published by Springer, Berlin Heidelberg 1995.

ROBERT LETTNER created the picture on the cover of this book. He is the person in charge of the central printing and reprography office at the Academy of Applied Arts in Vienna, Austria. In the past few years Robert Lettner, inspired by advancements in digital image processing, has developed new forms of aesthetics.
Sequences of pictures recently created by Robert Lettner include
Das Spiel vom Kommen und Gehen (Coming and Going),
Dubliner Thesen zur informellen Geometrie (Dublin Theses on Informal Geometry),
Bilder zur magischen Geometrie (Magical Geometry).
The illustration on the cover of this book was taken from the sequence
Das Spiel vom Kommen und Gehen.

Mathematics Subject Classification: 65-00, 65-01, 65-04, 65Dxx, 65Fxx, 65Hxx, 65Y10, 65Y15, 65Y20

Library of Congress Cataloging-in-Publication Data

Ueberhuber, Christoph W. 1946-
[Computer Numerik. English] Numerical computation : methods, software, and analysis /
Christoph W. Ueberhuber.   p.   cm.
Includes bibliographical references and index.
ISBN 3-540-62058-3 (v. 1 : soft : acid-free paper). - -
ISBN 3-540-62057-5 (v. 2 : soft : acid-free paper)
1. Numerical analysis - - Data processing. I. Title.
QA297.U2413 1997   96-46772
519.4' 0285'53 - -dc21   CIP

ISBN 3-540-62058-3 Springer-Verlag Berlin Heidelberg New York

© Springer-Verlag Berlin Heidelberg 1997
Printed in Germany

Typesetting: By the author using LaTeX.
SPIN 10543814   41/3143 - 5 4 3 2 1 0 - Printed on acid-free paper

Preface

This book deals with various aspects of scientific numerical computing. No attempt was made to be complete or encyclopedic. The successful solution of a numerical problem has many facets and consequently involves different fields of computer science. Computer numerics—as opposed to computer algebra—is thus based on applied mathematics, numerical analysis and numerical computation as well as on certain areas of computer science such as computer architecture and operating systems.

Applied Mathematics	
Numerical Analysis	Analysis, Algebra
Numerical Computation	Symbolic Computation

Operating Systems
Computer Hardware

Each chapter begins with sample situations taken from specific fields of application. Abstract and general formulations of mathematical problems are then presented. Following this abstract level, a general discussion about principles and methods for the numerical solution of mathematical problems is presented. Relevant algorithms are developed and their efficiency and the accuracy of their results is assessed. It is then explained as to how they can be obtained in the form of numerical software. The reader is presented with various ways of applying the general methods and principles to particular classes of problems and approaches to extracting practically useful solutions with appropriately chosen numerical software are developed. Potential difficulties and obstacles are examined, and ways of avoiding them are discussed.

The volume and diversity of all the available numerical software is tremendous. When confronted with all the available software, it is important for the user to have a broad knowledge of numerical software in general and to know where this software can be optimally applied in order for him to be able to choose the most appropriate software for a specific problem. Assistance in this complicated matter is offered in this book.

This book gives a comprehensive survey of the high quality software products available and the methods and algorithms they are based on. Positive properties and inherent weaknesses of specific software products are pointed out. Special subsections in this book, devoted to software, provide relevant information about commercially available software libraries (IMSL, NAG, etc.) as well as public domain software (such as the NETLIB programs) which can be downloaded from the Internet.

This book addresses people interested in the numerical solution of mathematical problems, who wish to make a good selection from the wealth of available software products, and who wish to utilize the functionality and efficiency of modern numerical software. These people may be students, scientists or engineers. Accordingly, this monograph may be used either as a textbook or as a reference book.

The German version *Computer-Numerik* was published in 1995 by Springer-Verlag, Heidelberg. The English version, however, is not merely a translation; the book has also been revised and updated.

Synopsis

Volume I starts with a short introduction into scientific model building, which is the foundation of all numerical methods which rely on finite models that replace infinite mathematical objects and techniques. This unavoidable finitization (found in floating-point numbers, truncation, discretization, etc.) is introduced in Chapter 2 and implications are discussed briefly.

The peak performance of modern computer hardware has increased remarkably in recent years and continues to increase at a constant rate of nearly 100 % every year. On the other hand, there is a steadily increasing gap between the potential performance and the empirical performance values of contemporary computer systems. Reasons for this development are examined and remedial measures are presented in Chapter 3.

Chapter 4 is dedicated to the objects of all numerical methods—numerical data—and to the operations they are used in. The main emphasis of this chapter is on standardized floating-point systems as found on most computers used for numerical data processing. It is explained as to how portable programs can be written to adapt themselves to the features of a specific number system.

Chapter 5 deals with the foundations of algorithm theory, in so far as this knowledge is important for numerical methods. Floating-point operations and arithmetic algorithms, which are the basic elements of all other numerical algorithms, are dealt with extensively.

Chapter 6 presents the most important quality attributes of numerical software. Particular attention is paid to techniques which provide for the efficient utilization of modern computer systems when solving large-scale numerical problems.

Chapter 7 gives an overview of readily available commercial or public domain software products. Numerical programs (published in TOMS or in other journals or books), program packages (LAPACK, QUADPACK etc.), and software libraries are dealt with. Particular emphasis is placed on software available on the Internet (NETLIB, ELIB etc.).

Chapter 8 deals with modeling by approximation, a technique important in many fields of numerical data processing. The applications of these techniques range from data analysis to the modeling processes that take place in numerical programs (such as the process found in numerical integration programs in which the integrand function is replaced by a piecewise polynomial).

The most effective approach to obtaining model functions which approximate given data is interpolation. Chapter 9 gives the theoretical background necessary to understanding particular interpolation methods and programs. In addition, the algorithmic processing of polynomials and various spline functions is presented.

Volume II begins with Chapter 10 which contains best approximation techniques for linear and nonlinear data fitting applications.

Chapter 11 is dedicated to a very important application of approximation. the Fourier transform. In particular, the fast Fourier transform (FFT) and its implementations are dealt with.

The subject of Chapter 12 is numerical integration. Software products which solve univariate integration problems are systematically presented. Topical algorithms (such as lattice rules) are covered in this chapter to enable the user to produce tailor-made integration software for solving multivariate integration problems (which is very useful since there is a paucity of ready made software for these problems).

The solution of systems of linear equations is undoubtedly the most important field of computer numerics. Accordingly, it is the field in which the greatest number of software products is available. However, this book deals primarily with LAPACK programs. Chapter 13 answers many questions relevant to users: How should algorithms and software products appropriate for a given problem be chosen? What properties of the system matrix influence the choice of programs? How can it be ascertained whether or not a program has produced an adequate solution? What has to be done if a program does not produce a useful result?

Chapter 14 deals with nonlinear algebraic equations. The properties of these equations may differ greatly, which makes it difficult to solve them with black box software. Chapter 15 is devoted to a very special type of nonlinear equations: algebraic eigenproblems. There is a multitude of numerical software available for these problems. Again, this book deals primarily with LAPACK programs.

Chapter 16 takes a closer look at the topics found in previous chapters, especially problems with large and sparse matrices. Problems of this type are not covered using black box software; a solution method has to be selected individually for each of them. Accordingly, this chapter gives hints on how to select appropriate algorithms and programs.

Monte Carlo methods, which are important for solving numerical problems as well as for empirical sensitivity studies, are based on random numbers. The last chapter of the book, therefore, gives a short introduction into the world of random numbers and random number generators.

Acknowledgments

Many people have helped me in the process of writing this book. I want to thank all of them cordially.

Arnold Krommer, with whom I have worked closely for many years, has had a very strong influence on the greater part of this book. Arnold's expertise and knowledge has particularly enhanced the chapter on numerical integration, a field of interest to both of us.

Roman Augustyn and Wilfried Gansterer have contributed significantly to the chapter on computer hardware.

Winfried Auzinger and Hans Stetter reviewed parts of early manuscripts. Their comments have helped to shape the final version.

Many students at the Technical University of Vienna have been of great assistance. Their suggestions and their help with the proofreading contributed in turning my lecture notes on numerical data processing and a subsequent rough draft into a manuscript. In particular the contributions of Bernhard Bodenstorfer, Ernst Haunschmid, Michael Karg and Robert Matzinger have improved and enriched the content of this book. I wish to express my thanks to all of them.

Wolfgang Moser contributed considerably to the process of translating the German version of this book. He organized the technical assistance of many students who produced a rough draft of the English version. He also translated several chapters himself.

Yolanda Goss was a competent and helpful collaborator in linguistic matters. She brought the rough draft of the English manuscript a giant step nearer to the final version. We even managed to squeeze some fun out of the dry business of improving the grammar, style and form of the draft version.

Robert Lynch (Purdue University) has made a great many valuable suggestions for improving the comprehensibility and the formulation of several chapters.

Robert Bursill (University of Sheffield) and Ronnie Sircar (Stanford University) have proofread and revised the manuscript competently.

Jeremy Du Croz (NAG Ltd.) and Richard Hanson (Visual Numerics Inc.) have checked the software sections and made numerous suggestions for improvements.

Christoph Schmid and Thomas Wihan deserve particular appreciation for the appealing text lay out and the design of the diagrams and illustrations. Christoph Schmid created the final LaTeX printout with remarkable dedication.

In addition, I would like to acknowledge the financial support of the Austrian Science Foundation.

January 1997   Christoph W. Ueberhuber

Contents

Chapter 1

Scientific Modeling

Will man Schweres bewältigen,
muß man es sich leicht machen.[1]

BERTOLT BRECHT

The term *model* is a commonly used term with meanings which vary greatly from situation to situation. In science and mathematics, where models and model based investigation play an important role, a model is defined as follows (Ören [304]):

*A **model** is an artificial object which reflects and reproduces essential features, relationships (the structure), and functions of a concrete object or phenomenon (reality) in a simplified way and can therefore be used as a tool for examining and analyzing reality.*

The interrelationship among reality, the model, and the *model subject* (i.e., the person who develops or uses the model) characterizes the concept of modeling:

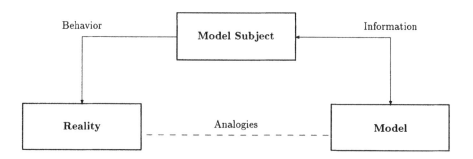

1.1 Reality Versus Model

Reality and the model must be similar with respect to certain features and functions.

The fact that a model is obtained by simplifying something is the reason it is easier to work with than the original object or phenomenon: The model is a

---

[1]If you want to achieve something that is difficult, you must first make it easy.

reduction (an *abstraction*) of reality. The purpose of the model must be taken into account when choosing the features to be represented (see Section 1.2). Models are thus reductions of the original objects to simpler ones. At the same time they frequently have features which do not correspond to anything in the original object.

Example (Neural Networks)   Neural networks—a class of nonlinear approximation functions—are coarse-grained models of the transmission and processing of stimuli in relatively small ensembles of neurons (nerve cells) in the human brain (Arbib, Robinson [89]). In these models many properties of human nerve cells are *ignored*, for example, the regeneration of neurons, the stochastic variation of the transmission speed of nerve fibers, or the temporal changes in the structure of the human nervous system. On the other hand, there are certain features in the model which do not correspond to any property found in the original, in this case the human nervous system, for example, the *infinitely fast* transmission and processing of stimuli.

This makes it clear that there is usually more than one model of a specific phenomenon. In fact there is often an infinite number of *partial models* which are more or less well suited to the analysis and description of particular features of the original object or phenomenon.

There is a distinction between structural models and functional models.

Structural Models

With structural models there is an analogy between the original object and the model with respect to the relationship between the elements of the system. However, the concrete realization of the elements of the model does not necessarily correspond with the original object.

Example (Neural Networks)   Neural networks, the mathematical models of small ensembles of human nerve cells, are roughly similar to the original with respect to structure as well as with respect to the way in which they transmit information. Neural networks in the form of computer programs, however, have nothing in common with the underlying biological system, the human brain.

Functional Models

Functional models represent the behavior of a system on the basis of input-output relations. As a result the model ignores the inner structure of the original object or phenomenon and aims only to reproduce functional relationships. The original system is regarded as a *black box*.

Example (Air Pollution Prediction)   The following simple time-series model can be used to predict the atmospheric concentration, y_{k+1}, of sulfur dioxide at time $k + 1$ in Vienna (Bolzern, Fronza, Runca, Ueberhuber [121]):

$$\hat{y}_{k+1} := 0.26y_k + 38900(\hat{T}_{k+1} + 23.1)^{-2} + 443(\hat{v}_{k+1} + 1.16)^{-1}. \tag{1.1}$$

The predicted SO_2 concentration $\hat{y}_{k+1}$ is modeled in terms of the measured SO_2 concentration y_k on the kth day, the predicted average temperature $\hat{T}_{k+1}$ (provided it is above $-20°C$) and the predicted average wind speed $\hat{v}_{k+1}$ of the following day. Equation (1.1) models the functional relationship between an increasing concentration of SO_2 and a decreasing outdoor temperature (increased domestic heating etc.) and decreasing SO_2 concentration for increasing wind speed (due to the intensified intermixing of pollutants and air).

1.2 The Model Subject and the Model

As a rule, models are used by people for particular reasons (the *property of finality*). The person who develops a model incorporates certain information about the original phenomenon into the model, and the user of that model[2] gains knowledge about the original phenomenon from the model. For both of them the model is the medium which conveys information and acts as a substitute for the original object or phenomenon.

Example (Storage of Information)   For the Austrian national standard ÖNORM M 9440 "Dispersion of pollutants in the atmosphere; determination of stack heights and computation of emission concentrations", experts have developed a model of the dispersion of gaseous pollutants in the atmosphere in the form of step-by-step computing directives. This model comprises information used to decide which exhaust control measures must be imposed, for example, in governmental and municipal licensing procedures.

Example (Acquisition of Information)   When designing new cars, crash tests which must be performed add significantly to development costs. Nowadays finite element programs (such as LS-Dyna 3D) can simulate crashes. This reduces, considerably, the number of crash tests, and thus the overall cost. Moreover, the safety properties of the cars are improved due to more comprehensive testing.

Simulations like this require considerable computing resources. On a Cray Y-MP4 of the Daimler-Benz Inc., it takes between 10 and 30 hours to simulate that split second (around 0.1 s) during which the crash takes place.

1.2.1 The Purpose of a Model

Models can be divided into three categories, according to what they are meant to do: acquire information, impart information, or be used in technical applications.

Information Acquisition

When certain information cannot be obtained from the original object or when it is too costly to obtain, a model can be used.

Example (Petri Nets)   A Petri net provides the means for modeling a process (in computer science a process is an action executed in the form of an algorithm). The static structure of a process is described by a directed graph which has two types of nodes: places and transitions. A place can be interpreted as a process state, a transition as a process action. The edges describe the possible transitions from one state to another (Peterson [313]).

Useful applications of Petri nets are, for example, deadlock detection and tests for the termination conditions of a process. *Program Resource Mapping* (PRM) nets are Petri nets with additional time components, which were developed in order to model and analyze the performance of parallel computer systems, either during the design phase of new systems or in order to assess existing systems (Ferscha [189]).

---

[2]The designer and the user of a model are often different persons.

Knowledge Dissemination

A model may be intended to teach and inform newcomers to a certain field about well established relationships.

Example (Visualization)   *Graphical models* can be used to visualize numerical data. For instance, every geographical map is a graphical model of geodetic-numerical data. Bar diagrams can be used to visualize economic data. For doctors, the graphical visualization of the numerical results of computed tomography (i.e., the results of the numerical inversion of the Radon transformation) is an indispensable diagnostic tool. Multicolor representations of the results of a simulation in the field of civil engineering enable, at a glance, a qualitative overview of stresses and strains and other phenomena of interest.

Example (Flight Simulator)   For the purpose of training and testing pilots and other crew members, a flight simulator imitates the process of flying an aircraft or spacecraft.

Technical Applications

Sometimes models can be used to replace original systems.

Example (Autopilot)   Based on models of aerodynamic properties, engine behavior and other properties of an aircraft, the autopilot, which is a real time computer system with special tailor-made programs, takes over the control of an aircraft according to specified parameters such as the heading and the altitude.

1.3 The Model Subject and Reality

Normally the user applies the knowledge and the information gained from the model to the original, i.e., the user gathers information from the model in order find out how to handle the original object or phenomenon.

However, *no* model analysis, experiments, or simulations can replace empirical experiments. Each assertion about the original object gained from the model has to be verified within the context of the original object.

Example (Weather Forecasting)   Any contemporary weather forecast is based on meteorological data and mathematical models of the atmosphere and is meant to predict, roughly, the weather, for example, in order to plan agricultural work. However, experience shows that the weather may be quite different from the forecast (the knowledge gained from the model).

Example (Medical Data Processing)   Various graphical-mathematical methods used in medical data processing, such as computed tomography (CT), magnetic resonance imaging (MRI), positron emission tomography (PET), digital subtraction angiography (DSA), etc. which are nowadays very important diagnostic tools, produce special graphical models for parts of the human body.

However, even when these methods are improved, they will still only be able to deliver *incomplete* information on the state and the course of events inside the human body. Other forms of diagnosis have to be used as well.

For example, in certain cases, a cerebral apoplexy cannot be distinguished from a cerebral tumor by using only a computed tomogram. A reliable diagnosis is possible only in connection with other examinations.

1.4 Model Building

To a certain extent the process of building and using models resembles the development process of computer software as expressed, for instance, in the *waterfall model* or the *spiral model* of software engineering.[3]

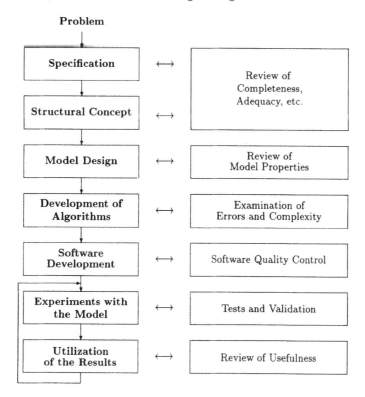

Figure 1.1: Evolutionary steps of model building and corresponding quality checks.

1.4.1 Problem Specification

The first step in building models is to give a clear statement of the problem. The purpose of the investigation is determined, the scope of the problem is defined and the required accuracy is fixed. In the course of the model building process, the problem is often modified and reformulated in a more precise way.

1.4.2 Creating a Structural Concept

With the help of a *structural concept (paradigm)*, the unstructured initial information is conceptually grouped and organized.

---

[3]In fact, many techniques frequently used in software engineering can be used (after slight modification perhaps) for building models.

Example (Transistor Simulation)   There are many ways to describe and investigate the motion of electrons in a semiconductor device. One way is to use a structural concept, which assumes that there is a continuum of electrons satisfying a system of partial differential equations (a numerical solution of which is used to obtain actual results). A second way is to use the concept of particle tracing, i.e., simulating the movement of all relevant electrons through an electric field in a crystal grid.

Example (Computer Models)   The structural concept used in examining the parts and the functions of a computer system and the models used to represent these can either be based on automata theory (e.g., Turing machines), graph theory (e.g., Petri nets), or queuing theory.

The creation of structural concepts also requires the choice of suitable modeling tools. There is an interrelationship between the choice of tools and the choice of models.

Representational Models

Representational models are physical images of objects and phenomena.

Example (Analog Computers)   General purpose analog computers use electric circuits as models of phenomena (e.g., mechanical vibrations). In mathematical terms the analog computer provides particular solutions of linear or nonlinear systems of differential equations.

Formalized Models

The relevant properties of an object can be described using a natural language (*verbal models*) or a formal language (*formalized models*). Mathematical models, like systems of algebraic equations, as well as the models in the form of computer programs which are dealt with in this book, are formalized models.

1.4.3 Choosing and Designing a Model

The methodological criteria as well as the cost must be taken into account when selecting and designing a model. The two most important criteria for selecting a particular model are suitability and simplicity.

Suitability

The qualitative and quantitative description of the object or phenomenon to be modeled must be sufficiently accurate with respect to the intended purpose of the model.

- The model must enable the correct *qualitative* description of a problem.

  Example (Circuit Design)   During a computer aided design process, the maximal possible clock rate of a particular electric circuit has to be determined. For such a task a *dynamic* (i.e., time dependent) model has to be used because otherwise the time dependent behavior of the electric circuit could not be investigated. A *steady state* model would only allow for the characterization of a static state of the circuit.

- The quantitative description of the object must meet pre-determined criteria which are part of the problem specification.

More complex problems tend to conceal the unsuitability of a model. The use of an inadequate model can lead to the neglect or distortion of existing properties of the object, such that the investigation concentrates on something which does not exist or is of no interest. Thus, checking and verifying the function of a model is crucial.

Simplicity

Out of two equally suitable models, the one which requires fewer assumptions and fewer resources is the better.[4]

More complex models are seemingly more *suitable* than simpler ones, as the use of more parameters makes it possible to take more factors into account. However, if there are too many degrees of freedom in the model, a misleading correspondence between the results of the model and the experiment can result. A close agreement between the model and the data is often erroneously taken as proof of the suitability of the model, which possibly does not reveal its unsuitability until it is used under different conditions.

Example (Forecasting) The time-series data ($\circ$) in Fig. 1.2 is described by a model which is intended for the short term prediction of the future development of the time-series.

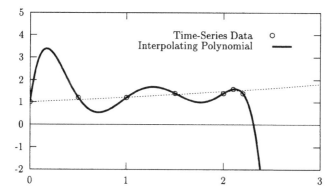

Figure 1.2: *Interpolation* (—) versus *regression* ($\cdots$) of time-series data.

If a polynomial $P \in \mathbb{P}_k$, which interpolates all data points $(x_0, y_0), (x_1, y_1), \ldots, (x_k, y_k)$, i. e.,

$$P(x_0) = y_0, \quad P(x_1) = y_1, \ldots, \quad P(x_k) = y_k,$$

is used (curve —), then completely absurd and worthless predictions are obtained. Instead, a regression function with fewer parameters, for instance, a polynomial whose degree is smaller than what was used and which does *not* perfectly agree with the values $y_0, y_1, \ldots y_k$ at the points $x_0, x_1, \ldots, x_k$, might be much more adequate for prediction purposes (curve $\cdots$).

---

[4]This *principle of minimality* was formulated in the Middle Ages by William Occam; it is also known as *Occam's razor*.

The principle of minimality can be facilitated using different numerical methods, e. g., a singular value analysis (see Chapter 13) or a factor analysis.[5]

1.4.4 Establishing Parameter Values

When deciding on a model type, the number and the types of parameters are usually established at the same time. In order to derive an actual model, *suitable values* of these parameters have to be determined. In this book it is always assumed that the data necessary for determining this are available. In this case parameter estimation usually involves solving systems of linear or nonlinear equations or optimization (minimization) problems.

1.4.5 Tests and Validation

Before a model can be used, its validity or its invalidity must be established. The techniques used in this context are called verification and falsification.

To achieve this the following questions must be answered: Does the behavior of the model correspond to a sufficient degree with the behavior of the corresponding system? Is the structure of the model too sketchy, does it have to be further refined? What is the behavior of the model near the limits of the range of its applicability?

The *verification* of a model usually follows one of the following strategies:

Indirect verification demonstrates that falsification is impossible, or at least that falsification fails in spite of serious efforts: *verification by the lack of falsification*. As with the testing of programs (which also is only an indirect verification), this is called *model testing*.

Statistical verification is based on statistical tests which examine (within the limits of a certain error probability) whether or not the residuals (the part of the data not covered by the model) are random.

A *sensitivity analysis* investigates how sensitive the model is to changes in the parameters or the structure.

The features and behavior of a *validated model* correspond with the features and behavior of the given system within the limits of the required accuracy bounds. The validated model is then used—e. g., in computer simulations—to find a solution of the original problem.

In practice, the steps outlined above for model building are often taken in a different order. These steps indicate a general scheme for the quite complex and interdependent process of building models.

---

[5] *Factor analysis* is a method which analyzes the relationship between independent random phenomena (features) by relating them to as small a number of causes (factors) as possible.

Chapter 2

Fundamental Principles of Numerical Methods

Ein Ideal der Genauigkeit ist nicht vorgesehen;
wir wissen nicht, was wir uns darunter vorstellen sollen –
es sei denn, du selbst setzt fest, was so genannt werden soll.[1]

LUDWIG WITTGENSTEIN

2.1 From Application Problems to their Numerical Solution

There are two different approaches to describing, analyzing, and controlling technical, scientific or economic objects and processes: experiment or simulation.

Experiment: Information is gathered by examining the objects or phenomena themselves.

Simulation: Experiments and examinations are performed by computer programs based on mathematical models. The models are derived from scientific knowledge (*theories*) which is used to explain certain phenomena and their underlying laws.

When deciding which strategy to use, both the *cost* and the *feasibility* of a project have to be taken into consideration. For example, experiments on real objects are obviously beyond consideration when determining the optimal design of an aircraft wing, finding out the breaking load of a bridge or determining the consequences of uncontrolled nuclear fusion. In such cases mathematical, scientific models, and computational solutions must form the basis for dealing with these issues.

Fixing the desired *level of accuracy* is an important aspect of problem specification. With experiments it is, among other things, the arrangement and the quality of the instruments used that determine the level of accuracy. In the case of simulation studies, the quality of the models and the calculations based on them are assessed with respect to the desired level of accuracy. In order to make such an assessment possible, all factors which influence the accuracy of the results must be known and quantified.

---

[1] There is no consensus about the meaning of accuracy because a general definition of accuracy is impossible; unless, you decide for yourself what you mean by it.

2.1.1 Case Study: Pendulum

In special cases the period of oscillation T of a pendulum with fixed pivot can be determined by conducting an experiment (by performing measurements on real pendulums). However, the inverse problem—of determining the required pendulum length l for a given period of oscillation—would require much more experimental effort. It is therefore more profitable to use a mathematical model instead of a real pendulum to determine the period of oscillation as a function of relevant factors.

The process of building a mathematical model for the pendulum is aimed at creating as simple a tool as possible whose numerical properties are well understood (at least by experts) and which is suitable for calculating approximate, yet sufficiently accurate, values for the period of oscillation. To do this the following questions have to be asked and answered:

1. What quantities are relevant to the motion of the pendulum?

2. How are these quantities to be combined into a model so that the motion is described quantitatively?

The First Specification of the Problem

In this example, the actual process to be investigated is the oscillation of a real pendulum. Due to dissipative forces (friction in the bearing, air resistance) oscillations will cease to be visible after a certain time. Hence, the notion of a "period of oscillation" has to be redefined more precisely. For the rest of this case study, "period" denotes the interval of time, T, between the moment that the pendulum is released to the moment it next returns to that point; i.e., the time required for the pendulum to make its first complete swing.

The accuracy requirement for this problem may, for example, be formulated as

$$|T_{\text{computed}} - T_{\text{actual}}| < \varepsilon, \tag{2.1}$$

where the tolerance ε depends on the specific application.

Mathematical Pendulum

The concept of a *mathematical pendulum* (see Fig. 2.1) is the result of the following idealizations:

1. The oscillation is assumed to be *undamped*: All dissipative forces (bearing friction and air resistance) are ignored.

2. The mass of the pendulum is assumed to be concentrated in a single point and is suspended by a weightless rod.

The motion of the mathematical pendulum is described by the solution, φ, of the second order nonlinear ordinary differential equation

$$\varphi''(t) = -\omega^2 \sin(\varphi(t)) \qquad \text{where} \quad \omega^2 = g/l \tag{2.2}$$

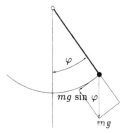

Figure 2.1: Mathematical pendulum.

with initial conditions

$$\varphi(0) = \varphi_0 \quad \text{and}$$
$$\varphi'(0) = 0, \quad \text{i.e., release of the pendulum at } t = 0.$$

The solution φ, a function of time, models the angular displacement of the pendulum from the rest position ($\varphi = 0$).

The *parameters* of the model are

1. the initial angular displacement φ_0 at $t = 0$,

2. the length l of the pendulum (the distance between the point mass and the axis), and

3. the gravitational constant g (with the standard value $g = 9.80665 \, \text{ms}^{-2}$ used for technical applications).

Linearization

When φ_0 is sufficiently small, the nonlinear differential equation (2.2) can be replaced by the simpler *linear* differential equation

$$\varphi''(t) = -\omega^2 \varphi(t) \tag{2.3}$$

since $\sin \varphi \approx \varphi$ for small φ. The advantage of the linear equation over the nonlinear equation (2.2) is that its general solution

$$\varphi(t) = A \cos \omega t + B \sin \omega t$$

is much simpler than the solution of the nonlinear differential equation. The constants A and B can be determined by solving a system of linear algebraic equations which follow from the initial conditions $\varphi(0) = \varphi_0$ and $\varphi'(0) = 0$. They turn out to be $A = \varphi_0$ and $B = 0$.

On the other hand, the solution of the nonlinear problem (2.2) can only be represented analytically with the aid of a particular class of special functions: the Jacobi elliptic functions.

Rigid Body Pendulum

Like the mathematical pendulum, the rigid body pendulum is assumed to oscillate in an undamped way for the sake of simplicity. However, the physical shape of the pendulum is fully taken into account. It is assumed that the pendulum body is rigid, i.e., the relative positions of its constituent parts are fixed.

The oscillation of a rigid body pendulum can be described by the same differential equations as the mathematical pendulum. Taking into account the moment of inertia, the *radius of gyration* l^* has to be used instead of l.

More Complex Models

Losses due to friction, which have been ignored up to this point, can be incorporated into the model (2.2) by adding a term F to model friction:

$$\varphi''(t) = -\omega^2 \sin(\varphi(t)) - F(\varphi'(t)). \qquad (2.4)$$

Its solutions model *damped* oscillations if F is positive. The actual form of the frictional term F can only be obtained as the result of a model building process which has to be based on experiments relevant to the given system.

If none of the models discussed so far is sufficiently accurate for a particular application, more refined and complex models can be constructed. For example, rather than using the simple frictional term in equation (2.4), models based on partial differential equations are needed to model the elastic deformation or the effect of aerodynamic forces on the pendulum.

Determining the Period of Oscillation

Once a suitable model for the oscillation is chosen, the given problem can be solved: the calculation of the period of oscillation T.

Since the general solution of the differential equation (2.3) is

$$\varphi(t) = \varphi_0 \cos \omega t,$$

the period of oscillation is

$$T_L = \frac{2\pi}{\omega} = 2\pi \sqrt{l/g}. \qquad (2.5)$$

Note that the expression for the period of oscillation T_L obtained in this way is independent of the initial angular displacement φ_0. The result (2.5) is accurate only for small initial values of angular displacement because $\sin \varphi$ was approximated with φ and friction was ignored. The maximum value of φ_0 for which the result (2.5) is still acceptable depends on the accuracy requirement (2.1).

The period which results from the solution of the nonlinear differential equation (2.2) can be expressed in terms of the *complete elliptic integral of the first kind*

$$K(k) = \int\limits_0^{\pi/2} \frac{1}{\sqrt{1 - k^2 \sin^2 t}} \, dt \qquad \text{where} \quad k = \sin(\varphi_0/2). \qquad (2.6)$$

The period of oscillation is

$$T = \frac{4}{\omega} K(k).$$

A solution of the original problem is thus derived by building a model and solving a mathematical problem (see Fig. 2.2).

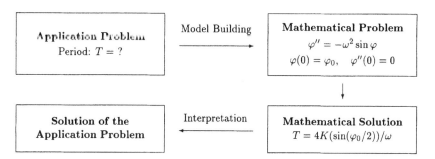

Figure 2.2: Solving a practical problem using a mathematical model.

If the value of the period, as modeled by the nonlinear differential equation, is required, then the integral (2.6) has to be calculated for given values of φ_0. However, elliptic integrals cannot be represented by elementary functions; they require particular computational methods. One such method was developed by C. F. Gauss (Hancock [218], Abramowitz, Stegun [1]). For given initial values a_0 and b_0, two sequences $\{a_i\}$ and $\{b_i\}$ are calculated whose $(i+1)$st elements are the geometric and the arithmetic means of their elements, respectively. The first sequence, defined by

$$
\begin{aligned}
a_0 &:= \sqrt{1 - k^2} \\
a_{i+1} &:= \sqrt{a_i b_i}; \qquad i = 0, 1, 2, \ldots,
\end{aligned}
\tag{2.7}
$$

increases monotonically; the second sequence, defined by

$$
\begin{aligned}
b_0 &:= 1 \\
b_{i+1} &:= (a_i + b_i)/2; \qquad i = 0, 1, 2, \ldots,
\end{aligned}
\tag{2.8}
$$

decreases monotonically. Both sequences converge to the same limit c. The value of the elliptic integral $K(k)$ is related to c by

$$K(k) = \frac{\pi}{2c}. \tag{2.9}$$

Numerical example: For $\varphi_0 = 4.4°$, $k = \sin(2.2°)$, the following numerical values are obtained:[2]

$$
\begin{array}{ll}
a_0 = 0.999\,262916\,41062 & b_0 = 1. \\
a_1 = 0.9996\,31390\,26874 & b_1 = 0.9996\,31458\,20531 \\
a_2 = 0.9996\,31424\,23703 & b_2 = 0.9996\,31424\,23703
\end{array}
$$

---

[2]The leading digits of the numbers in the table, that are identical to the corresponding digits of the result are written in *italics*.

and 0.9996 31424 23703 is the exact value of c rounded to 14 decimal digits. Hence, with the value

$$K(\sin(2.2\,\pi/180)) = 1.5713\,75497\,71788$$

of the elliptic integral and a pendulum length $l = 0.85\,\mathrm{m}$, the value $T = 1.8505$ seconds is obtained. Using the linear model for the same length the value $T_L = 1.8498$ is obtained, i.e.,

$$T_L - T \approx -7 \cdot 10^{-4}.$$

Therefore, with an accuracy requirement of one millisecond, the less accurate value T_L is acceptable as an approximate result, provided the effect of all other disturbing factors is also below the level of accuracy $\varepsilon = 10^{-3}\,\mathrm{s}$.

For a large initial angular displacement φ_0, the linear model cannot be used any longer. Even if the required accuracy allows for larger errors (see Fig. 2.3), the use of the nonlinear model (2.2) is often unavoidable.

If the frictional losses *cannot* be ignored, the required period of oscillation has to be determined by finding a numerical solution of a nonlinear differential equation, e.g.,

$$\varphi'' = -\omega^2 \varphi - Q\,\mathrm{sgn}(\varphi')|\varphi'|^2, \tag{2.10}$$

and the subsequent numerical determination of the length of the first period (see Fig. 2.4 and Fig. 2.5 for $l = 0.85\,\mathrm{m}$, $\varphi(0) = 4.4°$, $\varphi'(0) = 0°\mathrm{s}^{-1}$).

2.1.2 Qualitative and Quantitative Problems

The evaluation of accurate models frequently leads to mathematical problems which are *not* solvable in a simple manner. This means that the exclusive use of elementary operations (including differentiation and integration) is not sufficient for deriving a solution from the given information (the problem data). It may be possible to establish a *qualitative* relationship between the given data and the desired values, but this is rarely enough to solve *quantitative* problems.

Example (Transistor Simulation) The internal processes of a semiconductor device can be modeled by the solution of the system of partial differential equations (Selberherr [345]):

$$\varepsilon\Delta\Psi = q(n - p - C)$$
$$\nabla \cdot J_n - q\frac{\partial n}{\partial t} = qR$$
$$\nabla \cdot J_p + q\frac{\partial p}{\partial t} = -qR$$
$$J_n = q\mu_n(U_T\,\nabla n - n\,\nabla\Psi)$$
$$J_p = -q\mu_p(U_T\,\nabla p + p\,\nabla\Psi).$$

Qualitative statements which describe the electrostatic potential Ψ, the vectorial electron and hole-current densities J_n and J_p as well as other characteristics of the transistor can be derived from this system of equations.

However, if specific properties, for example, of an MOS transistor for a new VLSI chip are to be predicted and possibly optimized, then numerical calculations are inevitable.

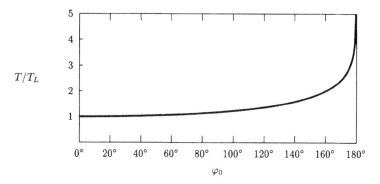

Figure 2.3: The ratio T/T_L of the period of oscillation T of the nonlinear model (2.2) and $T_L = 2\pi\sqrt{l/g}$ of the linear model (2.3) tends to infinity when φ_0 approaches $180°$.

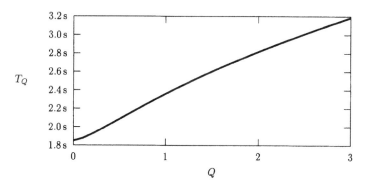

Figure 2.4: The period of oscillation T_Q for a damped oscillation modeled by (2.10) shows irregular behavior.

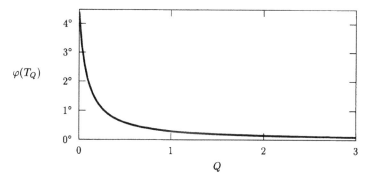

Figure 2.5: The angular displacement $\varphi(T_Q)$ of the pendulum after the first cycle of a damped oscillation modeled by (2.10) decreases as the frictional resistance is increased.

Mathematical models frequently take the form of equations (algebraic equations, differential equations and the like) and inequalities with known or unknown parameters. The mathematical problems related to them may be either qualitative or quantitative:

Qualitative problems occur, for instance, in examinations of the stability of solutions (determined by finding out if or how fast the effect of perturbations is damped), of their asymptotic behavior (determined by finding out if the solutions increase or decrease after a long period of time) and so on.

Quantitative problems occur when actual solution functions or values are determined. They also occur when values are determined for variables of state, constructional variables and control variables.

2.2 Numerical Problems

Mathematical problems which arise in a model based investigation often require the determination of the values of variables of state. The specification of constraints transforms such mathematical problems into *numerical problems*: Computational methods can be used to determine the desired values of all variables if mathematical relations, equations, inequalities, numerical data, etc., are given.

Numerical mathematics and *numerical data processing* deal with the design, the analysis and the implementation of methods for the computer aided solution of numerical problems. Numerical mathematics includes the development and analysis of mathematical tools and methods (see e. g., Deuflhard and Hohmann [44], Hämmerlin and Hoffman [52], Hamming [54], Isaacson, Keller [61], Schwarz [72], Stoer, Bulirsch [75]). Numerical data processing mainly concentrates on the design, the implementation, and the assessment of numerical software (see e. g., Cowell [10], [144], Mason and Cox [285], Patterson and Hennessy [68] or Rice [70], [71], [326]). However, there is no rigid separation between the two disciplines.

The following diagram schematically depicts the position of numerical mathematics and numerical data processing in a layer model indicating the scientific fields which relate the user and the computer:

User

Applied Mathematics	
Numerical Mathematics	Symbolic Mathematics
Numerical Data Processing	Symbolic Data Processing

System Software
Hardware

Restrictions on Numerical Problem Solving

For the non-symbolic numerical solution of mathematical problems, one must use the numbers available on the particular computer in use. These numbers constitute only a *finite* set of *rational* numbers; these are called *machine representable numbers*, or *floating-point numbers*. Any real number which does not happen to be a machine number itself has to be approximated by one of the machine representable numbers.

On a computer, the result of applying the arithmetic operations $+$, $-$, $\cdot$, and $/$ can only be approximated. This is due to the fact that the set of machine numbers is not closed with respect to these operations: Generally, the exact result of such an operation is not a machine number and has to be *approximated* by one.

The situation is the same for the evaluation of the elementary functions *sin*, *cos*, *exp*, *log*, and the like. Most programming languages developed for numerical applications provide means for evaluating such functions with built-in function procedures SIN, COS, EXP, LOG, and so on, i. e., mathematical software which is an integral part of the language implementation. These function procedures calculate a machine representable value of the exact result for a given value of the function's argument. At best this approximation is the machine number closest to the exact result, but under certain circumstances the machine number returned from the procedure may differ from the exact number by a wide margin. Nevertheless, the built-in function procedures are used in the same way as the corresponding elementary functions are used in calculus.

Example (Pendulum) Using a scientific programming language (here Fortran 90), the iterative computation of $K(\sin(\varphi_0/2))$ may look like

```
INTEGER  ::  i
REAL     ::  a, arithm, b, c, geom, k, phi_0, phi
...
degree_to_radian = ATAN(1.)/180.
phi = degree_to_radian * phi_0
a = COS(phi/2.)          ! a = SQRT(1. - SIN(phi/2.)**2)
b = 1.
DO i = 1, 5              ! Gaussian iteration:
   geom   = SQRT(a*b);   !     geometric mean
   arithm = (a + b)/2.   !     arithmetic mean
   a      = geom
   b      = arithm
END DO
c = (a + b)/2.
k = 2.*ATAN(1.)/c         ! complete elliptic integral of the first kind
...
```

With $\varphi_0 \in [0, 179.99°]$, this implementation of the Gaussian iteration (2.7) returns approximate values of the elliptic integral of the first kind with around 7 correct decimal digits.

The Need for Finiteness

In principle, many operations cannot be used in numerical methods as they have been defined in analysis: Due to the *finiteness* of the set of machine numbers there are neither arbitrarily small nor arbitrarily large numbers, nor arbitrarily close neighbors. Hence, the fundamental concepts of limit and continuity are not applicable in numerics.

Moreover, the execution of any computational operation takes time, so calculations on a computer can only comprise a *finite number of steps*. Hence, it is impossible, for example, to calculate the sum of an infinite series by calculating each term. A partial sum (the sum of a finite number of terms in the series) must suffice. This fundamental flaw cannot be overcome by using ultra fast CPUs or multiprocessor computers: on a computer most methods and techniques developed in analysis have to be replaced by *finite methods*.

The truncation of infinite series as well as the replacement of analytical operations (differentiation, integration etc.) with a finite number of evaluations of arithmetic expressions is referred to as *discretization*. As a result of this discretization, this restricted range of numbers, this inexact arithmetic and other reasons, there is usually no way to calculate the exact result of a mathematical problem on a computer. Only *approximations* of the desired result can be determined.

The quality, or accuracy, of a numerical approximation can often be enhanced by increasing the computational effort; for example, by executing more iterations, using more sophisticated methods or refining partitions. However, sooner or later, an insurmountable barrier is reached due to the limited accuracy of the computer and the limited time which can be spent on solving a problem.

Example (Pendulum) Consider again the Gaussian iteration (2.7), (2.8). If the initial value $b_0 = 1$ is used as an approximation of c, the approximation $K(0) = \pi/2 = 1.5707\,9632$ of $K(\sin(2.2°))$ results; the corresponding period of oscillation is $1.8498\,1796$ s, deviating from the exact result $1.8505\,00017\ldots$ s by about $-0.68 \cdot 10^{-3}$ s. For many technical applications an approximation of this accuracy is sufficient.

If the experimental procedure is to use a pendulum to determine the gravitational constant g with an accuracy of 10^{-6} ms^{-2} (provided the values of l and φ_0 are given with sufficient accuracy), then at least 6 decimal digits of K must be correct.

If, in the example above, ($l = 0.85$ m, $\varphi_0 = 4.4°$), b_1 is used as an approximation of c, then the error of the approximation of K is reduced to $\approx -5 \cdot 10^{-8}$, and the determined period of oscillation $T = 1.8504\,99954$ s matches the exact value to $-0.63 \cdot 10^{-9}$ s. This accuracy is more than sufficient for an experimental determination of the gravitational constant.

Additional iterations could increase the accuracy of the approximation of K. However, the machine arithmetic of the computer is in any case the ultimate restriction as to what accuracy can be attained.

The Separation of Numerical Methods from Specific Applications

As can be seen even from such a simple example as the pendulum, determining which numerical value to accept as a solution of the problem is a decision which depends on the particular application. Does that imply that the numerical solution of a problem cannot be isolated from a specific application?

Such a separation is indeed desirable—it is a prerequisite for the universal applicability of numerical methods and numerical software. In the case of the pendulum, for example, the numerical approximation with the *maximum number* of correct digits can easily be determined. However, this strategy only works with simple examples like the pendulum where the required surplus effort is not significant.

Things change as soon as a frictional term is added to the mathematical model of the pendulum, as in equation (2.4). Then in general the differential equation (2.4) has to be solved numerically and the distance between the zeros of that solution has to be determined. In this case the computational effort for calculating all results with maximum accuracy can no longer be neglected. Thus, the aim is to determine numerical solutions which meet the accuracy requirements of the user and which require as little computational effort as possible. A solution of optimal cost in this sense is only as accurate as necessary, but not as accurate as possible.

2.2.1 Numerical Problems

It is possible to make numerical methods independent of specific application problems by making the accuracy requirements part of the problem. In the case of the pendulum, one of the mathematical problems is to determine the period of the solutions of the differential equation (2.2). In this case the newly defined problem is to determine an *approximation* of the period which does not differ from the actual value by more than a specified *error bound (tolerance)*.

The combination of a

mathematical problem (of constructive nature)   *and*

accuracy requirements for the results

is called a *numerical problem* from this point on.

Example (Pendulum)   For the complete elliptic integral (2.6), a particular numerical problem could be formulated as follows:

Find an approximation $K_{\mathrm{num}}(k)$ of $K(k)$ which satisfies the inequality

$$|K_{\mathrm{num}}(k) - K(k)| \leq \varepsilon,$$

where $\varepsilon = 10^{-3}$, provided that $k \in [0, \sin 89.99°]$.

The smaller the chosen error bound ε, the more computational effort is needed to solve the numerical problem. For example, on most current computers an error bound $\varepsilon = 10^{-9}$ can only be achieved by using double precision arithmetic.

This example demonstrates an important difference between mathematical and numerical problems: However different the representations of the solution of a uniquely solvable mathematical problem may look, they are all equivalent in the mathematical sense. On the other hand, a multitude of different numerical problems can generally be set up in correspondence with one particular mathematical problem, and each of them possesses a multitude of solutions. There are many possible ways to obtain one of those numerical solutions, ranging from the development and implementation of self-devised algorithms to the use of commercial software products (see Fig. 2.7).

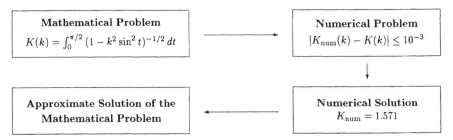

Figure 2.6: Numerical approach to solving a mathematical problem.

The Need for Error Estimates

The exact solution of a given problem is of course not known execpt for textbook examples, otherwise the effort of finding a numerical solution would be wasted. Since an accuracy requirement is part of any numerical problem, numerical methods must contain mechanisms for estimating the accuracy of approximate solutions. The resulting error estimates are used for control purposes and to inform the user about the accuracy of results.

Example (Pendulum) Due to the monotone convergence of the sequences $\{a_i\}$ and $\{b_i\}$,

$$\frac{\pi}{2a_i} \downarrow K(k) \quad \text{and} \quad \frac{\pi}{2b_i} \uparrow K(k),$$

a numerical method based on equations (2.7), (2.8), and (2.9) can use the difference

$$e_i := \frac{\pi}{2a_i} - \frac{\pi}{2b_i}$$

as an *error estimate* for both approximations:

$$\left|\frac{\pi}{2a_i} - K(k)\right| < e_i, \qquad \left|\frac{\pi}{2b_i} - K(k)\right| < e_i.$$

If a numerical integration program is used to determine the integral (2.6), the error tolerance has to be passed as a parameter to the integration program, which must contain mechanisms for error estimation and accuracy control.

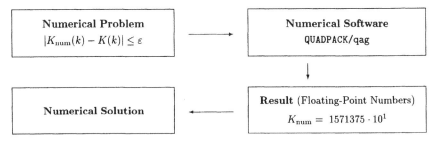

Figure 2.7: Solving a numerical problem using numerical software.

Methods without Accuracy Control

For some classes of problems, such as systems of linear equations, there are numerical methods which return the result of the mathematical problem after performing a finite number of steps. They do so, however, only if the computation is
exact[3]. Depending on the properties of such algorithms and the actual problem
data, the computer implementation of these algorithms may produce numerical
results with errors which are so large that the accuracy requirements of certain
numerical problems are no longer met. Unfortunately, in such cases—e.g., the
direct (*non*-iterative) solution of a system of linear equations—the accuracy requirements usually cannot be made a control parameter of the method in use.
This would require a program-driven adjustment of the machine arithmetic to the
needs of particular numerical problems, which is usually unacceptably inefficient
or completely impossible.[4]

An essential element of such methods is therefore the determination of certain
characteristic numbers of the problem (e. g., condition number estimates) or of the
solution (error estimates). They help the user to decide if the obtained numerical
result can be used for the particular application.

2.2.2 Categories of Numerical Problems

Most numerical problems can be assigned to one of the following categories:

The evaluation of a functional $l : \mathcal{F} \to \mathbb{R}$, e. g., the calculation of function
  values $f(x)$, derivatives $f'(x)$, $f''(x)$, ... (numerical differentiation), definite
  integrals $\int_a^b f(t)\, dt$ (numerical integration) and norms $\|f\|_p$, etc.

The solution of algebraic equations: the determination of unknown values
  in algebraic relationships by solving systems of linear or nonlinear equations.

The solution of analytic equations: the determination of functions (or function values) given operator equations such as ordinary or partial differential
  equations, integral equations, functional equations, etc.

The solution of optimization problems: the determination of particular numerical values or functions, subject to specified constraints, which optimize
  (maximize or minimize) a given objective function.

Constructive mathematical problems, which numerical problems originate from,
can be regarded as abstract mappings $F : X \to Y$ between two normed linear
spaces (see Chapter 8). Depending on which of the quantities y, x, or F is
unknown in the equation

$$Fx = y,$$

---

[3]If real numbers instead of machine numbers and exact arithmetic operations instead of
approximate numerical operations were used.

[4]The situation changes with *iterative* methods for solving systems of linear equations: In
this case it is reasonable to introduce the accuracy requirement as a control parameter of the
numerical method.

this is a direct problem, an inverse problem, or an identification problem:

	F	x	y
direct problem	given	given	desired
inverse problem	given	desired	given
identification problem	desired	given	given

There is no compelling criterion for classifying actual problems with respect to the framework represented in the chart. As to how a particular problem is formulated is often a matter of convenience.

Example (Integration)   The evaluation of a definite integral

$$Fx = \int_a^b x(t)\,dt$$

is a *direct problem*. In this problem, $F : \mathcal{F} \to \mathbb{R}$ is a functional which maps, for example, elements of the space $\mathcal{F} = C[a,b]$ of all continuous functions on $[a,b] \subset \mathbb{R}$ into $\mathbb{R}$. The integrand—in this case a function x defined on the interval $[a,b]$—is given, and the integral $y = Fx$ is to be determined.

Example (System of Linear Equations)   The solution of a system of linear equations

$$Ax = b, \qquad A \in \mathbb{R}^{n \times n}, \quad x, b \in \mathbb{R}^n$$

is a typical *inverse problem*. The linear mapping F is characterized by the n^2 coefficients of the matrix A. The vector b (the image) is given, and the vector x (which is mapped onto b) is to be determined.

Example (Analysis of a Mixture)   The composition of a mixture of evaporating substances is to be determined. Provided that a fixed, characteristic quantity of each substance evaporates per time unit, the following time-dependent model of the whole mixture can be used:

$$y(t) = \sum_{i=1}^m y_{i0} e^{-k_i t}. \tag{2.11}$$

In addition to the initial quantities $y_{i0}, i = 1, 2, \ldots, m$, and the evaporation rates $k_1, \ldots, k_m$ of the unknown substances, the number m of different substances is also often unknown. If at the times $t_1, \ldots, t_n$ the respective quantities $y_1, \ldots, y_n$ are measured, the parameters $y_{i0}, k_i,\ i = 1, 2, \ldots, m$ and m of the mapping

$$F : (t_1, \ldots, t_n) \to (y_1, \ldots, y_n) = \left(\sum_{i=1}^m y_{i0} e^{-k_i t_1}, \ldots, \sum_{i=1}^m y_{i0} e^{-k_i t_n} \right)$$

are to be determined; whence, this is an *identification problem*. For each *fixed* $m \in \{1, 2, \ldots, n/2\}$ (there are at most $2 \cdot n/2 = n$ unknown variables in this case) one can try to minimize the distance between the model function (2.11) and the given data, which could be defined as the sum of the squared deviations

$$r_m = \sum_{j=1}^n \left(y_j - \sum_{i=1}^m y_{i0} e^{-k_i t_j} \right)^2. \tag{2.12}$$

This is an exceptionally difficult *inverse problem* (nonlinear estimation problem). One has to determine the respective minima with (constrained) optimization techniques.

The value of the minimal residual r_m^* makes it possible to determine the value of m which can be used for the best adaptation of the model to the given data. However, this does not simply mean minimizing r_m without constraints on m, rather the aim is to choose m no larger than necessary in order to obtain the simplest possible model .

2.2.3 The Accuracy of Numerical Results

In order to investigate the accuracy of numerical results quantitatively, suitable and precise systems of measurement and assessment are needed.

Terminology (Errors and Effects)   In order to provide a clear and easy-to-read representation, often, in this book, no distinction is made between errors and their *effects*. The word
error is used as a collective term.

Forward Errors and Backward Errors

In numerical analysis a distinction is made between forward errors and backward errors. The *forward error* is the deviation of the computed solution Y from the exact solution X of a mathematical problem $\mathcal{P}$. The *backward error* gives information on the extent to which the problem $\mathcal{Q}$, of which Y is the exact solution, deviates from the original problem $\mathcal{P}$ (see Fig 2.8), provided that $\mathcal{Q}$ is in fact unique.

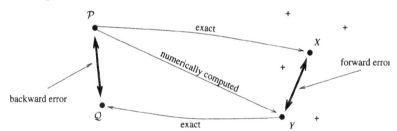

Figure 2.8: The forward and the backward errors are deviations in the solution space or the problem space respectively. Solutions derived using other methods than $\mathcal{P}$ are indicated by +. Their deviation from the exact solution X may be used to assess the magnitude of the forward error of the numerical solution Y.

Terminology (Error = Forward Error)   Unless stated otherwise, from this point on the
term *error* always denotes a forward error.

Absolute Accuracy

The (*absolute*) *error* will always denote the quantity

$$absolute\ error\ :=\ approximation - exact\ value. \qquad (2.13)$$

Terminology (The Sign of the Error)   Literature on the subject does not adhere to a
uniformly defined *sign* of the error. Sometimes the reverse difference "exact value − approximation" is used instead of (2.13). The term *error* may also denote the *absolute value* of the
error (2.13).

Relative Accuracy

The value of an error taken out of the specific context of a numerical problem is often not very useful in practical applications.

Example (Pendulum) Using $K(0)$ and the linear model, the period of oscillation T was approximated by T_L, the error in which amounted to $-0.68 \cdot 10^{-3}$ s $= -0.68$ ms. This quantity, if taken out of its practical context, is hardly suitable for assessing the usefulness of the numerical result. Since the period of oscillation of this pendulum is about 1.85 s, the error is tolerable for most technical applications. However, for an oscillating quartz crystal with a frequency of 20 kHz and a corresponding period of oscillation of 50 μs, the same error would make any numerical result totally useless.

This observation suggests that, instead of the *absolute* error, the *relative* error with respect to the exact result should be considered. However, since the exact value is usually unknown, other values of the same size as the exact result have to be used as a reference:

$$relative \; error \quad := \quad \frac{absolute \; error}{value \; of \; reference}. \tag{2.14}$$

The error definitions (2.13) and (2.14) play an important role in specifying a numerical problem—i.e., for the a priori decision as to what to accept as a solution—as well as for the assessment of obtained numerical results.

The two cases are dealt with in different ways. For fixing a tolerable error of a calculation in advance, the user will prefer to have the exact result as the value of reference. If it is known, it can be used as a value of reference, e.g., in the process of validating a model. Numerical methods which calculate an estimation of the error, usually employ the obtained (intermediate or final) approximation as the value of reference for estimating the relative error since the exact result is not known.

The definition of the relative error assumes a non-zero value of reference. If it is known (e.g., in test cases) that the exact result is zero, then one has either to be content with the *absolute* error or use other quantities as a reference such as values derived from the given problem data.

At the beginning of a calculation, the user of a numerical method normally does not know if the exact result is zero or just close to zero. In order to define an error bound which also works for approximations $R_{num} \approx 0$, a hybrid form such as

$$|R_{num} - R_{exact}| \leq \max\{\varepsilon_{abs}, \varepsilon_{rel}R_{num}\}, \tag{2.15}$$

is often preferred, a special case of which, namely $\varepsilon_{abs} = 0$, constitutes an accuracy requirement for the *relative* error. With $\varepsilon_{rel} = 0$ (2.15) constitutes an accuracy requirement for the *absolute* error.

Correct Decimal Places

In conjunction with the relative error of a numerical quantity, the number of *correct decimal places* is often used to indicate accuracy. This measure of accuracy indicates the maximum number of significant decimal places

$$\max\{m \in \mathbb{N}_0 : \text{rnd}(R_{\text{num}}; m) = \text{rnd}(R_{\text{exact}}; m)\}$$

in which R_{num} and R_{exact} are identical after rounding them to m decimal places:

$$\text{rnd}(x; m) \quad := \quad \text{sgn}(x) \lfloor |x|/10^{e(x)-m} + 1/2 \rfloor \, 10^{e(x)-m} \quad \text{where}$$

$$e(x) \quad := \quad \lfloor \log_{10} |x| \rfloor + 1.$$

Correct decimal places does not mean equality of decimal *digits*.

Example (Correct Decimal Places)   The approximation $R_{\text{num}} = 0.09996$ of $R_{\text{exact}} = 0.1$ has four correct decimal places, whereas $\bar{R}_{\text{num}} = 0.09994$ has only three correct decimal places. There is no agreement of significant decimal digits in this case at all; only the leading zero is identical for all three numbers.

The Accuracy of Vectors, Matrices and Functions

If the result of a computation is a vector or a function rather than a single number, then the absolute or the relative error is also a vector or a function. In this case a suitable norm $\| \, \|$ for the absolute or of the relative error is used. However, important pieces of information may be lost if a norm is used. The choice of a suitable norm is therefore critical to the assessment of numerical results.

Example (Function Approximation)   The task is to approximate numerically a differentiable function f defined on the interval $[a, b]$ by a piecewise polynomial p with degree d such that on the interval $[a, b]$

$$\|p - f\| < \varepsilon.$$

Two applications, in which the choice of d as well as the choice of a suitable norm are critical for obtaining a reasonable and useful result, are discussed below.

Graphical Representation

The function p is required in order to draw the function f on a plotter. Since plotters can basically draw only polygons the choice $d = 1$ is reasonable. One norm is the maximum norm $\| \, \|_\infty$, which guarantees that

$$|p(t) - f(t)| < \varepsilon \qquad \text{for all} \quad t \in [a, b]. \tag{2.16}$$

The resulting function p is thus acceptable for the intended purpose and can be used in practice (see Fig. 2.9) provided that the tolerance ε is less than the resolution of the plotter. On the other hand, choosing the norm $\| \, \|_2$ with

$$\left(\int_a^b |p(t) - f(t)|^2 \, dt \right)^{1/2} < \varepsilon$$

would only guarantee that the approximation requirement is satisfied in the quadratic mean (see Fig. 2.10).

Numerical Differentiation

The desired function p is to model f so that p' can be used as a convenient approximation for f'. In this case the choice of d has to be made according to the required smoothness (differentiability) properties of p. Moreover, it is an indispensable requirement that the transition from

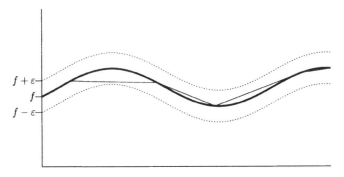

Figure 2.9: Approximation of the function f (——) by a polygon p (—) with $\|p - f\|_\infty < \varepsilon$.

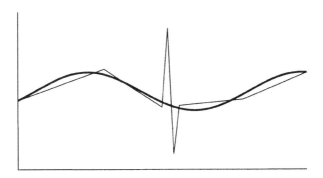

Figure 2.10: Approximation of f (——) by a polygon p (—) with $\|p - f\|_2 < \varepsilon$.

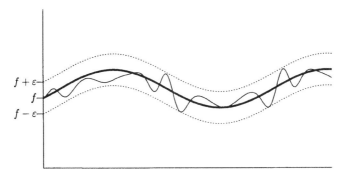

Figure 2.11: Approximation of the function f (——) by a mathematically smooth function g (—) which satisfies $\|g - f\|_\infty < \varepsilon$ whereas the point-wise deviation $g' - f'$ is large.

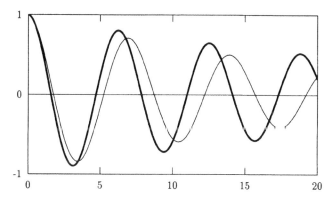

Figure 2.12: The exact function f (——) and its approximation $\tilde{f}$ (—).

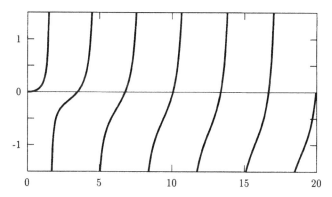

Figure 2.13: The relative error $e_{\text{rel}}(x) := [\tilde{f}(x) - f(x)]/f(x)$, $x \in [0, 20]$. This function has poles at the zeros of f (cf. Fig. 2.12).

one polynomial piece to another is differentiable, too. The maximum norm is not suitable to meet these requirements.

Even though (2.16) guarantees that p does not differ from f by a larger value than ε, the derivatives f' and p' may differ by an arbitrarily large margin (see Fig. 2.11). By choosing

$$\|p - f\| := \max\{|p(t) - f(t)| + |p'(t) - f'(t)| : t \in [a, b]\}$$

one gets the desired accuracy for f', as well.

The question thus arises as to how to define the relative error of functions and a corresponding norm. An obvious approach is to define an *error function* based on the *point-wise* application of (2.14) (see Fig. 2.12 and Fig. 2.13):

$$e_{\text{rel}}(t) := \frac{\tilde{f}(t) - f(t)}{f(t)}, \qquad t \in [a, b]. \tag{2.17}$$

Evidently, only if $f(t) \neq 0$ for all $t \in [a, b]$ is this error function meaningful over the whole interval $[a, b]$.

By forming the norm $\|e_{\mathrm{rel}}\|$ of the function (2.17), a single (scalar) error index is obtained. In order to attain the accuracy requirement

$$\|e_{\mathrm{rel}}\|_\infty < \varepsilon_{\mathrm{rel}},$$

$|\tilde{f}(t) - f(t)|$ must be extremely small wherever $|f(t)|$ is small. However, this requirement is not easy to satisfy. Thus, in order to provide a reference value around the same size as the solution at points t where f is zero or close to zero, often only *one* value (a norm of f), rather than the function value $f(t)$, is used to define an error function:

$$e_{\mathrm{rel}}(t) := \frac{\tilde{f}(t) - f(t)}{\|f\|}, \qquad t \in [a, b];$$

$$\|e_{\mathrm{rel}}\| := \left\| \frac{\tilde{f}(t) - f(t)}{\|f\|} \right\| = \frac{\|\tilde{f}(t) - f(t)\|}{\|f\|}.$$

For vectors $x = (x_1, \ldots, x_n)^\top, \tilde{x} = (\tilde{x}_1, \ldots, \tilde{x}_n)^\top \in \mathbb{R}^n$ the relative error may be characterized analogously either by

$$e_{\mathrm{rel}} := \left\| \left(\frac{\tilde{x}_1 - x_1}{x_1}, \ldots, \frac{\tilde{x}_n - x_n}{x_n} \right)^\top \right\| \qquad \text{or by} \qquad e_{\mathrm{rel}} := \frac{\|\tilde{x} - x\|}{\|x\|}.$$

For matrices the only characterization of the relative error of practical importance is

$$e_{\mathrm{rel}} := \frac{\|\tilde{A} - A\|}{\|A\|}.$$

Estimates as Values of Reference

If the tolerance parameter of a numerical computer program is specified in terms of the *relative error*, then the computed solution itself is used as the reference value for determining the relative error because the exact value is not available:

$$\tilde{e}_{\mathrm{rel}} = \frac{\|\tilde{f} - f\|}{\|\tilde{f}\|} \quad \text{or} \quad \tilde{e}_{\mathrm{rel}} = \frac{\|\tilde{x} - x\|}{\|\tilde{x}\|} \quad \text{or} \quad \tilde{e}_{\mathrm{rel}} = \frac{\|\tilde{A} - A\|}{\|\tilde{A}\|}.$$

In practice, the values $\|\tilde{f} - f\|$, $\|\tilde{x} - x\|$ and $\|\tilde{A} - A\|$ also have to be replaced by numerically determined estimates.

2.3 Types of Errors in Numerics

In numerical mathematics and numerical data processing, there are four types of errors: *model errors*, *data errors*, *algorithm errors* and *rounding errors*. These errors are *not* the consequences of unsound reasoning or poorly thought out decisions. Unlike, for example, programming errors, they are not avoidable and correctable; they are inevitable. They can be anticipated, and accuracy requirements can be imposed on them, i. e., they can be kept below given error bounds.

The error bound is part of the specification of the numerical problem. In order to make sure that the sum of all error effects does not exceed the error bound, i. e., that

‖effect of model error + effect of data error

  + effect of truncation error + effect of rounding error‖ ≤ tolerance,

all relevant errors have to be recognized, and their effects on the numerical result have to be assessed.

The following four sections discuss and characterize the four basic types of errors and their effects. Later chapters deal with related issues in more detail.

2.3.1 Model Errors

As an integral part of any model building process, several quantities are neglected (see Chapter 1). The result is a simplified depiction of reality. The inevitable deviation of the model from the original is denoted the *model error*.

It would be necessary to estimate at least the magnitude of the effect of the model error in order to guarantee that the given error tolerance is met. However, such an estimate is usually not obtainable since there are *unknown* and unquantifiable factors involved.

Example (Unknown Model Error Effects) Relativistic effects were *unknown* to nineteenth century scientists when they developed their models in the field of mechanics. However, these effects no doubt make some contribution to model error, since they cause a deviation from reality.

The effect of known model errors can often be interpreted as the effect of data errors, which can be estimated, for example, by analyzing the condition of the model (see Section 2.4).

Example (Pendulum) Neglect of frictional forces in mathematical models of the pendulum, constitute a model error. If the size of the frictional term F in the differential equation (2.4) can be estimated, the effects of neglecting frictional forces can be analyzed. This requires an analysis of the condition of the initial value problem for ordinary differential equations.

2.3.2 Data Errors

In general, models not only cover a specific application, but a whole class of similar applications. Any special case is identified by the values of model parameters. For example, the length l of the pendulum, the initial angular displacement φ_0 and the constant of gravitation g (which depends on the geographic location) are parameters of the mathematical pendulum model. In order to investigate a specific pendulum, these parameters have to be specified.

Due to inaccurate measurements and other factors, the values used for the model parameters usually deviate from the true values; this is referred to as the *data error*. Accordingly, the impact of data errors on the results are called data error *effects*. They can be estimated using condition number analysis.

In addition to scalar quantities (as in the case of the pendulum model), vectors, matrices and all kinds of *functions* may also belong to the data of a problem. An example of such a function is the integrand function in integration problems. Such data are also known as *analytical data* as opposed to *algebraic data* (scalars, vectors, matrices etc). The effects of errors in analytical data can also be analyzed using condition number estimates.

Example (Pendulum)   The analytical data of the pendulum problem include the functions, $f : \mathbb{R} \times \mathbb{R} \to \mathbb{R}$, on the right sides of the differential equations (2.2), (2.3), and (2.4)

$$\varphi''(t) = f(\varphi(t), \varphi'(t)) = \begin{cases} -\omega^2 \varphi(t) \\ -\omega^2 \sin(\varphi(t)) \\ -\omega^2 \sin(\varphi(t)) - F(\varphi'(t)). \end{cases}$$

In this sense the differences among these functions can be regarded as data errors, and their effects can be quantified using condition number analysis.

2.3.3 Algorithm Errors

If a mathematical problem cannot be solved using symbolic manipulations, then it must be solved using numerical algorithms. When developing numerical algorithms, it is necessary to simplify a number of things before a finite problem formulation can be derived or the required computational effort can be reduced to a reasonable level. The resulting deviation of the algorithmically obtained results from the solution of the mathematical problem is denoted the *algorithm error*.

Example (The Iterative Solution of a System of Linear Equations)   Iterative schemes have been used for a long time to formulate approximate methods for solving large systems of linear equations (see Chapter 16)

$$Ax = b, \qquad A \in \mathbb{R}^{n \times n}, \quad b, x \in \mathbb{R}^n.$$

This was done to reduce the required computational effort. Direct methods (e.g., Gaussian elimination) require a computational effort which is proportional to n^3. If an approximate solution $\tilde{x}$ satisfying

$$\|A\tilde{x} - b\| < \varepsilon$$

is sufficient, the computational cost can sometimes be reduced significantly. If, for example, the matrix A does not have a special structure, then only

$$k \approx \sqrt{\kappa_2}\, \frac{\ln(2/\varepsilon)}{2}, \qquad \kappa_2 := \|A\|_2 \|A^{-1}\|_2$$

matrix-vector multiplication operations are necessary for the iterative solution (Traub, Wozniakowski [368]). The number κ_2 is the Euclidean condition number of the matrix A (see Section 13.8).

Example (One-Dimensional Optimal Cutting Problem)   Parts $\mathcal{T}_n = \{T_1, T_2, \ldots, T_n\}$ of lengths $l(T_i) \in (0, 1]$ are cut out of given pieces $B_1, B_2, \ldots$ of length 1 each. The task is to organize the cutting so that the overall number L of required B-pieces is *minimal*. The computational effort of all known algorithms for an *exact* solution of this problem grows (at least) exponentially with n. Even the fastest computers fail to compute the exact solution even for a moderately sized n. However, if an ε-approximation suffices, i.e., a value L which is just slightly larger than the optimal value L_{opt},

$$\max_{\mathcal{T}_n} \frac{L}{L_{\text{opt}}} \leq 1 + \varepsilon \qquad \text{for all} \quad n \in \mathbb{N},$$

then a solution of the problem which requires *considerably* less computational effort—as low as $O(n \log n)$ (Karmarkar, Karp [248], Johnson, Garey [245])—can be calculated.

Truncation Errors

Numerical algorithms implemented on a computer can only comprise a *finite* sequence of arithmetic operations (addition, subtraction, multiplication, division, and logic). The number and order of the computations are generally not fixed in advance but are determined by the computer at runtime; they depend on the data and the intermediate results. For predefined function procedures evaluating the elementary functions *sin, exp*, ..., as well, only a finite sequence of arithmetic operations is executed on the computer.

The error due to replacing an infinite process by a finite sequence of arithmetic operations is referred to as the *truncation error*.

Example (Standard Functions) In calculus the standard mathematical functions, such as the exponential, sine, or cosine functions, are defined in terms of infinite power series:

$$e^x = \sum_{k=0}^{\infty} \frac{x^k}{k!}, \quad \sin x = \sum_{k=0}^{\infty} \frac{(-1)^k}{(2k+1)!} x^{2k+1}, \ \ldots \ .$$

Many programming languages include intrinsic function procedures for mathematical standard functions. To implement such representations on a computer, the infinite series must be replaced by finite expressions, e.g., Taylor polynomials. For example, the replacement of the infinite cosine series

$$\cos x = \sum_{k=0}^{\infty} \frac{(-1)^k}{(2k)!} x^{2k} \quad \text{with} \quad P_8(x) = 1 - \frac{x^2}{2!} + \frac{x^4}{4!} - \frac{x^6}{6!} + \frac{x^8}{8!},$$

yields the truncation error

$$e_{\text{trunc}}(x) := P_8(x) - \cos(x),$$

which grows with the distance between x and zero (see Fig. 2.14).

Discretization Errors

The error that results from the replacement of continuous information with discrete information is referred to as the *discretization error*.

Example (Numerical Integration) In the evaluation of an integral

$$If = \int_a^b f(t)\, dt, \tag{2.18}$$

complete information is not usually available. The integrand f is often given in terms of an algebraic expression, the evaluation of which returns $f(t)$ at *any* point $t \in [a, b]$. However, a

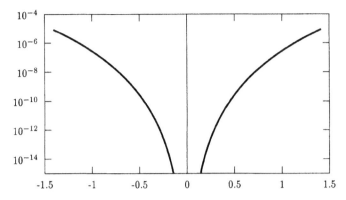

Figure 2.14: Truncation error $e_{\text{trunc}}(x) = P_8(x) - \cos(x)$ for $x \in [-1.5, 1.5]$.

computer program can only work with a *finite* set of values $\{f(t_1), \ldots, f(t_N)\}$. The continuous information, which is theoretically available, cannot be exploited except for a finite set of discrete points used to compute an approximation Q_N:

$$Q_N f = \sum_{i=1}^{N} c_i f(t_i) \approx \text{I}f.$$

Algorithm Errors

All simplifications made in the course of formulating a numerical algorithm are referred to as *algorithm errors*. The result of these errors is the deviation of the numerical approximation (which is, in this context, considered without rounding errors) from the exact solution of the mathematical problem.

Terminology (Algorithm Errors)   Publications written in English do not normally use the general term "algorithm errors". Only the terms "truncation errors" and "discretization errors" are used.

The Reduction of the Algorithm Error

If the algorithm error can be reduced to an arbitrarily small size by increasing the computational effort, then the algorithm is referred to as a *convergent* approximation method. Since algorithmic computations always have to terminate after a finite number of steps, the algorithm error is usually *unavoidable*.

Example (Pendulum)   The Gaussian iteration—(2.7) to (2.9)—for computing values of the complete elliptic integral of the first kind is a convergent method, i.e., additional computation narrows the interval containing the desired value.

Methods which by their very nature require only a finite number of algebraic operations are exceptions. The most famous example of such an exception is Gaussian elimination and other direct methods for the solution of linear algebraic equations.

Bounds for the Algorithm Error

For many numerical approximation methods, *error bounds* can be derived. This makes it possible to guarantee a specified level of algorithm accuracy.

Example (Cosine Series)   The cosine function is defined by the infinite series

$$\cos x = 1 - \frac{x^2}{2!} + \frac{x^4}{4!} - \frac{x^6}{6!} + \cdots .$$

For $x \in (-\sqrt{2}, \sqrt{2})$ the series is alternating with the absolute values of its terms converging strictly monotonically to zero. For such a series the estimate

$$|s_k(x) - S(x)| \le |a_{k+1}(x)|$$

holds if the finite sum $s_k(x) = a_0(x) + a_1(x) + \cdots + a_k(x)$ is used as an approximation for

$$S(x) = \sum_{i=0}^{\infty} a_i(x).$$

Accordingly, for

$$p_k(x) = 1 - \frac{x^2}{2!} + \frac{x^4}{4!} - \cdots + (-1)^k \frac{x^{2k}}{(2k)!}$$

the rigorous estimate of the algorithm error

$$|p_k(x) - \cos x| < \frac{x^{2k+2}}{(2k+2)!} =: e_k(x), \qquad x \in (-\sqrt{2}, \sqrt{2})$$

holds (neglecting rounding error).

Example (Recursive Approximation of 2π)   The sums of the sides of regular polygons circumscribed about the unit circle and inscribed in the latter can be used in order to derive upper and lower bounds, respectively, for 2π, the circumference of the unit circle. Continued doubling of the number k of vertices leads to the recursion for the length s_k of a side of the inscribed regular polygon:

$$s_4 = \sqrt{2}, \qquad s_{2k} = \sqrt{2 - \sqrt{4 - s_k^2}}. \tag{2.19}$$

The length c_k of a side of the circumscribed regular polygon can be determined according to the following rule based on s_k:

$$c_k = \frac{2s_k}{\sqrt{4 - s_k^2}}. \tag{2.20}$$

Both the sequences $\{ks_k\}$ and $\{kc_k\}$ are *monotonically* convergent to the limit 2π:

$$ks_k \uparrow 2\pi, \qquad kc_k \downarrow 2\pi. \tag{2.21}$$

Hence, $kc_k - ks_k$ is a *bound* for the two algorithm errors $|ks_k - 2\pi|$ and $|kc_k - 2\pi|$ respectively.

Often the explicit knowledge of characteristic numbers associated with a problem (such as upper bounds on certain derivatives) is a prerequisite for the calculation of useful algorithm error bounds. However, such characteristic numbers are often *not* available or can only be determined with an unreasonable amount of effort. Nevertheless, detailed information about the algorithm error plays an important role in the *qualitative* assessment of numerical algorithms when their efficiency has to be assessed with respect to whole *classes* of problems.

Example (Numerical Integration) In order to determine a numerical approximation of the definite integral (2.18), the interval $[a, b]$ can be divided into N subintervals of equal length

$$h = \frac{b-a}{N}.$$

Both the formulas

$$R_N^l f = h[f(a) + f(a+h) + \cdots + f(a + (N-1)h)], \quad \text{and} \tag{2.22}$$

$$T_N f = h[f(a)/2 + f(a+h) + \cdots + f(a + (N-1)h) + f(b)/2] \tag{2.23}$$

return approximations for If. The algorithm error of the simple Riemann sum (2.22) can be estimated by (see Chapter 12)

$$|If - R_N^l f| \le h \frac{b-a}{2} M_1 \quad \text{where} \quad M_1 := \max\{|f'(x)| : x \in [a, b]\}.$$

For the compound trapezoidal rule (2.23), the error estimate (see Chapter 12)

$$|If - T_N f| \le h^2 \frac{b-a}{2} M_2 \quad \text{where} \quad M_2 := \max\{|f''(x)| : x \in [a, b]\}$$

holds. These two error bounds make it clear that there are qualitative differences between the rates of convergence of the two methods. The convergence as $h \to 0$ of the simple Riemann sum is only linear, whereas the discretization error of the compound trapezoidal rule converges to zero as h^2. These rates of convergence provide an intuitive feel for the efficiency of the two methods.

In order to make sure that the error lies within a specified tolerance, estimates of the actual algorithm error are needed. Most numerical programs calculate such estimates during the computation of a result. For that purpose the algorithm error is estimated on the basis of information obtained during the computation. Although this way of estimating the algorithm error is, in many cases, not suitable for guaranteeing error bounds, it does yield useful information about the actual error which is reliable enough for many practical problems.

2.3.4 Rounding Errors

A computer provides only a *finite* set of numbers: *integers*, and *floating-point numbers* with fixed mantissa lengths. This is why the operations involved in a computer program cannot generally be executed exactly. Each computational step maps its result onto one of the available floating-point numbers—usually the closest one. The difference between the exact result and the rounded result of an operation is referred to as the *round-off error* or the *rounding error*. The effect of the accumulated rounding errors on the final result of the approximation method is called the rounding error effect.

Example (Systems of Linear Equations) The numerical solution of the system of linear equations $Ax = b$ with complex coefficients

$$A = \begin{pmatrix} 100 - 100\,i & -100 & 100\,i & 0 \\ -100 & 150 + 150\,i & -50 & 150\,i \\ 100\,i & -50 & 250 - 100\,i & 200 \\ 0 & 150\,i & 200 & 200 + 150\,i \end{pmatrix}, \quad b = \begin{pmatrix} 50 \\ 0 \\ 0 \\ 50 \end{pmatrix}$$

in complex, single precision, floating-point arithmetic using Gaussian elimination yields the following answer:

$$x = \begin{pmatrix} 0.2437 + 0.2083\,i \\ 0.0111 - 0.0638\,i \\ -0.0283 - 0.0591\,i \\ 0.1719 - 0.0781\,i \end{pmatrix}.$$

During the computation neither an exponent overflow, nor a division by zero, nor exceptionally large intermediate results occurred. It is thus tempting to believe that x is the unique solution of a system of linear equations with a non-singular matrix of coefficients $(\det(A) \neq 0)$. However, the opposite is true: The matrix A is *singular* because its rows r_i satisfy the relation

$$r_4 = r_1 + r_2 + r_3.$$

In this example there are neither data errors (i. e., all coefficients are integers) nor algorithm errors. Gaussian elimination either returns the exact result or terminates prematurely, provided *exact arithmetic* is used. Only the accumulated effect of rounding errors is responsible for a seemingly reasonable but incorrect result. Thus, the user of this program is led to believe that the situation is normal, because the subroutine used to compute the solution did not include appropriate tests to inform the user that the system is singular and that the computed answer might not be correct.

Example (Cosine Series)   Using a suitable truncation criterion, the Taylor expansion

$$\cos x = 1 - \frac{x^2}{2!} + \frac{x^4}{4!} - \frac{x^6}{6!} + \cdots$$

can be used in algorithms for the numerical evaluation of the cosine function. Because the Taylor series converges for all $x \in \mathbb{R}$, for *each* argument x and *each* $\varepsilon > 0$ there exists a $k(x;\varepsilon)$ such that the Taylor polynomial (the truncated Taylor series)

$$p(x;\varepsilon) = 1 - \frac{x^2}{2!} + \cdots + (-1)^{k(x;\varepsilon)} \frac{x^{2k(x;\varepsilon)}}{[2k(x;\varepsilon)]!}$$

has the required approximation accuracy

$$|p(x;\varepsilon) - \cos(x)| < \varepsilon.$$

This result is based on equations and data used in calculus, i. e., on *exact computation* in the field of real numbers. However, this result cannot be transferred to the computer without difficulty, as is demonstrated by the following implementation of the Taylor polynomial algorithm.[5]

```
cosine = 0.;   term = 1.;   k2 = 0;   xx = x*x
DO WHILE (ABS(term) > ABS(cosine)*EPSILON(cosine))
   cosine = cosine + term
   k2     = k2 + 2
   term   = (-term*xx) / REAL(k2*(k2 - 1))
END DO
```

For the parameter value x = 3.1415 9265 $\approx \pi$, the algorithm (using single precision floating-point arithmetic) returns the correct result

$$\text{cosine} = -1.000000E + 00$$

---

[5] The Fortran 90 function EPSILON($\cdot$) used in this program fragment returns a bound for the relative rounding error in the floating-point data type of its argument (see Section 4.8).

in all decimal places. For $x = 31.415\,9265 \approx 10\pi$, however, the same program does not return
the expected result $\cos 10\pi = 1$, but

$$\text{cosine} = -8.593623\text{E} + 04,$$

which is an absurd value for a function whose range is the interval $[-1, 1]$. The reason for the
utter collapse of the algorithm is *cancellation* (a secondary effect of rounding errors). This
effect increases as x increases: In the case of $x \approx 10\pi$, the absolute values of the terms of the
series reach a peak at

$$\text{term} = -\frac{x^{30}}{30!} \approx -3.09 \cdot 10^{12}$$

and then decrease down to zero (see Fig. 2.15). Some of the intermediate sums are therefore
extremely large although the exact result is found in the interval $[-1, 1]$. Continued summation
inevitably cancels out all significant digits.

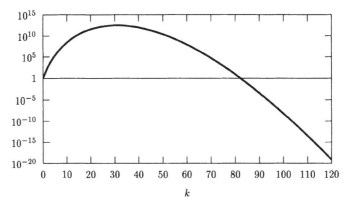

Figure 2.15: Size of the terms $x^k/k!$, $k = 0, 2, 4, \ldots, 120$ of the cosine series for $x = 10\pi$.

Example (Recursive Approximation of 2π) The practical computation of the recursions
(2.19) and (2.20) show—as could be expected because of the convergence (2.21)—a monotonic
decrease of the algorithm error bound $kc_k - ks_k$ down to the value $2.38 \cdot 10^{-7}$ after 11 doublings
of the number of vertices, i.e., for $k = 8192$. Beginning with the next iteration, the calculated
error bound constantly remains *zero*. It is tempting to conclude that approximations ks_k and
kc_k were exact within the limits of the respective floating-point arithmetic. But in fact for 11,
12, and 13 doublings of the number of vertices the approximations of $2\pi \approx 6.28318531$

$$
\begin{aligned}
8192\, s_{8192} &= 6.324556 \\
16384\, s_{16384} &= 5.656854 \\
32768\, s_{32768} &= 0.
\end{aligned}
$$

become worse and completely useless in the end. The reason for the failure of the recursive
approximation of 2π is again cancellation which occurs in the subtraction operations in formula
(2.19).

Fig. 2.16 depicts the relative error

$$e_{\text{rel}}^{(k)} := \left| \frac{ks_k - 2\pi}{2\pi} \right|.$$

It also demonstrates that double precision arithmetic does not prevent the disturbing effects of
cancellation: it only postpones their occurrence.

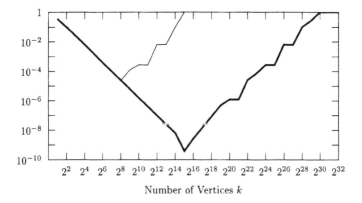

Number of Vertices k

Figure 2.16: The relative error of the recursive approximation of 2π in single precision arithmetic (—) and double precision arithmetic (**—**).

In literature on numerical mathematics, techniques are presented for estimating the rounding error effect. However in practice, such procedures often require a prohibitive amount of work when applied to complicated algorithms. Moreover, they tend to return error bounds which are too large. As a rule, methods which involve a large number of operations often partially compensate for individual rounding errors. Nevertheless, it is nearly impossible to take such compensatory effects into consideration when deriving guaranteed error bounds.

2.3.5 Error Hierarchy

In order to meet the accuracy requirement which is part of the numerical problem, each of the error sources can be influenced within certain limits. The leading principle should be to strive for *balanced error levels*. For example, if a model reproduces the corresponding original process with an accuracy of only 10 %, then it would be unreasonable to use a double precision arithmetic with 15 decimal places for numerical calculations or to use high-precision instruments for data procurement.

The following is a list of what can be done concerning specific error effects:

Model error effects can be controlled in the process of model building. In the pendulum example, the need to consider frictional forces stems from accuracy requirements. Estimates of the condition number may aid decision making in such cases, though a basic uncertainty remains.

Data error effects can sometimes be influenced by increasing the accuracy of measurements. However, if the condition number estimate applied to the most accurate data returns too large a bound for the data error effect, then the required level of accuracy cannot be obtained at all. In such cases the accuracy of the result is not improved even by drastically increasing the algorithmic or the computational accuracy.

Algorithm error effects can be reduced for convergent methods to an arbitrary extent, by increasing the computational effort. However, for the sake of efficiency, there is no reason to push the algorithm error far below the level of other errors.

Rounding error effects can be controlled to some extent by choosing one of the available floating-point number systems (single, double, or perhaps even multiple precision).

Due to the effects of algorithm errors and rounding errors, the solution $\tilde{x}$ of a numerical problem is generally not equal to the solution x of the corresponding mathematical problem. It should nevertheless comply with the specified accuracy requirement

$$\|\tilde{x} - x\| < \varepsilon.$$

Due to model errors and data errors, the solution x of the mathematical problem does not describe exactly the process that the model is based on. If the estimate

$$\|x - x_{\text{real}}\| < \delta$$

is available in an actual situation then the *overall error effect*, can also be estimated:

$$\|\tilde{x} - x_{\text{real}}\| \leq \|\tilde{x} - x\| + \|x - x_{\text{real}}\| \leq \varepsilon + \delta.$$

This inequality makes it evident that it is *not* enough to choose a small tolerance ε for the numerical problem in order to guarantee a given overall accuracy for the approximate solution $\tilde{x}$. Moreover, the effect of the model and of the data errors $\|x - x_{\text{real}}\|$ must be small enough to allow for an appropriate control of the overall error effect $\|\tilde{x} - x_{\text{real}}\|$.

Example (Pendulum) The parameters g, l and φ_0 constitute the numerical data of the linear differential equation problem (2.3). Moreover, the φ'-term (the frictional term) in the right side $f : \mathbb{R} \times \mathbb{R} \to \mathbb{R}$ also belongs to the data of the differential equation (2.4)

$$\varphi''(t) = f(\varphi(t), \varphi'(t)).$$

In fact, the notion of *data* comprises: numbers, assumptions, formulas, and the like. All changes in these entities are changes in the data. Hence, the omission of the frictional term, which has been classified as a model error so far, can also be interpreted as a data error. Its effect on the result must be investigated prior to any assessment of the overall error effect.

The reliable estimation of the various error effects is often very difficult, and in certain cases it turns out to be completely impossible. For a practical assessment of the computational results, carefully designed experiments can be performed in order to *validate* whether or not the model, the algorithms and the programs in use comply with the required accuracy of a given problem (cf. Section 2.7).

If the validation implies that the achieved level of accuracy does not meet the requirements it is necessary to try to isolate the factors responsible for the deviation. Condition numbers can often be a very useful tool to quantify the influence of such factors.

2.4 The Condition of Mathematical Problems

The impact of data errors (and of some model errors) $\|x - x_{\text{real}}\|$ can be estimated using a *data error analysis*. Such an analysis investigates the extent to which the solution x is changed if the data $\mathcal{D}$ it is based on is altered. The transition from the unperturbed data $\mathcal{D}$ of the original problem to the perturbed data $\overline{\mathcal{D}}$ gives rise to the transition from the exact solution x to the perturbed solution $\bar{x}$.

2.4.1 Perturbed and Unperturbed Problems

In a data error analysis which is meant to estimate $\|x - x_{\text{real}}\|$, the following correspondence is obvious: The exact solution of the mathematical problem corresponds to the result of the unperturbed problem, and the value x_{real} corresponds to $\bar{x}$, the result of the perturbed problem. This means, for example, that the frictional term R can be regarded as a perturbation in the pendulum example. For specific applications more or less exact values (estimates) of $\|\overline{\mathcal{D}} - \mathcal{D}\|$ are known. In the pendulum example, the size of the data errors which occur in the values of φ_0, g and l are easy to obtain. A bound for $\|\bar{f} - f\|$, the error caused by the omission of friction on the right side of the differential equation (2.4), is usually much more complicated to determine.

In general, the user is free to decide which of the two problems is classified as perturbed or unperturbed. For example, an original system of linear equations (which corresponds to the object being modeled) is the obvious choice of the unperturbed problem. The linear system stored in the computer (the coefficients of which are affected by measurement errors and conversion inaccuracies) represents the perturbed problem. However, in practice it is often reasonable to interchange the two roles since the system stored in the computer is actually available, whereas the original system is generally unknown. The condition number (depending on the matrix of coefficients) can only be determined for a problem whose information is actually stored in the computer. As in the context of data error analysis, a condition number always refers to the unperturbed problem; in this case, the system actually stored in the computer is identified as the *unperturbed* problem.

2.4.2 The Absolute Condition Number

Error bounds of the form

$$\|\bar{x} - x\| \le l\|\overline{\mathcal{D}} - \mathcal{D}\|, \tag{2.24}$$

are desirable for characterizing the sensitivity of the solution to changes in the data. The function $\| \; \|$ on the right side of the inequality (2.24) represents a suitable norm in the data space, which conveys a measure of the distance between $\mathcal{D}$ and $\overline{\mathcal{D}}$. The smallest possible factor

$$k = \inf\{l : \|\bar{x} - x\| \le l\|\bar{\mathcal{D}} - \mathcal{D}\|; \; \bar{\mathcal{D}}, \mathcal{D} \in \text{data set}\} \tag{2.25}$$

is called the *absolute condition number* of a mathematical problem with respect to a particular data set.

2.4.3 The Relative Condition Number

The condition number estimate (2.25) is based on *absolute* changes in the data; the condition number k is therefore referred to as the *absolute* condition number. If relative changes in the result are related to relative changes in the data by inequalities of the form

$$\frac{\|\bar{x} - x\|}{\|x\|} \leq K \frac{\|\overline{\mathcal{D}} - \mathcal{D}\|}{\|\mathcal{D}\|},$$

then the *relative condition number* of the mathematical problem is obtained as the smallest factor K compatible with a particular data set.

In order to quantify the relative sensitivity of a mapping

$$F \; : \; \mathbb{R}^m \to \mathbb{R}^n, \quad a \mapsto y$$

to perturbations, one can start by examining *one* particular data variable $a_i \in \mathbb{R}$ and the impact of its perturbations on a particular result variable $y_j \in \mathbb{R}$. Comparing the *relative changes in the result*

$$\frac{|y_j(a_1, \ldots, a_{i-1}, \tilde{a}_i, a_{i+1}, \ldots, a_m) - y_j(a_1, \ldots, a_m)|}{|y_j(a_1, \ldots, a_m)|}$$

to the *relative changes in the data*

$$\frac{|\tilde{a}_i - a_i|}{|a_i|}$$

leads to a quantification of the relative disturbance sensitivity of the result variable y_j, at a particular point a_i, in the form of the formula

$$\frac{\dfrac{|y_j(a_1, \ldots, a_{i-1}, \tilde{a}_i, a_{i+1}, \ldots, a_m) - y_j(a_1, \ldots, a_m)|}{|y_j(a_1, \ldots, a_m)|}}{\dfrac{|\tilde{a}_i - a_i|}{|a_i|}}. \tag{2.26}$$

If the quantities $a_1, \ldots, a_i, \ldots, a_m$ and $\tilde{a}_i \neq a_i$ are varied in such a way that

$$(a_1, \ldots, a_i, \ldots, a_m) \quad \text{and} \quad (a_1, \ldots, \tilde{a}_i, \ldots, a_m)$$

are elements of a given data set $\mathcal{D}$, then the *relative condition number* $K_{y_j \leftarrow a_i}(\mathcal{D})$ (which represents the sensitivity of y_j with respect to a_i) is obtained as the supremum of all compound fractions (2.26).

Note (Impact on the Result)   A relative condition number $K_{y_j \leftarrow a_i} = 5$ for determining the impact of a_i on y_j means that changing a_i by 1 % might change the result y_j by 5 %. Clearly the condition number $K_{y_j \leftarrow a_i}$ must cover the worst case, such that the sensitivity of y_j with respect to *arbitrary* changes in a_i is represented correctly.

In order to characterize the *overall error sensitivity* of a mapping with m scalar data values and n result values with *one* condition number, the relative modifications of the various data and results have to be condensed into two scalar values: *data changes* and *result changes*. The best way to achieve this depends on the particular application.

Example (System of Linear Equations) The solution of the linear system $Ax = b$ with the data

$$A = \begin{pmatrix} 4 & 6 & 4 & 1 \\ 10 & 20 & 15 & 4 \\ 20 & 45 & 36 & 10 \\ 35 & 84 & 70 & 20 \end{pmatrix}, \qquad b = \begin{pmatrix} 602 \\ 2012 \\ 4581 \\ 8638 \end{pmatrix} \tag{2.27}$$

is the vector x with

$$x_1 = 22, \quad x_2 = 57, \quad x_3 = 36, \quad x_4 = 28.$$

A perturbation of the right side b by the relatively small vector

$$\Delta b = (-0.31, 0.72, -0.59, 0.17)^T,$$

changes the exact and rounded solution to

$$\tilde{x}_1 = 13.91, \quad \tilde{x}_2 = 84.03, \quad \tilde{x}_3 = -25.54, \quad \tilde{x}_4 = 144.03.$$

If the Euclidean norm is used as the measure for deviations in the data and the result, the relative data changes are characterized by

$$\frac{\|\Delta b\|_2}{\|b\|_2} \approx 0.0001 \quad (0.01\,\%)$$

and the relative changes of the result by

$$\frac{\|\Delta x\|_2}{\|x\|_2} = 1.76 \quad (176\,\%).$$

An overall condition number for the system of linear equations above must thus amount to at least $1.76/0.0001 = 17\,600$. In fact, the actual condition number (cf. Chapter 13) is

$$\mathrm{cond}_2(A) = \|A\|_2 \|A^{-1}\|_2 = 18\,200.$$

The approximate size of a condition number is more important than its exact value. Condition numbers are only supposed to give a general idea of the accuracy of the result since the accuracy of the data itself is only known roughly. For example, the table below gives an overview of what accuracy is to be expected in the result for a relative accuracy 10^{-4} in the data of a problem (i.e., 4 to 5 correct decimal places of the data values can be safely determined):

relative condition number	1	10	100	1000	10 000
correct decimal places in the result	4–5	3–4	2–3	1–2	none

If the inverse of the condition number is of the same order of magnitude as the relative data accuracy, then even the most significant digit in the result is unreliable.

Example (System of Linear Equations) The condition number of the system (2.27) of linear equations $Ax = b$ in fact exceeds 10 000. Hence, changes in the data of only 0.01 % can modify the leading digit of the result.

NOTE: Condition numbers, as defined above, characterize the sensitivity of the *mathematical problem* to data perturbations. Consequently any condition number is independent of the numerical technique used to solve the problem since it is only changes in the exact solution of the mathematical problem which are taken into consideration.

2.4.4 Narrowing the Problem Class

In previous sections it has been implicitly assumed that the application at hand leaves no doubt as to which set of problems particular representatives are to be chosen from when comparing their results. Such an assumption implies a limitation or restriction of the data set, as is expressed, for example, in (2.25).

Example (System of Linear Equations) With a system of linear equations $Ax = b$, perturbations may change the matrix of coefficients A into another matrix $\bar{A}$ or change the right side b into another vector $\bar{b}$. In either case, the perturbed system $\bar{A}x = \bar{b}$ remains of order n.

In the following chapters formulas of the form (2.26) are derived for different classes of problems (integration problems, systems of linear equations etc.). These formulas can be used to determine the condition numbers (or at least the upper bounds) for the various problem classes.

With nonlinear problems it is useful to confine the condition analysis to a subset of the class of problems in question, which usually consists of a neighborhood of the unperturbed problem.

Example (Initial Value Problem) For the initial value problems in ordinary differential equations, the set of all perturbed problems

$$\begin{aligned} y' &= \bar{f}(t,y), \qquad t \in [0,T] \\ y(0) &= \bar{y}_0 \end{aligned}$$

can be dealt with, where $\bar{f}$ and $\bar{y}_0$ are restricted by perturbation bounds

$$\begin{aligned} \|\bar{f}(t,y) - f(t,y)\| &\leq \varrho_1 \qquad \text{for all} \quad (t,y) \in G \\ \|\bar{y}_0 - y_0\| &\leq \varrho_0. \end{aligned}$$

G is a suitable region around the solution of the original (unperturbed) problem

$$\begin{aligned} y' &= f(t,y), \qquad t \in [0,T] \\ y(0) &= y_0. \end{aligned}$$

The absolute condition number is the smallest possible factor in (2.24), where all perturbed problems of the subclass are considered.

The smaller the subset of the class of nonlinear problems can be made, the more precise the condition estimate and the smaller the corresponding condition number is. Clearly, the subclass must nevertheless be large enough to cover all data errors which may occur in practice and in the model building process.

2.4.5 Calculating Condition Numbers Using Differentiation

The field of *differential calculus* and the notion of *derivatives* have been developed in attempts to quantify the impact on function values of changes in the function arguments: Assume that f is a continuously differentiable function which maps the argument a to the result y

$$y = f(a).$$

Changing a to $\tilde{a} = a + \Delta a$ leads, according to the mean-value theorem, to a modification of y to

$$\tilde{y} = f(\tilde{a}) = f(a + \Delta a) = f(a) + f'(\bar{a})\Delta a = y + \Delta y,$$

where $a \leq \bar{a} \leq a + \Delta a$. Hence, the change in the result is given by

$$\Delta y = \tilde{y} - y = f'(\bar{a})\Delta a.$$

The relation between the *relative* changes $\Delta y/y$ and $\Delta a/a$ (where $y = f(a)$) can be derived using simple transformations:

$$\frac{\Delta y}{y} = \frac{a f'(\bar{a})}{f(a)} \cdot \frac{\Delta a}{a}.$$

The value corresponding to (2.26) is thus given by

$$\left| \frac{a f'(\bar{a})}{f(a)} \right|.$$

Because $\bar{a} \approx a$, the formula

$$K_{y \leftarrow a} = \left| \frac{a f'(\bar{a})}{f(a)} \right| \approx K_{y \leftarrow a}(B), \tag{2.28}$$

provides a satisfactory approximation of the relative condition number on a sufficiently small data domain B.

For multidimensional data (2.28) can either be interpreted by analogy as a multidimensional formula, or the partial derivatives of the individual data elements can be used to derive *individual* condition numbers.

2.4.6 Case Study: Quadratic Equation

The quadratic equation

$$y^2 + a_1 y + a_0 = 0$$

contains two data elements: a_0 and a_1. If $a_1^2 - 4a_0 > 0$ then the formulas

$$y_1 = (\sqrt{a_1^2 - 4a_0} - a_1)/2,$$
$$y_2 = (-\sqrt{a_1^2 - 4a_0} - a_1)/2$$

relate the two data parameters to the two results y_1 and y_2. In order to determine the relative condition number $K_{y_j \leftarrow a_i}$, the partial derivatives $\partial y_j/\partial a_i$ are needed:

$$\frac{\partial y_j}{\partial a_1} = \frac{1}{2}\left(\pm \frac{a_1}{\sqrt{a_1^2 - 4a_0}} - 1 \right) = \frac{\pm a_1 - \sqrt{a_1^2 - 4a_0}}{2\sqrt{a_1^2 - 4a_0}},$$

that is

$$\frac{\partial y_1}{\partial a_1} = -\frac{y_1}{\sqrt{a_1^2 - 4a_0}}, \qquad \frac{\partial y_2}{\partial a_1} = \frac{y_2}{\sqrt{a_1^2 - 4a_0}}.$$

Similarly,

$$\frac{\partial y_1}{\partial a_0} = -\frac{1}{\sqrt{a_1^2 - 4a_0}}, \qquad \frac{\partial y_2}{\partial a_0} = \frac{1}{\sqrt{a_1^2 - 4a_0}}.$$

Hence, from (2.28) the relative condition numbers $K_{y_j \leftarrow a_i}$ are

$$K_{y_j \leftarrow a_1} = \frac{|a_1|}{\sqrt{a_1^2 - 4a_0}}, \qquad j = 1, 2 \qquad (2.29)$$

and

$$K_{y_j \leftarrow a_0} = \frac{|a_0|}{|y_j|\sqrt{a_1^2 - 4a_0}} = \frac{|y_{3-j}|}{\sqrt{a_1^2 - 4a_0}}, \qquad j = 1, 2. \qquad (2.30)$$

The last transformation uses the identity $y_1 y_2 = a_0$. If the condition number is to be determined outside the immediate vicinity of one particular point, then a_1 and a_0 have to be varied over the whole data domain, and an upper bound on (2.29) and (2.30) has to be found.

Considering the case $a_0 < 0$ it can be immediately seen that

$$|y_j| < \sqrt{a_1^2 - 4a_0} \qquad \text{and} \qquad |a_1| < \sqrt{a_1^2 - 4a_0}.$$

All four condition numbers are therefore smaller than unity! A relative change in a data element of a given percentage affects both solutions of the quadratic equation to an even smaller extent. The problem is hence well-conditioned for $a_0 < 0$.

Numerical Example: With $a_1 = -5$ and $a_0 = -0.1$ the results are

$$y_1 = 5.01992\ldots \qquad \text{and} \qquad y_2 = -0.0199206\ldots.$$

Changing the data to $\tilde{a}_1 = -4.9$ and $\tilde{a}_0 = -0.098$, i.e., by 2%, leads to the modified results

$$\tilde{y}_1 = 4.91992\ldots \qquad \text{and} \qquad \tilde{y}_2 = -0.0199190\ldots,$$

i.e., y_1 is changed by 2% and y_2 by less than 0.01%.

2.4.7 Ill-Conditioned Problems

If the condition number of the mathematical problem is large ($K \gg 1$), then even for small changes in the data, large deviations of $\bar{x}$ from x have to be expected. This is why problems with large condition numbers are referred to as *ill-conditioned*, as opposed to *well-conditioned* problems with small condition numbers.

It is not possible to draw a sharp line between well-conditioned and ill-conditioned problems. The condition number of a problem can only be assessed with respect to the desired level of overall accuracy.

The accuracy requirement for the overall error effect

$$\|x - x_{\text{real}}\| < \tau,$$

where τ is the tolerance specified by the user, can only be met if it is guaranteed that

$$\|\tilde{x} - x_{\text{real}}\| \le \|\tilde{x} - x\| + \|x - x_{\text{real}}\| \le \varepsilon + \delta < \tau$$

or, the even more stringent requirement, that

$$\varepsilon + K\|\overline{\mathcal{D}} - \mathcal{D}\| < \tau.$$

If the size of the data error $\|\overline{\mathcal{D}} - \mathcal{D}\|$ is known, a problem with a condition number K is considered to be ill-conditioned whenever

$$K\|\overline{\mathcal{D}} - \mathcal{D}\| \ge \tau$$

or even when $K\|\overline{\mathcal{D}} - \mathcal{D}\| \approx \tau$, since in these cases even the choice of an extremely small ε (which usually entails a great deal of computational effort) cannot guarantee that $\tilde{x}$ is sufficiently accurate.

On the other hand, it is generally useless to choose the tolerance parameter ε of a numerical problem much smaller than $K\|\overline{\mathcal{D}} - \mathcal{D}\|$ since, in this case, the *overall* level of accuracy is almost entirely determined by the effects of model and data errors. For ill-conditioned problems, keeping the error of the numerical solution $\tilde{x}$ small is normally not worth the necessary computational effort. This extra work does not help to improve the overall level of accuracy.

Thus, only if the condition number of the mathematical problem, the desired overall error tolerance τ specified by the user, and information about the size $\|\overline{\mathcal{D}} - \mathcal{D}\|$ of the data error is available can a reasonable tolerance ε for the numerical problem be specified. The tolerance parameters, which frequently occur as input for numerical programs, refer to the error $\tilde{x} - x$ and, hence, play the role of ε. This is due to the fact that such a program is, at best, able to control the deviation of internal numerical results from the ideal mathematical result. The discrepancies between the model, or the program itself, and reality can be assessed or influenced from the outside of the system only by the user.

2.4.8 Ill-Posed Problems

If the result of a mathematical problem depends *discontinuously* on continuously varying data, then a numerical solution of the problem is generally impossible if the given data is in the neighborhood of the discontinuity. In such cases the result can be perturbed substantially even for extremely accurate data and despite the use of a multiple-precision arithmetic. Situations like this lead to the notion of *ill-posed* or *improperly posed* problems.

Ill-posed problems may occur, for example, if an integer result has to be calculated from real (i. e., continuously varying) data, e. g., the number of real zeros of a function or the rank of a matrix.

Example (The Number of Real Zeros of a Polynomial)   The cubic polynomial

$$P_3(x; c_0) = c_0 + x - 2x^2 + x^3$$

has

$$\left.\begin{array}{r} \text{one} \\ \text{two} \\ \text{three} \end{array}\right\} \text{real zeros if } c_0 \left\{\begin{array}{l} > 0 \\ = 0 \\ < 0 \end{array}\right.$$

(see Fig. 2.17). Hence, for values c_0 close to 0, the determination of the *number* of real zeros of P_3 is an ill-posed problem.

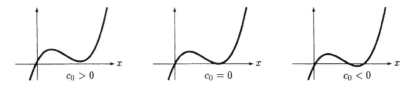

Figure 2.17: The polynomial $P_3(x; c_0) = c_0 + x - 2x^2 + x^3$ for c_0 close to 0.

2.5 The Condition of Application Problems

Analogously to the condition of mathematical problems, application problems are said to be either well-conditioned or ill-conditioned depending on the extent to which small changes in the experimental arrangement or the initial state affect the result x_{real}.

An important property of *relevant* mathematical models is the similarity between the application and the mathematical problem regarding sensitivity to perturbations. If there is *no* such similarity, it is very likely that the model is not realistic.

Example (Pushing a Bicycle)   A bicycle can be described using a simple model consisting of a steerable front wheel F connected to a fixed back wheel B by a rod of length a. The bicycle is pushed forward so that the front wheel moves along a line—the x-axis. The behavior of the back wheel is to be investigated after a perturbation which forces it away from the x-axis employing a lateral displacement δ (see Fig. 2.18).

The center of the back wheel B moves along a tractrix, which is the solution of the differential equation

$$y' = -\frac{y}{\sqrt{a^2 - y^2}}. \tag{2.31}$$

When moving the bicycle forward, B rapidly approaches the x-axis, and the effect of the initial displacement δ decreases. However, moving the bicycle backward increases the effect of the perturbation, and the back wheel deviates more and more rapidly from the x-axis.

With a wheel base of $a = 1$ m in (2.31), the impact of the initial displacement $\delta = 0.01$ m = 10 mm on the deviation of B in the direction of the y-axis can be seen from the table:

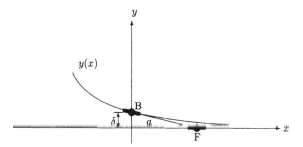

Figure 2.18: Model of the wheeled bicycle: The front wheel F is moved along the x-axis; the distance y of the back wheel B from the x-axis initially amounts to δ.

			forward	backward
after	1 m	movement	3.7 mm	27 mm
after	2 m	movement	1.4 mm	74 mm
after	3 m	movement	0.5 mm	197 mm
after	4 m	movement	0.2 mm	509 mm

The exponentially decreasing and increasing effect of the perturbation indicates that the forward movement is well-conditioned, whereas the backward movement is ill-conditioned.

The solution of very ill-conditioned problems may appear to be undetermined.

Example (Pendulum)   Usually all efforts fail to bring the pendulum to rest in the unstable, upper position of equilibrium. The initial condition $\varphi(0) = \pi$, $\varphi'(0) = 0$ cannot be exactly complied with, and even a very small disturbance $\varphi(0) = \pi + \alpha_0$, $\varphi'(0) = \alpha_1$ eventually makes the pendulum leave its initial position. Since the side to which the pendulum falls depends on extraordinarily small disturbances, this process can be practically regarded as undetermined.

For such ill-conditioned problems it is usually wise to do without a mathematical solution. However, it may be that a process turns out to be ill-conditioned only after a model has been derived and the corresponding mathematical problem has been analyzed. In some cases the user *knows* that the actual problem is well-conditioned although the corresponding mathematical problem proves to be ill-conditioned. This gives rise to doubts about the relevance of the mathematical model since the character of the original problem has changed considerably. It may be possible with a more diligent modeling to obtain a better-conditioned mathematical problem.

In practice, however, a different approach is usually chosen: Additional information about the expected solution (e.g., the monotonic behavior or the convexity of the solution) is added to the mathematical model and improves the condition number significantly. This strategy is referred to as *regularization* and is discussed in the chapters on interpolation and systems of linear equations.

2.6 The Mathematical Elements of Condition Estimation

Having introduced the notion of condition in Section 2.4, this section deals with the condition numbers of mathematical problems in more detail. Only the condition of direct and inverse problems are considered since identification problems are usually reduced to inverse problems.

2.6.1 The Condition of Direct Mathematical Problems

For direct mathematical problems, i.e., the evaluation of F at the point x

$$y = Fx,$$

the most general type of data perturbation is made up of a combined perturbation of x and F (see Fig. 2.19). However, this does not play a role in the chapters ahead and is therefore not discussed in this section any further.

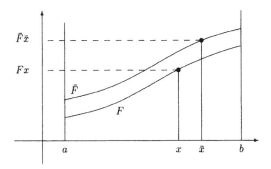

Figure 2.19: The combined perturbation of x and F with $\Delta x = \bar{x} - x$ and $\Delta F = \bar{F} - F$ in the case of a function $F : [a,b] \to \mathbb{R}$.

The impact of changes in x on the result y are quantified using the error bound (2.24), which, in the case of direct mathematical problems has the form

$$\|\bar{y} - y\| = \|F\bar{x} - Fx\| \le k\|\bar{x} - x\|.$$

It is therefore reasonable to characterize the sensitivity of direct problems to changes in the data using Lipschitz constants and the Lipschitz norm of F.

Definition 2.6.1 (Lipschitz Constant) *The function $F : X \to Y$ is said to be Lipschitz-continuous on a region $B \subseteq X$ if there is a number $l(F,B) \in \mathbb{R}_+$ such that*

$$\|Fx_1 - Fx_2\| \le l(F,B)\|x_1 - x_2\| \qquad \text{for all} \quad x_1, x_2 \in B. \tag{2.32}$$

Each number $l(F,B)$ is called a Lipschitz constant of the mapping F on the region B.

Clearly, the most realistic characterization of the sensitivity is given by the smallest Lipschitz constant $\mathrm{Lip}(F, B)$, which is referred to as the Lipschitz norm of F on B.

Definition 2.6.2 (Lipschitz Norm) *The quantity*

$$\mathrm{Lip}(F, B) \quad := \quad \inf\{l \in \mathbb{R} : \|Fx_1 - Fx_2\| \leq l\|x_1 - x_2\|, \quad x_1, x_2 \in B\}$$

$$= \quad \sup\left\{\frac{\|Fx_1 - Fx_2\|}{\|x_1 - x_2\|} : \quad x_1, x_2 \in B, \, x_1 \neq x_2\right\} \tag{2.33}$$

is called the Lipschitz norm of the mapping $F : X \to Y$ on $B \subseteq X$.

The Lipschitz norm $\mathrm{Lip}(F, B)$ is the absolute condition number of the direct problem $y = Fx$ on the data set $B \subseteq X$.

The Dependence of the Condition Number on the Data Set

For a nonlinear function F the choice of the data set B has a critical influence on the size of the condition number.

Example (Varying Condition Numbers) A condition number for the evaluation of a function $F : \mathbb{R} \to \mathbb{R}$ is to be determined. F is depicted in the following graph:

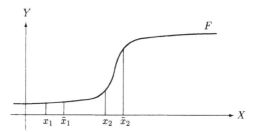

The direct problem of evaluating F is very well-conditioned in a neighborhood of the point x_1 or $\tilde{x}_1$, whereas the same problem is ill-conditioned in a neighborhood of x_2 and $\tilde{x}_2$.

Nevertheless, the monotonicity condition

$$B_1 \subseteq B_2 \quad \Longrightarrow \quad \mathrm{Lip}(F, B_1) \leq \mathrm{Lip}(F, B_2) \tag{2.34}$$

always holds.

The Condition Number of Linear Problems

Consider a linear mapping F on X with Lipschitz norm $\mathrm{Lip}(F, B)$ on $B \subseteq X$, where B contains

$$S_\delta(x_0) := \{x \in X : \|x - x_0\| < \delta\} \qquad (\delta > 0),$$

i. e., a δ-neighborhood of x_0. Then for all $x_1, x_2 \in X$

$$
\begin{aligned}
\|Fx_1 - Fx_2\| &= \|F(x_1 - x_2)\| \\
&= \left\| cF\left(\frac{x_1}{c} - \frac{x_2}{c}\right) \right\| \\
&= \left\| cF\left(\left(x_0 + \frac{x_1}{c}\right) - \left(x_0 + \frac{x_2}{c}\right)\right) \right\| \\
&\leq c\operatorname{Lip}(F, B) \left\| \left(x_0 + \frac{x_1}{c}\right) - \left(x_0 + \frac{x_2}{c}\right) \right\| \\
&= c\operatorname{Lip}(F, B) \left\| \frac{x_1 - x_2}{c} \right\| \\
&= \operatorname{Lip}(F, B)\|x_1 - x_2\|
\end{aligned}
$$

holds. In such cases $c > 0$ has to be chosen such that

$$\|x_1/c\| < \delta \quad \text{and} \quad \|x_2/c\| < \delta$$

in order to guarantee

$$x_0 + \frac{x_1}{c} \in S_\delta(x_0) \quad \text{and} \quad x_0 + \frac{x_2}{c} \in S_\delta(x_0),$$

e. g., $c = (1 + \|x_1\| + \|x_2\|)/\delta$.

In the linear case therefore, the Lipschitz norm does not depend on the region $B \subseteq X$—which is expressed by the notation $l(F)$—provided that B contains a ball $S_\delta(x_0)$ where $x_0 \in X$ and $\delta > 0$.[6]

For a set B which does not contain a ball, the quantity $\operatorname{Lip}(F, B)$ is generally not a Lipschitz norm on the whole of X, as is demonstrated not only in the trivial cases $B = \emptyset$ or $B = \{x_0\}$ but also by the following example:

Example (Linear Function) The function $F(x, y) = x + 2y$ defined on $X = \mathbb{R}^2$ has the Lipschitz norm $\operatorname{Lip}(F, B) = 1$ on

$$B = \{(x, 0) : x \in \mathbb{R}\} \subset \mathbb{R}^2$$

since

$$\|F(x_1, y_1) - F(x_2, y_2)\| = |x_1 - x_2|$$

holds for all $(x_1, y_1), (x_2, y_2) \in B$. However, for $(x_1, y_1) = (0, 0) \in B$ and $(x_2, y_2) = (0, 1) \notin B$ the inequality

$$\|F(x_1, y_1) - F(x_2, y_2)\| = 2 > 1 = \|(x_1, y_1) - (x_2, y_2)\|$$

holds. Heuristically, this means that the set B does not contain those paths along which the mapping is particularly steep.

Because

$$\|Fx_1 - Fx_2\| = \|F(x_1 - x_2)\| \leq \|F\|\|x_1 - x_2\|$$

the smallest constant possible still satisfying (2.32) is the *norm* $\|F\|$ of the linear mapping F (see Chapter 13). Hence, the definition of the Lipschitz norm in (2.33) leads to

$$\operatorname{Lip}(F, B) = \|F\| \qquad \text{for linear functions } F.$$

The Lipschitz constant $\operatorname{Lip}(F, B)$ can therefore be thought of as a norm for non-linear operators.

---

[6]Such a set B is called a *neighborhood* of x_0.

Condition Determination by Differentiation

For the mathematical problem of evaluating a univariate real function $F : B \subseteq \mathbb{R} \to \mathbb{R}$, each number that constitutes an upper bound of the gradients of all chords between (x_1, Fx_1) and (x_2, Fx_2) where $x_1, x_2 \in B$ is a Lipschitz constant $l(F, B)$. For the convex function

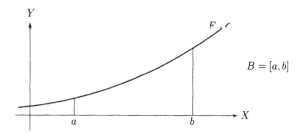

$\mathrm{Lip}(F, B)$ is given by $F'b$, the gradient of the steepest tangent over B. This is because there cannot be any chord whose gradient is steeper than $F'b$. This fact, which is clear in the one-dimensional case, can be generalized:

Theorem 2.6.1 *If $F : X \to Y$ is continuously differentiable on the convex region $B \subseteq X$, then*

$$\mathrm{Lip}(F, B) = \sup\{\|F'x\| : x \in B\},$$

where $\| \ \|$ denotes the operator norm for linear mappings corresponding to the norms on the linear vector spaces X and Y.

Proof: From the mean-value theorem of integral calculus it follows that, for all $x_1, x_2 \in B$:

$$Fx_1 - Fx_2 = \int_0^1 F'(x_2 + \lambda(x_1 - x_2))(x_1 - x_2)\, d\lambda,$$

$$
\begin{aligned}
\|Fx_1 - Fx_2\| &\leq \int_0^1 \|F'(x_2 + \lambda(x_1 - x_2))(x_1 - x_2)\|\, d\lambda \\
&\leq \sup\{\|F'(x_2 + \lambda(x_1 - x_2))\| : \lambda \in [0, 1]\}\, \|x_1 - x_2\| \\
&\leq \sup\{\|F'x\| : x \in B\} \cdot \|x_1 - x_2\|.
\end{aligned}
$$

On the other hand, because of

$$Fx_1 - Fx_2 = F'x_2\,(x_1 - x_2) + R(x_1, x_2) \quad \text{where} \quad \|R(x_1, x_2)\| = o(\|x_1 - x_2\|),$$

the quotient in (2.33) can be arbitrarily close to $\sup\{\|F'x\| : x \in B\}$.   □

Example (Derivative of a Function as a Condition Number)   The condition of the mathematical problem of evaluating the function $F : \mathbb{R}^2 \to \mathbb{R}^2$ defined by

$$
\begin{aligned}
F_1(x_1, x_2) &= (\sin x_1 + \cos x_2)/2 \\
F_2(x_1, x_2) &= (\cos x_1 - \sin x_2)/2
\end{aligned}
$$

is to be estimated. The mean-value theorem gives:

$$
\begin{aligned}
F_1(x_1^1, x_2^1) - F_1(x_1^2, x_2^2) &= \frac{\partial F_1}{\partial x_1}(\vartheta_{11}, \vartheta_{12})(x_1^1 - x_1^2) + \frac{\partial F_1}{\partial x_2}(\vartheta_{11}, \vartheta_{12})(x_2^1 - x_2^2) \\
&= \frac{1}{2}\cos\vartheta_{11}(x_1^1 - x_1^2) - \frac{1}{2}\sin\vartheta_{12}(x_2^1 - x_2^2) \\
F_2(x_1^1, x_2^1) - F_2(x_1^2, x_2^2) &= \frac{\partial F_2}{\partial x_1}(\vartheta_{21}, \vartheta_{22})(x_1^1 - x_1^2) + \frac{\partial F_2}{\partial x_2}(\vartheta_{21}, \vartheta_{22})(x_2^1 - x_2^2) \\
&= -\frac{1}{2}\sin\vartheta_{21}(x_1^1 - x_1^2) - \frac{1}{2}\cos\vartheta_{22}(x_2^1 - x_2^2),
\end{aligned}
$$

where $(\vartheta_{11}, \vartheta_{12})$ is located on the straight line from (x_1^1, x_2^1) to (x_1^2, x_2^2).

$$
Fx^1 - Fx^2 = \frac{1}{2}\begin{pmatrix} \cos\vartheta_{11} & -\sin\vartheta_{12} \\ -\sin\vartheta_{21} & -\cos\vartheta_{22} \end{pmatrix}(x^1 - x^2)
$$

is then obtained. Estimating the quantities $\cos\vartheta_{11}, \sin\vartheta_{12}, \sin\vartheta_{21}$, and $\cos\vartheta_{22}$ using their maximum absolute value 1 gives the inequality

$$
\|Fx^1 - Fx^2\|_\infty \le \|x^1 - x^2\|_\infty.
$$

Thus, $l = 1$ is a Lipschitz constant $l(F, \mathbb{R}^2)$ on the whole of $\mathbb{R}^2$.

The First Order Condition Number

Under suitable circumstances the condition number $\mathrm{Lip}(F, B)$ can be expected to converge to $\|F'x_0\|$ with $B \to \{x_0\}$. Then $\|F'x_0\|$ can be used as an approximation of the condition number if perturbations are small. In what follows this idea is verified and formulated more precisely.

The mathematical formulation of the transition from B to the set $\{x_0\}$ involves a characterization problem based on the first order condition.

Definition 2.6.3 (First Order Condition Number)   *Let $F : X \to Y$ and $x_0 \in X$; then*

$$
k_{F \leftarrow x}(x_0) := \lim_{\delta \to 0+} \mathrm{Lip}(F, S_\delta(x_0))
$$

is called the absolute first order condition number of F at x_0.

Due to the monotonicity (2.34), the upper bound

$$
k_{F \leftarrow x}(x_0) \le \mathrm{Lip}(F, B)
$$

is valid for all neighborhoods[7] B of x_0.

Because $\mathrm{Lip}(F, S_\delta(x_0)) \ge 0$ for all $S_\delta(x_0)$, the lower bound $k_{F \leftarrow x}(x_0) \ge 0$ holds. If $k_{F \leftarrow x}(x_0) > 0$ then the first order condition number is a suitable estimate of $\mathrm{Lip}(F, B)$ on small neighborhoods B of x_0, as the following theorem shows:

---

[7]It is conventional to consider *neighborhoods* of a point because the consequences of changing x in an *arbitrary* direction are then analyzed.

Theorem 2.6.2 *If $k_{F\leftarrow x}(x_0) > 0$, then for all $\rho > 0$ there exists $\delta > 0$ such that the condition number $\mathrm{Lip}(F, B)$ for all neighborhoods B of x_0 where $B \subseteq S_\delta(x_0)$ can be estimated by the first order condition number $k_{F\leftarrow x}(x_0)$:*

$$k_{F\leftarrow x}(x_0) \leq \mathrm{Lip}(F, B) \leq (1 + \rho)k_{F\leftarrow x}(x_0).$$

Proof: The considerations above imply that the inequality $k_{F\leftarrow x}(x_0) \leq \mathrm{Lip}(F, B)$ on the left side holds for all neighborhoods B of x_0.

In order to prove the inequality on the right side, the quantity

$$\varepsilon := \rho k_{F\leftarrow x}(x_0) > 0$$

is defined according to the assumption of the theorem. Then, there exists $\delta > 0$ with

$$\mathrm{Lip}(F, S_\delta(x_0)) \leq k_{F\leftarrow x}(x_0) + \varepsilon$$

because

$$k_{F\leftarrow x}(x_0) := \lim_{\delta \to 0+} \mathrm{Lip}(F, S_\delta(x_0)).$$

The proof is completed by substituting ε and applying the monotonicity property (2.34) on $B \subseteq S_\delta(x_0)$. □

Theorem 2.6.2 says that the absolute condition number of F in a sufficiently small neighborhood of x_0 is of the same size as $k_{F\leftarrow x}(x_0)$.

In comparison with the Lipschitz norm $\mathrm{Lip}(F, B)$, the first order condition number $k_{F\leftarrow x}(x_0)$ is much easier to compute, making it a convenient tool for sensitivity analyses. In particular, for continuously differentiable F, Theorem 2.6.1 and the monotonicity property (2.34) imply

$$k_{F\leftarrow x}(x_0) = \|F'x_0\|. \tag{2.35}$$

However, if the first order condition number vanishes, i.e., $k_{F\leftarrow x}(x_0) = 0$, it does not follow that F is absolutely insensitive to changes in x in a neighborhood of x_0.

Example (Vanishing First Order Condition Number) For the function $Fx = x^2$, the first order condition number vanishes at $x_0 = 0$, i.e. $k_{F\leftarrow x}(0) = 0$. However, $|x_1^2 - x_2^2|$ does not vanish for all $x_2 \notin \{x_1, -x_1\}$. This can be illustrated by the chord between two points in $B = (a, b) \ni 0$,

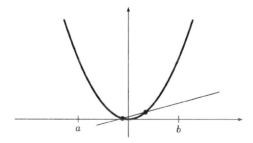

which is hardly ever horizontal, even for very small neighborhoods B of 0. In that case the first order condition number $k_{F\leftarrow x}(x_0) = 0$ does not provide all the information about the sensitivity of F to perturbations around x_0.

For linear mappings F the equation

$$k_{F\leftarrow x}(x_0) = \mathrm{Lip}(F, B) = \|F\| \tag{2.36}$$

holds for all $x_0 \in X$ and for all $B \subseteq \mathrm{interior}(X)$.

The Relative First Order Condition Number

In order to obtain local estimates of the relative condition number, the absolute first order condition number can be related to the data x of the problem and to the corresponding results Fx. By doing so the *relative first order condition number*

$$K_{F\leftarrow x}(x_0) := \frac{\|x_0\|}{\|Fx_0\|} k_{F\leftarrow x}(x_0)$$

is defined. The assumption $\|Fx_0\| > 0$ has to be made in order to obtain a useful definition of a relative condition number.

For continuously differentiable F, equation (2.35) leads to the formula

$$K_{F\leftarrow x}(x_0) = \frac{\|x_0\|}{\|Fx_0\|} \|F'x_0\|$$

for multivariate functions and to

$$K_{F\leftarrow x}(x_0) = \frac{|x_0|}{|Fx_0|} |F'x_0|$$

in the case of univariate functions.

The condition analyses in the following chapters are often based on the relative first order condition number. The argument "(x_0)" is usually omitted unless there is the danger of ambiguity.

2.6.2 The Condition of Inverse Mathematical Problems

Consider the inverse mathematical problem $Fx = y$, where F and y are given and x is to be determined. The perturbed problem

$$\bar{F}\bar{x} = \bar{y};$$

where both y *and* F are affected by errors, i.e.,

$$\bar{F} = F + \Delta F, \qquad \bar{x} = x + \Delta x, \qquad \bar{y} = y + \Delta y,$$

is to be investigated. As a prerequisite of the following analyses, F and $\bar{F}$ are assumed to have a unique inverse in a neighborhood of the solution, and the Lipschitz norm is assumed to be available either for F^{-1} or $\bar{F}^{-1}$:

$$\|x_1 - x_2\| \leq \mathrm{Lip}(F^{-1}, B)\|Fx_1 - Fx_2\| \tag{2.37}$$

or

$$\|x_1 - x_2\| \leq \mathrm{Lip}(\bar{F}^{-1}, B)\|\bar{F}x_1 - \bar{F}x_2\|. \tag{2.38}$$

It is further assumed that the data set $B \subseteq Y$, which the Lipschitz norm is based on, is the same in both cases.

The Estimation of Absolute Condition Numbers

Under the above assumptions the following holds:

$$\begin{aligned}
F\bar{x} - Fx &= F\bar{x} - y \\
&= F\bar{x} - y - \Delta y + \Delta y \\
&= F\bar{x} - \bar{y} + \Delta y \\
&= F\bar{x} - \bar{F}\bar{x} + \Delta y \\
&= -\Delta F\bar{x} + \Delta y.
\end{aligned}$$

Substituting $x_1 = \bar{x}$ and $x_2 = x$ in (2.37) leads to the inequality

$$\begin{aligned}
\|\Delta x\| &\leq \mathrm{Lip}(F^{-1}, B)\|F\bar{x} - Fx\| \\
&\leq \mathrm{Lip}(F^{-1}, B)(\|\Delta F\bar{x}\| + \|\Delta y\|).
\end{aligned} \tag{2.39}$$

By analogy,

$$\bar{F}\bar{x} - \bar{F}x = -\Delta Fx + \Delta y,$$

and from (2.38) it can be concluded that

$$\begin{aligned}
\|\Delta x\| &\leq \mathrm{Lip}(\bar{F}^{-1}, B)\|\bar{F}\bar{x} - \bar{F}x\| \\
&\leq \mathrm{Lip}(\bar{F}^{-1}, B)(\|\Delta Fx\| + \|\Delta y\|).
\end{aligned} \tag{2.40}$$

According to (2.39) and (2.40) either $\mathrm{Lip}(F^{-1}, B)$ or $\mathrm{Lip}(\bar{F}^{-1}, B)$ can be chosen as the absolute condition number of the inverse problem with respect to perturbations of the right-hand side and of the operator. From the monotonicity (2.34), it follows that the bounds (2.39) and (2.40) become more precise as B is made smaller. Ideally, for an analysis of the condition number, the region B is selected to be as small a neighborhood of $y \in Y$ as possible which still includes all perturbed values $y + \Delta y$ which can occur due to data errors Δy.

If the Lipschitz norm $\mathrm{Lip}(\Delta F, B)$ of the perturbation ΔF is known then it is possible to eliminate $\mathrm{Lip}(\bar{F}^{-1}, B)$ from formula (2.40) by estimating the latter by $\mathrm{Lip}(F^{-1}, B)$ and $\mathrm{Lip}(\Delta F, B)$. From

$$\|\Delta Fx_1 - \Delta Fx_2\| \leq \mathrm{Lip}(\Delta F, B)\|x_1 - x_2\| \qquad \text{for all} \quad x_1, x_2 \in B$$

and

$$\begin{aligned}
Fx_1 - Fx_2 &= (Fx_1 - \bar{F}x_1) + (\bar{F}x_1 - \bar{F}x_2) + (\bar{F}x_2 - Fx_2) \\
&= -\Delta Fx_1 + (\bar{F}x_1 - \bar{F}x_2) + \Delta Fx_2
\end{aligned}$$

it follows that

$$\|Fx_1 - Fx_2\| \leq \|\bar{F}x_1 - \bar{F}x_2\| + \text{Lip}(\Delta F, B)\|x_1 - x_2\|.$$

It further follows that

$$\|x_1 - x_2\| \leq \text{Lip}(F^{-1}, B)(\|\bar{F}x_1 - \bar{F}x_2\| + \text{Lip}(\Delta F, B)\|x_1 - x_2\|)$$

and finally

$$\|x_1 - x_2\| \leq \frac{\text{Lip}(F^{-1}, B)}{1 - \text{Lip}(F^{-1}, B)\text{Lip}(\Delta F, B)}\|\bar{F}x_1 - \bar{F}x_2\|,$$

provided the denominator is greater than zero, i.e., if the perturbation ΔF as measured by its Lipschitz norm $\text{Lip}(\Delta F, B)$ is sufficiently small. The last inequality gives the upper bound

$$\text{Lip}(\bar{F}^{-1}, B) \leq \frac{\text{Lip}(F^{-1}, B)}{1 - \text{Lip}(F^{-1}, B)\text{Lip}(\Delta F, B)}. \tag{2.41}$$

Using (2.41), the error bound (2.40) can be recast into

$$\|\Delta x\| \leq \frac{\text{Lip}(F^{-1}, B)}{1 - \text{Lip}(F^{-1}, B)\text{Lip}(\Delta F, B)}(\|\Delta Fx\| + \|\Delta y\|). \tag{2.42}$$

The error bounds (2.39) and (2.40) are complementary in a remarkable way: In (2.39) the Lipschitz norm $\text{Lip}(F^{-1}, B)$ is found together with $\|\Delta F\bar{x}\|$, whereas in (2.40) $\text{Lip}(\bar{F}^{-1}, B)$ and $\|\Delta Fx\|$ are combined. As explained in the introductory remarks, the user of these formulas is free to decide which of the two problems he wants to regard as the perturbed one and which as the unperturbed one. In this sense (2.40) is readily derived from (2.39) by simply interchanging the expressions "perturbed problem" and "unperturbed problem" (i.e., by interchanging barred and non-barred quantities).

As with direct problems, the norm of a derivative—$\|(F^{-1})'\|$ or $\|(\bar{F}^{-1})'\|$—can be used as a satisfactory approximation of the Lipschitz norms for F^{-1} and $\bar{F}^{-1}$, provided the perturbations are not too large.

The Relative Condition Number of Linear Problems

In this section relative condition numbers of *linear* inverse problems (e.g., systems of linear equations) are derived. From equation (2.36), which holds for linear operators F and ΔF and for a set B containing a non-empty open ball $S_\delta(x_0)$ centered at $x_0 \in B$, the following hold:

$$\text{Lip}(F, B) = \|F\|, \qquad \text{Lip}(F^{-1}, B) = \|F^{-1}\| \quad \text{and} \quad \text{Lip}(\Delta F, B) = \|\Delta F\|.$$

Thus, in this case the Lipschitz norms can be replaced by the operator norms; the dependence on B disappears; and relative condition number estimates can be derived without difficulty.

The Effect of a Perturbed y

Suppose only y is perturbed and F is given without any errors ($\Delta F = 0$);

$$\|\Delta x\| \leq \text{Lip}(F^{-1}, B)\|\Delta y\| = \|F^{-1}\|\|\Delta y\|$$

follows from (2.39). This inequality and

$$\|y\| = \|Fx\| \leq \|F\|\|x\|$$

lead to a bound on the relative error:

$$\frac{\|\Delta x\|}{\|x\|} \leq \|F\|\|F^{-1}\|\frac{\|\Delta y\|}{\|y\|}.$$

The Effect of a Perturbed F

If only a perturbation of the function F is taken into consideration and y remains unperturbed ($\Delta y = 0$), then

$$\begin{aligned}
\|\Delta x\| &\leq \text{Lip}(F^{-1}, B)\|\Delta F\bar{x}\| \\
&= \|F^{-1}\|\|\Delta F\bar{x}\| \\
&\leq \|F^{-1}\|\|\Delta F\|\|\bar{x}\|
\end{aligned}$$

follows from (2.39). Hence,

$$\frac{\|\Delta x\|}{\|\bar{x}\|} \leq \|F\|\|F^{-1}\|\frac{\|\Delta F\|}{\|F\|}. \tag{2.43}$$

From the error bounds of both types of perturbations, the condition number $\|F\|\|F^{-1}\|$ for linear inverse problems is derived; however, in (2.43) the relative error is related to the perturbed solution $\bar{x}$. A formula analogous to (2.43), which relates the relative error to x rather than $\bar{x}$ can be derived from (2.42) for $\Delta y = 0$:

$$\frac{\|\Delta x\|}{\|x\|} \leq \frac{\|F\|\|F^{-1}\|}{1 - \|F^{-1}\|\|\Delta F\|}\frac{\|\Delta F\|}{\|F\|}.$$

Relative error estimations are also possible for nonlinear inverse problems, provided suitable bounds

$$\|\Delta Fx\| \leq M(\Delta F)\|x\| \qquad \text{and} \qquad \|Fx\| \leq M(F)\|x\|$$

are available, i.e., if the behavior of the functions F and ΔF in the region B is similar to that of linear mappings.

2.7 Validation of Numerical Computations

Numerical computations are subject to uncertainty which originates from a variety of sources. For example, inadequate scientific models or data seriously afflicted with inaccuracies can occur. Hence, finding out whether or not the accuracy of numerical results meets specific requirements is crucial. However, there is no way of knowing this for sure!

Although the inherent uncertainties of numerical computations can never be completely disposed of, this chapter briefly discusses a number of techniques which may reduce the degree of uncertainty (to what degree is very hard to quantify in practice).

Methods for reducing the degree of uncertainty require a great deal of effort. An examination of all the factors of uncertainty usually increases the overall effort by at least an order of magnitude.

Example (Systems of Linear Equations)   Solving a system of linear equations using an implementation of the Gaussian algorithm involves a considerable number of uncertainties: In general, the program does not recognize ill-conditioned problems. There may thus be cases in which a completely useless result is returned. However, the additional determination of condition number estimates requires much larger and more complicated programs and, depending on the method used, from a 30 % increase in computation time (needed for determining an estimate or bound of the condition number) to 200 % or more (needed for a singular value decomposition).

Examining the validity of a numerical computation, i.e., examining the extent to which the results comply with the accuracy requirements, is referred to as *validation*. The most common method of validation is *testing*—a verification by lack of falsification, i.e., falsification fails in spite of serious efforts to find flaws.

2.7.1 The Uncertainty of Numerical Computations

The result of any numerical computation is afflicted with some degree of uncertainty. This uncertainty may be due to model, data, algorithm, and computational errors; but may also be caused by a poor choice and incorrect use of numerical software.

Model Uncertainty

Models only cover some aspects of reality (cf. Chapter 1 and Chapter 8). Knowledge acquired from models can therefore only be transferred to reality with caution. The significance of every model based statement concerning the properties of the original system must be carefully checked.

Moreover, inadequate models may lead to unstable algorithms and thus contribute to model uncertainty.

Numerical Uncertainty

Developing a mathematical model is always the first step in the numerical solution of application problems. The model may represent an ideal approach, but may not be the basis for developing a useful numerical algorithm. As a result, it may be necessary to modify this model and create new algorithms. The uncertainties arising in the construction of implementable algorithms occur in the form of *algorithm errors* (see Section 2.3.3).

Another source of numerical uncertainties is *rounding error* (see Section 2.3.4) caused by the inevitable shortcomings of machine numbers and the computer arithmetic.

Software Uncertainty

Other reasons for the uncertainty of numerical computations result from the use of numerical software:

1. The choice of ready-made software involves the risk of selecting a program which is unsuitable for the problem at hand and may lead to inaccurate or useless results.

2. The inadequate use of ready-made software may also lead to incorrect results or may be the basis of an incorrect interpretation of results.

3. In practice it is not possible to eliminate software bugs completely; such bugs include design errors, coding errors, shortcomings in the documentation, and so on. They pervade all types of numerical software: ready-made software (commercial software as well as public domain programs on computer networks) and, last but not least, numerical software designed by the user himself.

4. System software, compilers and optimizers may also contain bugs or may be used incorrectly.

There is no comprehensive method for validating and assessing numerical computation. Model uncertainty, numerical uncertainty and software uncertainty are therefore discussed separately in the following sections.

2.7.2 Validation of Mathematical Models

The *validation of models* is one of the main topics of applied mathematics. In validating a model, the computational results obtained from the model and a particular set of parameters are compared to the real phenomena from which the model was derived. Suitable experiments have to be performed to determine reference values.

The fundamental questions involved in model validation are related to the conformity between the model and the real object or phenomenon: To what extent does the behavior of the model correspond with the behavior of the original system? How reliable and accurate is the model in as far as the reproduction of empirically determined quantities of the original system is concerned? Is the structure of the model too coarsely grained, and does it need to be refined? Which data sets cause instability in the model behavior?

There is no certainty concerning the compliance of the model with the required level of accuracy, irrespective of the validation effort. The computational results may always deviate significantly from the situation being modeled. In particular, values of parameters which were not taken into consideration when validating the results may well involve unacceptably large discrepancies.

2.7.3 Sensitivity Analysis and Error Estimation

Empirical sensitivity analysis is a qualitative as well as quantitative investigation into the sensitivity of computational results. It is based on deliberately made changes to the model structure, the input parameters, the solution algorithms, the machine arithmetic, and other influential factors. The types and sizes of these deliberately produced perturbations depend greatly on the individual problem. If these perturbations only cause small changes in the computational result which can be tolerated in the context of the given numerical problem, then the degree of uncertainty decreases. Although empirical sensitivity analyses must be designed and executed within the context of the given problem, there are still general aspects which can be discussed.

The Impact of Model Errors

Sensitivity analyses, which are an important part of the model building process, can be quite complicated since changes in models have to be made and analyzed. In numerical data processing, tools are available which make the task of empirical sensitivity analysis much easier.

The Impact of Data Errors

Monte Carlo methods (see, for instance, Chapter 12) can be used to analyze the impact of data errors on results: Elements of the actual data set (representative of the application) are exposed to random perturbations, the distribution and correlation properties of which correspond as close as possible to the data errors that actually occur. The experimentally determined distribution characteristics of the computational results gives partial information about the sensitivity of algorithms and programs to data and rounding errors at the same time.

If there are large inaccuracies in the numerical results and a Monte Carlo analysis rules out certain input data as the cause of these inaccuracies, and the algorithm error is also sufficiently small, then the model itself is inadequate. In this case the inappropriate model must be replaced by a more appropriate one, and the whole validation process must begin all over again.

Software (Sensitivity Analysis) The *Software for Rounding Analysis* in TOMS/532 can be used to analyze data and rounding error sensitivity. The programs contained in this package analyze Fortran 77 programs (in particular, those for linear algebra computations) and detect input data sets which cause the most serious instability.

The Impact of Rounding Errors

If a computer program is executed once using single precision arithmetic and then again using double precision arithmetic, then in many cases the identical decimal places of the two results are correct with respect to the machine arithmetic.

Example (Iterative Solution of a Nonlinear Equation) In order to solve the equation

$$\cosh x + \cos x = a \quad \text{where} \quad a \in \mathbb{R} \tag{2.44}$$

on the basis of Newton's method

$$x_{n+1} = x_n - \frac{f(x_n)}{f'(x_n)},$$

a program was written for $f(x) = \cosh x + \cos x - a$ using the termination criterion

if $|x_n - x_{n+1}| < \varepsilon$ **then exit**.

With the data values $a = 2$, $x_0 = 1$ and $\varepsilon = 10^{-6}$, the program returned the following results:

$x_{13} = 0.0310\,800$ for *single* precision arithmetic;
$x_{32} = 0.0001\,307$ for *double* precision arithmetic.

The large discrepancy between the two values suggests that the result is inaccurate and that effects of the floating-point arithmetic are responsible for this inaccuracy. In this particular case, the program encounters some difficulty because of the fact that $x^* = 0$ is a multiple zero of equation (2.44).

Moreover, this example demonstrates that program failure is *not* detected by varying the tolerance parameter ε: in single precision arithmetic. For all values

$$\varepsilon = 10^{-2}, 10^{-3}, 10^{-4}, \ldots, 10^{-8}$$

the same result, $x_{13} = 0.0310\,800$, is returned by the program. This is completely unexpected and is contrary to what should be returned from the program. Such behavior indicates a mistake in the program.

The Impact of Algorithm Errors

There are an enormous number of algorithms implemented in numerical software available for the solution of standard mathematical problems. In many cases it is thus possible to examine the dependence of numerical results on the underlying algorithm by simply replacing one library routine with another.

Numerical algorithms are usually controlled by input parameters which specify the required accuracy, the maximum number of iterations, and so on. A sensitivity analysis must investigate the dependence of the numerical results on those parameters.

Example (Numerical Integration) If a globally adaptive integration program based on Gauss-Kronrod formulas, e.g., QUADPACK/qag, is used to solve an integration problem, then the dependence of the results on the algorithm can be examined by selecting different pairs of

Gauss-Kronrod formulas (ranging from 7/15-point formulas to 30/61-point formulas) via the input parameter key.

Moreover, the dependence of the results on the parameters specifying the required accuracy (epsabs and epsrel for the program QUADPACK/qag) should be examined. To do so, at least three program runs are advisable: one with the originally selected parameters and two more with accuracy parameters scaled by factors of 10 and 0.1. However, such an experiment can never rule out anomalies, e.g., an undetected *peak* of the integrand that can still cause inaccurate results.

Empirical Condition Analysis

In Sections 2.4 and 2.6 it was implicitly assumed that condition numbers such as $\mathrm{Lip}(F, B)$ and $\mathrm{Lip}(F^{-1}, D)$ can be computed or at least estimated. However, for complicated problems which occur in practice, efforts to determine condition numbers for particular constellations often fail. As some kind of information on the sensitivity of the results to data errors is indispensable for assessing the accuracy of results, this knowledge has to be acquired by simulating the perturbations. In doing so, the usual scheme

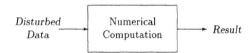

is replaced by an extended scheme

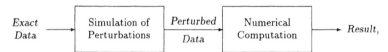

beginning with the exact data. However, the exact data of an application problem is often not known. Thus, it is necessary to have an alternative data set which should be as much like the actual data as possible. Ideally, the exact result for this data set is known in advance. This artificial input data set is perturbed in a way similar to the way data are actually perturbed (by errors in measurement etc.). (Pseudo-) random numbers with a distribution similar to that of an actual perturbation can be used for that purpose (see Chapter 17). A random number generator can be used to produce several perturbed data sets used as inputs for the investigated numerical method. This leads to a set of results (belonging to the original *exact* result), the variability of which offers insight into how sensitive the method is to data errors. The statistical analysis of the values thus obtained is too large a subject to be handled in this book, but the user is warned not to jump to conclusions: a small variance (or standard deviation) of the result values does *not* imply small errors (cf. the density functions in Fig. 2.20).

The empirical condition analysis is an effective tool for reducing uncertainty. However, its use is limited, in particular, in a *systematic* analysis of *all* data variations. Suppose, for example, that a numerical problem has 100 scalar input data (e.g., the elements of a 10×10 matrix), and all perturbed data sets which

Figure 2.20: The results to the left have a large variance, their mean value, however, coincides with the exact value. This is contrasted by the results to the right which have a small variance, but are afflicted with a significant systematic error.

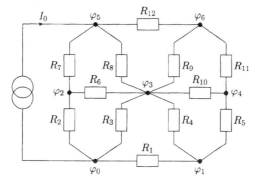

Figure 2.21: Electric circuit with 12 resistors.

result from the variation of each individual data element by $+1\,\%$ or $-1\,\%$ are to be examined. In this case $3^{100} \approx 5 \cdot 10^{47}$ computations would be needed.

Statistical methods are the only key to a manageable sensitivity analysis: In a statistical approach, a subset of the 3^{100} possible data sets which is easy to handle is selected using a random number generator. The number of computations to be performed in experiments designed in this way is determined according to statistical criteria.

Example (Empirical Condition Analysis) An electric resistor circuit (see Fig. 2.21) was subject to an experimental condition analysis. It was based on the following parameter values:

$$R_1 = 60\Omega, \quad R_2 = 10\Omega, \quad R_3 = 5\Omega, \quad R_4 = 70\Omega, \quad R_5 = 5\Omega, \quad R_6 = 100\Omega,$$
$$R_7 = 30\Omega, \quad R_8 = 50\Omega, \quad R_9 = 30\Omega, \quad R_{10} = 10\Omega, \quad R_{11} = 5\Omega, \quad R_{12} = 80\Omega,$$
$$I_0 = 100A.$$

Using the node-potential method a system of linear equations[8]

$$
\begin{pmatrix}
a_{11} & 0 & a_{13} & a_{14} & 0 & 0 \\
0 & a_{22} & a_{23} & 0 & a_{25} & 0 \\
a_{31} & a_{32} & a_{33} & a_{34} & a_{35} & a_{36} \\
a_{41} & 0 & a_{43} & a_{44} & 0 & a_{46} \\
0 & a_{52} & a_{53} & 0 & a_{55} & a_{56} \\
0 & 0 & a_{63} & a_{64} & a_{65} & a_{66}
\end{pmatrix}
\begin{pmatrix}
\varphi_1 \\
\varphi_2 \\
\varphi_3 \\
\varphi_4 \\
\varphi_5 \\
\varphi_6
\end{pmatrix}
=
\begin{pmatrix}
0 \\
0 \\
0 \\
0 \\
I_0 \\
0
\end{pmatrix}
\tag{2.45}
$$

---

[8]The electrical units V, A and Ω are omitted in the mathematical formulation of the problem.

can be derived for the circuit of Fig. 2.21. The coefficients $a_{11}, \ldots, a_{66}$ are derived from the resistors $R_1, \ldots, R_{12}$:

$$
\begin{aligned}
a_{11} &= G_1 + G_4 + G_5 &&\approx 0.231 & a_{13} &= -G_4 &&\approx -0.014 \\
a_{22} &= G_2 + G_6 + G_7 &&\approx 0.143 & a_{14} &= -G_5 &&= -0.200 \\
a_{23} &= -G_6 &&= -0.010 & a_{25} &= -G_7 &&\approx -0.033 \\
a_{31} &= -G_4 &&\approx -0.014 & a_{32} &= -G_6 &&= -0.010 \\
a_{33} &= G_3 + G_4 + G_6 + G_8 + G_9 + G_{10} &&\approx 0.378 & a_{43} &= -G_{10} &&= -0.100 \\
a_{34} &= -G_{10} &&= -0.100 & a_{35} &= -G_8 &&= -0.020 \\
a_{36} &= -G_9 &&\approx -0.030 & a_{41} &= -G_5 &&= -0.200 \\
a_{44} &= G_5 + G_{10} + G_{11} &&= 0.500 & a_{56} &= -G_{12} &&\approx -0.013 \\
a_{63} &= -G_9 &&\approx -0.033 & a_{64} &= -G_{11} &&= -0.200 \\
a_{66} &= G_9 + G_{11} + G_{12} &&\approx 0.246 & a_{65} &= -G_{12} &&\approx -0.013
\end{aligned}
$$

$$
\text{Conductance values} \quad G_i := \frac{1}{R_i}, \quad i = 1, 2, \ldots, 12.
$$

In an empirical condition analysis, the resistors were perturbed in a particular way: The distribution of the perturbations corresponds to a *sliced* normal distribution, such that the perturbations deviate from the exact value by 5 % to 10 % (see Fig. 2.22).

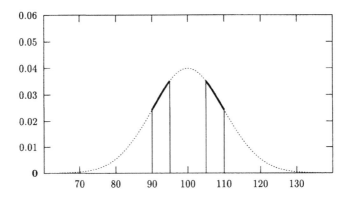

Figure 2.22: Distribution of the perturbations applied to the resistors (related to 100%).

A suitably constructed random number generator was then used to perturb the resistor values $R_1, \ldots, R_{12}$, and each of the resulting systems of linear equations (2.45) was solved numerically. The statistical analysis of the results showed amazing phenomena, such as the distribution of the value I_8, i.e. the electric current through R_8, as depicted in Fig. 2.23.

In this example, an unusually large number of simulation runs was performed: The distribution of the current I_8 depicted in Fig. 2.23 is based on 10 000 runs. In practice, the number of runs has to be chosen according to the properties of the parameters examined, the accuracy requirements, and the computing time available.

Theoretical Error Analysis

Error analysis is an alternative to experimental sensitivity analysis. It is a *mathematical* examination of various uncertainty factors and is qualitative as well as quantitative. Different types of error analyses may lead to different results for the same problem. For example, there are different mathematical methods for

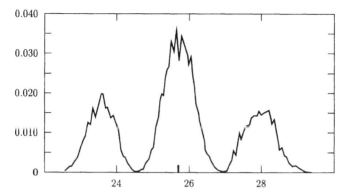

Figure 2.23: Frequency distribution of I_8 (10 000 simulation runs).

examining the effect of data errors on the solution vector of a system of linear equations, which may lead to different assessments.

There may be a number of difficulties connected with error analyses:

1. The necessary mathematical investigations may be too complicated to make an error analysis feasible.

2. The estimation of the impact of the different factors may be far too pessimistic (i.e., the effects may be overestimated); as a result, the findings may be useless.

3. The prerequisites may be restrictive to such an extent that an error analysis cannot be done: For example, the mathematical analysis of the accuracy of numerical integration schemes requires upper bounds on certain derivatives of the integrand function which are often not available in practice.

In other words, neither empirical sensitivity analysis nor mathematical error analysis supply unfailing mechanisms for the validation of numerical results. Nevertheless, both techniques are suitable for improving the reliability of numerical computations. Complicated problems with high reliability requirements always necessitate that a considerable amount of work be done in order to validate them.

2.7.4 Validation of Numerical Software

Roughly speaking, a software bug is usually manifested in software which exhibits undesirable or unexpected behavior. Many typical non-numerical data processing problems make it easy to decide whether or not the results comply with specifications. For example, the output of a sorting program can easily be checked. On the other hand, it is sometimes completely impossible to tell if there is a bug in numerical software. This dilemma is caused by the use of heuristic approaches in the individual routines, rather than by flaws in the documentation. In this context the term "heuristic" means components of algorithms that are not derived from mathematical or scientific knowledge but rather stem from the programmer's intuition. Numerical software is rarely free of heuristic elements.

Example (Numerical Integration)   Many programs in the numerical integration package QUADPACK [22] contain heuristic sections for detecting integrands contaminated by stochastic disturbances. These programs are designed to avoid unnecessary computational effort and to stop computation as soon as the iterative algorithm cannot improve the result any more (see Section 12.5.6). However, in some cases, e. g., for heavily oscillating integrands, an undisturbed integrand may be classified as disturbed. It is not possible to know if this is caused by a program bug or not: *All* numerical integration programs are based on the principle of point-wise sampling, whence distinguishing between stochastically disturbed functions and highly oscillating functions is not possible with a finite amount of effort. Nevertheless, the so-called *noise detector* is an indispensable part of the algorithm, which makes a heuristic approach unavoidable.

Software Testing

For many mathematical problems the potential data set is considerably large. For example, the data set of the integration problem comprises all Riemann integrable functions (defined on a particular region).

Tests usually have to confine themselves to finite (and generally small) random samples, so the quality of a numerical algorithm or a numerical program cannot even be assessed statistically with much certainty.

Example (The Cardinality of the Set of Functions)   On a computer the continuum $\mathbb{R}$ of real numbers is replaced by a *finite* set $\mathbb{F}$ of rational numbers, floating-point numbers (see Chapter 4).

Accordingly only functions $\tilde{F} : \mathbb{F}^n \to \mathbb{F}^m$ can be defined on a computer, and functions $F : \mathbb{R}^n \to \mathbb{R}^m$ do not exist. The set of *all* functions $\tilde{F}$ that can be defined for a particular set of machine numbers $\mathbb{F}$ is *finite*. However, even in the simplest case, $m = n = 1$, i. e., $\tilde{f} : \mathbb{F} \to \mathbb{F}$, the number of different functions is enormous: about $10^{40\,000\,000\,000}$ for single precision IEC/IEEE floating-point numbers. Even extensive experiments, therefore, cannot examine a representative random sample of this set.

Software Verification

The following criteria must be met in order to perform a software verification:

1. There must be a formal (mathematical) specification of the tasks the program is assumed to accomplish, i. e., a precise description of the program output as a function of the program input is a necessity.

2. There must be analytical mechanisms which allow for the verification of the correctness of the rules which transform the input values into the desired output.

Programs which use heuristic approaches do *not* meet the first condition. It is of course possible to construct a formal specification of the heuristic components based on the actual behavior of the respective program fragments after they have been implemented. However, this kind of specification does not convey any useful information concerning the correctness of these program components.

Example (Numerical Integration) The *noise detector* is always based on heuristic considerations, i.e., prior to actual coding, a subalgorithm which fulfills the task of detecting stochastically disturbed functions is established. The verification of such a program cannot guarantee the expected behavior but can only identify whether or not the algorithm is implemented exactly as it was specified.

The first condition is also violated when there is no precise mathematical description of the scope that a software product is expected to cover.

Example (Numerical Integration) In the program package QUADPACK [22] there is, among many other codes, a non-adaptive program QUADPACK/qng which is intended to be used for smooth integrand functions (functions without singularities, oscillations, etc.). However, there is no precise specification of the set of *all* integrand functions $\mathcal{F}$ for which this program is reliable and efficient.

The second condition is generally *not* met due to the size and the complexity of numerical software. Even automatic verification systems cannot deal with numerical software because of the complexity of most numerical problems. Further obstacles to the use of such verification systems are the effects of rounding errors or special mathematical properties required of the numerical solution (e.g., the orthogonality of a solution matrix).

Chapter 3

Computers for Numerical Data Processing

This apparatus was regarded as a prodigy of nature, because it captured in a machine an art which altogether dwells in man's mind, and because thus means were discovered to execute all operations of this art with perfect confidence without exerting one's mind.

Report on the calculator developed by PASCAL (1642)

Both the simulation and optimization of processes in science, engineering or economics usually involve a huge number of computational steps which are to be performed within a given time limit.

For complex, large-scale computations, there has always been a demand for higher performance, i.e., a larger number of computational steps executed per time unit. Hardware developers and manufacturers have indeed been able to meet this demand to a high degree: since 1984 the peak performance of microprocessors has nearly doubled every year.

The stunning performance increase of modern computer systems is mainly due to the continued miniaturization of electronic modules and innovations in computer architecture.

Modern Integrated Circuit Technology

Primarily, improvements in the field of semiconductor technology are advanced by the incessant miniaturization of electronic components. Over the last ten years, the number of transistors per unit area on a chip (its *packing density*) has increased by about 60 % every year. In 1961 when the first commercial, integrated circuits appeared they contained no more than four transistors (in addition to other components). Today, up to nine million transistors are accommodated on a chip of 100 to 300 mm^2.

The increase in packing density allows for the development of still more complex processors as well as memory chips of increased capacity. Also, the switching speed of transistors, too, is inversely proportional to their physical size. This makes higher processor clock rates and hence performance improvements possible. Another way to boost the processor clock rate is to use new semiconductor technology such as BiCMOS.[1]

---

[1] BiCMOS integrated circuits combine CMOS (complementary metal oxide semiconductor) technology with bipolar transistors on one chip.

A technological limitation to clock rate increases is imposed by the electric power consumption which is entailed. The heat emission of a high performance processor makes sophisticated cooling devices necessary.

Advanced Computer Architecture

The most promising way to increase modern computer performance is to modify computer architecture, i. e., to introduce new organizational and functional principles. Among other things, *parallelism* has been the most important concept for increasing computer performance. Conventional computers tackle tasks step by step, while modern computers are capable of dealing with several subtasks concurrently. The following is a list of landmarks in computer architecture development in the past three decades:

in the seventies: the development of vector processors,
in the eighties: the "RISC revolution",
in the nineties: the widespread use of multiple-CPU computer systems.

However, these achievements have been followed by an unwelcome phenomenon which troubles the users of modern high-performance computers: The gap between the hardware peak performance and the actual performance achieved by application programs is becoming wider and wider. It is increasingly difficult for application programmers to utilize the available hardware resources to a satisfactory degree.

Example (Matrix-Matrix Multiplication) The multiplication of two matrices is a basic operation required by many linear algebra algorithms. Hence, the performance of computers when executing this operation is of particular importance in scientific computing.

The simplest algorithm for the matrix-matrix multiplication operation is derived directly from the definition of the matrix dot product. An implementation of

$$C = AB \quad \text{where} \quad A \in \mathbb{R}^{m \times p}, \ B \in \mathbb{R}^{p \times n}, \ C \in \mathbb{R}^{m \times n}$$

is given by the following code fragment:

```
c = 0
DO i = 1, m
   DO j = 1, n
      DO k = 1, p
         c(i,j) = c(i,j) + a(i,k)*b(k,j)                    (3.1)
      END DO
   END DO
END DO
```

However, this *classical* method turns out to be unsuitable for yielding satisfactory performance rates on contemporary computer systems. As experiments with square matrices $(m = n = p)$ on a RISC workstation with 16 MB main memory and a theoretical peak performance of 50 Mflop/s demonstrate, the performance of that program is no more than 13 Mflop/s for square matrices of small order $(n \leq 300)$ (see Fig. 3.1). For matrices of medium order $(300 \leq n \leq 800)$, performance was generally below 5 Mflop/s, i. e., less than 10 % of peak performance. For matrices of higher order, with $n \geq 800$, performance collapses completely.

Hence, even for such a simple algorithm, exploiting the hardware potential is by no means straightforward.

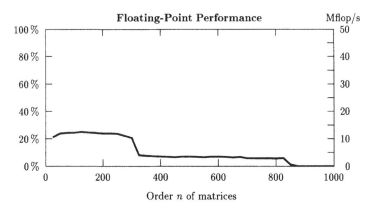

Figure 3.1: Performance and efficiency of a RISC workstation when executing the *classical* matrix-matrix multiplication (3.1) in double precision arithmetic.

As shown by this example (which is typical for modern computer systems), actual performance rates are poor, especially for large-scale problems where the amount of computational work would require outstanding performance over and above anything else.

Increased peak performance due to innovative computer architecture can only be exploited to a satisfactory degree if application programs are individually re-structured and modified. Analyzing and optimizing the performance of numerical programs is impossible without insight into the structure and function of modern computers.

The primary purpose of this chapter is to discuss the fundamentals of processor architectures insofar as they are relevant to numerical data processing. In particular, the most important developments which have triggered the rapid performance increases in the past several years are dealt with.

The other sections of this chapter deal with the performance assessment of computer systems (both hardware and software). Optimization of numerical programs is discussed in Chapter 6.

3.1 Processors

There are two main approaches to improving processor performance: increasing the clock rate and parallel processing.

Increasing the Clock Rate

The performance of a processor can be increased in proportion to its clock rate unless interfaces to other units, especially to the memory system (see Section 3.2), impose restrictions.

Example (DEC Alpha Processors) The single chip implementation of the DEC Alpha architecture of the RISC processor Digital 21064 is operated at a clock rate of 100 to 133 MHz, the 21064A processor at 225 to 275 MHz, the 21164 processor at 250 to 333 MHz and the 21164A processor at 417 MHz, i.e., a performance boost by a factor of four is possible by using a processor with a higher clock rate.

Increased clock rates are made possible by innovations in the field of circuit technology, as mentioned earlier, as well as measures affecting the processor architecture (e.g., superpipelining[2] as discussed in Section 3.1.2). But even the most remarkable improvements in this field reach a limit: the speed of light. In one nanosecond (1 ns $= 10^{-9}$ s) an electric impulse travels only 100 to 250 mm inside and/or between computer components (depending on the conducting material)[3], and the clock rates of today's fastest computers are of this order of magnitude.

Parallelization

When all known technical means of increasing the clock rate are exhausted, further improvements of processor performance are possible if the value CPI $= 1$ (one clock cycle required per instruction; see Section 3.4.2) which is typical for RISC processors, can be reduced to a value below one. This can only be achieved if the processor can be made to execute more than just a single instruction at a time. The *parallel* (i.e., concurrent and independent) execution of tasks and/or subtasks, gives rise to new means of enhancing performance. *Pipelining*, the *superscalar principle* and *vectorization* are examples of processor speedup achieved using parallelization. Such processors no longer follow the ideas of the von Neumann machine, which require the serial, i.e., sequential execution of single instructions.

Note (Multiprocessor Hardware)   Performance improvements gained by using more than one CPU, which is implemented in *multiprocessor computers*, are not discussed in this book.

3.1.1 Pipelining

RISC workstations have been subject to the most significant performance boost of late. For each year since the introduction of such computer systems, their peak performance has nearly doubled. However, the clock rate has doubled only every three years!

The performance improvement of RISC machines was primarily due to the ever more extensive application of pipelining, i.e., the overlayed, simultaneous execution of different substeps of instructions (Kogge [255]).

Terminology (RISC)   The acronym RISC (*Reduced Instruction Set Computer*), which was initially intended to distinguish computer architectures with reduced instruction sets from *Complex Instruction Set Computers* (CISC), is largely misleading nowadays. Many so-called RISC computers have an instruction set which is anything but simple. Indeed, the *RISC movement* of the eighties made a number of changes in processor architecture in addition to reducing the

---

[2]Note that superpipelining is, above all, a strategy for concurrently processing different steps of single instructions. However, this partitioning into small subtasks also make it possible to increase the clock rate.

[3]In a *vacuum* light travels about 300 mm in one nanosecond.

instruction set. Among these were the elimination of the *microprogram layer* (the implementation of instructions of a machine language within the microprogram unit of the processor) and the introduction of pipelining. The reduction of the instruction set turned out to be of minor significance for performance improvements compared to other modifications. Hence, in the course of time, the policy of reducing the instruction set, which once gave this processor class its name, was gradually abandoned (Hennessy, Patterson [55]).

The execution of an instruction consists of a continually repeated sequence of actions, the *instruction cycle*. Typically, an instruction cycle comprises the following five phases:[4]

Instruction fetch phase (IF): moving an instruction from a slower layer of the memory hierarchy to an instruction register.

Decoding phase (D): decoding and interpretation of the instruction; extracting operand addresses.

Operand fetch phase (OF): moving the required operands from slower layers of the memory hierarchy to data registers.

Execution phase (EX): processing the operands in the arithmetic unit.

Save phase (SV): moving the result to a register or to a slower layer of the memory hierarchy.

The chronological order of phases during the sequential processing of instructions is depicted in Fig. 3.2.[5] When data or instructions are retrieved from memory during the instruction or operand fetch phase, all functional units of the processor remain unused. At the end of the operation, when the result is being stored, these units are also inactive. Hence, in a processor executing all phases of an instruction sequentially, all but one of the functional units are idle at any time. This is indeed a waste of resources.

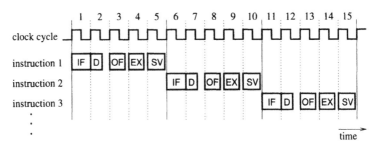

Figure 3.2: Sequence of phases for the serial execution of machine instructions.

---

[4]The order of phases as described here should be looked upon as a generic instruction cycle. The exact partition into phases varies from processor to processor.

[5]It is assumed here that each phase of an instruction does not take more than one clock cycle, as commonly found in RISC machines. In practice, particular phases, e.g., fetching operands from the cache or main memory, may extend over several clock cycles.

This unsatisfactory situation can be improved by using a technique long known in industry: the assembly line. To this end the manufacturing process is partitioned into many small steps which are carried out concurrently and independently.

If the phases of a sequence of instructions can be executed independently at the same time, then the execution of consecutive instructions can be overlayed. Similar to industrial assembly-line production, the execution is said to follow the principle of *instruction pipelining*. Fig. 3.3 depicts the resulting time diagram for the concurrently executed phases.

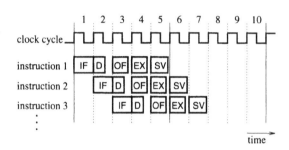

Figure 3.3: Concurrent execution of machine instructions in a five-stage pipeline.

The overlapped execution of instructions does not reduce the number of clock cycles required for the execution of a single instruction; hence the execution time of a single instruction remains the same.

However, the number of clock cycles between the completion of successive machine instructions is reduced by a factor corresponding to the number of pipeline stages (the *pipeline depth*). Hence, the throughput, the number of instructions completed per time unit, is increased by the same factor. The speed gain approaches this factor as the number of concurrently executed instructions increases.

Example (RISC Processors)   The sequential execution of machine instructions according to Fig. 3.2 requires at least a five-clock-cycle interval between the completion of instructions. In the case of a five-stage pipeline as depicted in Fig. 3.3, the minimal interval between the completion of consecutive instructions is just one clock cycle. This reduction to 1/5 of the sequential completion time corresponds to the pipeline depth.

As a prerequisite for the concurrent execution of different instruction phases there must be an autonomous processing unit for each phase, which requires an appropriate processor design.[6]

Pipeline Stalls

So far it has been assumed that all necessary memory access operations may be performed within one clock cycle. RISC processors achieve this one-cycle

---

[6]In practice the same processing unit may be used for the execution of different phases of particular instruction sequences. Obviously, such phases can then no longer be executed concurrently.

memory access by fetching operands from registers and writing results back to them. The goal is to access the slower levels of the memory hierarchy—the cache memory and the even slower main memory—as seldom as possible (see Chapter 3.2.1) by exploiting data locality and data reuse. In practice this ideal of exclusively accessing registers cannot be maintained; from time to time slower levels of the memory hierarchy have to be referenced. In such cases the execution of all instructions in the pipeline is delayed until the necessary memory access is completed (see Fig. 3.4). This is called a *pipeline stall*. Pipeline stalls also occur, for example, if an instruction requires the result of one of its predecessors which is still being executed—this is said to be a logical dependence of phases, which makes parallelism impossible.

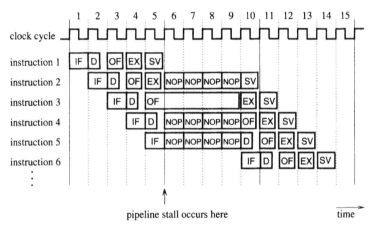

Figure 3.4: *Pipeline stall* due to delay during the operand fetch phase of instruction 3. The instruction cycle is extended by empty instructions (NOP = *no operation*).

It is the compiler's task to arrange program instructions so that pipeline stalls occur as infrequently as possible. Programming techniques such as loop unrolling (see Section 6.6) can assist the compiler in this task. By using such means it is possible, for example, to place an unrelated instruction between a loading operation from memory and the matching processing instruction so that a pipeline stall is averted.

Further Applications of Pipelining

Generally, pipelining can be applied to accelerate any process which can be decomposed into a sequence of *independent* phases. Any speed gain in execution time is due to the layered execution of those phases. An upper limit on the speedup factor is given by the maximum number of phases which can be executed concurrently.

Modern computers also employ the principle of pipelining in order to accelerate memory access and floating-point operations (addition, multiplication, comparison, type conversion—sometimes square root and division as well).

A necessary requirement for applying pipelining is the independence of the different phases of consecutive instructions. In practice, this requirement is often not fully met. Frequently, sequences occur which do not fully utilize the pipeline, causing a type of pipeline stall, the reasons for which are discussed in detail in Section 6.2.

3.1.2 Superpipeline Architecture

Pipelining primarily aims at reducing the overall time needed to perform a task by increasing the number of instructions being executed concurrently per clock cycle. The clock rate itself is not affected. The increasingly finer decomposition of instructions results, however, in shorter and shorter execution times for individual phases. It is thus possible to further improve performance by increasing the clock rate for sufficiently refined decompositions.

Pipelines with more than five stages are called *superpipelines*. Older RISC processors with three, four or five stages are called *underpipelined.*

However, the number of pipeline stages is limited, firstly because the number of phases into which an instruction may be properly decomposed is limited, secondly because increasing the number of pipeline stages results in more complex hardware, which adversely affects signal times and hence impedes clock rate increments, and thirdly because the startup time (which arises, for instance, after each jump instruction) of the pipeline is proportional to the number of stages. For complex operations more stages are usually suitable than for simple ones whence pipelines for floating-point operations tend to be longer than those for integer operations.

Example (DEC Alpha Processor)   The DEC Alpha processor 21164 contains five parallel superpipelines: two 9-stage pipelines in the floating-point unit, one 7-stage pipeline in the integer unit, and two 12-stage pipelines for load and store operations.

3.1.3 Superscalar Architectures

The best that can be achieved by the application of pipelining—provided that no pipeline stalls occur—is one machine instruction to be completed per clock cycle. Superscalar technology yields an even larger increase in the number of instructions processed per clock cycle.

The superscalar approach takes into account the fact that the instruction set of a processor can be partitioned into several classes which are fundamentally different, i.e., floating-point, integer and memory management operations. These classes generally correspond to independent hardware modules. For example, any computer suitable for extensive numerical computations has a floating-point unit which is completely separate from the integer unit. The floating-point unit itself may be further divided, for example into an addition unit and a multiplication unit.

The *concurrent* execution of instructions belonging to different classes of hardware units is a very straightforward approach. The idea behind the *superscalar principle* is to employ several concurrently operating base-pipeline processors. A number of modifications to the computer architecture as well as higher demands on the compilers are necessary in order to achieve this:

Concurrent instruction load: In order to execute more than one instruction per clock cycle, it must be possible to fetch the appropriate number of instructions from the memory hierarchy.

Increasing data throughput: When several instructions are executed concurrently more data per time unit passes between processor and memory hierarchy. Increase in data throughput may be achieved by overlaying the execution of load/store instructions and by implementing several independent load/store units.

Instruction synchronization: When executing instructions concurrently, conditional dependencies have to be considered (cf. Section 6.2). The sequence of operations has to be carefully controlled in order to prevent conflicts, such as one operation needing the result of another. In particular, different relative execution speeds have to be taken into account. For example, launching a floating-point operation before an integer operation does not mean that both operations will terminate in the same order.

Instruction scheduling: The compiler is responsible not only for preventing pipeline stalls but also for arranging program instructions in such a way that the concurrent execution of consecutive instructions is implemented as far as possible. An unfavorable case would be for example, a sequence of floating-point operations; this leaves the integer unit idle.

Newly developed processors often use several hardware units of the *same* class. This makes it possible for two floating-point multiplications, for example, to be executed at the same time.

Example (IBM POWER2 Processor)   The IBM POWER2 RISC processor contains two parallel pipelines for integer operations in the integer unit and two parallel pipelines for floating-point operations in the floating-point unit (Weiss, Smith [379]).

Superscalar and VLIW Processors

The superscalar principle comes in static and dynamic hardware variants:

Superscalar processors are designed to allocate instructions to the appropriate execution units at runtime without any support from the compiler (i. e., *dynamically*).

VLIW processors (*very long instruction word processors*) cannot perform this allocation, so it has to be done by the compiler at compile time (i. e., *statically*).

Furthermore, there are hybrids of these two variants. With the Intel processor i860, for example, the allocation of instructions to processing units is done dynamically by default. However, there is a mode which requires the compiler to arrange integer and floating-point operations in alternating order.

For both processor variants—superscalar and VLIW processors—performance gains achieved by applying the superscalar principle are limited by the extent to which instruction sequences can be parallelized.

3.1.4 Vector Processors

Many numerical algorithms allow those parts which consume the majority of computation time to be expressed as vector operations. This holds especially for almost all linear algebra algorithms (Golub, Van Loan [50], Dongarra et al. [164]). It is therefore a straightforward strategy to improve the performance of processors used for numerical data processing by providing an instruction set tailor-made for vector operations as well as suitable hardware.

This idea materialized in vector architectures comprising specific *vector instructions*, which allow for componentwise addition, multiplication and/or division of vectors as well as the multiplication of the vector components by a scalar. Moreover, there are specific load and store instructions enabling the processor to fetch all components of a vector from the main memory or to move them there.

The hardware counterparts of vector instructions are the matching *vector registers* and *vector units*. Vector registers are memory elements which can contain vectors of a given maximum length. Vector units performing vector operations, as mentioned above, usually require the operands to be stored in vector registers. The high performance of floating-point operations in such vector units is mainly due to the concurrent execution of operations (as in a very deep pipeline).

But there are further advantages to vector processors as compared with other processors capable of executing overlayed floating-point operations:

- The number of instructions to be executed is dramatically reduced, as a single vector instruction represents a large number of arithmetic operations. This reduces memory traffic.

- As vector components are usually stored contiguously in memory, the access pattern to the data storage is known to be linear. Vector processors exploit this fact for a very fast vector data fetch from a massively interleaved main memory space (see Section 3.2.6).

- There are no memory delays for a vector operand which fits completely into a vector register.

- There are no delays due to branch conditions as they might occur if the vector operation were implemented in a loop.

In addition, vector processors may utilize the superscalar principle by executing several vector operations per time unit (Dongarra et al. [166]).

Fig. 3.5 schematically summarizes the way different concepts for improving performance—increasing the clock rate and reducing the number of clock cycles per instruction—are accomplished for different types of processors.

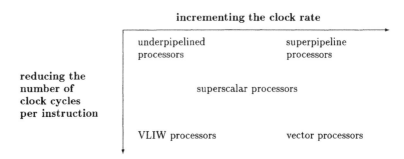

Figure 3.5: Realization of the two performance improving concepts of computer architecture for different types of processors (Hennessy, Patterson [55]).

3.2 Memory

One of the most significant properties of the von Neumann computer model is that memory contains the code of the executable program as well as the data. In the course of different instruction execution phases, the code and one or more operands are loaded from memory, and afterwards the result of the operation is stored in memory. *Several* memory access operations may be thus necessary for the execution of a single instruction. Whenever new developments in circuit technology or improvements in processor architecture speed up a processor, memory access time has to be reduced as well, otherwise memory delays could nullify all possible performance gains.

One of the fundamental problems of computer development in the past decade has been the growing gap between the dramatic increase in processor performance on the one hand and the comparatively small improvement of memory access times on the other. One of the reasons for this discrepancy arises from the following state of affairs.

The main advance in semiconductor technology is the increase in circuit density per chip. For the same physical dimensions, the capacity of a memory chip grows quadratically with transistor packing density. Since the latter has been the main demand so far (and still is), every opportunity to increase memory capacity has been taken. The capacity of RAM chips has multiplied by nearly a factor of four every three years in the past couple of decades, which implies an average capacity increase of about 60 % every year. At present, 64 Mbit chips constitute the maximum capacity.

However, every increase in the capacity of a memory chip adversely affects its access time. It boosts overheads such as the cost of *address decoding*, i.e., referencing the desired memory cell. The increased decoding effort nullifies the gains in access time which are due to lower transistor switch times. This is reflected by the overall gains in memory access time, which used to be no more than a modest 6 % per year.

A number of measures have been taken in recent years to narrow the widening gap between the annual 100 % processor performance growth rate and the poor 6 % increase in memory speed per year. For example, the data bus width between the CPU and the memory hierarchy, has been increased steadily, so that now several memory words can be transferred at a time. The pipelined execution of memory access increases the data throughput as well.

However, the effects of these innovations on software development engineers wishing to optimize the performance of programs for numerical data processing remain marginal. The crucial stimuli for the optimization of program performance are the newly developed concepts of *memory hierarchy* and *interleaved memory*, which are discussed in the following sections.

3.2.1 Memory Hierarchy

As mentioned above, the unsatisfactory gain in main memory speed is, above all, due to the fact that any improvements in integrated circuit technology have been used for increasing memory capacity rather than access speed. On the other hand, if capacity is kept more or less invariant while improving circuit technology, then memory access time may indeed be reduced (approximately inversely proportional to the physical dimensions of a single transistor) due to faster transistor switch times. Moreover, low capacity memory units may employ faster circuit technology which would be too expensive for large-scale storage schemes. Hence, it is easier to economically implement short access times in low capacity memory units than in larger ones.

A straightforward approach to avoiding performance drops due to delayed memory access is to use small but fast *caches*. By storing regularly referenced pieces of code and data in a fast cache instead of in the main memory, the memory access delay may be reduced significantly. To do so it is crucial to decide which data and/or pieces of code should be stored in the cache. Hardware development engineers employ the empirically established principle of *temporal locality of reference*, which states that a machine word which has been referenced recently is likely to be referenced again soon afterwards. According to this principle, every machine word which has been referenced is stored in the cache and remains there until—due to its limited capacity—space is needed for a more recently referenced word.

Using not only a single cache but a graded system of several memory levels leads to the notion of a *memory hierarchy*.

Definition 3.2.1 (Memory Hierarchy) *A memory hierarchy consists of one or more memory levels. Each level is of lower capacity but faster accessibility than the one below it. Generally, each level contains a full copy of the data stored in the level above it.*

A machine word is stored in the highest level once it has been referenced. Each word remains in the memory level it has been stored in as long as the capacity of this level is not exhausted.[7]

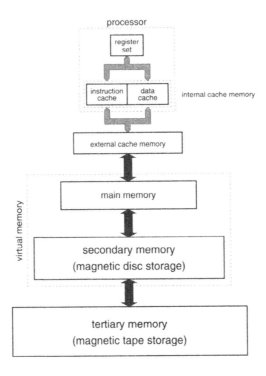

Figure 3.6: A typical memory hierarchy consisting of six levels.

Typically, the memory hierarchy of contemporary computers for numerical data processing consists of six levels (see Fig. 3.6). The registers with the shortest access times but the smallest capacities constitute the highest level. One or more levels of cache memory are located between the registers and the increasingly slower but larger levels of main, secondary and tertiary memory.

The difference in access times to different levels of the memory hierarchy amounts to several orders of magnitude (powers of 10). For example, registers, the fastest memory elements, provide access times in the range of nanoseconds, whereas magnetic disks take milliseconds to be accessed. However, capacity also varies greatly: registers cannot hold more than a few kilobytes, whereas the capacity of secondary memory is in the range of gigabytes; tertiary memory may even be in the range of terabytes ($1\,\mathrm{TB} = 2^{40}$ bytes $\approx 10^{12}$ bytes).

---

[7]Exceptions to this rule are discussed in Section 3.2.4 (Cache Memory).

3.2.2 Addressing Schemes

A data element is localized in memory according to its *address*. Usually a group of basic *memory elements* (which can hold one bit each) is assigned to an address rather than individual elements. Such groups are of a certain size which is characteristic of the memory system, typically 8 bits = 1 byte. They are the smallest memory units which can be addressed and are called *memory cells*.

A collection of memory cells (typically four or eight) is called a *memory word* or simply a *word*. The addressing scheme makes it possible to identify the particular memory cell or word which contains a certain data element.

Physical and Logical Address

Operand registers, which constitute the fastest level of the memory hierarchy to hold program data, are usually addressed explicitly by the compiler. Thus, code generated by a compiler contains register names which correspond uniquely and invariably to physical memory words, i.e., the processor registers. This is called *physical* address allocation.

For cache memory units the choice between physical and logical address allocation is left to the discretion of hardware developers. It does not affect software development in any way, as the addressing scheme for cache memory is transparent (i.e., invisible and uncontrollable) to the software.

All remaining levels of the memory hierarchy use only *logical* address allocation: addresses generated by the compiler do *not* refer to physical addresses in the main memory, secondary memory and so on. Hence, *address transformation* is necessary in order to determine the physical address given a logical address. This task is performed by the hardware and/or the system software of the computer.

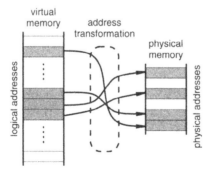

Figure 3.7: Transformation of logical into physical addresses.

As a rule, physical address allocation allows for faster memory access than logical address allocation, since the transformation effort is disposed of. This is why physical address allocation is used for the higher levels of the memory hierarchy—registers and sometimes the cache as well, because short access times are crucial there.

For all other levels of the memory hierarchy, access time is nonetheless important, but the advantages of segregating the program related address space from the physical address space by far outweighs the disadvantage of additional transformation costs. For example, the use of logical addresses makes the concurrent execution of different programs in one single physical address space (as necessary for multitasking and multiuser systems) much more convenient.

Moreover, logical address allocation makes it easier to store program data in different physical layers (address spaces) of the memory hierarchy at the same time. There are two main methods of logical address allocation: *paging* and *segmentation*. Often a combination of both is used.

Paging

The overhead which stems from address transformation has to be kept within tolerable limits. This transformation is thus not performed on the word level: There is no list containing the physical addresses corresponding to each logical address. Rather the address space is partitioned into blocks of *invariant* size, so-called *pages*, which are allocated to physical memory areas of equal size.

For the mapping of a logical address to a physical one, it is sufficient to determine the physical page which corresponds to the logical page of this particular address. The location relative to the first element of the page is the same for both physical and logical pages.

In comparison to other methods of logical address allocation, such as segmentation, the main advantage of paging is its simplicity and speed. To illustrate this, the page size may be adapted to the properties of the respective storage medium: cache blocks are usually about a hundred times smaller than main memory pages which are typically 4 KB.

Segmentation

From the programmer's point of view, segmentation is an obvious approach to partitioning the address space according to its *contents*; so-called *segments* are managed instead. Segments are independent, protected address areas of *variable* size, which can also be changed dynamically. Segments may be assigned a type identification (code segments containing program fragments, data segments containing program data) and access privileges (system segments may be referenced by the operating system, user segments by an application program only and so on). Thus a high degree of data safety is guaranteed.

In order to benefit from the advantages of both segmentation and paging, the two techniques may be combined by dividing segments into pages. Address transformation is then performed in two stages.

3.2.3 Registers

An integral part of the processor, a *register*, is a special memory cell with a very short access time. The collection of all registers is often referred to as a *register set* or a *register file*. In order to keep the transmission time of electrical signals short, the register set is located close to the operational units of the processor. More than one register may be referenced at a time.

The registers of a processor are divided into classes according to their use. In particular, there is a distinction between *address registers*, *instruction registers*, and *operand registers*. With superscalar processors, an additional distinction is made between floating-point and integer instruction registers as well as floating-point and integer operand registers because of their independent operational units for the concurrent execution of floating-point and integer operations.

Since the access time of register files is proportional to their size and long register access times adversely affect the clock rate, only a limited number of registers is feasible. Typically, the total capacity of a register file does not exceed 1 KB (except for vector registers).

Example (Register) Many workstations are equipped with 32 integer and 32 floating-point registers, each of which can store a 64 bit word and, therefore, have an overall capacity of 512 bytes.

Vector computers are equipped with *vector registers* which contain vectors of data. Vector registers are of a specified length and number which on some machines may be user-defined.

Example (Vector Register) The NEC SX-3/14R vector computer has

- 8 vector registers for storing the operands of the vector pipeline and for the data transfer to and from main storage,

- 64 vector registers for the storage of intermediate results.

Each of these vector registers can store 128 words of 64 bits.

3.2.4 Cache Memory

A *cache* is a fast memory level (memory buffer) inserted between processor and main memory (Handy [219]). Modern microprocessors usually have an *internal* cache integrated into the processor chip (*on-chip cache*), the size of which is, for the time being, restricted to a maximum of 112 KB for technical as well as economic reasons. The internal cache may be spread across *two* levels in the memory hierarchy (see Fig. 3.6); the *level-1 cache* (which is closer to the processor) and the *level-2 cache*.

External cache memory, on the other hand, which is not placed on the processor chip, may be much larger—at the moment up to 4 MB. As with registers, separate cache areas are often used for different classes of data (e.g., program data and instructions), especially in the case of internal caches.

Example (Workstations) The following table gives a survey of the caches (data-only and instruction-only caches) and the corresponding cache sizes of some workstations:

Processor	Internal cache memory		External cache memory
	Data	Instructions	
HP-PA 7200	2 KB fully associative	—	typically 256 KB/512 KB data separated from instructions
HP-PA 8000	no *on-chip* cache[8]		up to 2 MB
IBM POWER[9] IBM POWER2[9]	32/64 KB 64/128/256 KB	32 KB 32 KB	— typically 512 KB/1 MB
DEC Alpha 21164	8 KB	8 KB	variable, e.g., 512 KB
	96 KB *level-2 cache*		
MIPS R4400	16 KB	16 KB	variable, e.g., 1 MB
MIPS R8000	16 KB[10]	16 KB	variable, e.g., 4 MB[11]
UltraSPARC-I	16 KB	16 KB	512 KB – 4 MB

As processors are becoming faster and faster and main memory sizes larger and larger, the ideal cache is becoming both faster and larger. Since these two aims are incompatible for *a single* cache level, systems with *two* or more cache levels have come into existence: Being smaller and faster, the internal cache does not access the main memory directly, but a larger, albeit slower (external) cache instead, which then loads data from the main memory.

One difference between cache and register management is that load/store operations for the cache usually do not have to be generated explicitly by the compiler. Rather, such operations are generated by a hardware device whenever they are required. For the purposes of the compiler, cache load/store operations are usually transparent (invisible). Moreover, the processor can access several registers concurrently, whereas only one cache or main memory word may be stored or loaded at a time.

Another notable distinction between cache memory and registers concerns the amount of data moved by a load/store instruction. Whereas only one machine word is transmitted during a register load/store operation, for efficiency's sake a whole page is passed during a cache load/store (denoted *cache line* or *cache block*). The typical length (size) of a cache line varies from 32 bytes to 128 bytes, depending on the computer being used.

---

[8]The processor HP-PA 8000 is a single chip implementation without *on-chip* cache; however, its *off-chip* cache is exceptionally fast.

[9]The POWER and the POWER2 architecture were implemented until 1996 as multiple chip sets. Consequently the size of the "internal" cache memory can be chosen within technical limits. Since 1996 *single chip* implementations of the POWER2 architecture have become available.

[10]for integer data only

[11]*level-1* cache for floating-point numbers, *level-2* cache for integer data and instructions

Load and Update Operations

Referencing not only the cache but the storage of the first slower memory level in parallel as well may reduce unnecessary read delays during load operations. However, this is not a useful strategy for multiprocessor machines, as it causes too much traffic on the memory bus (Handy [219]).

When an addressed word is found in the cache, this is referred to as a *cache hit*. A *cache miss*, on the other hand, means that the referenced word could not be identified as a valid cache entry. In such a case the program execution has to be halted until the required data has been loaded from a lower level of the memory hierarchy to the cache.

For cache load/store operations, it is common to handle cache lines as integral wholes. This not only makes address transformation easier, but also utilizes the principle of *spatial locality of reference*: Whenever a cache memory cell is referenced, memory cells nearby, i. e., locations with similar addresses, are likely to be referenced soon as well. Thus, if a machine word has to be fetched from the main memory and a certain number of words in adjacent memory locations are fetched as well, it can be assumed that they, too, will be needed. The appropriate reference will then require no additional main memory loads and therefore cause only marginal delays.

It is one of the most important methods of program performance optimization for modern single processor computers to reorganize memory access in accordance with the actual cache resources. So for the developer of numerical software, it is necessary not only to know the cache size, but also to have insight into the memory management schemes.

Placement of Cache Lines

In what is to follow, the number of physical cache lines is denoted by Z, and the lines are enumerated $z = 0, 1, \ldots, Z - 1$. Also the number of logical cache lines is denoted by N, and the lines are enumerated $n = 0, 1, \ldots, N - 1$ respectively.

One problem which emerges during assigning the cache lines is determining which of the Z physical lines a new logical line n can be allocated to and to check whether or not this logical line n is already stored in the cache. To that end, all physical lines which might contain a logical line n have to be checked. Hence, the smaller the set $\mathcal{Z}$ of eligible lines, the easier this test and the less hardware effort required. Which of the eligible physical lines the logical line n will actually be stored in depends on the *replacement strategy* of the cache.

There are different strategies for placing data blocks in cache lines, which correspond to cache organization and cache types. The three most important cache types are:

Directly mapped cache: This cache type assigns *exactly one* physical line $\mathcal{Z} := \{z\}$ to each logical line; its number can be calculated according to $z = n \bmod Z$ (see Fig. 3.8).

Example (Workstations)   The HP-PA 8000, DEC 21164 (Alpha) and MIPS R8000 processors use directly mapped level-1 caches.

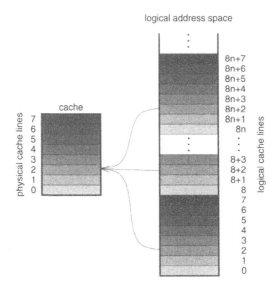

Figure 3.8: Direct mapping of cache lines in a model cache with $Z = 8$ lines.

Fully associative cache: In complete contrast to a directly mapped cache, in a fully associative cache a logical line may be allocated to *any* physical line (see Fig. 3.9), i.e., $\mathcal{Z} = \{0, 1, \ldots, Z - 1\}$.

The memory word is stored in the cache together with its main memory address, which functions as the reference key for cache words. This is a flexible placement strategy indeed but it requires considerable computational effort and, consequently, results in a reduced cache capacity.

Example (HP Workstations)   The HP-PA 7200 processor is equipped with a small fully associative internal cache (of only 2 KB) to supplement the external cache.

Set associative cache: This cache type is somewhere between directly mapped and fully associative caches: Although the physical cache line is not uniquely determined by the number of the logical line (as with directly mapped caches), not all physical lines are eligible (as with fully associative caches). The set associative cache determines an eligibility set $\mathcal{Z}$ from the factor k of Z by
$\mathcal{Z} = \{z \; : \; z \equiv n \bmod k\}$.

A logical cache line may thus be stored in $l = Z/k$ physical lines (see Fig. 3.10). This is also called an l-fold set associative cache.

Example (Workstations)   The IBM POWER and POWER2 processors have internal 4-fold set associative data caches. Their instruction caches are also 4-fold set associative.

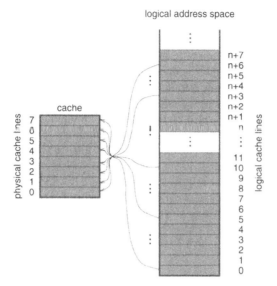

Figure 3.9: Fully associative cache line placement.

The external cache of the MIPS R8000 processor is 4-fold set associative.

The internal level-2 cache of the DEC Alpha 21164 processor is 3-fold set associative.

The internal instruction cache of the UltraSPARC-1 processor is 2-fold set associative.

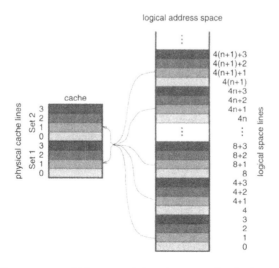

Figure 3.10: 2-fold set associative placement of cache lines.

Cache Line Replacement

If all physical lines $\mathcal{Z}$ eligible for storing a logical cache line are already occupied, then one of these entries has to be replaced with the new logical line. Obviously the best line to replace would be one whose entry is not needed any longer or at least in the near future. However, since hardware cannot perform a prediction of that kind, the following heuristic approach is taken[12]:

The *random* strategy, which choses (pseudo-) randomly one line of the eligibility set $\mathcal{Z}$ to be overwritten, is the simplest method. The *least recently used* strategy (LRU), on the other hand, selects the line of $\mathcal{Z}$ which has not been accessed for the longest time. Since a strict LRU strategy is very costly, an *approximate* LRU strategy is usually employed instead: Each physical line has a status flag, which is set whenever the line is referenced; periodically all flags are reset. For the replacement, one of the lines whose status flag is not set is chosen randomly. The *first-in first-out* (FIFO) strategy replaces the cache line which has been resident in the cache for the longest time.

3.2.5 Virtual Memory

The simulation of a larger main memory capacity than is actually present, which can be done by means of backing storage such as hard disks, is denoted *virtual memory*. Program instructions and data are loaded into physical memory at runtime whenever they are required; they are eventually relocated to virtual memory in order to make room for other data. The actual memory structure is hidden from the software, which can only access the virtual memory. This transparency is made possible by the use of logic address spaces, which decouple logical and physical addresses entirely.

The relationship between main memory and backing storage is basically the same as that between cache and main memory: For secondary storage, main memory plays the role of the cache.

The page placement and replacement strategies of virtual memory are of crucial importance to the performance optimization of numerical software. The method of address transformation—referred to as *page identification strategy*—may influence program performance severely.

For virtual memory the effects of those strategies are much more serious than for caches. A virtual memory miss causes a secondary storage access which is at least a 100 000 times slower than a main memory access. In contrast, reference times of cache and main memory usually differ by one order of magnitude (a factor of 10) only.

Page Placement and Replacement

In principle, every logical page of virtual memory may be loaded to any physical page frame. This guarantees the exploitation of all empty page frames in the main

---

[12]For directly mapped caches a replacement strategy is redundant.

memory unit and avoids unnecessary accesses to backing storage. This reduction is crucial, as accessing backing storage is slower by many orders of magnitude in comparison to main memory.

For page replacement, virtual memory also employs an approximate LRU strategy. The approximation method simply picks up a page that has not been used recently (NUR).

Address Transformation

Address transformation from logical pages to physical addresses in the main memory or backing store requires a *page table* . It contains, roughly speaking, the addresses of physical pages for all matching logical pages of a segment. Hence address translation basically means table lookup.

Due to the principle of locality of reference, one single logical page is often referenced several times in a row. In order to allow fast address translation, recently used parts of the page table are stored in a fast memory buffer. To that end, a separate, fully associative cache called *translation lookaside buffer* (TLB) is used, which—depending on the system—may contain 20 to 200 entries.

If the required virtual page is not found in the TLB, the page table fraction stored in main memory has to be referenced, which slows down address translation due to additional main memory references. For relocated pages in the backing store, the respective fraction of the page table itself may not even be in main memory. In this case, the backing store has to be referenced several times, which accordingly increases the access time.

3.2.6 Interleaved Memory

Interleaving (bank phasing) is a method to accelerate access to a memory level which is used primarily for main memory. To that goal, the main memory is divided into B independent *banks (modules)* of equal size which can each be read or written to independently.

Given the length w of a memory word in bytes, each physical address a can be allocated to one of the memory banks b according to the formula $b = \lfloor a/w \rfloor \bmod B$, i.e., consecutive machine words are written to cyclically ordered banks (see Fig. 3.11).

This method is effective due to the discrepancy between access time and *cycle time* of a main memory module. The access time is the amount of time required for loading or storing a machine word, whereas the cycle time of a memory module[13] is the time which elapses between the beginning of a memory access and the next possible memory access. The gap between these two values for main memory modules is a result of the *memory regeneration* which takes place after each access. The cycle time of a typical memory module is three to four times the access time.

---

[13] *Memory* cycle time must not be confused with the smaller *processor* cycle time.

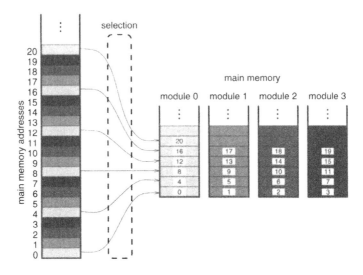

Figure 3.11: Interleaved memory unit with $B = 4$ memory banks.

The throughput of non-interleaved memory units is determined by the cycle time, not the access time. For interleaved memory, however, access time (which is much shorter) together with the number of modules used is critical for the overall throughput, provided the modules are accessed in a suitable order. The maximum speed gain is achieved if consecutive memory references access different modules. For example, if contiguous machine words are referenced in a row, every bank is affected once in B accesses only whence the interval between two accesses of a specific bank is B times the access time. It follows that unless the cycle time is more than B times the access time, the bank is granted enough time for regeneration and no delays occur.

3.3 Performance Quantification

The assessment and choice of computer hardware and/or scientific software requires the use of numerical values, which may be determined analytically or empirically, for the quantitative description of performance. As to which parameters are used and how they will be interpreted depends on *what* is to be assessed.

The user of a computer system who waits for the solution of a particular problem is mainly interested in the time it takes for the problem to be solved. This time depends on two parameters, workload and performance:

$$\text{time} = \frac{\text{workload}}{\text{performance}_{\text{effective}}} = \frac{\text{workload}}{\text{performance}_{\text{maximum}} \, \text{efficiency}} \, .$$

The computation time is therefore influenced by the following quantities:

1. The amount of work (*workload*) which has to be done. This depends on the nature and complexity of the problem as well as on the properties of the

algorithm used to solve it (see Chapter 5). For a given problem complexity, the workload is a characteristic of the algorithm. The workload (and hence the required time) may be reduced by improving the algorithm.

2. *Peak performance* characterizes the computer hardware independently of particular application programs. The procurement of new hardware with high peak performance usually results in reduced time requirements for the solution of the same problem.

3. *Efficiency* is the percentage of peak performance achieved for a given computing task. It tells the user what share of the potential peak performance is actually exploited and thus measures the quality of the implementation of an algorithm. Efficiency may be increased by optimizing the program (see Chapter 6).

The correct and comprehensive performance assessment requires answers to a whole set of complex questions: What limits are imposed, independently of specific programming techniques, by the hardware? What are the effects of the different variants of an algorithm on performance? What are the effects of specific programming techniques? What are the effects on efficiency of an optimizing compiler?

3.3.1 The Notion of Performance (Power)

The notion of *performance* is frequently used in computer science, although its definition is rather vague. There are several problems regarding the analytical assessment as well as the empirical testing of computer performance. The most important is finding a satisfactory measure of computer power.

In physics the *average power* P is defined as the ratio between the overall amount of work ΔW and the time Δt required to do it: *Power equals work accomplished per unit of time.* By analogy, the average power of a computer system over a period of time $[t_1, t_2]$ may be defined by the difference quotient

$$P_{[t_1,t_2]} := \frac{W(t_2) - W(t_1)}{t_2 - t_1} = \frac{\Delta W}{\Delta t}. \tag{3.2}$$

In contrast to the average power during a time interval there is the *instantaneous power* $P(t)$ at a single point in time. It is defined as the limiting average power as $\Delta t \to 0$, i. e., the derivative of work with respect to time:

$$P(t) := \frac{dW}{dt} = \lim_{\Delta t \to 0} P_{[t,t+\Delta t]} = \lim_{\Delta t \to 0} \frac{W(t + \Delta t) - W(t)}{\Delta t}. \tag{3.3}$$

In practice, the instantaneous power of a computer system is calculated using the difference quotient (3.2), not the differential quotient (3.3), for a sufficiently short period of time $[t_1, t_2]$, e. g., one clock cycle.

Notation (Period of Time, Workload) Periods of time are denoted T in the following, e. g., T_c for the processor cycle time. Workloads will be denoted W, e. g., W_I for the entirety of all instructions to be executed for a particular program.

Thus, the computing power of a data processing system depends, by definition, on two factors:

1. the workload imposed on the system, and
2. the time required by the system to do this work.

In order to discuss the performance of a computer system, information on *both* factors must be available! A common mistake occurs as a result of the inaccurate quantification of the workload, ΔW.

There are two basic concepts for the assessment of computer performance: Usually a particular workload, such as a certain problem to be solved, is given and the time required by the computer system (the hardware, the system software and the application software) to do this work is measured. But it is also reasonable to make the assessment the other way round: A period of time is given and the amount of work which is done in this period is determined. This is called the *throughput* of the system, or the *bandwidth* for the assessment of memory systems (see Section 3.4.4).

Benchmarks are standardized computer programs for the comparative performance assessment of computer systems. For numerical data processing, the average number of floating-point operations executed per second is usually determined and rated. Accordingly, the *floating-point performance* is measured in flop/s (*floating-point operations per second*) (see Section 3.4.1).

Example (LINPACK Benchmark)   The LINPACK benchmark (Dongarra [160]) uses the programs `LINPACK/sgefa` and `LINPACK/sgesl` for the solution of systems of linear equations with $n = 100$ variables. The effective floating-point performance is determined by the workload as given by the number of executed floating-point operations $W_F \approx 2n^3/3 + 2n^2$ and the required computing time T

$$P = \frac{W_F}{T} \quad [\text{flop/s}].$$

Relating this value to the peak performance of the computer gives the *efficiency*, i. e., the degree to which two particular LINPACK programs exploit computer resources.

The peak performance of NEC SX-3/14R computers is 6400 Mflop/s. For a 100×100 system of linear equations, effective performance only amounts to 368 Mflop/s, which means that efficiency is a mere 6 %.

However, a mathematically equivalent program which makes use of the properties of the computer as far as possible can be developed; this program, applied to a system of linear equations with $n = 1000$ variables, yields a 5199 Mflop/s floating-point performance, which implies a satisfactory efficiency index of 81 %.

3.3.2 Time as a Factor in Performance

Further distinctions and definitions have to be made for the exact explanation of workload and time, which so far have only been used intuitively, so that meaningful and reproducible performance analyses will be possible. This section deals only with the time factor. A detailed discussion of workload, the other factor of performance can be found in Chapter 5.

Response Time

For the user almost nothing other than the time required for the computer to complete a task is relevant. This time, referred to as *response time*, *elapsed time*, or sometimes *wall-clock time*, passes between the launching of the command which starts a particular computational task and its completion, the *response* of the computer.

However, this number, which is crucial to the user, may depend on factors that cannot be determined through reproducible experiments. *Multitasking*, for instance, may be the reason a program is not active all the time from its beginning to its completion, because the CPU handles other programs as well.

The difference between the response time experienced by the user and the time the system actually spends solving the problem can be particularly large for a *multiuser* system. Other users, who consume their share of computer resources, remain unseen. The user is thence led to the illusion that he has the computer *all to himself*, whereas the response time in fact depends on the total system load and the processor scheduling scheme. In such circumstances, experiments designed for performance assessment cannot be executed without creating arrangements which are difficult to realize, such as running the computer system in *single user mode*.

CPU Time

The fact that only a small share of computer resources are spent on a particular *job* in a multiuser environment is taken into account by measuring *CPU time*. This quantity specifies the amount of time the processor was actually engaged in solving a particular problem and neglects the time spent on other jobs or on waiting for input/output.

CPU time itself is divided into user CPU time and system CPU time. *User CPU time* is the time spent on executing an application program and its linked routines. *System CPU time* is the time consumed by all system functions required for the execution of the program, such as accessing virtual memory pages in the backing store or executing I/O operations.

However, CPU time does *not* comprise times during which the program is idle (e. g., times when the program waits for data to be loaded from the backing store), even if the program happens to be the only job on the system!

Before any performance assessment is carried out, the decision as to whether performance should be characterized in terms of response time or CPU time has to be made. Accordingly, *system performance* is based on response time and *CPU performance* on user CPU time (Hennessy, Patterson [55]). As discussed in the following sections, the type of time measurement influences all performance characteristics. This stresses the importance of documenting the type of time measurement on which a performance assessment is based.

3.4 Analytical Performance Assessment

The purpose of Sections 3.1 and 3.2 was to impart a basic understanding of contemporary computer architectures. To that end, it was sufficient to consider just the *qualitative* aspect of performance. For example, it is obvious even without a detailed analysis, that the concurrent execution of instructions reduces expenditure of time and hence increases performance.

However, often an exact *quantitative* description of computer performance is required. For example, in order to determine to what degree one computer is "faster" or "slower" than another, quantitative performance analysis cannot be renounced, even if the complexity of parameters and the variety of computer architectures make such analyses difficult.

The *analytical assessment* of hardware performance starts with the technical data and the physical features of the computer, such as

- processor type,
- cycle time T_c or its reciprocal, the clock rate f_c,
- the capacities and access times of the different levels in the memory hierarchy,
- the structure and hierarchy of the different memory modules as well as the interconnections between them, and
- the bandwidth of data buses (number of bits which can be transmitted in parallel within the CPU, or between the CPU and the memory hierarchy or the periphery respectively).

From these hardware features a number of indices can be derived which form a basis for the assessment of system performance.

The actual execution of a program usually gives a lower performance index than the one calculated. It is thus necessary to distinguish between analytically determined *ideal performance indices* (like the ones in this section) and *actual performance indices* determined using empirical means as discussed in Section 3.5.

3.4.1 Peak Performance

An important hardware characteristic, the *peak performance* $P_{\max}$ of a computer, specifies the maximum number of floating-point (or other) operations which can theoretically be performed per time unit (usually per second).

The peak performance of a computer can be derived from its cycle time T_c and the maximum number N_c of operations which can be executed during a clock cycle:

$$P_{\max} = \frac{N_c}{T_c}.$$

If $P_{\max}$ refers to the number of floating-point operations executed per second, then the result states the *floating-point peak performance*. It is measured in

flop/s   (*floating-point operations per second*)

or Mflop/s (10^6 flop/s), Gflop/s (10^9 flop/s), or Tflop/s (10^{12} flop/s). Unfortunately, the fact that there are different classes of floating-point operations, which take different amounts of time to be executed, is neglected far too often (see Section 5.5.6).

Notation (flop/s)   Some authors use the notation flops, Mflops etc. instead of flop/s, Mflop/s etc. As the first notation suggests a plural form rather than a performance index it will not be used in this book.

It is most obvious that no program, no matter how efficient, can perform better than peak performance on a particular computer. In fact, only specially optimized parts of a program may come close to peak performance. One of the reasons for this is that, in practice, address calculations, memory operations, and other operations which do not contribute directly to the result are left out of the operation count. Thus, peak performance may be looked upon as a kind of *speed of light* for a computer.

Example (Cray T916)   The cycle time of a CRAY T916 is $T_c = 2.2\,\text{ns} = 2.2 \cdot 10^{-9}\,\text{s}$. Each processor can execute at most $N_c = 4$ floating-point operations—two additions *and* two multiplications—per clock cycle. Hence the floating-point peak performance of such a processor results in

$$P_{\max} = \frac{N_c}{T_c} = \frac{4 \text{ operations}}{2.2 \cdot 10^{-9} \text{ seconds}} = 1.8 \cdot 10^9 \text{ flop/s} = 1800 \text{ Mflop/s}.$$

Accordingly, the peak performance of a CRAY T916 with four processors is 7200 Mflop/s, which could theoretically be achieved if all four processors executed four floating-point additions and multiplications *per clock cycle* over a given period of time.

Example (IBM Workstation)   The cycle time of the IBM-POWER2 workstation model 591 is $T_c = 13\,\text{ns}$. At most, four floating-point operations can be executed per clock cycle—a multiply-*and*-add instruction in each of the two floating-point pipelines (White, Dhawan [379]). The peak performance can thus be determined as follows:

$$P_{\max} = \frac{N_c}{T_c} = \frac{4 \text{ operations}}{13 \cdot 10^{-9} \text{ seconds}} = 308 \text{ Mflop/s}.$$

The smaller the ratio of multiply-and-add instructions to all floating-point operations in the program, the bigger the difference between actual and peak performance on this workstation.

3.4.2 Instruction Performance

The CPI (*cycles per instruction*) value of a processor states the number of clock cycles required for the execution of an instruction. As different instructions usually require different numbers of cycles, only the *average* CPI value can be determined. This mean value should, of course, be a weighted mean: accurate knowledge of the frequency of different instructions (which can be obtained by hardware monitoring, for example) leads to a reliable weight function.

The (average) CPI value not only depends on the program, but also on the nature of the instruction set as well. Complex instructions (as in CISC architectures) usually require longer execution times and hence increase the CPI value. At the same time, however, they usually reduce the overall number of program instructions.

Average performance (instructions per time unit) can be derived from formula (3.2) if the number N of executed instructions (often referred to as *path length*) is taken as the workload:

$$P_I = \frac{W_I}{T} = \frac{N}{NT_c\text{CPI}} = \frac{1}{T_c\text{CPI}}. \tag{3.4}$$

This performance measure (*instruction performance*) is used in the design process of new computer architectures as well as for empirical performance assessment (see Section 3.5.2). If P_I refers to the number of instructions per microsecond, the unit of instruction performance is *Mips* (*million instructions per second*); for nanoseconds, it is *Bips* (*billion instructions per second*, i.e., 10^9 instructions per second).[14]

Example (RISC—CISC) Formula (3.4) makes clear that instruction performance can be increased by reducing either the cycle time or the CPI value. There are technological limits to reducing the cycle time, which are due to the finite velocity of signal propagation. Given the cycle time T_c, the formula

$$T = \frac{W_I}{P_I} = NT_c\text{CPI}$$

shows that the execution time not only depends on the CPI value, but also on the path length N, which is itself influenced by the nature of the instruction set.

The conventional CISC instruction set consists of rather complex instructions. Each of those can take on a large workload. As a result, such an instruction set reduces the path length N of a program. On the other hand, more complex instructions require more clock cycles for their execution, which increases the CPI value and reduces instruction performance.

The RISC approach is reversed: The instruction set consists of comparatively simple instructions. This reduces the CPI value—often down to CPI = 1, the *single cycle instructions* characteristic of RISC processors. Further reductions in the CPI value can be obtained, although they necessitate the use of several parallel units, as can be found in superscalar architectures (see Section 3.1.3). However, the simplicity of the instructions boosts the path length of a program as each instruction performs a smaller task. Thus for a RISC instruction set, much depends on the quality of the optimizing compiler which has to make sure that the potential gains due to small CPI values are not lost.

Better (higher) instruction performance may result in worse (longer) execution times and vice versa. The following example illustrates that extreme caution is advisable when using instruction performance as a basis for hardware and/or software comparisons:

Example (Optimizing Compiler) The following table shows empirically determined frequencies of instructions as generated by the GNU C compiler and the respective clock cycles (Hennessy, Patterson [55]):

Operation	Frequency	Required clock cycles
arithmetic-logical operations	43 %	1
branch operations	24 %	2
load operations	21 %	2
store operations	12 %	2

---

[14] As with Mflop/s, Gflop/s etc., the notation Mi/s, Gi/s etc. would be more reasonable. However, it is not customary.

The corresponding average CPI value amounts to 1.57. If the clock frequency of the processor is $f_c = 50$ MHz, which corresponds to cycle time $T_c = 20$ ns. The instruction performance is determined by

$$P_I = \frac{1}{T_c \mathrm{CPI}} = \frac{1}{20 \cdot 10^{-9} \cdot 1.57} = 31.8 \,\mathrm{Mips}.$$

The CPU time consumed by such a program is given by

$$T = \frac{W_I}{P_I} = \frac{W_I}{31.8 \cdot 10^6} = 31.4 \cdot 10^{-9} W_I,$$

where W_I denotes the number of required instructions. If a better optimizing compiler halves the number of arithmetic-logical operations, then the CPI value is increased to

$$\mathrm{CPI}_{\mathrm{optimized}} = \frac{(0.43/2) \cdot 1 + 0.24 \cdot 2 + 0.21 \cdot 2 + 0.12 \cdot 2}{1 - (0.43/2)} = 1.73,$$

and the instruction performance of the optimized code is *reduced* to

$$\overline{P}_I = \frac{1}{T_c \mathrm{CPI}} = \frac{1}{20 \cdot 10^{-9} \cdot 1.73} = 28.9 \,\mathrm{Mips}.$$

However, the optimization improves (shortens) execution time, although the instruction performance of the processor deteriorates (is reduced):

$$\overline{T} = \frac{\overline{W}_I}{\overline{P}_I} = \frac{\overline{W}_I}{28.9 \cdot 10^6} = \frac{0.785 W_I}{28.9 \cdot 10^6} = 27.2 \cdot 10^{-9} W_I,$$

where $\overline{W}_I$ denotes the number of operations required after optimization.

3.4.3 Performance of Vector Processors

In this section indices suitable for a rough assessment of vector processors, which are then compared with scalar processors, are discussed. Only *arithmetic pipelines*, i.e., pipelines which perform arithmetic operations on data vectors of length n, are discussed. For analytic performance formulas, it is assumed that the pipeline fetches operands from vector registers only. The influence of the vector load operations from main memory as well as other performance decreasing factors are considered for the empirical performance assessment of vector processors (see Section 3.5.4).

Pipeline Throughput

For the performance analysis of arithmetic pipelines, it is usually sufficient to employ a linear model of the execution time $T_v(n)$ of vector operations:

$$T_v(n) = T_0 + nT_c.$$

Fig. 3.12 depicts the execution scheme of a vector operation for vectors of length n in a five-stage pipeline ($s = 5$).

With this scheme in mind, it is easy to understand the following formula, which models the execution time of a given vector operation for vector length n ($n \geq s$):

$$T_v(n) = (k + s + (n - 1))T_c. \tag{3.5}$$

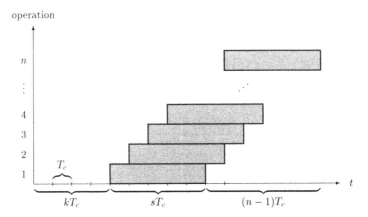

Figure 3.12: Overlayed vector operation in a 5-stage pipeline.

At the beginning of the vector operation, k clock cycles are needed for initialization, which corresponds to a fixed time kT_c. During this period of time, the addresses of the first and the last components of the vector operands are calculated, and data elements are transported from the vector registers to the arithmetic pipeline. After another s clock cycles, the pipeline returns the first result. The pipeline is now full, and all further components of the vector of results take only one clock cycle to be returned. Hence, the remaining $n - 1$ components are calculated within the time span $(n - 1)T_c$.

The *start-up time* T_0 can be derived from formula (3.5):

$$T_0 = (k + s - 1)T_c.$$

The performance which depends on the vector length n is called the *throughput* $r(n)$ of an arithmetic pipeline. It specifies the number of results returned by the pipeline per time unit (for $n \geq s$):

$$r(n) = \frac{n}{T_v(n)} = \frac{n}{T_0 + nT_c}. \tag{3.6}$$

The *asymptotic throughput* is the (3.6) throughput as $n \to \infty$:

$$r_\infty = \lim_{n \to \infty} r(n) = \frac{1}{T_c} = f_c.$$

The theoretical peak performance of an arithmetic pipeline can be derived from the clock rate f_c and the maximum number of floating-point operations per result.

Example (Asymptotic Throughput)   The clock rate of the NEC SX-3/14R vector computer is $f_c = 400\,\text{MHz}$. Hence, its asymptotic throughput is

$$r_\infty = f_c = 400 \cdot 10^6\,\text{s}^{-1}.$$

This computer has four independent groups of vector pipelines, each of which comprises two pipelines for additions and two pipelines for multiplications. The peak performance is therefore

$$4(2 + 2)400\,\text{Mflop/s} = 6.4\,\text{Gflop/s}.$$

Characteristic Vector Length

An important index for vector pipelines is the vector length $n_{1/2}$ for which half the asymptotic throughput is achieved:

$$r(n_{1/2}) = \frac{r_\infty}{2}. \tag{3.7}$$

Formula (3.6) gives $r(n_{1/2})$:

$$n_{1/2} = \frac{T_0}{T_c} = k + s - 1. \tag{3.8}$$

This quantity helps to estimate which vector length n is necessary to make vectorization efficient.

Not surprisingly, $n_{1/2}$ is proportional to the required start-up time T_0. It is inversely proportional to the time T_c which elapses between component operations: the reason for this is that for larger T_c the value r_∞ is smaller and hence shorter vectors are sufficient for achieving half the r_∞ value.

The time $T_v(n)$ required to perform a vector operation can also be expressed in terms of the following formula:

$$T_v(n) = \frac{n + n_{1/2}}{r_\infty} = (n + n_{1/2})T_c. \tag{3.9}$$

This leads to another interpretation of the parameter $n_{1/2}$: It specifies the number of internal pipeline cycles but one which pass until the first component of the result vector is returned from the pipeline. Hence, a reasonable model for the start-up time is the assumption that a fictitious vector of length $n_{1/2}$ is processed in addition to the actual data vector. In practice $n \gg n_{1/2}$ should be true, otherwise the start-up time would influence the throughput rate too much.

Applying (3.9) to (3.6) gives the formula

$$r(n) = \frac{r_\infty}{1 + n_{1/2}/n}, \tag{3.10}$$

which specifies the degree to which the difference between the actual throughput $r(n)$ and the theoretical maximum r_∞ is influenced by the ratio $n_{1/2}/n$.

The cross-over point n_v helps to compare a vector processor to a scalar processor by specifying the minimal vector length for which vector arithmetic is faster than scalar arithmetic: the execution of n scalar operations is compared to *one* appropriate vector operation (implemented on an s-stage pipeline) on a vector of length n. For the execution time $T_1(n)$ of n scalar operations,

$$T_1(n) \leq sT_c n. \tag{3.11}$$

In practice, a single scalar operation takes less time than that required for the processing of scalar operands in a pipeline, since the pipeline steps work synchronously for the different suboperations. However, for the sake of simplicity, it is assumed that equality holds in (3.11), i.e., that the scalar operation takes as long as the processing of s pipeline stages.

If $T_1(n) > T_v(n)$, then it is faster to process a vector of length n than to perform n scalar operations. It results from (3.5) and (3.11) that

$$n > \frac{k}{s-1} + 1 \qquad \text{and hence} \qquad n_v = \left\lceil \frac{k}{s-1} \right\rceil + 1.$$

3.4.4 The Influence of Memory on Performance

As discussed in Section 3.2, technical progress in the past years has mainly concentrated on the expansion of the capacity of memory chips. The speed of memory chips has not been able to keep pace with the increases in processor speed.

Memory, i.e., the memory hierarchy as a whole, is of crucial importance to performance. If, for example, there is a high miss rate due to unskillful programming on the fast levels of the memory hierarchy, dramatic performance deterioration is to be expected (see Section 6.2.3).

Transfer Rate

The *transfer rate* describes the throughput of the interface between processor and memory; it is often referred to as *bandwidth*. It specifies the amount of data which can be transferred per time unit between the CPU and the memory and is calculated as the product of the width of the data bus in bytes and the frequency, at which single memory words are transferred, which is usually proportional to the clock rate f_c of the processor.

Access Time

The *average memory access time* is specified by the statistical mean of the times required for accessing data stored in a memory hierarchy. Referencing a higher level of the memory hierarchy is not always successful—misses occur with a certain probability and delay the memory access operation.

The probability p_m of a cache miss with a subsequent main memory hit and the probability p_p of a main memory miss after a cache miss are empirically determined or estimated. Moreover, the cache access time T_h, the time T_m required for a cache miss with a subsequent main memory hit, and finally the time T_p required for a main memory miss after a cache miss are needed. Properties of a memory hierarchy dictate that $T_p \gg T_m \gg T_h$. From these quantities the average memory access time can be derived:

$$\overline{T} = p_p T_p + p_m T_m + (1 - p_p - p_m) T_h. \tag{3.12}$$

The probabilities p_m and p_p depend not only on the capacities of the different levels of the memory hierarchy, but also on the memory organization scheme. They are difficult to determine, either with the help of hardware monitors or by means of rather complicated simulations (Gee et al. [208]).

Level-2 caches and translation lookaside buffers (TLBs) can be modeled through suitable modifications of formula (3.12).

Performance Parameters

A most important case is that when a whole data vector of length n is referenced. By analogy to the model of the execution time of a vector operation as given by formula (3.9), the following formula can be used to describe this case:

$$T_s(n) = \frac{n + n_{1/2}^s}{r_\infty^s}. \tag{3.13}$$

The parameters r_∞^s and $n_{1/2}^s$ reflect memory access time as well as the memory transfer rate.

3.5 Empirical Performance Assessment

In contrast to analytical performance assessment obtained from the technical data from the computer system, empirical performance assessment is based on experiments and surveys conducted on given computer systems or abstract models (using simulation).

3.5.1 Temporal Performance

In order to compare different algorithms for solving a given problem on a single computer, either the *execution time*

$$T := t_{\text{end}} - t_{\text{start}}$$

itself or its inverse, referred to as *temporal performance* $P_T := T^{-1}$, can be used. To that end, the execution time is measured—as opposed to calculated as in Section 3.4. The workload is normalized by $\Delta W = 1$, since only *one* problem is considered.

This kind of assessment is useful for deciding which algorithm or which program solves a given problem fastest (Addison et al. [81]). The execution time of a program is the main performance criterion for the user. He only wants to know how long he has to wait for the solution of *his* problem. For him, the most powerful and most efficient algorithm is determined by the shortest execution time or the largest temporal performance. From the user's point of view the workload involved and other details of the algorithm are usually irrelevant.

3.5.2 Empirical Instruction Performance

For the experimental determination of the number W_I of instructions (as a measure of workload) executed over a time span T, the *empirical instruction performance* can be calculated as

$$P_I = \frac{W_I}{T} = \frac{\text{number of executed instructions}}{\text{required time}}.$$

As in Section 3.4.2, its unit is *Mips* (million instructions per second or instructions per microsecond):

$$P_I\,[\text{Mips}] \;=\; \frac{\text{number of executed instructions}}{\text{duration in microseconds}}$$

$$=\; \frac{N}{N \cdot \text{CPI} \cdot T_c}$$

$$=\; \frac{f_c}{\text{CPI} \cdot 10^6}. \qquad\qquad (3.14)$$

High performance computers execute more instructions per time unit and, therefore, normally have a higher Mips rate. Nonetheless, one has to be careful when interpretating Mips values, as the instruction performance depends heavily on the instruction set of the respective computer. Mips values of machines with different instruction sets *cannot* be compared.

Example (RISC—CISC)   As the Mips rate depends only on the clock rate f_c of the processor, according to formula (3.14), RISC processors *always* achieve better Mips values than CISC processors: The instruction set of RISC processors consists of relatively simple instructions which take fewer clock cycles than the more complex instructions of CISC computers (cf. the example on page 96).

However, a higher Mips value does not necessarily mean a shorter execution time for a given program. This is because for the completion of a given task, the number of simple RISC instructions which have to be executed is often much larger than the number of the more powerful CISC instructions. It has to be examined as to which effect prevails.

3.5.3 Empirical Floating-Point Performance

The *floating-point performance* characterizes the workload completed over the time span T as the number of floating-point operations executed in T:

$$P_F\,[\text{flop/s}] = \frac{W_F}{T} = \frac{\text{number of executed floating-point operations}}{\text{time in seconds}}.$$

In contrast to the peak performance calculated as in Section 3.4, this empirical quantity is obtained by measuring executed programs in real life situations. The results are expressed in terms of Mflop/s, Gflop/s or Tflop/s as with analytical performance indices.

Floating-point performance is more suitable for the comparison of different machines than instruction performance, because it is based on *operations* instead of *instructions*. This is because the number of instructions related to a program differs from computer to computer but the number of floating-point operations will be more or less the same.

Floating-point performance indices based simply on counting floating-point operations may be too inaccurate unless a distinction is made between the different classes of floating-point operations and their respective number of required clock cycles. If these differences are neglected, a program consisting only of floating-point *additions* will have considerably better floating-point performance than a program consisting of the same number of floating-point *divisions*. On the POWER processor, for instance, a floating-point division takes around twenty times as long as a floating-point addition.

The *standardization* of floating-point performance may be a way to make this performance index more reliable for the comparison of different computer systems (see Section 5.5.6).

Interpretation of Empirical Performance Values

In contrast to peak performance, which is a hardware characteristic, the empirical floating-point performance analysis of computer systems can only be made with real programs, i. e., algorithm implementations. However, it would be misleading to use floating-point performance as an absolute criterion for the assessment of *algorithms*.

A program which achieves higher floating-point performance does not necessarily achieve higher temporal performance, i. e., shorter overall execution times. In spite of a better (higher) flop/s value, a program may take *longer* to solve the problem if a larger workload is involved. Only for programs with equal workload can the floating-point performance indices be used as a basis for assessing the quality of different *implementations*. Measures suitable for the assessment of algorithms, regardless of their implementation, are discussed in Chapter 5.

Example (Performance and Time) To solve a particular diffusion problem on a parallel computer, an iterative, local implementation based on the Jacobi algorithm achieved 5.6 times the Mflop/s rate of a multigrid method. However, the first implementation required 317 times the execution time of the multigrid method. In such cases only temporal performance is relevant for the comparison of the two programs.

For the benchmark assessment of computer systems, empirical floating-point performance is also suitable and is frequently used (for instance in the LINPACK benchmark or the SPEC[15] benchmark suite).

Empirical Efficiency

Sometimes it is of interest to obtain information about the degree to which a program and its compiler exploit the potential of a computer. To do so, the ratio between the empirical floating-point performance and the peak performance of the computer is considered.

This *empirical efficiency* is usually significantly lower than 100 %, a fact which is in part due to simplifications in the model for peak performance.

Example (LINPACK Benchmark) (Dongarra [162]) For the LINPACK benchmark the empirical performance and efficiency values listed in Table 3.1 are determined along with other values. For systems of order $n = 100$, the time measurements are based on the original LINPACK routines, LINPACK/sgefa and LINPACK/sgesl.

For $n = 1000$, mathematically equivalent programs, which were designed to make the best use of hardware resources, were developed. Obviously, machine dependent program optimization significantly improves efficiency.

---

[15]SPEC is the abbreviation of *Systems Performance Evaluation Cooperative*.

Table 3.1: Empirical performance and efficiency for the LINPACK benchmark.

Computer		LINPACK benchmark			
Type	Peak performance	$n = 100$		$n = 1000$	
	[Mflop/s]	[Mflop/s]	[%]	[Mflop/s]	[%]
NEC SX-3/14R	6400	368	6	5199	81
CRAY T916	1800	522	29	1576	88
IBM RS/6000-590 (POWER2)	264	130	49	236	89
HP 9000/735	198	41	21	120	61
IBM RS/6000-580 (POWER)	125	38	30	104	83

3.5.4 Empirical Performance of Vector Processors

With analytical performance indices, as discussed in Section 3.4.3, only the clock rate f_c of a vector processor is relevant for its throughput rate r_∞. In practice, however, several other factors are also of significance for the performance of a vector processor. For example, if a vector exceeds a particular maximum length, a decrease in performance results, as such a vector has to be split into smaller pieces before being processed. Such *stripmining*, which is usually performed by a hardware device, causes additional overhead (depending on the vector length n), which reduces the asymptotic throughput rate. An even bigger overhead is caused by load operations of vector operands which have to be fetched from the main memory and moved into vector registers.

Moreover, r_∞ is heavily influenced by the *type* of vector operation (which is a construct of a high-level language) and the implementation by the compiler, which has to take into account peculiarities of the respective computer (such as instruction set and memory hierarchy). For example, performance is adversely affected by indirectly addressed vectors or by an unfavorable ratio between memory access operations and executed arithmetic operations.

Effects like these can be taken into consideration by empirically determining the estimate $\bar{r}_\infty$, which can be done as follows:

1. Measure the runtimes of the vector construct for several n;
2. Calculate $\bar{r}_\infty$ for $n \to \infty$ by means of curve fitting in accordance with the least squares method.

The empirical values $\bar{r}_\infty$, thus obtained, are usually significantly below the peak performance r_∞. Of course they are only valid for the vector construct and computer system being studied.

Example (Loops) (Dongarra et al. [164]) The peak performance of a CRAY 2 processor is 488 Mflop/s. For the loop

```
DO i = 1, n              ! update loop
   vector_1(i) = vector_1(i) + a*vector_2(i)
END DO
```

the throughput rate obtained experimentally is $\bar{r}_\times = 71$ Mflop/s. Hence, even in the best case, the efficiency rate cannot exceed 14 %. For the loop

```
DO i = 1, n                  ! bidiagonal loop
   vector_1(i) = vector_2(i) - vector_3(i) * vector_1(i-1)
END DO
```

the asymptotic throughput rate $\bar{r}_\infty$ amounts to only 8.5 Mflop/s, which is less than 2 % of the peak performance.

For the *update loop*, $\bar{n}_{1/2} = 41$ is obtained, and $\bar{n}_{1/2} = 10$ for the *bidiagonal loop*. The latter value, in particular, demonstrates that vectorization as such is reasonable for relatively small vector lengths, even if the respective performance indices (in relation to the peak performance) are anything but favorable.

Example (LINPACK Benchmark)   The empirical indices $\bar{r}_\infty$ and $\bar{n}_{1/2}$ can also be obtained using the subroutines LINPACK/sgefa and LINPACK/sgesl of the LINPACK benchmark.

For the CRAY 2 the values

$$\bar{r}_\infty = 138\,\text{Mflop/s}, \qquad \bar{n}_{1/2} = 305$$

are determined experimentally (Dongarra et al. [164]).

Thus, even for very large systems of equations, the efficiency of the two programs LINPACK/sgefa and LINPACK/sgesl (i.e., the unmodified LINPACK routines) on a CRAY 2 processor cannot exceed

$$100\,\frac{\bar{r}_\infty}{P_{\text{max}}} = 100\,\frac{138\,\text{Mflop/s}}{488\,\text{Mflop/s}} = 28\,\%.$$

For $n = 100$, the LINPACK benchmark produces a value of 44 Mflop/s, i.e., an efficiency of only 9 %. For $n = \bar{n}_{1/2} = 305$, efficiency climbs to 14 %, which is still no more than half the theoretical maximum efficiency rate for this program of 28 %.

Chapter 4

Numerical Data
and Numerical Operations

> *Well-informed people require only as much accuracy*
> *as the nature of the subject allows.*
>
> ARISTOTLE

4.1 Mathematical Data

Mathematical methods cannot be applied directly to physical objects or any other actual objects. That is why mathematical models of concrete situations are indispensable prerequisites for the use of mathematical methods. Similar objects which are mapped to the same abstract object are distinguished by their respective data. All numerical methods are eventually applied to those data, leading to the notion of numerical data processing.

4.1.1 Fundamental Mathematical Data

Numbers are the fundamental units of mathematical data. Classical calculus is based on *real numbers*, an uncountable set, the members of which span the real line from $-\infty$ to $+\infty$. For example, the following are valid real numbers:

$$0, \quad 1997, \quad 10^{10^{24}}, \quad -273.15, \quad 1/3, \quad \sqrt{2}, \quad \pi, \quad e, \quad \dots .$$

Real numbers are basic elements in science and technology. Practically all quantitative descriptions of technical and scientific processes are based on real numbers, even if certain subsets would be sufficient. Such subsets (consisting, for example, of integers or rational numbers) are embedded in the set of real numbers.

4.1.2 Algebraic Data

Real numbers can be used for the creation of compound data objects of higher or lower complexity—*data structures*—using aggregation (the creation of tuples), recursion (the creation of linear lists, trees etc.), and other techniques. Some examples of such data structures are listed in the following, where lower case letters $x, y, z, \dots$ denote real numbers:

Complex Numbers: $(x_1, x_2) := x_1 + i\, x_2 \in \mathbb{C}$, where i denotes the imaginary unit, with $i^2 + 1 = 0$;

Vectors: one-dimensional arrays[1] of real numbers $(x_1, x_2, \ldots, x_n) \in \mathbb{R}^n$;

Matrices: rectangular two-dimensional arrays of real numbers $A \in \mathbb{R}^{m \times n}$;

Sparse Matrices: matrices $A \in \mathbb{R}^{m \times n}$, with coefficients a_{ij} which are non-zero only for a few index pairs (i, j); all other elements are implicitly zero. *Condensed storage* is used for such matrices in order to reduce memory requirements, i.e., only the non-zero elements are stored.

Homogeneous Coordinates: vectors $P = (x_1, x_2, x_3, w) \in \mathbb{R}^4$, which represent coordinates in a projective coordinate system and which therefore denote points in $\mathbb{R}^3$ (or ideal points).

Homogeneous coordinates are used, for example, in graphical data processing (Reisenfeld [323]). They enable a uniform description of geometric transformations in terms of 4×4 matrix operations which are supported by graphical standards (PHIGS) and the hardware devices of graphical workstations.

These data structures themselves can be used in order to create more complicated data structures, e.g., complex vectors and matrices, three-dimensional or higher-dimensional arrays and the like. All such data objects are referred to in this book as *algebraic data*.

4.1.3 Analytic Data

A real function $f : D \subseteq \mathbb{R}^n \to \mathbb{R}$ maps each element in its domain to exactly one real number in its range. The domain typically consists of *infinitely* many numbers, usually an interval or a collection of intervals.

Example (Real Functions of One Variable)

$$
\begin{aligned}
f_1 : x &\mapsto 1 + 2x + 5x^3 & D &= \mathbb{R} \\
f_2 : x &\mapsto \sin x & D &= \mathbb{R} \\
f_3 : x &\mapsto \ln x & D &= \mathbb{R}_+ := \{x \in \mathbb{R} : x > 0\} \\
f_4 : x &\mapsto 2x/(x^2 - 1) & D &= \mathbb{R} \setminus \{-1, 1\} = \{x \in \mathbb{R} : x \neq \pm 1\}.
\end{aligned}
$$

Instead of a subset of $\mathbb{R}^n$, the domain or codomain (range) of a function may also be an algebraic set, making the definition of more complicated functions possible. This strategy can be continued recursively (functions of functions, and so on). Among these, *functionals* are of outstanding practical importance. They are mappings from a space of functions into $\mathbb{R}$. In particular, the definite integral of a real function is the functional

$$
\mathrm{I}(f; [0, 1]) \; : \; f \to \int_0^1 f(x)\, dx.
$$

In this book, the various functions or mappings of one or more variables are referred to as *analytic data*.

---

[1] In computer science, an *array* is an ordered collection of elements of the same type. Each component is referenced via its index.

4.2 Numerical Data on Computers

The functioning of a computer can be understood from different levels of abstraction: from the point of view of semiconductor technology, transistors, flip-flops, switching circuits, processors, and so on up to the computer as a whole. Each of those hardware levels has its own data and operations.

Information hiding, one of the fundamental principles of computer science, implies that none of the levels of higher abstraction has to handle the details of less abstract levels. It must be possible to handle a data object without any knowledge about its internal structures and processes. There must be a sufficiently detailed abstract description of the properties and the performance of such an object which enables its correct use and application. Such a description is usually referred to as an *interface specification.*

The advantage of this abstraction in layers is that modifications at lower levels do not necessarily entail changes at higher levels[2]. Moreover, this method makes it easier to concentrate on the original problem without being distracted by insignificant details.

Example (Rounding Error)   (Kulisch, Miranker [261]) In order to understand why the program

```
REAL  ::  x, y, z
...
x = 192119201.
y =  35675640.
z = (1682.*X*(Y**4) + 3.*(X**3) + 29.*X*(Y**2) - 2.*(X**5) + 832.)/107751.
PRINT *,  "z = ", z
...
```

using double precision IEC/IEEE arithmetic does not return the correct solution $z = 1783$, but instead the completely useless result z = 7.18056E20, it is necessary to know something about the rounding behavior of the arithmetic operations +, -, *, / and the exponential operator ** when applied to floating-point data objects. This requires insight into the model of the internal representation of those data objects (cf. for example model (4.4) in Section 4.4.3) rather than detailed knowledge of the hardware platform on which the program is executed.

4.2.1 Fundamental Numerical Data

The fundamental data unit at the flip-flop and switching-circuit level of a computer is the *bit*[3], the only possible values of which are 0 (binary zero) = *false* and 1 (binary one) = *true*. Manipulations at the bit-level is performed by the binary operations (Boolean functions) disjunction ($\vee$, *or*), conjunction ($\wedge$, *and*) and a unary operation, namely negation ($\neg$, *not*):

---

[2]For example, the functionality of the UNIX operating system is the same no matter what computer architecture is used.

[3]Bit is an abbreviation for *binary digit.*

x	y	$x \lor y$	$x \land y$	$\neg x$
0	0	0	0	1
0	1	1	0	1
1	0	1	0	0
1	1	1	1	0

At processor level, another elementary data type for the representation of information is added: the *character*. A character is an element of a finite set, the *character set*. A character set for which an order relation is defined is referred to as an *alphabet*. The most important alphabets in the field of data processing are the ASCII code[4], comprising 128 characters, and the EBCDIC code[5], comprising 256 characters (Engesser, Claus, Schwill [15]). The character set of these alphabets contains capital letters and lower case letters, numbers and control characters. A sequence of 8 bits (one *byte*) is used for coding, allowing for at most $2^8 = 256$ characters.

In this context, sequences of digits (used as characters) frequently represent the integer numbers they symbolize, e. g., set declarations, dates, assessments, and the like. However, arithmetic operations and other operations which do not confine themselves to simple character manipulations of the sequence of digits, are prudently not executed in terms of this character representation of numbers. Special bit sequences are used for the internal representation of numbers instead, such as the binary coded decimals (BCD) used for the representation of integers. This code represents each decimal digit by a combination of four bits, known as a *nibble*:

decimal digit	0	1	2	3	4	$\cdots$	9
BCD bit combination	0000	0001	0010	0011	0100	$\cdots$	1001

Decimal numbers are coded place by place. For example, the BCD representation of the decimal number 1997 is 0001 1001 1001 0111.

Efficient numerical computations—a fundamental demand of numerical data processing—are made possible by the particular internal representations of numbers on computers: There are INTEGER number systems (see Section 4.4.1) for integer numbers and floating-point number systems which use numbers of the form $x = \pm m \cdot b^e$ (see Section 4.4.3) for the approximate representation of real numbers. Simply speaking, m denotes the sequence of digits in base b of the number and e denotes the position of the unit's place, and hence the order of magnitude of the number.

---

[4] ASCII is an abbreviation for *American Standard Code for Information Interchange*.
[5] EBCDIC is an abbreviation for *Extended Binary Coded Decimal Interchange Code*.

4.2.2 Algebraic Data

Non-elementary data types can be derived from elementary data types (e.g., the Boolean data type or the set of characters). This can be accomplished by using aggregation (e.g., by creating tuples), by using generalization (e.g., by forming the disjoint union of ranges), by the creation of power sets, and other constructive methods. The manipulation of these data is an important part of non-numerical data processing (Aho, Hopcroft, Ullman [82], King [252]). Some of these derived data types are of universal importance and hence are implemented in programming languages designed for numerical data processing.

For algebraic data each element of a *finite* index set (i.e., a k-tuple of integers) is allocated to exactly one real number. For example, for complex numbers the index set consists of the numbers 1 and 2, for a vector $x \in \mathbb{R}^n$ of the numbers $1, 2, \ldots, n$, and for a (sparse) matrix $A \in \mathbb{R}^{m \times n}$ of (a subset of) the set of pairs

$$\{(i, j) : \ i = 1, 2, \ldots, m, \ j = 1, 2, \ldots, n\}.$$

There are different ways to perform this index mapping, which is crucial for the efficient implementation of algebraic data types and their internal processing by computers.

4.2.3 Analytic Data

The implementation and the processing of functions—analytic data—is an important task in numerical data processing (see Section 4.11 and Chapter 8).

However, the concept of function spaces (sets of functions, e.g., the set $C^2[a, b]$ of all twice continuously differentiable functions $f : [a, b] \to \mathbb{R}$) has not as yet impinged upon programming languages as there are serious difficulties encountered in its implementation. Imperative programming languages (e.g., Fortran or C) only allow for *unchangeable* functions, which are defined in the form of procedures (subprograms). Those programming languages do not allow the specification of subroutines which return other subroutines as results.

4.2.4 Numerical Data Types

The concept of *data types* is of fundamental importance in all branches of data processing. It denotes the grouping of data (data structures) and their respective operations. In order to stress that the focus is on the properties of data structures and operations rather than on their implementation on computers, they are frequently referred to as *abstract data types*. This chapter deals with abstract *numerical* data types.

Programming languages designed for numerical data processing implement abstract numerical data types using *actual* data types REAL (or FLOAT etc.) and INTEGER (or INT etc.).

The range of the INTEGER data type is the subset $\{i \in \mathbb{Z} : i_{\min} \leq i \leq i_{\max}\}$ of the integers. The range of the REAL data type is a subset of the real numbers, which is discussed in detail in Section 4.4.3. Some programming languages (e. g., Fortran 90), moreover, contain a COMPLEX data type for the representation of complex numbers.

For all actual data types there are particular arithmetic operations, the details of which are discussed in Section 4.7.

Vectors and matrices can be derived from data objects of the simple types IN-TEGER, REAL or COMPLEX using aggregation. Some programming languages allow for the application of arithmetic operations to such arrays, as well

Example (Fortran 90) In Fortran 90 operations on arrays are defined *element by element*. For example, the multiplication of two matrices A*B results in the elementwise product $(a_{ij}b_{ij})$ rather than the matrix-matrix multiplication $(c_{ij} = a_{i1}b_{1j} + a_{i2}b_{2j} + \cdots + a_{in}b_{nj})$; this operation is implemented in Fortran 90 by the intrinsic function MATMUL (Adams et al. [2]).

Example (Fortran XSC) FORTRAN XSC, a language extension of Fortran 90 especially designed for numerical data processing, contains the data type INTERVAL. With $\underline{x} < \overline{x}$, data of this type can be used such that a pair of real numbers represent an infinite set of real numbers $[\underline{x}, \overline{x}] := \{x \in \mathbb{R} : \underline{x} \le x \le \overline{x}\}$. Moreover, this data type allows the combination of intervals by arithmetic operations.

It is possible for the programmer to define derived data types in many imperative programming languages. In doing so, the ranges and their respective operations e. g., for homogeneous coordinates or intervals, may be defined.

4.3 Operations on Numerical Data

In this section, the most important operations with numerical operands are discussed without going into their actual implementation. In accordance with the classification of the data, there is a difference between arithmetic, algebraic and analytic operations.

4.3.1 Arithmetic Operations

The *rational operations* addition, subtraction, multiplication, and division are the basis for manipulating numerical data. In analysis these operations satisfy fundamental laws: commutativity and associativity of addition and multiplication, distributivity of multiplication over addition etc. One of the most serious difficulties of numerical data processing is that some of those elementary relations are not maintained in computer implementations.

In addition to rational operations, there are several *standard functions* of extraordinary importance, which are frequently considered *arithmetic operations* as well:

sign inversion	$-$ (a unary prefix operator)
absolute value function	$|x|$
power function	x^m
square root function	$\sqrt{x}$

exponential function	$\exp x$
logarithmic functions	$\log x$, $\log_{10} x$
trigonometric functions	$\sin x$, $\cos x$, $\tan x$, ...

All programming languages which are suitable for numerical data processing contain intrinsic procedures (e. g., -, ABS, **, SQRT, EXP, LOG, LOG10, SIN, COS, TAN in Fortran 90) which implement these standard functions. This, to some extent, makes it possible to handle them as rational operations. In Fortran 90, the set of standard functions is even more comprehensive than the list above and comprises the hyperbolic functions SINH, COSH, and TANH as well as the inverse trigonometric functions ASIN, ACOS, and ATAN.

Moreover, there are standard functions with an arbitrary number (two or more) of real arguments (operands), e. g., the minimum function MIN and the maximum function MAX.

In a broader sense, the *comparison operations*, can also be considered to be arithmetic operations. They attribute a Boolean value to a pair of real operands and are used e. g., for conditional jumps depending on numerical values; they are represented by

$$< \quad \geq \qquad > \quad \leq \qquad = \quad \neq \, .$$

From a strictly logical point of view one operator in each of these pairs could be disposed of as it simply returns the opposite Boolean value of the other operator. Nevertheless, for the sake of improved program readability, all six operators are part of programming languages used in numerical data processing, e. g., `<` `>=` `>` `<=` `==` `/=` in Fortran 90.

4.3.2 Algebraic Operations

Since the implementation of algebraic data types is based on a finite set of real numbers, it must be possible to reduce all operations applied to such data to arithmetic operations acting on the real components, elements etc. which constitute the algebraic data type.

For example, the *rational operations of complex operands* can be represented as the composition of rational operations on the real and imaginary parts of the operands $z_1 = x_1 + iy_1, z_2 = x_2 + iy_2 \in \mathbb{C}$:

$$
\begin{aligned}
z_1 \pm z_2 &= (x_1 \pm x_2) + i\,(y_1 \pm y_2) \\
z_1 z_2 &= (x_1 x_2 - y_1 y_2) + i\,(x_1 y_2 + y_1 x_2) \\
z_1/z_2 &= \frac{x_1 x_2 + y_1 y_2}{x_2^2 + y_2^2} + i\,\frac{y_1 x_2 - x_1 y_2}{x_2^2 + y_2^2} \qquad (z_2 \neq 0).
\end{aligned}
$$

For convenient and efficient programming it is essential that the programmer has powerful algebraic operations at his disposal. These operations should *not* require the use of the internal components of the data objects.

Example (Complex Numbers)   The data type COMPLEX for complex numbers, for which all algebraic operations and standard functions are implemented, is part of Fortran 90.

```
COMPLEX  ::  z_1, z_2, z_sum, z_mult, z_div, z_sin
...
z_sum  = z_1 + z_2
z_mult = z_1 * z_2
z_div  = z_1 / z_2
...
z_sin  = SIN (z_1)
...
```

Example (Intervals) In FORTRAN XSC data of the type INTERVAL can also be combined by arithmetic operators. For two intervals $[\underline{x}, \overline{x}]$, $[\underline{y}, \overline{y}]$ there are the following rational operations:

$$
\begin{aligned}
[\underline{x},\overline{x}] + [\underline{y},\overline{y}] &:= [\underline{x}+\underline{y}, \overline{x}+\overline{y}] \\
[\underline{x},\overline{x}] - [\underline{y},\overline{y}] &:= [\underline{x}-\overline{y}, \overline{x}-\underline{y}] \\
[\underline{x},\overline{x}] \cdot [\underline{y},\overline{y}] &:= [\min\{\underline{x}\underline{y}, \underline{x}\overline{y}, \overline{x}\underline{y}, \overline{x}\overline{y}\}, \max\{\underline{x}\underline{y}, \underline{x}\overline{y}, \overline{x}\underline{y}, \overline{x}\overline{y}\}] \\
1 / [\underline{y},\overline{y}] &:= [1/\overline{y}, 1/\underline{y}] \\
[\underline{x},\overline{x}] / [\underline{y},\overline{y}] &:= [\underline{x},\overline{x}] \cdot (1/[\underline{y},\overline{y}])
\end{aligned}
\left.\right\} \quad \text{for} \quad \underline{y}, \overline{y} > 0 \quad \text{or} \quad \underline{y}, \overline{y} < 0.
$$

As in the case of operations with complex numbers or intervals, *basic linear algebra operations* can be represented by (compound) rational operations on the components of the vector or matrix operands, e.g., with $x, y \in \mathbb{R}^n$, $A \in \mathbb{R}^{m \times n}$:

$$
\begin{aligned}
\langle x, y \rangle &= \sum_{k=1}^{n} x_k y_k \\
(Ax)_j &= \sum_{k=1}^{n} a_{jk} x_k, \qquad j = 1, 2, \ldots, m.
\end{aligned}
$$

Array processing is a very important branch of numerical data processing. Hardware components which are particularly efficient for such tasks, namely *vector computers* have thus been developed.

4.3.3 Array Processing in Fortran 90

Current multi-purpose programming languages—which were largely developed for sequential algorithms on conventional computers—do *not* allow for satisfactory exploitation of vector computer resources (see e.g., Dongarra et al. [164], Ling [268], Sekera [344]). The manufacturers of vector computers thus developed non-portable (non-standardized) language extensions to Fortran 77, which allow for much better exploitation of those computers, e.g., *Cray Fortran* for the various Cray models or *VECTRAN*, developed for IBM machines.

Many elements of these hardware-specific language extensions (sometimes in modified form) became part of the standardized, hardware-independent Fortran 90 language specification (Adams et al. [2], Ueberhuber, Meditz [78]).

Array Operations

In Fortran 90 arithmetic operations are defined elementwise, e.g., the operations $C = A + B$ or $C = A/B$ are allowed provided the arrays A, B and C have the same form (e.g., matrices with the same number of rows and columns).

Scalar constants are automatically expanded to constant arrays of compatible form, such that e. g., the expression $1/B$ corresponds to A/B, where $A = (a_{ij}) = (1)$.

Array Composition and the Extraction of Array Subsections

Fortran 90 supports the recursive definition of n-dimensional arrays from $(n-1)$-dimensional ones, e. g., the construction of an $n{\times}n$ matrix given an $(n-1){\times}(n-1)$ matrix by adding a row vector and a column vector. The extraction of an array subsection from a given larger array can be coded in a simple manner.

In Fortran 90 standard functions are applicable to arrays as well; e. g., LOG $(A) := (\log a_{ij})$ is defined elementwise for the array $A = (a_{ij})$.

Intrinsic Functions

There are intrinsic functions for matrix-matrix multiplication (cf. the Fortran 90 example on page 111) and matrix transposition. *Vector* operations are part of any *matrix* algebra; a vector is interpreted as a $(n \times 1)$ matrix.

Example (Inner Product)   The Fortran 90 function DOT_PRODUCT (vector_u,vector_v) returns the inner product

$$u^\top v = u_1 v_1 + \cdots + u_n v_n$$

of the two vectors u and v. On the other hand, the array operation vector_u * vector_v returns the *vector*

$$(u_1 v_1, u_2 v_2, \ldots, u_n v_n)^\top.$$

In addition to the matrix algebra of general dense matrices, there are also intrinsic functions for the condensed storage of sparse matrices: PACK and UNPACK. However, in Fortran 90 there are no intrinsic functions for manipulating and combining sparse matrices.

Power of Array Operations and Array Functions

The Fortran 90 language elements for array processing are remarkably comprehensive to the point that the coding of algorithms may be significantly simplified and shortened.

Example (Vector and Matrix Norms)   The following simple code fragment returns different norms of vectors and matrices.

```
REAL, DIMENSION (n)    ::  x, y
REAL, DIMENSION (n,n)  ::  a, b
...                                          ! VECTOR NORMS:
x_norm_1   = SUM (ABS (x))                   !   1-norm
x_norm_2   = SQRT (SUM (x**2))               !   Euclidean norm (2-norm)
x_norm_max = MAXVAL (ABS (x))                !   maximum (infinity-) norm
                                             ! MATRIX NORMS:
a_norm_1   = MAXVAL (SUM (ABS (a), DIM = 1)) !   column-sum norm
a_norm_max = MAXVAL (SUM (ABS (a), DIM = 2)) !   row-sum norm
a_norm_f   = SQRT (SUM (a**2))               !   Frobenius norm
```

Example (Statistical Data Analysis)   The $m \times n$ matrix $T = (t_{ij})$ contains the test results (classified by scores $t_{ij} \geq 0$) of m students after n tests. The following Fortran 90 statements return various analyses of the test results.

1.   What is the maximal score of each student?

```
MAXVAL (t, DIM = 2)   ! result is a vector of length m
```

2.   How many test results were above average (with respect to all tests and all students)?

```
above_average = t > SUM (t)/SIZE (t)            ! matrix
number_of_good_tests = COUNT (above_average)    ! scalar
```

3.   Increase each score above average by 10 %.

```
WHERE (above_average) t = 1.1*t
```

4.   What was the *lowest* score of all results *above* average?

```
min_above_average = MINVAL (t, MASK = above_average)
```

5.   Was there at least one student who scored above average in all tests?

```
response_boolean = ANY (ALL (above_average, DIM = 2))
```

Example (Chi-Square Statistics)   The value χ^2 is derived from the empirical frequency p_{ij} and the expected frequency e_{ij} as follows:

$$\chi^2 := \sum_{i,j} \frac{(p_{ij} - e_{ij})^2}{e_{ij}},$$

where the elements of the matrix $E = (e_{ij})$ are given by

$$e_{ij} := \frac{\left(\sum_k p_{ik} \right) \left(\sum_k p_{kj} \right)}{\sum_{k,l} p_{kl}}.$$

As few as *four* Fortran 90 statements are sufficient for computing χ^2:

```
REAL                   ::  chi_square
REAL, DIMENSION (m,n)  ::  p, e
REAL, DIMENSION (m)    ::  r
REAL, DIMENSION (n)    ::  c
...
r = SUM (p, DIM = 2)
c = SUM (p, DIM = 1)
e = SPREAD(r, DIM=2, NCOPIES=n) * SPREAD(c, DIM=1, NCOPIES=m) / SUM(p)
chi_square = SUM (((p-e)**2)/e)
```

4.3.4 Analytic Operations

Operations using functions as the data of a problem are frequently used in the analysis and evaluation of mathematical models. From a mathematical point of view, even the evaluation of a function at a given point is an operation with the function as a data element. However, in the context of (single or multiple) differentiation problems and (definite or indefinite) integration problems, it is easier to regard the function itself as an operand.

Contemporary programming languages designed for numerical data processing allow for operations with functions only as long as they are based on the pointwise evaluation of functions—a serious restriction, as will be shown later.

Computer algebra systems—MAPLE, MATHEMATICA, AXIOM etc.—allow for the direct manipulation of the defining terms of a function (Davenport et al. [148]).

Example (Mathematica) In the following example possible ways of manipulating formulas with the use of the computer algebra system MATHEMATICA are demonstrated.

Symbolic Differentiation: All elementary mathematical functions can be derived using MATHEMATICA. For example, the input/output

```
In[1]:= -3x^4 + 12x^2 + 25x^3 +2
                2         3      4
Out[1]= 2 + 12 x  + 25 x   - 3 x

In[2]:= D[%,x]
                 2       3
Out[2]= 24 x + 75 x   - 12 x
```

is used to find the derivative

$$p(x) = P'(x) = -12x^3 + 75x^2 + 24x$$

of the polynomial

$$P(x) = -3x^4 + 25x^3 + 12x^2 + 2.$$

Symbolic Integration: MATHEMATICA can be used for symbolic integration as well. If no integration limits are specified, an indefinite integral of the given function is determined. For the above function $p(x)$, an indefinite integral is given by

$$\overline{P}(x) = -3x^4 + 25x^3 + 12x^2.$$

The user dialogue of MATHEMATICA is given by:

```
In[3]:= Integrate[24x +75x^2 -12x^3,x]
              2       3      4
Out[3]= 12 x  + 25 x   - 3 x
```

If MATHEMATICA cannot come up with an analytical evaluation of a function, then the function is simply returned in its original form.

Determination of zeros: MATHEMATICA can be used to determine the zeros of elementary functions. In order to do so, it has to be specified as to which variable of the equation is sought; all other variables are assumed to be constants.

```
In[4]:= Solve[4 a x + 25 a x^2 - 4 x^3==0, x]
```

$$
Out[4]= \{\{x \rightarrow 0\}, \{x \rightarrow \frac{25\ a\ -\ Sqrt[128\ a\ +\ 625\ a^2]}{8}\},
$$

$$
> \quad \{x \rightarrow \frac{25\ a\ +\ Sqrt[128\ a\ +\ 625\ a^2]}{8}\}\}
$$

Hence, the three zeros of the function $g(x) = -4x^3 + 25ax^2 + 4ax$, $a \in \mathbb{R}$ are

$$
x_1 = 0, \quad x_{2,3} = \frac{25a \pm \sqrt{128a + 625a^2}}{8}.
$$

Solution of Differential Equations: The symbolic derivation of solutions of ordinary differential equations is of particular interest and importance in science. As an example, for the differential equation $y'(x) = ay(x) + 4x$ MATHEMATICA derives the function

$$
y(x) = \frac{-4}{a^2} - \frac{4x}{a} + C_1 e^{ax}
$$

as a solution in response to the following instruction:

```
In[6]:= DSolve[y'[x]  == a y[x] +4x, y[x],x]
                      -4    4 x      a x
Out[6]= {{y[x] -> -- - --- + E     C[1]}}
                   2    a
                  a
```

If initial values are specified by the user, MATHEMATICA also determines the constant C_1.

4.4 Number Systems on Computers

For the sake of efficiency, the implementation of real numbers—the elementary data of mathematics—must be based on a fixed code or, at the very most, a small number of different codes.

This code must map each real number to a particular *bit sequence of fixed length* (the *format width*) N. However, this only allows for the coding of 2^N different real numbers at the most, compared to the uncountable continuum of real numbers. This striking disproportion can neither be disposed of by the choice of a particularly large format width nor by the use of a larger number of different codes. Moreover, there is a natural bound for N, as the code of a real number should fit into one register.[6]

Whatever the parameters of such a code may be, only a finite set of real numbers can be represented. The question arises as to how to choose the set of numbers so that the demands of the user are complied with as far as possible. Moreover, the efficiency of coded numbers must always be considered.

Example (32 Bit Numbers) A format width of $N = 32$ bits implies that

$$
2^{32} = 4\,294\,967\,296
$$

numbers can be coded at the most. Usually not all possible bit sequences are actually used to represent real numbers.

---

[6]Customary register sizes are $N = 32$ and $N = 64$ bits (see Section 3.2.3).

4.4.1 INTEGER Number Systems

The most natural way to interpret a bit sequence $d_{N-1}d_{N-2}\cdots d_2 d_1 d_0$ of length N as a real number is to regard it as the digitwise representation of a non-negative integer in a decimal place code with the base $b = 2$, i.e.,[7]

$$d_{N-1}d_{N-2}\cdots d_2 d_1 d_0 \;\doteq\; \sum_{j=0}^{N-1} d_j 2^j, \qquad d_j \in \{0,1\}. \tag{4.1}$$

Thus all numbers ranging from 0 to $2^N - 1$ are covered:

$$
\begin{aligned}
0000\cdots 0000 &\;\doteq\; 0 \\
0000\cdots 0001 &\;\doteq\; 1 \\
&\;\;\vdots \\
1111\cdots 1111 &\;\doteq\; 2^N - 1.
\end{aligned}
$$

In order to represent negative integers, too, the range of non-negative integers is restricted, and all bit sequences thus obtained are interpreted as negative numbers. Three different methods are customary: Separate sign coding, one's complement and two's complement.

Separate Sign Coding

For the representation of integers the sign may be coded separately. If a bit $s \in \{0,1\}$ is used for representing the sign then the bit sequence of length N is interpreted as follows:

$$s d_{N-2} d_{N-3} \cdots d_2 d_1 d_0 \;\doteq\; (-1)^s \sum_{j=0}^{N-2} d_j 2^j.$$

This covers the integers $[-(2^{N-1} - 1), 2^{N-1} - 1] \subset \mathbb{Z}$:

$0000\cdots 0000$	$\doteq$	0	$1000\cdots 0000$	$\doteq$	-0
$0000\cdots 0001$	$\doteq$	1	$1000\cdots 0001$	$\doteq$	-1
$\vdots$			$\vdots$		
$0111\cdots 1111$	$\doteq$	$2^{N-1} - 1$	$1111\cdots 1111$	$\doteq$	$-(2^{N-1} - 1)$.

Here the code for the number zero is not unique: there being $+0$ and -0.

One's Complement Coding

The *one's complement coding* uses

$$\bar{x} = \sum_{j=0}^{N-1}(1 - d_j)2^j$$

---

[7]Here the symbol $\doteq$ is used to mean "corresponds to".

as the representation of $-x$:

$0000\cdots0000$	$\doteq$	0	$1111\cdots1111$	$\doteq$	-0
$0000\cdots0001$	$\doteq$	1	$1111\cdots1110$	$\doteq$	-1
$\vdots$			$\vdots$		
$0111\cdots1111$	$\doteq$	$2^{N-1}-1$	$1000\cdots0000$	$\doteq$	$-(2^{N-1}-1)$.

Again, the representation of the number zero is not unique. Positive numbers are represented as in equation (4.1).

Two's Complement Coding

The *two's complement coding* uses the representation

$$\bar{\bar{x}} = 1 + \sum_{j=0}^{N-1}(1-d_j)2^j$$

for the coding of $-x$:

$0000\cdots0000$	$\doteq$	0			
$0000\cdots0001$	$\doteq$	1	$1111\cdots1111$	$\doteq$	-1
$\vdots$			$\vdots$		
$0111\cdots1111$	$\doteq$	$2^{N-1}-1$	$1000\cdots0001$	$\doteq$	$-(2^{N-1}-1)$
			$1000\cdots0000$	$\doteq$	-2^{N-1}.

This way of coding yields a unique representation of the number zero, but the range of integers $[-2^{N-1}, 2^{N-1}-1]$ is unsymmetric.

Example (Intel)   The Intel microprocessors use two's complement coding of integers with format widths $N = 16, 32$, and 64 bits, referred to as *short integer*, *word integer*, and *long integer* respectively. The range of *word-integer* numbers is

$$[-2^{31}, 2^{31}-1] = [-2\,147\,483\,648,\ 2\,147\,483\,647] \subset \mathbb{Z}.$$

INTEGER Data Type

In most programming languages subsets of the integers—together with the arithmetic operations $+$, $-$, $\times$ and

div	(integer division)
mod	(remainder of integer division)
abs	(absolute value) ,

are combined into an INTEGER data type, leading to the notion of an *INTEGER number system*. Such a set of numbers can be characterized by its smallest number $i_{min} \le 0$ and its largest number $i_{max} > 0$. For a set of non-negative integer numbers $i_{min} = 0$ holds.

The coding is said to be symmetric if $i_{min} = -i_{max}$ and unsymmetric otherwise. In particular, with the two's complement coding $i_{min} = -(i_{max}+1)$ holds.

Example (C) The implementation of the programming language C on HP workstations provides the following INTEGER number systems:

Data Type	$i_{\min}$	$i_{\max}$
unsigned short int	0	65 535
short int	-32 768	32 767
unsigned int	0	4 294 967 295
int	-2 147 483 648	2 147 483 647
unsigned long int	0	4 294 967 295
long int	-2 147 483 648	2 147 483 647

Both of the data types `int` and `long int` are coded in format width $N = 32$. Hence, `long int` does *not* have more elements than `int`.

Modulo Arithmetic

If the result of an arithmetic operation is outside of the range $[i_{\min}, i_{\max}]$, it is referred to as an *integer overflow*. The occurrence of an integer overflow is either indicated by the computer as an error or a value x_{wrap} in the range $[i_{\min}, i_{\max}]$, and congruent with the actual result x, is returned instead, *without* an error message. This value is specified by the following relation:

$$x_{\text{wrap}} \equiv x \bmod m, \qquad m := i_{\max} - i_{\min} + 1$$

$$x_{\text{wrap}} \in [i_{\min}, i_{\max}].$$

This concept is referred to as *modulo arithmetic*.

Example (C) The implementation of C on HP workstations uses modulo arithmetic for all integer formats. This can neither be changed to an overflow mode at the program level nor at the operating system level.

Example (Random Number Generator) Modulo arithmetic is used for most generators of uniformly distributed pseudo-random numbers. For example, a set of 17 odd 32-bit integer numbers can be used as the initial values $x_{-16}, x_{-15}, \ldots, x_0$, of the iteration

$$x_k := x_{k-17} x_{k-5} \quad \bmod 2^{32}, \qquad k = 1, 2, 3, \ldots \tag{4.2}$$

which generates a sequence of pseudo-random numbers with a very long period of around $7 \cdot 10^{13}$ (see Fig. 4.1). The modulo function in (4.2) is implemented quite effectively with modulo arithmetic.

4.4.2 Fixed-Point Number Systems

The interval of the real number line covered[8] by codable numbers can be enlarged or reduced by a scaling factor 2^k where k is an arbitrary integer. This

---

[8]This does not mean a complete covering of *all* elements in a set. Finite sets can cover an interval of real numbers only by leaving out spaces in between. In the context of numerical data processing, points of a finite set covering real numbers are often referred to as a *grid*.

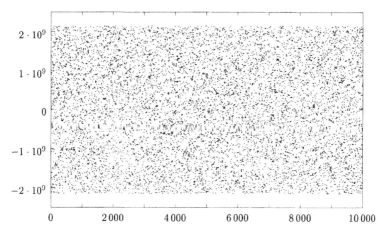

Figure 4.1: 10 000 uniformly distributed (pseudo-) random numbers.

enlarges or reduces the *density* of the represented numbers respectively. Scaling with 2^k, where $k < 0$, corresponds to the interpretation of the bit sequence $s d_{N-2} d_{N-3} \cdots d_2 d_1 d_0$ as a signed binary fraction with $-k > 0$ places behind the binary point.

$$s \, d_{N-2} d_{N-3} \cdots d_2 d_1 d_0 \; \dot{=} \; (-1)^s 2^k \sum_{j=0}^{N-2} d_j 2^j \qquad (4.3)$$

$$\dot{=} \; (-1)^s d_{N-2} \cdots d_{-k} . d_{-k-1} \cdots d_0$$

A set of numbers with fixed scaling of that kind is called a *fixed-point number system*.

Example (Fixed-Point Number System) A fixed-point number system with a format width of $N = 16$ bits and a scaling factor of 2^k, $k = -4$ specifies the set of numbers

$$s d_{14} d_{13} \ldots d_1 d_0 \; \dot{=} \; (-1)^s 2^{-4} \sum_{j=0}^{14} d_j 2^j$$

An example of such a fixed-point number is

$$1000\,0001\,1011\,0101 \; \dot{=} \; -1\,1011.0101_2 \; = \; -27.3125_{10}.$$

For $k = -(N-1)$ the binary point is assumed to be positioned directly in front of the leading digit d_{N-2} of the sequence of binary digits. As with INTEGER numbers one bit $s \in \{0, 1\}$ can be used to represent the sign $(-1)^s$. For $k = -(N-1)$ the interval $(-1, 1)$ is covered uniformly with a constant distance $2^k = 2^{-N+1}$ between the points.

The use of bit sequences suggest an interpretation as numbers of base $b = 2$. However, alternative number systems may be used as well, such as *hexadecimal numbers*, which condense four binary digits into one digit of base 16 (in the range $\{0, \ldots, 9, A, \ldots, F\}$).

Example (Hexadecimal Numbers) The number 2842_{10} with the binary representation $1011\,0001\,1010_2$ is written $B1A_{16}$ in the hexadecimal system.

Obviously the hexadecimal code is far more compact than the corresponding binary representation.

Alternatively, four bits can be used for the coding of a decimal digit $d_i \in \{0, 1, \ldots, 9\}$; however, this results in significant redundancy. This *decimal fixed-point code* mimics well the usual way in which the traditional decimal number system is handled. Such codes are frequently used for pocket calculators, either exclusively or in conjunction with other codes.

Example (Intel) In addition to the binary INTEGER number system discussed above, Intel microprocessors are supplied with an 18-place *decimal* INTEGER code of format width $N = 73$ bits (18×4 bits and one sign bit).

The main disadvantage of fixed-point number systems in numerical data processing is that the size of the real numbers which can be covered is fixed once and for all whereas the orders of magnitude of data involved in numerical methods may vary over a wide range which is rarely known in advance. Moreover, in the course of solving a problem, very large and very small real numbers may occur as intermediate results.

4.4.3 Floating-Point Number Systems

The disadvantage of the fixed-point codes discussed above can be overcome if the scaling parameter e is allowed to vary and information as to its size is stored in the code. This leads to a *floating-point* code, which is used for the vast majority of applications in numerical data processing.

A bit sequence is split into three parts so as to be interpreted as a floating-point code: the *sign s*, the *exponent e* and the *mantissa m* (the *significant figures*). The *format* of such a bit sequence, which specifies the length of each of the three parts, is fixed for each floating-point code.

Example (IEC/IEEE Floating-Point Numbers) The standard IEC 559:1989 for binary floating-point arithmetic specifies two *basic formats*:

Single Precision Format: format width $N = 32$ bits

1	8 bits	23 bits
s	e	m

Double Precision Format: format width $N = 64$ bits

1	11 bits	52 bits
s	e	m

The IEC/IEEE code as outlined above is discussed in full detail in Section 4.6.

The mantissa m is a non-negative real number in a fixed-point code (*without sign*)

$$m = d_1 b^{-1} + d_2 b^{-2} + \cdots + d_p b^{-p}$$

with respect to a fixed base b with the binary point (decimal point etc.) directly in front of the leading digit.

Notation (Position of Binary or Decimal Points) In this book the binary or decimal point of all floating-point number systems is deliberately assumed to be in front of the leading digit of the mantissa. Moreover, in correspondence with technical conventions the *most significant digit* is denoted by d_1 and the *least significant digit* is denoted by d_p. This notation does not correspond to the notation of the indices of a fixed-point number system (4.3).

Both conventions comply with the proposed international standard *Language Independent Integer and Floating-Point Arithmetic* [239] as well as with the specifications of the ANSI-C and the Fortran 90 standards.

The exponent e is an integer. The real number represented by x is given by

$$x = (-1)^s \, b^e \, m,$$

where the bit $s \in \{0,1\}$ specifies the sign of x. The set of real numbers which can be represented by the floating-point code is denoted by $\mathbb{F}$, and varies with the choice of base b, the fixed-point code m, and the permissible values of e.

For numerical data processing, it is irrelevant as to how the floating-point code is implemented by the underlying hardware of a particular computer (see e. g., Swartzlander [363]). What really counts is the distribution of the floating-point numbers on the real line and the arithmetic in $\mathbb{F}$. The discussion of computer arithmetic is therefore based on the floating-point *number system* rather than on their respective *codes*.

A system of floating-point numbers with base b, precision p, and range of exponents $[e_{\min}, e_{\max}] \subset \mathbb{Z}$ contains the following real numbers:

$$x = (-1)^s \, b^e \sum_{j=1}^{p} d_j b^{-j} \tag{4.4}$$

where

$$s \in \{0,1\} \tag{4.5}$$

$$e \in \{e_{\min}, e_{\min}+1, \ldots, e_{\max}\}, \tag{4.6}$$

$$d_j \in \{0, 1, \ldots, b-1\}, \quad j = 1, 2, \ldots, p. \tag{4.7}$$

The figures d_j denote the *digits* of the mantissa $m = d_1 b^{-1} + \cdots + d_p b^{-p}$.

Example (Representation of Numbers) In a decimal system of floating-point numbers with six decimal places ($b = 10$, $p = 6$), the real number 0.1 is represented as

$$.100000 \cdot 10^0 \quad .$$

On the other hand, in a system of binary floating-point numbers, the number 0.1 *cannot* be represented exactly; the rounded representation, e. g., with 24 binary digits

$$.110011001100110011001101 \cdot 2^{-3}$$

is used instead.

In addition to the so-called *single precision* floating-point number system, most computers provide further number systems with longer mantissas and possibly larger ranges of exponents. Such number systems are called *double* or *k-fold precision systems* although the number of digits is not necessarily an integral multiple of p_1, the length of the mantissa of single precision numbers.

Normalized and Denormalized Floating-Point Numbers

Clearly, each choice of values for the parameters s, e, and $d_1, d_2, \ldots, d_p$ which complies with the restrictions (4.5), (4.6) and (4.7) corresponds to a particular real number. However, different parameter values may represent the same real number x.

Example (The Ambiguous Representation of a Real Number)   In a decimal floating-point number system with precision $p = 6$ the number 0.1 can be represented by

$$.100000 \cdot 10^0 \quad \text{or} \quad .010000 \cdot 10^1 \quad \text{or}$$
$$.001000 \cdot 10^2 \quad \text{or} \quad .000100 \cdot 10^3 \quad \text{or}$$
$$.000010 \cdot 10^4 \quad \text{or} \quad .000001 \cdot 10^5,$$

whence the representation of 0.1 as a floating-point number is *not* unique.

In order to make the representation of numbers unique without considerably reducing the range of the floating-point number system, the additional condition $d_1 \neq 0$ can be introduced. The numbers thus obtained, with $b^{-1} \leq m < 1$, are called *normalized* floating-point numbers. The set of normalized floating-point numbers extended by the number zero (specified by the mantissa $m = 0$) is denoted $\mathbb{F}_N$ in the following.

The restriction $d_1 \neq 0$ leads to a one-to-one correspondence between the permissible values

$$s \in \{0, 1\},$$
$$e \in \{e_{\min}, e_{\min} + 1, \ldots, e_{\max}\},$$
$$d_1 \in \{1, 2, \ldots, b - 1\},$$
$$d_j \in \{0, 1, \ldots, b - 1\}, \quad j = 2, 3, \ldots, p,$$

and the numbers in $\mathbb{F}_N \setminus \{0\}$ (see Table 4.1).

The numbers lost as a result of the normalization condition $d_1 \neq 0$, namely those of such small absolute values that they cannot be represented as normalized floating-point numbers, can be regained without destroying the uniqueness of the representation by letting $d_1 = 0$ only for the smallest exponent $e = e_{\min}$. The numbers with $m \in [b^{-p}, b^{-1} - b^{-p}]$, which are all elements of the interval $(-b^{e_{\min}-1}, b^{e_{\min}-1})$ are called *denormalized* (or *subnormal*) numbers and are denoted $\mathbb{F}_D$.

The number zero (which corresponds to the mantissa $m = 0$) is a member of $\mathbb{F}_N$ according to the definition above. To bring about the uniqueness of this representation, some choice of the sign bit has to be made, e. g., $s = 0$.

Table 4.1: The one-to-one correspondence between the non-negative numbers in $\mathbb{F}$ and the tuples (e, m). An analogous relation holds for the negative numbers in $\mathbb{F}$ which are characterized by $s = 1$.

	exponent	mantissa
normalized floating-point numbers	$e \in [e_{min}, e_{max}]$	$m \in [b^{-1}, 1 - b^{-p}]$
denormalized floating-point numbers	$e = e_{min}$	$m \in [b^{-p}, b^{-1} - b^{-p}]$
zero	$e = e_{min}$	$m = 0$

Parameters of a Floating-Point Number System

Four integer parameters and a Boolean value:

1. base (radix) $b \geq 2$,
2. precision $p \geq 2$,
3. smallest exponent $e_{min} < 0$,
4. largest exponent $e_{max} > 0$,
5. normalization indicator $denorm \in Boolean$

characterize each floating-point number system.

The Boolean value *denorm* indicates whether or not the floating-point number system contains denormalized numbers (*denorm* = *true* in the positive case, *denorm* = *false* otherwise).

In the following, for floating-point number systems the short notation

$$\mathbb{F}(b, p, e_{min}, e_{max}, denorm)$$

is used. It satisfies the relations

$$\mathbb{F}(b, p, e_{min}, e_{max}, true) = \mathbb{F}_N(b, p, e_{min}, e_{max}) \cup \mathbb{F}_D(b, p, e_{min}, e_{max})$$
$$\mathbb{F}(b, p, e_{min}, e_{max}, false) = \mathbb{F}_N(b, p, e_{min}, e_{max}).$$

Contemporary computers use base 2, 10, or 16 exclusively, i.e., only binary, decimal, or hexadecimal floating-point number systems.

Example (Intel) In compliance with the IEC/IEEE-Norm (see Section 4.6), the number systems usually employed on Intel microprocessors are $\mathbb{F}(2, 24, -125, 128, true)$ and $\mathbb{F}(2, 53, -1021, 1024, true)$ for single precision and double precision floating-point number representation respectively. Intel microprocessors also have the extended precision floating-point number system $\mathbb{F}(2, 64, -16381, 16384, true)$.

Example (IBM System/390) IBM System/390 mainframes are provided with three hexadecimal floating-point number systems: *short precision* $\mathbb{F}(16, 6, -64, 63, false)$, *long precision* $\mathbb{F}(16, 14, -64, 63, false)$ and *extended precision* $\mathbb{F}(16, 28, -64, 63, false)$.

Example (Cray) Cray computers are provided with the two floating-point number systems: $\mathbb{F}(2, 48, -16384, 8191, false)$ and $\mathbb{F}(2, 96, -16384, 8191, false)$.

Example (Pocket Calculators)   Scientific pocket calculators are usually provided with just *one* floating-point number system, namely the decimal system $\mathbb{F}(10, 10, -98, 100, \textit{false})$. Some pocket calculators use higher precision systems internally while the display shows only 10 decimal places.

In practice the exponent is rarely coded in a symmetric (or nearly symmetric) INTEGER number system, as introduced in Section 4.4.1, but a *biased notation* is used instead, which adds a fixed bias to the exponent such that only positive exponents are used internally.

Hidden Bit

For a binary system of normalized floating-point numbers, $d_1 = 1$ necessarily holds as $d_1 \in \{0, 1\}$. It thus follows that d_1 does *not* have to be coded explicitly for the numbers of the binary system $\mathbb{F}_N(2, p, e_{\min}, e_{\max})$—the mantissa can have a *hidden bit*. For all denormalized numbers $\mathbb{F}_D(2, p, e_{\min}, e_{\max})$, as well, the first bit can be disposed of as it is always zero ($d_1 = 0$). However, if it is omitted in the internal representation, then there must be a mechanism for distinguishing between normalized and denormalized numbers. Most implementations of binary floating-point systems with hidden bits, e.g., IEC/IEEE numbers (see Section 4.6), distinguish denormalized numbers by storing some specific value not found in $[e_{\min}, e_{\max}]$ (usually $e_{\min} - 1$) in the exponent field.

Example (IBM System/390)   Because IBM mainframes use $b = 16$, they do *not* have a hidden bit. The mantissa of single precision floating-point numbers consists of six hexadecimal digits. For a normalized number the only requirement is $d_1 \neq 0000$. Thus, in the extreme $d_1 = 1_{16} = 0001_2$, it can happen that the first *three* digits of a normalized number are zero.

4.5 Structure of Floating-Point Systems

This section describes in detail the structure of the floating-point number system $\mathbb{F}(b, p, e_{\min}, e_{\max}, \textit{denorm})$, i.e., the number of floating-point numbers and their distribution.

4.5.1 The Number of Floating-Point Numbers

Due to the one-to-one correspondence between the numbers in $\mathbb{F}$ and the tuples $(s, e, d_1, \ldots, d_p)$, depicted in Table 4.1, the set $\mathbb{F}$ can only be a *finite* subset of all real numbers. The cardinality of normalized numbers in a floating-point system $\mathbb{F}(b, p, e_{\min}, e_{\max}, \textit{denorm})$ is

$$2\,(b-1)\,b^{\,p-1}(e_{\max} - e_{\min} + 1).$$

This quantity does not take into account the number zero and, if *denorm = true*, the set of denormalized numbers. For the purpose of numerical computation, this finite set of floating-point numbers has to replace the uncountable continuum of real numbers. In this book, therefore, the "numbers in $\mathbb{R}$ covered by $\mathbb{F}$" denote those real numbers which are relatively close to elements of $\mathbb{F}$ (compare with the footnote on page 120).

Example (IEC/IEEE Floating-Point Numbers) In their basic formats the IEC/IEEE floating-point number systems $\mathbb{F}(2, 24, -125, 128, \textit{true})$ and $\mathbb{F}(2, 53, -1021, 1024, \textit{true})$ contain

$$2^{24} \cdot 254 \approx 4.26 \cdot 10^9 \quad \text{and} \quad 2^{53} \cdot 2046 \approx 1.84 \cdot 10^{19}$$

normalized numbers respectively.

4.5.2 Largest and Smallest Floating-Point Numbers

In contrast to real numbers and due to the finiteness of $\mathbb{F}$, each floating-point number system contains a *largest number* $x_{\max}$ and a *smallest positive* normalized number $x_{\min}$.

With $m = m_{\max}$ and $e = e_{\max}$, the *largest floating-point number*

$$x_{\max} := \max\{x \in \mathbb{F}\} = (1 - b^{-p})b^{e_{\max}}$$

is obtained; there are no larger numbers in the floating-point number system $\mathbb{F}$. With $e = e_{\min}$ and $m = m_{\min}$ the smallest positive *normalized* floating-point number

$$x_{\min} := \min\{x \in \mathbb{F}_N : x > 0\} = b^{e_{\min} - 1}$$

is obtained.

Example (IEC/IEEE Floating-Point Numbers) The largest and the smallest positive normalized single precision floating-point numbers, i. e., the minimum and the maximum of the number set $\{x \in \mathbb{F}_N(2, 24, -125, 128) : x > 0\}$ are

$$
\begin{aligned}
x_{\min} &= 2^{-126} &\approx 1.18 \cdot 10^{-38} \quad \text{and} \\
x_{\max} &= (1 - 2^{-24})\, 2^{128} &\approx 3.40 \cdot 10^{38}
\end{aligned}
$$

respectively. The scope of double precision numbers $\mathbb{F}_N(2, 53, -1021, 1024)$ is much larger:

$$
\begin{aligned}
x_{\min} &= 2^{-1022} &\approx 2.23 \cdot 10^{-308}, \\
x_{\max} &= (1 - 2^{-53})\, 2^{1024} &\approx 1.80 \cdot 10^{308}.
\end{aligned}
$$

Example (IBM System/390) The IBM single precision floating-point numbers cover a considerably larger range of real numbers than the IEC/IEEE single precision floating-point numbers. The minimum and the maximum of $\{x \in \mathbb{F}(16, 6, -64, 63, \textit{false}) : x > 0\}$ are given by

$$
\begin{aligned}
x_{\min} &= 16^{-65} &\approx 5.40 \cdot 10^{-79} \quad \text{and} \\
x_{\max} &= (1 - 16^{-6})16^{63} &\approx 7.24 \cdot 10^{75}.
\end{aligned}
$$

The double precision numbers $\mathbb{F}(16, 14, -64, 63, \textit{false})$ nearly cover the same range of real numbers. On IBM mainframes, therefore, the use of double precision numbers enhances the accuracy of a result rather than the scope of a problem.

Due to the separate representation of the sign by $(-1)^s$ the distribution of floating-point numbers is symmetric with respect to zero:

$$x \in \mathbb{F} \Longleftrightarrow -x \in \mathbb{F}.$$

Accordingly $-x_{\max}$ and $-x_{\min}$ are the *smallest* and the *largest negative* normalized numbers in $\mathbb{F}$ respectively. In the case *denorm* = *true* there is also a smallest positive denormalized number

$$\bar{x}_{\min} = b^{e_{\min}-p}$$

and a largest negative denormalized number $-\bar{x}_{\min}$ in $\mathbb{F}$.

4.5.3 Distances Between Floating-Point Numbers

For each choice of $e \in [e_{\min}, e_{\max}]$, the largest and the smallest mantissa of normalized floating-point numbers are characterized by the figures

$$d_1 = 1, \quad d_2 = \cdots = d_p = 0 \qquad \text{and} \qquad d_1 = d_2 = \cdots = d_p = \delta := b - 1$$

respectively. So the mantissa m runs between

$$m_{\min} = .100\ldots00_b = b^{-1} \qquad \text{and}$$
$$m_{\max} = .\delta\delta\delta\cdots\delta\delta_b = \sum_{j=1}^{p}(b-1)b^{-j} = 1 - b^{-p},$$

with a constant increment b^{-p}. This *basic increment* which corresponds to the absolute value of the last position is often referred to as *ulp* (**u**nit of **l**ast **p**osition). In this chapter *ulp* is denoted by u:

$$u := 1\,\mathrm{ulp} = b^{-p}.$$

The distance between neighboring numbers from $\mathbb{F}_N$ in the interval $[b^e, b^{e+1}]$ is *constant*,

$$\Delta x = b^{e-p} = ub^e;$$

so each interval is characterized by a *constant (absolute) density* of normalized floating-point numbers. The transition from this exponent e to the next smallest one reduces this constant distance by a factor b and increases the density of numbers in $\mathbb{F}_N$ by the same factor. Similarly, for a transition to the next largest exponent, the distance is increased and the density is reduced by the factor b. Hence there is a repetition of sequences of $(b-1)b^{p-1}$ equidistant numbers, with each sequence being an image of the previous one scaled by the factor b.

The Gap around Zero

The normalization of floating-point numbers causes a gap in the range of $\mathbb{R}$ covered by $\mathbb{F}_N$ around zero (see Fig. 4.2).

In the case *denorm* = *false*, there are just *two* members of $\mathbb{F}_N$ in the interval $[0, x_{\min}]$, namely the two boundary values 0 and $x_{\min}$. However, the next interval $[x_{\min}, bx_{\min}]$ (which is of equal length if $b = 2$) contains $1 + b^{p-1}$ numbers in $\mathbb{F}_N(b, p, e_{\min}, e_{\max})$. For the IEC/IEEE system $\mathbb{F}_N(2, 24, -125, 128)$ this amounts to 8 388 609 numbers.

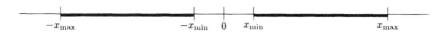

Figure 4.2: Real numbers covered by $\mathbb{F}_N$.

On the other hand, if $denorm = true$ then the interval $(0, x_{\min})$ is covered uniformly by $b^{p-1} - 1$ denormalized numbers (see Fig. 4.3) with the constant distance $ub^{e_{\min}}$ between two neighboring numbers. Hence, the smallest positive denormalized number

$$\bar{x}_{\min} := \min\{x \in \mathbb{F}_D : x > 0\} = ub^{e_{\min}} = b^{e_{\min}-p}$$

is much closer to zero than the smallest positive normalized number $x_{\min} = b^{e_{\min}-1}$. The negative numbers in $\mathbb{F}_D$ are obtained by reflection relative to the origin so the denormalized numbers $\mathbb{F}_D$ close the gap around zero.

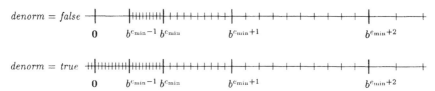

Figure 4.3: $\mathbb{R}$ covered by normalized and denormalized floating-point numbers.

4.5.4 Relative Distances
Between Floating-Point Numbers

The absolute distance $|\Delta x|$ where $\Delta x = x_{\text{nearest}} - x$ between a number x in $\mathbb{F}_N$ and the next largest number x_{nearest} in $\mathbb{F}_N$ increases with the exponent e. By contrast, the *relative distance* $\Delta x/x$ remains (nearly) unchanged over the range of floating-point numbers because it only depends on the mantissa $m(x)$ and not on the exponent of x:

$$\frac{\Delta x}{x} = \frac{(-1)^s \, u \, b^e}{(-1)^s \, m(x) \, b^e} = \frac{u}{m(x)} = \frac{b^{-p}}{m(x)}.$$

With $b^e \leq x \leq b^{e+1}$, the relative distance $\Delta x/x$ decreases for increasing x from bu nearly down to u because $b^{-1} \leq m(x) < 1$. For $x = b^{e+1}$, where $m = b^{-1}$, it jumps back to bu and then starts decreasing again. This behavior is the same for each interval $[b^e, b^{e+1}]$ (see Fig. 4.4). In this sense the *relative density* of the numbers in $\mathbb{F}_N$ is *nearly constant*. The variation of the density over the interval $[b^e, b^{e+1}]$ is called a *wobble*. Clearly, as b becomes larger, the wobble increases (e.g., if $b = 10$ or $b = 16$ the wobble is larger than that in a binary system with $b = 2$).

For *denormalized* numbers $\mathbb{F}_D$, the uniformity of the relative density is lost. Because $m(x) \to 0$, the relative distance $\Delta x/x$ increases rapidly as $x \to 0$ while the absolute distance $\Delta x = b^{e_{\min}-p}$ remains constant.

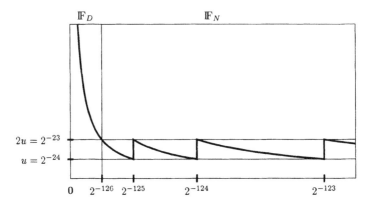

Figure 4.4: Relative distances between the IEC/IEEE single precision floating-point numbers.

In a floating-point number system $\mathbb{F}(b, p,\ e_{min}, e_{max},\ true)$, the number distribution is different in each of the regions[9]

$$\mathbb{R}_N \quad := \quad [-x_{max}, -x_{min}] \cup [x_{min}, x_{max}]$$

$$\mathbb{R}_D \quad := \quad (-x_{min}, x_{min})$$

$$\mathbb{R}_{overflow} \quad := \quad (-\infty, -x_{max}) \cup (x_{max}, \infty).$$

This suggests a decomposition of the real axis into these three regions with respect to any floating-point number system $\mathbb{F}$:

1. $\mathbb{R}_N$ is that subset of real numbers which is covered by floating-point numbers with nearly uniform *relative* density.

2. $\mathbb{R}_D$ is covered by zero and the denormalized numbers (provided there are any in $\mathbb{F}$) with uniform *absolute* density.

3. $\mathbb{R}_{overflow}$ does not contain any members of $\mathbb{F}$.

The (nearly) constant relative density of members of $\mathbb{F}_N$ reflects well the nature of data in numerical data processing. These data are frequently results of measurements in which the absolute value of a measurement depends on the unit being used. The *relative accuracy of measurement*, however, is usually independent of the unit and is therefore independent of the order of magnitude of data; it corresponds to a constant number of significant figures.

Example (Digital Measuring Instruments)   The accuracy of digital electronic measuring instruments (e.g., digital voltmeters) is specified with respect to the decimal places of the display. Some voltmeters return e.g., a value with 4 digits. The accuracy of this value is independent of the actual voltage (e.g., 220 V or 10 mV); it only depends on the quality of the instrument.

---

[9]The decomposition $\mathbb{R} = \mathbb{R}_N \cup \mathbb{R}_D \cup \mathbb{R}_{overflow}$ is chosen deliberately to emphasize the nature of the covering of $\mathbb{R}$ and, in particular, the different densities. It thus deviates from the decomposition $\mathbb{F} = \mathbb{F}_N \cup \mathbb{F}_D$, where zero is regarded as a member of $\mathbb{F}_N$.

4.5.5 Case Study: $\mathbb{F}(10, 6, -9, 9, true)$

The floating-point number system $\mathbb{F}(10, 6, -9, 9, true)$ is used as a prototype in the following case study in order to illustrate the ideas mentioned in the previous section on the structure of a floating-point number system. $\mathbb{F}$ contains all numbers

$$\pm . d_1 d_2 d_3 d_4 d_5 d_6 \cdot 10^e$$

where $d_j \in \{0, 1, 2, \ldots, 8, 9\}$ and $e \in \{-9, -8, \ldots, 8, 9\}$. For normalized numbers $d_1 \in \{1, 2, \ldots, 9\}$. Zero is characterized by the mantissa $m = 0$. In the case of denormalized numbers, $\mathbb{F}_D$ $d_1 = 0$ and $e = -9$. The smallest positive number in $\mathbb{F}_N$ is

$$x_{min} = .100000 \cdot 10^{-9} = 10^{-10},$$

and the largest number in $\mathbb{F}$ is

$$x_{max} = .999999 \cdot 10^9 = (1 - 10^{-6}) \cdot 10^9 \approx 10^9.$$

There are no numbers in $\mathbb{F}$ outside the interval $[-x_{max}, x_{max}]$.

The mantissa of normalized numbers runs through the values

$$m_{min} = b^{-1} = .100000 \qquad \text{to} \qquad m_{max} = 1 - b^{-p} = .999999,$$

with the basic increment $u = .000001 = 10^{-6}$. This corresponds to the distance $\Delta x = 10^{e(x)-6}$ between normalized numbers, where $e(x)$ is the exponent of $x \in \mathbb{F}$. For example, the interval $[100, 1000]$ contains $900\,001$ numbers:

$$100.000, \quad 100.001, \quad 100.002, \quad \ldots \quad 999.998, \quad 999.999, \quad 1000.00.$$

These numbers are equidistant, separated by a distance of 10^{-3}. The next largest number is 1000.01, and the distance between it and 1000.00 amounts to 10^{-2}. The *relative* distance of neighboring numbers in the interval $[100, 1000)$ decreases from

$$\frac{10^{-3}}{100} = 10^{-5} = 10u = bu \qquad \text{to} \qquad \frac{10^{-3}}{999.999} \approx 10^{-6} = u$$

by the factor $b = 10$ and when it reaches 1000 it jumps back to its maximal value

$$\frac{10^{-2}}{1000} = 10^{-5} = 10u = bu.$$

The relative distance between the two smallest normalized floating-point numbers, $0.100000 \cdot 10^{-9}$ and $0.100001 \cdot 10^{-9}$, is thus the same as that between relatively large neighboring numbers $0.100000 \cdot 10^9$ and $0.100001 \cdot 10^9$, while the *absolute* distance between the respective numbers amounts to 10^{-15} in the first case and 10^{+3} in the second case.

Without the denormalized numbers $\mathbb{F}_D$, the gap between 0 and x_{min} mentioned on page 128 would be substantial since the numbers in $\mathbb{F}_N$ close to $x_{min} = 10^{-10}$ are separated by the distance 10^{-15}. This is indeed small in comparison to the size of the gap. The $10^5 - 1$ positive, subnormal numbers in $\mathbb{F}_D$

cover this gap with constant distance 10^{-15}. This means that the relative distance grows to the value 1 when approaching the smallest positive subnormal number $\overline{x}_{min} = 10^{-15}$. Similar results apply to the region between $-x_{min}$ and 0 containing the negative denormalized numbers.

As a whole, $\mathbb{F}_N$ consists of 34 200 000 normalized floating-point numbers including zero. $\mathbb{F}(10, 6, -9, 9, true)$ contains $\mathbb{F}_N$ and as well as 199 998 denormalized numbers in $\mathbb{F}_D$. The half-axis $[0, \infty)$ is partitioned into

$$[0, 10^{-10}) \qquad\qquad \ldots \quad \mathbb{R}_D,$$
$$[10^{-10}, (1 - 10^{-6})10^9] \qquad \ldots \quad \mathbb{R}_N, \text{ and}$$
$$((1 - 10^{-6})10^9, \infty) \qquad \ldots \quad \mathbb{R}_{overflow}.$$

4.6 Standardization of Floating-Point Number Systems

Standards are introduced for the sake of simplicity, uniformity, and efficiency in professional activities of various kinds. All organized procedures in industry, economics, and administration are based on standards. They constitute the basis of the useful development of any discipline.

Only technical standards are dealt with in this book. They are highly qualified recommendations by experts combining their knowledge and experience in any field, not laws in a legal sense[10]: Standards are worked out by those who are in need of them and who are affected by them.

International standards are developed by the *International Standardization Organization* (ISO) to which around 95 national standardization organizations belong. For example, the standards of programming languages are worked out by the ISO.

A special case is the field of electrical engineering and electronics, in which standards are developed by the *International Electrotechnical Commission* (IEC).

4.6.1 The IEEE Standard for Floating-Point Numbers

In the seventies, efforts were made to develop standards for the arithmetic of binary floating-point numbers on microprocessors. As with any other standard, one of the main goals was to make this discipline more uniform so that programs could be portable and particular programming techniques used to deal with rounding error effects and arithmetic exceptions (e. g., exponent overflow) could be equally effective on different computer architectures.

As a result of lengthy negotiations, the American IEEE[11] computer society adopted the IEEE Standard 754-1985, the *IEEE Standard for Binary Floating Arithmetic* (often referred to as "IEEE 754"). In 1989 the IEC decided to let

---

[10]Nevertheless, technical standards or parts of them can be made part of law.

[11]IEEE is the abbreviation for *Institute of Electrical and Electronics Engineers*.

this national standard become an international standard, IEC 559 : 1989 *Binary Floating-Point Arithmetic for Microprocessor Systems*. W. Kahan (University of California, Berkeley) was awarded the *Turing Award*, the highest honor decoration in the field of computer science, by the *Association for Computing Machinery* (ACM) for his contribution to this standard.

Even prior to being officially adopted in 1985, IEEE 754 was a *de facto* standard in the field of floating-point numbers and arithmetic. It specifies

1. format widths and codes for two classes of floating-point number systems: *basic formats* and *extended formats* (Each class comprises formats for single precision and double precision.),

2. the elementary operations and rounding rules available,

3. the conversion between different number formats as well as between decimal numbers and binary numbers,

4. the handling of exceptional cases like exponent overflow or underflow, division by zero, and the like.

IEEE Formats

The basic formats consist of the single-precision floating-point number system $\mathbb{F}(2,24,-125,128,true)$ and the double-precision system $\mathbb{F}(2,53,-1021,1024,true)$.

For extended formats the standard only specifies lower or upper bounds for the parameters.

	extended IEEE formats	
	single precision	double precision
format width N	≥ 43	≥ 79
precision p	≥ 32	≥ 64
smallest exponent $e_{\min}$	≤ -1021	≤ -16381
largest exponent $e_{\max}$	≥ 1024	≥ 16384

Most implementations comprise only *one* extended format, which usually corresponds to the floating-point number system $\mathbb{F}(2,64,-16381,16384,true)$.

Example (HP Workstations) On HP workstations there is a double precision extended format which corresponds to the floating-point number system $\mathbb{F}(2,\mathbf{113},-16381,16384,true)$. In this case the precision significantly exceeds the standard's minimal requirement for p.

The American national ANSI/IEEE Standard 854-1987 *"A Radix Independent Standard for Floating-Point Arithmetic"* specifies the floating-point number systems $\mathbb{F}(b,p,e_{\min},e_{\max},denorm)$ without confining itself to binary systems ($b = 2$). Moreover, both the standards IEC 559 and ANSI/IEEE 854 comprise detailed requirements for the arithmetic in floating-point number systems, which is discussed in Section 4.7.

4.6.2 Hidden Bit

The implementation of IEC/IEEE floating-point numbers uses a hidden bit, i.e., the leading digits $d_1 = 1$ of normalized numbers and $d_1 = 0$ of denormalized numbers respectively are *not* coded explicitly; hence 23 bits and 52 bits respectively are sufficient for coding the mantissa.

In order to distinguish denormalized floating-point numbers from normalized numbers, the value $e_{min} - 1$ is stored in the exponent field of denormalized numbers.

4.6.3 Infinity, NaNs, and Signed Zero

The IEC/IEEE floating-point number systems provide the format widths $N = 32$, 64, and (usually) 80 bits respectively. Not all bit sequences correspond to a floating-point number; instead, there are the values ± 0, $\pm \infty$ and floating-point codes for symbolic quantities, so-called NaNs.[12]

value to be represented	value of the exponent field	value of the mantissa field
$(-1)^s \cdot 0 = \pm 0$	$e_{min} - 1$	zero
$(-1)^s \cdot \infty = \pm \infty$	$e_{max} + 1$	zero
NaN	$e_{max} + 1$	$\neq$ zero

These quantities are mainly used for handling exceptional operations such as attempting to take the square root of a negative number (Goldberg [211]). For processors which employ *all* bit sequences for coding valid floating-point numbers, the only reasonable reaction (towards an illegal operation) is breaking off (or interrupting) the computation in progress.

Example (IBM System/390) If an attempt is made to compute the square root of -9 on a computer in the IBM System/390 series, then either the program terminates with an error message or a result is calculated. Since each bit sequence represents a valid floating-point number, the square root function must return a valid number as well. In this case the value $\sqrt{|-9|} = 3$ is returned and the value 3 is used in the execution of the program from that moment on.

The IEC/IEEE arithmetic returns a NaN bit sequence as the result of an operation if an exceptional case appears. There is no need for immediate termination of the computation, as NaNs can be handled symbolically and may also appear in the program output.

4.6.4 Impact on Programming Languages

Program language and compiler construction specialists were *not* involved in the development of the IEEE standard. Due to this unfortunate circumstance, there

---

[12] NaN is short for *Not a Number* and symbolizes, for example, the result $0/0$.

is still no programming language which allows for the full utilization of the potential of the IEC/IEEE floating-point number system. The most comprehensive approach so far was undertaken in the Fortran 90 programming language with its mechanisms for the manipulation of floating-point numbers; this is discussed in Section 4.8.

Currently the IEC is drafting a standard which specifies programming language extensions aimed at the better utilization of the IEC/IEEE floating-point number system (ISO/IEC [239]). These extensions are explicitly formulated for Ada, Basic, C, Common Lisp, Fortran, Modula-2, Pascal, and PL/I.

Example (Fortran)   The proposed intrinsic functions DENORM and IEC_559 are designed to make it possible to test whether or not denormalized floating-point numbers are available and if the number system in use complies with the arithmetic of the IEC 559 norm respectively.

RND_NEAREST, RND_TRUNCATE, and RND_OTHER are designed to investigate the actual rounding mode. The proposed functions SET_INDICATORS, CLR_INDICATORS, TEST_INDICATORS, and SAVE_INDICATORS are used for the manipulation of the overflow and the underflow indicator.

4.7 Arithmetics for Floating-Point Systems

Numerical data processing can only use real numbers which are actually available on a computer. The fact that this is only a finite set of numbers would seem to jeopardize any possibility of useful numerical data processing. The consequences of the restrictions thus imposed are discussed in the following (cf. also Goldberg [211]). For the sake of simplicity, it is assumed that the numerical operations under investigation are based on only one floating-point number system $\mathbb{F}(b, p, e_{\min}, e_{\max}, denorm)$.

First of all a decision must be made: How should the *arithmetic operations* discussed in Section 4.3.1 be adapted to the floating-point number system $\mathbb{F}$? Since $\mathbb{F}$ is a subset of the real numbers $\mathbb{R}$ the rational and arithmetic operations are indeed defined for operands in $\mathbb{F}$, but their results are, in general, *not* members of $\mathbb{F}$ except in trivial cases such as $x \mapsto |x|$ and $x \mapsto (-x)$.

The results of arithmetic operations with operands in the floating-point number set $\mathbb{F}(b, p, e_{\min}, e_{\max}, denorm)$ often require more than p mantissa places and an exponent outside of the range $[e_{\min}, e_{\max}]$ for their representation as floating-point numbers. In general, for division, and most standard functions, there is no floating-point representation of results at all that has a *finite* mantissa.

Rounding of Results

The logical approach to overcoming this problem is well-known from handling decimal fractions: the exact result is *rounded* to a number in $\mathbb{F}$. In this book the term *exact result* denotes the result in $\mathbb{R}$ which would be obtained if the operation was executed in the field of real numbers. Because $\mathbb{F} \subset \mathbb{R}$, the exact result is always well-defined.

Example (Rounding of Results)   In the number system $\mathbb{F}(10, 6, -9, 9, true)$, the function $\square : \mathbb{R} \to \mathbb{F}$ maps each $x \in \mathbb{R}$ to the nearest number in $\mathbb{F}$.

$$
\begin{array}{rlll}
\textit{Arguments:} & x & = & .123456 \cdot 10^5 = 12345.6 \\
& y & = & .987654 \cdot 10^0 = .987654 \\[1ex]
\textit{Exact Computation:} & x + y & = & .1234658\,7654 \cdot 10^5 \\
& x - y & = & .12344612346 \cdot 10^5 \\
& x \cdot y & = & .121931\,812224 \cdot 10^5 \\
& x/y & = & .124999\,240624\ldots \cdot 10^5 \\
& \sqrt{x} & = & .111110\,755549\ldots \cdot 10^3 \\[1ex]
\textit{Rounding:} & \square(x + y) & = & .123466 \cdot 10^5 \\
& \square(x - y) & = & .123446 \cdot 10^5 \\
& \square(x \cdot y) & = & .121932 \cdot 10^5 \\
& \square(x/y) & = & .124999 \cdot 10^5 \\
& \square\sqrt{x} & = & .111111 \cdot 10^3
\end{array}
$$

4.7.1 Rounding Functions

A *rounding function* is a mapping (*reduction mapping*)

$$\square : \mathbb{R} \to \mathbb{F},$$

which maps each real number x to a number $\square x \in \mathbb{F}$ according to particular rules, discussed in this section.

With a given rounding function $\square$, a mathematical *definition of the arithmetic operations*[13] in $\mathbb{F}$ can be derived as follows: For each arithmetic operation in $\mathbb{R}$

$$\circ : \mathbb{R} \times \mathbb{R} \to \mathbb{R}$$

the *corresponding operation* in $\mathbb{F}$

$$\boxdot : \mathbb{F} \times \mathbb{F} \to \mathbb{F}$$

can be defined by

$$x \boxdot y := \square(x \circ y). \tag{4.8}$$

This definition corresponds to a two-step model: firstly, the exact result $x \circ y$ of the operation $\circ$ is determined and then it is mapped to a result in $\mathbb{F}$ by the rounding function $\square$; whence $x \boxdot y$ is again a number in $\mathbb{F}$.

Similarly, operations with only one operand—like most standard functions— can be defined in $\mathbb{F}$. For a function $f : \mathbb{R} \to \mathbb{R}$ the corresponding function

$$\boxed{f} : \mathbb{F} \to \mathbb{F}$$

is defined by

$$\boxed{f}(x) := \square f(x). \tag{4.9}$$

---

[13]The *implementation* of arithmetic operations is discussed in Section 4.7.4.

Of course the rounding function $\square$ cannot be defined by arbitrary rules if the operations based on definitions (4.8) and (4.9) are intended to be usable in practice. Required properties of practicable rounding functions are *projectivity*, i.e.,

$$\square x = x \qquad \text{for all} \quad x \in \mathbb{F}, \tag{4.10}$$

and *monotonicity*, i.e.,

$$x \leq y \quad \Longrightarrow \quad \square x \leq \square y \qquad \text{for all} \quad x, y \in \mathbb{R} \tag{4.11}$$

Clearly, these two requirements imply that each such function rounds all real numbers between two neighboring numbers x_1 and $x_2 \in \mathbb{F}$ which are smaller than a particular limit point $\hat{x} \in [x_1, x_2]$ to x_1, and all numbers larger than that limit point to x_2 (see Fig. 4.5). If $\hat{x}$ is neither equal to x_1 nor to x_2, it must be explicitly specified whether $\square\hat{x} = x_1$ or $\square\hat{x} = x_2$. The special situation in the range $\mathbb{R}_{\text{overflow}}$ is discussed later.

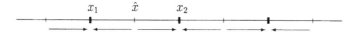

Figure 4.5: Definition of a rounding function using the limit point $\hat{x}$. The bold ticks (∎) symbolize numbers in $\mathbb{F}$.

A rounding function $\square$, in compliance with (4.10) and (4.11), is defined on the interval $[x_1, x_2]$ by the specification of a limit point $\hat{x}$ on the one hand and, when $\hat{x} \notin \{x_1, x_2\}$ by an additional rounding rule for $x = \hat{x}$ on the other hand. The rounding functions most frequently used are:

Round to Nearest (Optimal Rounding): This rounding function fixes the limit point $\hat{x}$ at midpoint of x_1 and x_2:

$$\hat{x} := \frac{x_1 + x_2}{2}.$$

Hence, $x \in \mathbb{R}_N \cup \mathbb{R}_D$ is always rounded to the *nearest* number in $\mathbb{F}$. For the cases where $x = \hat{x}$ is equidistant from x_1 and x_2, there are two alternatives.

Variant 1: The conventional strategy selects the number $\square x$ which is further away from zero (*round away from zero*).

Example (Round away from Zero) In $\mathbb{F}(10, 6, -9, 9, \textit{true})$ the number 0.1000005 is rounded to $.100001 \cdot 10^0$ as expected. This strategy is used by most electronic calculators.

Variant 2: When $x = \hat{x}$ the *round to even* strategy selects the neighboring number in $\mathbb{F}$ as the result $\square x$ of the rounding function containing an odd number as its last digit in the mantissa. This rounding strategy requires an even base b in the underlying number system.

Example (Round to Even) In $\mathbb{F}(10, 6, -9, 9, \textit{true})$ the number 0.1000005 is rounded to $.100000 \cdot 10^0$ rather than $.100001 \cdot 10^0$. However, the number 0.1000015 is rounded to $.100002 \cdot 10^0$ as usual.

Truncation: With truncation (*round toward zero*) the limit point of the rounding function depends on the sign[14] of the number x:

$$\hat{x} := \text{sign}(x)\ \max(|x_1|, |x_2|).$$

In this case the limit point $\hat{x}$ is the neighboring number x_1 or x_2 which has the largest absolute value; hence for $x \notin \mathbb{F}$ the value $\square x$ is always the number with the smaller absolute value among x_1 and x_2, and

$$x \in \mathbb{R} \setminus \mathbb{F}\quad \Rightarrow\quad |\square x| < |x|$$

always holds.

Truncation "away from zero" would also be a reasonable strategy, but it is of no practical importance.

Directed Rounding: For the *round toward plus infinity* and the *round toward minus infinity* strategies, the limit points are given by

$$\hat{x} := \min(x_1, x_2)\qquad \text{and}\qquad \hat{x} := \max(x_1, x_2).$$

With directed rounding the result is either given by the next smaller number in $\mathbb{F}$ or by the next larger one, no matter what the sign of $x \notin \mathbb{F}$ is. If $x \notin \mathbb{F}$ then it is always the case that $\square x > x$ and $\square x < x$ respectively.

Since the relation

$$\square(-x) = -(\square x)$$

holds for optimal rounding and truncation, these two rounding functions are also referred to as *symmetric rounding functions* as opposed to directed rounding. The latter approach is important for applications in rounding error analysis, particularly for interval arithmetic.

Example (Different Rounding Functions)   With $\mathbb{F}(10, 6, -9, 9, true)$ and $x = .123456\,789$
the respective results for the different rounding functions are

$$\square x = \begin{cases} .123457 & \text{for round to nearest and} \\ & \text{for round towards } +\infty, \\ .123456 & \text{for truncation and} \\ & \text{for round towards } -\infty. \end{cases}$$

However, the rounding rules can only be applied to an $x \in \mathbb{R} \setminus \mathbb{F}$ if there are actually two neighboring numbers $x_1 < x$ and $x_2 > x$ in $\mathbb{F}$. This is always the case for $x \in (\mathbb{R}_N \cup \mathbb{R}_D) \setminus \mathbb{F}$ but not for $x \in \mathbb{R}_{\text{overflow}}$.

Overflow: If $x \in \mathbb{R}_{\text{overflow}} = (-\infty, -x_{\max}) \cup (x_{\max}, \infty)$ the value $\square x$ is undefined. If such an x is the exact result of an operation with operands in $\mathbb{F}$, this is referred to as an *overflow*, or more precisely *exponent overflow*. Most computers interrupt the program execution with an error message in the case of an overflow; sometimes the particular activity of the computer can be specified by the user.

---

[14]Here, sign(x) denotes the function sign$(x) := -1$ for $x < 0$ and sgn$(x) := 1$ for $x \geq 0$.

Underflow: If $x \in \mathbb{R}_D = (-x_{\min}, x_{\min})$ the value $\Box x$ is always well-defined, regardless of whether denormalized numbers are available or not. This special case is referred to as (exponent) *underflow*. Usually the intermediate result zero is used for further computation after an underflow has occurred.

Example (Transformation of Coordinates) In the (fictitious) floating-point number system $\mathbb{F}(10, 6, -9, 9, true)$, the Cartesian coordinates $(x_1, y_1) = (10^{-8}, 10^{-8})$ and $(x_2, y_2) = (10^5, 10^5)$ are to be transformed to polar coordinates. In both cases $\tan \varphi = y_i/x_i = 1$, i.e., $\varphi = 45°$. In the determination of the radius $r = \sqrt{x_i^2 + y_i^2}$ of the first point an underflow occurs. If the computer proceeds with the value zero, the clearly unreasonable result $r = 0$ is obtained. For the second point an overflow occurs, which usually results in preliminary program termination. In both cases, however, suitable scaling may still lead to useful results.

4.7.2 Rounding Error

Definitions (4.8) and (4.9) make it evident that a knowledge of the maximal difference between $\Box x$ and x is crucial to the understanding of the arithmetic in $\mathbb{F}$. The quality of $\mathbb{F}$ as a model of the arithmetic in $\mathbb{R}$ largely depends on that difference.

The deviation of the rounded value $\Box x \in \mathbb{F}$ from the exact number $x \in \mathbb{R}$ is usually referred to as the (*absolute*) *rounding error* of x:

$$\varepsilon(x) := \Box x - x,$$

whereas

$$\rho(x) := \frac{\Box x - x}{x} = \frac{\varepsilon(x)}{x} \qquad (4.12)$$

denotes the *relative rounding error* of x.

Upper Bounds of the Absolute Rounding Error

The absolute value $|\varepsilon(x)|$ is clearly limited by Δx, the length of the smallest interval $[x_1, x_2]$ which contains x, $x_1, x_2 \in \mathbb{F}$; and, in the case of round to nearest, by half that length. To be more precise, for each $x \in \mathbb{R}_N$ the value $\Box x \in \mathbb{F}$ has the unique representation

$$\Box x = (-1)^s m(x) b^{e(x)},$$

where $m(x)$ denotes the mantissa and $e(x)$ the exponent of $\Box x$. Since the length of the smallest interval $[x_1, x_2]$ with $x_1, x_2 \in \mathbb{F}_N$ which contains $x \in \mathbb{R}_N \setminus \mathbb{F}_N$ is

$$\Delta x = u b^{e(x)},$$

the absolute rounding error of any $x \in \mathbb{R}_N$ is given by:

$$|\varepsilon(x)| \begin{cases} < u b^{e(x)} & \text{for directed rounding and truncation,} \\ \leq \dfrac{u}{2} b^{e(x)} & \text{for round to nearest.} \end{cases} \qquad (4.13)$$

For $x \in \mathbb{R}_D$ and round to denormalized numbers, the exponent $e(x)$ has to be replaced by $e_{\min}$.

Example (Absolute Rounding Error) For positive numbers in $\mathbb{F}_N(2,3,-1,2)$, the absolute rounding error for round to nearest has the behavior depicted in Fig. 4.7. In the case of truncation the rounding function is depicted in Fig. 4.6. The absolute rounding error is a piecewise linear function in both cases.

Upper Bounds of the Relative Rounding Error

For the relative rounding error $\rho(x)$ there are upper bounds which hold in the whole of $\mathbb{R}_N$. The existence of such upper bounds is due to the nearly constant relative density of floating-point numbers in $\mathbb{R}_N$. The smallest uniform bound of the relative rounding error of $x \in \mathbb{R}_N$ can be derived from (4.12) and the upper bound (4.13):

$$|\rho(x)| \begin{cases} < \dfrac{u}{|m(x)|} \leq bu & \text{for directed rounding} \\ & \text{and truncation,} \\ \leq \dfrac{u}{2|m(x)|} \leq bu/2 & \text{for round to nearest.} \end{cases} \tag{4.14}$$

eps frequently denotes this upper bound on the relative rounding error and is referred to as the *machine epsilon*. For *eps* the relation

$$eps = \begin{cases} bu = b^{1-p} & \text{for directed rounding} \\ & \text{and truncation,} \\ bu/2 = b^{1-p}/2 & \text{for round to nearest} \end{cases} \tag{4.15}$$

holds.

Example (Relative Rounding Error) The relative rounding error for the positive numbers in $\mathbb{F}_N(2,3,-1,2)$ behaves as depicted in Fig. 4.8. The variation of the relative error is comparably small in $\mathbb{R}_N$ for a floating-point number system with $b = 2$. For systems with bases $b = 10$ or $b = 16$ the variation is considerable.

Example (IEC/IEEE Arithmetic) For single precision IEC/IEEE arithmetic, the machine epsilon is determined as follows:

$$eps = \begin{cases} 2^{-23} & \approx & 1.19 \cdot 10^{-7} & \text{for directed rounding and truncation,} \\ 2^{-24} & \approx & 5.96 \cdot 10^{-8} & \text{for round to nearest,} \end{cases}$$

which corresponds to a relative precision of around seven decimal places. Corresponding bounds for a IEC/IEEE double precision arithmetic are

$$eps = \begin{cases} 2^{-52} & \approx & 2.22 \cdot 10^{-16} & \text{for directed rounding and truncation,} \\ 2^{-53} & \approx & 1.11 \cdot 10^{-16} & \text{for round to nearest,} \end{cases}$$

which corresponds to a relative precision of more than 15 decimal places.

Example (IBM System/390) On IBM mainframes with single precision arithmetic, the machine epsilon is determined as follows:

$$eps = \begin{cases} 16^{-5} & \approx & 9.54 \cdot 10^{-7} & \text{for directed rounding and truncation,} \\ 16^{-5}/2 & \approx & 4.77 \cdot 10^{-7} & \text{for round to nearest,} \end{cases}$$

which corresponds to a relative precision of only *six decimal places*. This poor accuracy (which is worse than that of any other current computer) has led to the frequent use of double-precision number systems in portable programs.

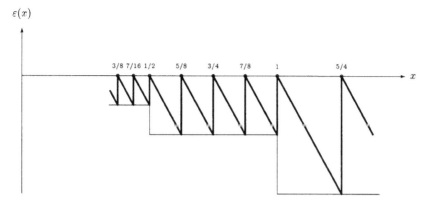

Figure 4.6: Absolute rounding error for truncation.

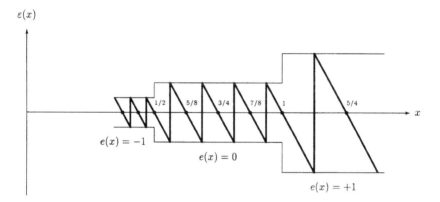

Figure 4.7: Absolute rounding error for round to nearest.

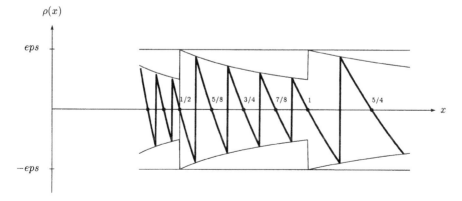

Figure 4.8: Relative rounding error for round to nearest.

Formula (4.12) is related to (4.14) and (4.15) by the following theorem:

Theorem 4.7.1 *For each $x \in \mathbb{R}_N$ there is a $\rho \in \mathbb{R}$ with*

$$\Box x = x(1 + \rho) \qquad and \qquad |\rho| \le eps. \tag{4.16}$$

However, when using the uniform bound thus derived for the relative rounding error

$$|\rho(x)| \le eps \qquad for \qquad x \in \mathbb{R}_N \tag{4.17}$$

it is necessary to remember that this bound may be far too *pessimistic*. The actual relative rounding error may be much smaller than the *eps* (cf. Fig. 4.8).

If additional information is available on the position of the mantissa $m(x)$ within the interval $[b^{-1}, 1 - b^{-p}]$, the estimate

$$|\rho(x)| \le \frac{u}{|m(x)|} \qquad or \qquad |\rho(x)| \le \frac{u}{2|m(x)|}$$

can be used; it is stricter than (4.17), particularly for larger values of the base b, like e. g., $b = 10$ or $b = 16$.

4.7.3 Rounding and Arithmetic Operations

Using definitions (4.8) and (4.9), the following characterization of the relative error of arithmetic operations and standard functions is a direct consequence of Theorem 4.7.1.

Theorem 4.7.2 *Provided that the exact result of an arithmetic operation $\circ$ is in $\mathbb{R}_N$, then for each $x, y \in \mathbb{F}_N$ there is a $\rho \in \mathbb{R}$ such that*

$$x \boxdot y = (x \circ y)(1 + \rho) \qquad and \qquad |\rho| \le eps.$$

Theorem 4.7.3 *Provided that, for the argument $x \in D \cap \mathbb{F}_N$, the exact value of $f : D \to \mathbb{R}$ is in $\mathbb{R}_N$, there is a $\rho \in \mathbb{R}$ such that*

$$\boxed{f}(x) = f(x)(1 + \rho) \qquad and \qquad |\rho| \le eps.$$

For multiple operations or operands a new ρ occurs each time.

Example (Sum of Three Numbers) Consider three floating-point numbers

$$x, y, z \in \mathbb{F}_N \qquad where \qquad x + y, \; y + z, \; x + y + z \in \mathbb{R}_N$$

are to be added up. The sum $x + y + z$ can be computed either as $(x \boxplus y) \boxplus z$ or as $x \boxplus (y \boxplus z)$. The first possibility can be expanded as follows:

$$
\begin{aligned}
(x \boxplus y) \boxplus z &= (x + y)(1 + \rho_1) \boxplus z \\
&= [(x + y)(1 + \rho_1) + z](1 + \rho_2) \\
&= x + y + z + (x + y)(\rho_1 + \rho_2 + \rho_1 \rho_2) + z\rho_2,
\end{aligned}
$$

and the second possibility as:

$$
\begin{aligned}
x \boxplus (y \boxplus z) &= x \boxplus (y + z)(1 + \rho_3) \\
&= [x + (y + z)(1 + \rho_3)](1 + \rho_4) \\
&= x + y + z + x\rho_4 + (y + z)(\rho_3 + \rho_4 + \rho_3\rho_4),
\end{aligned}
$$

where the relative rounding errors can be estimated using (4.16):

$$
|\rho_i| \leq eps, \qquad i = 1, 2, 3, 4.
$$

If the terms of order eps^2 are disregarded and only terms of order eps are taken into account, then the following error estimates are obtained for the two different approaches:

$$
\begin{aligned}
|(x \boxplus y) \boxplus z - (x + y + z)| &\leq K_1 \quad \text{where} \quad K_1 \approx (2|x + y| + |z|)\, eps \quad \text{and} \\
|x \boxplus (y \boxplus z) - (x + y + z)| &\leq K_2 \quad \text{where} \quad K_2 \approx (|x| + 2|y + z|)\, eps.
\end{aligned}
$$

For $|x| \gg |y|, |z|$ the *bound* is smaller in the second case; in fact the corresponding computation method often results in a smaller rounding error than in the first one. It follows that for the summation of a larger number of terms (e. g., series) the smallest error bound (and often the smallest actual error) is obtained by adding terms in order of increasing absolute value.

Pseudo-Arithmetic

An important consequence of the above example is that, in general,

$$
x \boxplus (y \boxplus z) \neq (x \boxplus y) \boxplus z;
$$

and, similarly,

$$
x \boxdot (y \boxdot z) \neq (x \boxdot y) \boxdot z.
$$

That is, the *associative law* for the addition and multiplication of real numbers *cannot* be transferred to the model of the floating-point arithmetic in $\mathbb{F}$. Moreover, the *distributive law* is no longer valid in $\mathbb{F}$: That is, in general,

$$
x \boxdot (y \boxplus z) \neq (x \boxdot y) \boxplus (x \boxdot z).
$$

However, because

$$
x \boxplus y = \Box(x + y) = \Box(y + x) = y \boxplus x
$$

the respective *commutative laws* of addition and multiplication still hold in $\mathbb{F}$.

Since both the associative law and the distributive law are invalid, it is clear that in $\mathbb{F}$ different (algebraically equivalent) representations of an arithmetic expression are *not* necessarily equivalent. The evaluation of different representations may yield different results. This leads to the notion of the *pseudo-arithmetic* of machine numbers.

Example (Pseudo-Arithmetic) Because $x^{n+1} - 1 = (x - 1)(x^n + x^{n-1} + \ldots + x + 1)$, the expansion into a sum leads to the result

$$
\frac{1.23456^4 - 1}{1.23456 - 1} = 1.23456^3 + 1.23456^2 + 1.23456 + 1 \approx 5.6403387 \tag{4.18}
$$

calculated in $\mathbb{R}$. A calculation in $\mathbb{F}(10, 6, -9, 9, true)$ with directed rounding (truncation), expansion of the powers into iterated multiplications, and summation from the left to the right, however, gives the results

$$5.64022$$

for the fraction but

$$5.64031$$

for the sum in (4.18).

Rounding and Comparison

As an important consequence of the monotonicity property (4.11) of rounding operations, the orientation of a *comparison operator* cannot be reversed after the transition from $\mathbb{R}$ to $\mathbb{F}$. Nevertheless, different real numbers can be mapped onto one and the same floating-point number in $\mathbb{F}$ by rounding.

For example, in order to decide whether the exact value of an arithmetic expression calculated in $\mathbb{F}$ is positive or not, the value must be far enough away from zero, i. e.,

$$expression \ \geq \ \alpha > 0$$

must hold, where α is an error bound for the evaluation of *expression* in $\mathbb{F}$. This means that, to assure the right decision it is accepted that many borderline cases are excluded.

Example (Large Number of Zeros)   The only zero of the polynomial

$$P_7(x) \ = \ x^7 - 7x^6 + 21x^5 - 35x^4 + 35x^3 - 21x^2 + 7x - 1 \ = \ (x - 1)^7$$

is $x^* = 1$. $P_7(x)$ is positive for $x > 1$ and negative for $x < 1$. The evaluation of P_7 in the form of a power $(x - 1.)**7$ yields a graph (see Fig. 4.9) which is satisfactory with respect to its accuracy[15]. However, evaluation of $P_7(x)$ using the Horner scheme

$$((((((x - 7.) * x + 21.) * x - 35.) * x + 35.) * x - 21.) * x + 7.) * x - 1.$$

yields a truly chaotic graph near $x = 1$ (see Fig. 4.10 and Fig. 4.11). Such an implementation $\tilde{P}_7$ of P_7 yields 128 749 zeros within single precision IEC/IEEE arithmetic, thousands of points $x > 1$ with $\tilde{P}_7(x) < 0$, and, similarly, thousands of points $x < 1$ with $\tilde{P}_7(x) > 0$. The cause of this phenomenon is the *cancellation of leading digits* (cf. Section 5.7.4):

When adding (subtracting) two numbers of nearly equal absolute value and different (equal) signs, the leading digits of the mantissa are canceled out; any inaccuracy of a less significant digit becomes an inaccuracy of a leading digit in the result.

Rounding and Overall Error

The error *bounds* (4.15) for directed rounding and round to nearest differ only by the factor 2. Nevertheless, for many problems involving a large number of arithmetic operations, the actual difference in the accuracy is much larger. This is due to the fact that a *compensation* of errors often takes place with round to nearest but rarely with directed rounding.

---

[15]In Fig. 4.9 and Fig. 4.10 *all* function values are depicted for $x \in [1 - 2^{-19}, 1 + 2^{-19}]$. The discrete structure of a floating-point function is clearly visible in these diagrams.

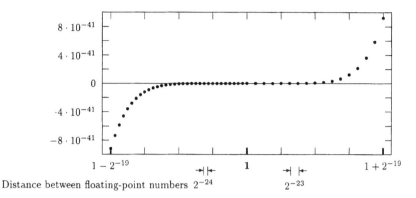

Figure 4.9: $P_7(x) = (x-1)^7$ evaluated in single-precision IEC/IEEE arithmetic for arguments $x \in [1 - 2^{-19}, 1 + 2^{-19}]$.

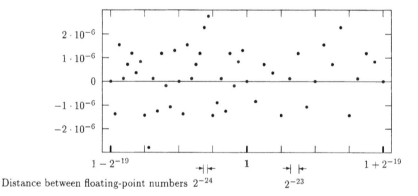

Figure 4.10: $P_7(x) = ((((((x - 7)x + 21)x - 35)x + 35)x - 21)x + 7)x - 1$ evaluated in single-precision IEC/IEEE arithmetic for $x \in [1 - 2^{-19}, 1 + 2^{-19}]$.

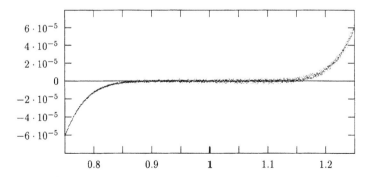

Figure 4.11: $P_7(x) = ((((((x - 7)x + 21)x - 35)x + 35)x - 21)x + 7)x - 1$ evaluated in single precision IEC/IEEE arithmetic for $x \in [0.75, 1.25]$. In this picture not *all* function values are depicted due to their large number.

Example (Cray)   The computers in the Y-MP series use directed rounding, unlike the series 2 machines. An actual application with a large sparse system of linear equations ($n = 16\,146$) once returned a result vector which was *100 times* less accurate on a Cray Y-MP than on a Cray 2 (Carter [135]).

4.7.4 Implementation of a Floating-Point Arithmetic

It is not a simple task to implement the mathematical definition (4.8) of the operation $\boxdot$ (i. e., the implementation of the rational operation $\circ$) in a floating-point number system $\mathbb{F}$ on a digital processor since it requires knowledge of the exact result found in $\mathbb{R}$. Up until a few years ago, most computers thus returned a machine number which usually complied with the definition (4.8) as the result of a rational operation. Nevertheless, in a considerable number of situations, substantial deviations from (4.8) occurred. As a consequence, there was no way to make a reliable assessment of the quality of the result of an arithmetic algorithm.

In fact the following procedure is necessary for deriving the result defined in (4.8): A longer register has to be used for creating a pseudo-result $\tilde{z}$ with the property that its rounding with respect to $\mathbb{F}$ results in the same number as the rounding of the exact result $z = x \circ y$:

$$x \boxdot y := \Box \tilde{z} = \Box z. \tag{4.19}$$

For a long time it was unclear what additional number of mantissa places was actually required for $\tilde{z}$ in order to guarantee (4.19) for each combination of x and $y \in \mathbb{F}$ and for each rational operation. It was also unclear as to how the generation of $\tilde{z}$ was actually to be accomplished. Since the end of the seventies, it has been accepted that for a suitable technique, two additional digits (a *guard digit* and a *round digit*) are sufficient for $\tilde{z}$ and that compliance with (4.8) does not create hardware difficulties. It is not within the scope of this book to discuss the particular technical details (see e. g., Hennessy, Patterson [55]).

The *IEC/IEEE floating-point standard* (cf. Section 4.6) requires of a binary processor that it *always* returns the machine number specified in definition (4.8) as the result of a rational operation of two machine numbers, unless the result is in $\mathbb{R}_{\mathrm{overflow}}$. The standard further specifies round to nearest as the default and requires truncation and directed rounding functions to be implemented as options.

Example (Intel)   The arithmetic processor Intel 8087, released in 1981, along with the 8086 and 8088 models, was the first to implement a "clean" floating-point arithmetic[16]. Meanwhile, practically all microprocessors conform to the requirements of the IEC 559 : 1989 standard.

Example (Cray)   The inital floating-point arithmetic of Cray computers does *not* comply with the requirements of the IEC/IEEE standard. The prevailing design principle was to optimize the speed of computation. For example, there are no guard digits, which results in

---

[16]To be more precise, the 8087 fulfilled the requirements of the then *proposed* American national standard IEEE 754-1985 since its design phase was completed before the publication of the final version of the IEC/IEEE standard. Only the Intel 80387 complied with the full IEC 559 : 1989 standard (IEEE 754-1985).

the fact that even addition is less accurate than on processors which conform to the IEC/IEEE standard; for division the situation is even worse. A division a/b is executed by multiplying a by $1/b$; for the calculation of $1/b$ an iteration is used which doesn't comply with the requirements of the IEC/IEEE standard. The overall error for the computation of $1/b$ and $a(1/b)$ may affect the last *three* bits of the mantissa of the result a/b. On Cray computers it is therefore difficult to implement programs which are based on the error bounds dealt with in this chapter.

Another drawback of Cray computers is their lack of uniformity as far as floating-point arithmetic is concerned. For example, the rounding properties of the Cray Y-MP are much worse than those of the Cray 2 (cf. the example on Page 146).

Increased Accuracy Requirements

Some of the most widespread floating-point arithmetics (e. g., the IEC/IEEE arithmetic or the IBM arithmetic) are *not* sufficiently accurate for very large problems involving a very large number of floating-point operations which could possibly accumulate rounding errors. When those standards were defined, no one conceived of the incredible improvements in computer performance (up to contemporary teraflop computers, which can execute 10^{12} floating-point operations per second). Actual problems will soon be so large that they will require a more accurate floating-point arithmetic (Bailey et al. [102]).

Example (System of Linear Equations)   For the numerical solution of a particular system of n linear equations, which occurs when solving certain partial differential equations (biharmonic differential equations), a mantissa with

$$p \geq 10 + 5\log_2 n$$

places is required when a binary arithmetic is used to calculate solutions with at least three correct decimal places. Even a double precision IEC/IEEE arithmetic can only guarantee this accuracy up to around $n = 390$; the extended double precision IEC/IEEE format with $p = 64$ further increases the possible order of the system to around $n = 1800$, whereas the double-precision Cray arithmetic allows for around $n = 100\,000$.

4.7.5 Multiple-Precision Software

For some applications the floating-point systems provided by a computer have been proved to have insufficient accuracy or range. For example, computations may require intermediate results which lead to an exponent overflow, although the final result itself is within the range of machine numbers (cf. Section 4.8.6). Moreover, it is possible—e. g., in the case of unstable algorithms (cf. Section 5.7.3)—for an algorithm to be particularly sensitive to changes in certain intermediate results. Such intermediate results should be computed in double-precision arithmetic in order to guarantee satisfactory accuracy in the overall result.

If the computations of an application are executed in single-precision arithmetic, then the use of double-precision floating-point numbers is a possible remedy to the difficulties mentioned above. If this is still not sufficient or if double-precision numbers are already used, an extended double-precision floating-point number system or one of k-fold precision may be used. However, this strategy is inconsistent with the portability of programs since not every computer system provides more than two levels of accuracy. Moreover, such a strategy is of limited use anyway: There are rarely more than four different levels, so the level of accuracy cannot be adapted to the requirements of a particular program.

A possible solution would be the use of special *multiple-precision software*, which enables computations of enhanced accuracy on any computer system, independent of the machine numbers provided by the hardware. Moreover, such software allows for a flexible and problem-oriented choice of the level of accuracy and of the required range of floating-point numbers.

Software Implementation of Floating-Point Numbers

In principle, the floating-point number systems used in multiple-precision software and the arithmetic operations defined there do *not* differ from the machine number systems such as the IEC/IEEE floating-point numbers, that are provided by the hardware. The striking difference is the way the number systems and the corresponding arithmetic operations are *implemented*. While machine number systems are implemented by processors which are suitable for numerical data processing, the number systems used for multiple-precision software are realized with the help of suitable data structures and algorithms. The numbers thus defined do not correspond to the elementary data types the processor can handle directly, but rather consist of several elementary data components. Similarly, the arithmetic operations of multiple-precision numbers are not identical to the built-in hardware operations but rather are implemented by suitable subroutines.

The software solution of number systems allows for independence of the accuracy and the range of a number system from the particular features of a processor. Moreover, advanced multiple-precision software allows for changes in the parameters of the emulated number system such as the precision, the smallest and the largest exponent, and others. The most serious drawback of a software implementation as opposed to a hardware number system is the significantly increased time requirement. The use of multiple-precision software with flexible accuracy specification causes an increase in execution times by a factor > 100 (Bailey [99]).

Quality Assessment of Multiple-Precision Software

To begin with, the effort necessary for adapting existing programs to multiple-precision software may be a critical issue. For example, when adding two floating-point numbers in extended precision, both operands have to be converted by suitable routines into multiple-precision numbers (unless they are already of the right format) before the multiple-precision add routine can be executed. Moreover, the multiple-precision numbers have to be declared in a suitable way. This conversion turns out to be lengthy and prone to errors if it has to be executed by hand for longer programs or for whole program packages.

Another important factor for the quality assessment of multiple-precision software is the scope of its applicability. If certain elementary functions or other frequently used functions are not available in multiple precision, then the full conversion of many programs is impossible.

There are basically two different ways to employ multiple-precision software. It can be employed either as a package of subroutines or as an integral part of a program system for symbolic computation. Multiple-precision software in the form of a subroutine package is normally used for extensive numerical computations, which demand high efficiency of the data structures and algorithms used for the representation and manipulation of the multiple-precision numbers. On the other hand, program systems for symbolic manipulations require multiple-precision calculations primarily for the (arbitrarily exact) evaluation of symbolically derived results. Large-scale computations in extended precision are rare in symbolic environments. In this case the efficiency of numerical computations in extended precision is not of the same importance as the user-friendly interface between the program system and the multiple-precision software. The use of multiple-precision software in program systems like MATHEMATICA, MACSYMA etc. is merely touched upon in this book, since this kind of software is currently not used for high performance *numerical* computation.

Software (Multiple-Precision Programs) In the IMSL/MATH-LIBRARY some subroutines for the exact multiplication of two double-precision floating-point numbers are provided. The exact result a floating-point number of quadruple precision which is internally represented by a double-precision array of length two. Double-precision floating-point numbers can be converted into floating-point numbers of four-fold precision using IMSL/MATH-LIBRARY/dqini, and the inverse conversion is executed by IMSL/MATH-LIBRARY/dqsto. The routine IMSL/MATH-LIBRARY/dqadd is used for adding up such quadruple precision floating-point numbers, and IMSL/MATH-LIBRARY/dqmul performs the exact multiplication of two double-precision numbers with subsequent addition to a number of quadruple precision. A corresponding set of routines (IMSL/MATH-LIBRARY/zqini etc.) provides the same functionality for complex double-precision numbers.

The subroutines SLATEC/xadd and SLATEC/dxadd add up floating-point numbers of single or double precision respectively over an enlarged range of exponents. Suitable initialization routines and conversion routines are also provided.

Software (Package MP) The software package MP (Brent [125]) provides a wide selection of more than a hundred different subroutines for the manipulation and processing of so-called *MP*-numbers. These are normalized floating-point numbers for which the base b, the precision p, and the range of the exponents can be varied. The precision—and hence the accuracy of a computation—is only limited by the available storage space. Some restrictions remain, however, due to certain details in the implementation: For the base b the number $8b^2 - 1$ must be representable by an INTEGER number of format width $N = 32$; and for the minimal exponent e_{min} and the largest exponent $e_{max} = -e_{min}$, the number $4e_{max}$ must be representable by an INTEGER number of the same kind. Hence, the range of MP-numbers is limited by restrictions on the base and the exponent range.

The programs implemented in the MP package set the standard by which all subsequent multiple-precision packages should be judged. In particular, MP contains subroutines for the following operations of MP-numbers:

- addition, subtraction, multiplication, and division;

- power function, root function;

- elementary functions;

- special functions and constants;

- conversion routines for INTEGER numbers and floating-point numbers of single and double precision.

MP can be procured as TOMS/524 (Brent [126]).

Software (Package FM) The multiple-precision package FM (Smith [353]) supports all rational operations as well as all Fortran 77 standard functions. Its advantage over comparable software (e. g., MP) is its enhanced accuracy of transcendental functions, which is obtained by computing intermediate results with higher accuracy. Only the final result is rounded to the accuracy desired by the user. Thus, even for transcendental functions, the goal of numerical computations is usually achieved, i. e., the numerical result is equal to the rounded exact result. FM is available as TOMS/693.

Software (Package MPFUN) Possibly the most advanced program package in multiple-precision software is MPFUN (Bailey [99]). Its advantages over comparable software— particularly MP and FM—are its ability to convert Fortran programs automatically and its performance oriented implementation which aims at maximal efficiency.

MPFUN mainly uses *MP*-numbers. This notation is in accordance with the original literature. Nevertheless, these numbers are denoted *MPF*-numbers in this book, in order to avoid confusion, since they are not identical to the MP-numbers in the MP package. What is really striking about an MPF-number is the representation of its *mantissa* using arrays of *floating-point numbers*; each floating-point number corresponds to a *single* digit in the mantissa with respect to a correspondingly large base for the floating-point number system.[17] At first sight this implementation hardly seems to be logical as the mantissa always contains *integers*, suggesting that these sizes should be implemented using INTEGER variables (as, for example, in MP). Surprisingly, the MPFUN approach turns out to be quite suitable for most computers designed to process numerical data. This is because the floating-point performance is optimized. In addition to MPF-numbers, MPFUN can also be used to process *MPFC*-numbers (complex MPF-numbers) and *DPE*-numbers (double-precision floating-point numbers with extended exponent range).

The functionality of MPFUN largely resembles that of MP. Interestingly, many operations are implemented in two versions, a normal one and an advanced one. The advanced version pays particular attention to the efficiency of high-precision computations. Moreover, the code is designed so that the innermost loop of iterative calculations can be vectorized, which further improves the performance on vector processors and, to a lesser extent, on RISC processors as well.

MPFUN derives a high degree of user-friendliness from its preprocessor (Bailey [100]), which automatically transforms conventional Fortran 77 programs into multiple-precision programs using suitable directives[18]. This makes most of the tiresome task of conversion, which is seriously prone to errors, redundant. All directives addressed to the MPFUN preprocessor begin with the string CMP+ in the first column of a line. For example, with the help of the directive CMP+ PRECISION LEVEL *nprec*, the level of accuracy of MPF-numbers is fixed to *nprec* significant digits. MPF- and MPFC-numbers are declared with the directives CMP+ MULTIP REAL and CMP+ MULTIP COMPLEX, respectively. On the basis of such declarations, the preprocessor globally replaces the predefined (internal) routines used for floating-point manipulations with the corresponding MPFUN subroutines. A description of all the directives of MPFUN can be found in Bailey [100].

Debugging is tricky for all programs generated by a preprocessor. Since the transformed program generally differs considerably from, and is usually a lot more complex than, the original program written by the programmer, no convenient source-code debugging is possible. The latest Fortran 90 version of MPFUN overcomes this problem (Bailey [101]) and allows for easy debugging: A Fortran 77 program can easily be converted into a Fortran 90 program which does not require a preprocessor when this MPFUN version and a Fortran 90 compiler are used. In

---

[17]For the IEC/IEEE floating-point number system, the base $b = 2^{24} = 16\,777\,216$ is chosen.

[18]A *directive* is a comment in a special format which is ignored by the compiler, but observed by the preprocessor.

these cases, it is sufficient to include the MPFUN module (with the directive USE MPMODULE) in all subroutines which are intended for enhanced precision. Moreover, suitable MPF-variables have to be declared as type MP_REAL.

More detailed information on the latest MPFUN software products are distributed via e-mail in response to the request send index sent to mp-request@nas.nasa.gov.

4.8 Inquiry Functions and Manipulation of Numbers in Fortran 90

The most important rule for the development of portable software is to avoid any machine dependent construct. For floating-point numbers and their corresponding arithmetic, this is only possible if the programming language provides suitable means for floating-point manipulations.

The parameters of the floating-point number system IF can be determined at runtime with the help of *numerical inquiry functions*. *Floating-point manipulation functions* can be used to analyze, synthesize, and scale floating-point numbers. These subroutines define operations whose results are adjusted to the computer system and its floating-point numbers, while the programs themselves are completely machine independent. The standardization of subroutines belonging to all Fortran 90 systems made it possible to develop portable programs which can adapt themselves to the features of a particular floating-point number system.

To ensure that a Fortran 90 programming system can be implemented on a wide range of computers (PCs, workstations, mainframes, supercomputers, etc.), special features of different types of computers had to be neglected. For example, *denormalized* floating-point numbers IF_D are implemented only on computers complying with the IEC/IEEE number system (i.e., on most workstations and PCs, but not on all mainframes and supercomputers). Hence, the Fortran 90 language definition does not utilize denormalized numbers explicitly.

In Fortran 90 the floating-point manipulation functions are only defined for a number set which constitutes a *simplified model* of the floating-point numbers found in *all* relevant computers or processors. The number model has parameters which are determined in such a way that the model best fits the actual machine on which a program is executed.

4.8.1 Parameters for Floating-Point Numbers

Floating-point numbers are represented by a model comprised of the following numbers:

$$x = \begin{cases} 0 & \text{or} \\ (-1)^s\, b^e\, [d_1 b^{-1} + d_2 b^{-2} + \cdots + d_p b^{-p}] \end{cases},$$

with

$$
\begin{aligned}
&\text{sign:} && s \in \{0, 1\} \\
&\text{base:} && b \in \{2, 10, 16, \ldots\} \\
&\text{exponent:} && e \in \{e_{\min}, e_{\min} + 1, \ldots, e_{\max}\} \\
&\text{digits:} && d_1 \in \{1, \ldots, b - 1\} \\
& && d_j \in \{0, 1, \ldots, b - 1\}, \quad j = 2, 3, \ldots, p.
\end{aligned}
$$

For $x = 0$ it is assumed that $d_1 = \cdots = d_p = 0$ and $e = 0$; this does *not* comply with the IEC/IEEE standard which specifies the exponent $e_{\min} - 1$ for the number zero.

The *model* floating-point numbers in Fortran 90 are comprised only of the numbers $\mathbb{F}_N$ of a floating-point number system $\mathbb{F}$. Hence the Fortran 90 model of floating-point numbers is characterized by only *four* integer parameters:

1. base (*radix*) $b \geq 2$,

2. precision $p \geq 2$,

3. smallest exponent $e_{\min} < 0$,

4. largest exponent $e_{\max} > 0$.

These parameters of the model number set $\mathbb{F}_N$ can be determined by any Fortran 90 program with the help of the intrinsic functions RADIX, DIGITS, MIN EXPONENT, and MAXEXPONENT.

Example (Workstation) On a workstation with IEC/IEEE floating-point numbers, the code fragment

```
b     = RADIX  (real_variable)
p     = DIGITS (real_variable)
e_min = MINEXPONENT (real_variable)
e_max = MAXEXPONENT (real_variable)
```

returned the values

2	for the base b,
24	for the precision p,
−125	for the smallest exponent e_min and
128	for the largest exponent e_max

for the model numbers corresponding to the data type of real_variable.

4.8.2 Characteristics of Floating-Point Numbers

Besides the four parameters in the number set $\mathbb{F}_N(b, p, e_{\min}, e_{\max})$, other characteristics, which are derived from those basic parameters, can also be determined using inquiry functions:

the *smallest positive* number in $\mathbb{F}_N$ $x_{\min} = b^{e_{\min} - 1}$,

the *largest* number in $\mathbb{F}_N$ (and $\mathbb{F}$) $x_{\max} = b^{e_{\max}}(1 - b^{-p})$

with the intrinsic functions TINY and HUGE respectively,

$$\text{the number of } \textit{decimal} \text{ places} \quad \begin{array}{ll} \lfloor (p-1)\log_{10} b \rfloor & \text{for} \quad b = 2, 16 \\ p & \text{for} \quad b = 10 \end{array}$$

$$\text{the exponent range} \qquad \lfloor \min\{-\log_{10} x_{\min}, \log_{10} x_{\max}\} \rfloor$$

with the functions PRECISION and RANGE respectively, and (a bound for)

$$\text{the machine epsilon} \quad b^{1-p}$$

with the function EPSILON.

Example (Workstation) On a workstation with IEC/IEEE floating-point numbers the program fragment

```
x_min     = TINY       (real_variable)
x_max     = HUGE       (real_variable)
digits_10 = PRECISION  (real_variable)
range_10  = RANGE      (real_variable)
eps       = EPSILON    (real_variable)
```

produced the values

$1.17549 \cdot 10^{-38}$	for the smallest positive model number x_min,
$3.40282 \cdot 10^{38}$	for the largest model number x_max,
6	for the decimal places digits_10,
37	for the exponent range range_10, and
$1.19209 \cdot 10^{-7}$	for the relative machine accuracy eps.

For the number of decimal places, the correct value 6 of the specification is determined, although binary floating-point numbers with $p = 24$ (and $(p-1)\log_{10} b \approx 6.92$) can represent almost 7 decimal places.

Example (Summation of the Series for the Hyperbolic Sine) For $|x| \le 1$ the series

$$\mathrm{sh}(x) = \sum_{i=0}^{\infty} \frac{x^{2i+1}}{(2i+1)!}$$

converges quite rapidly. A program fragment for the summation of this (truncated) series might look like

```
sh = x;    term = x;    i2 = 0;    xx = x*x
DO WHILE (ABS (term) > sh*EPSILON (sh))
    i2 = i2 + 2
    term = (term*xx)/(i2*(i2+1))
    sh = sh + term
END DO
```

The loop terminates precisely when the maximum possible accuracy for that type of series is obtained.

It has to be stressed that the value b^{1-p} returned by the function EPSILON is a bound for the relative rounding error which holds for *all* conventional rounding rules.

4.8.3 The Distance Between Floating-Point Numbers, Rounding

The following information on the distance between floating-point numbers can be obtained using the intrinsic functions SPACING, RRSPACING and NEAREST:

the absolute distance between machine numbers $\Delta x = b^{e-p}$,
the inverse relative distance between numbers $|x|/\Delta x$, and
next largest/next smallest machine number.

Example (Smallest Positive Floating-Point Number) The smallest positive floating-point number in $\mathbb{F}$ on the actual computer in use can be determined with the NEAREST function. The statement

```
real_min = NEAREST (0.,1.)
```

assigns the next largest number to zero in the direction of the number 1. On a computer with single-precision IEC/IEEE arithmetic, this is the value 1.40E-45 of the smallest positive *denormalized* floating-point number. It has to be noted that the function TINY, on the other hand, returns the smallest positive *normalized model* number:

```
real_min_model = TINY (1.).
```

For example, on a workstation with IEC/IEEE arithmetic the following values were obtained for double-precision floating-point numbers:

```
double_min       = NEAREST (0.D0,1.D0)   ! value: 4.94066E-324
! not a model number
! (denormalized number)
double_min_model = TINY (1.D0)                ! value: 2.22507E-308
```

In Fortran 90 there is *no* intrinsic function which provides information regarding the actual kind of rounding used at runtime.

Example (Rounding) The following program fragment can be used to determine whether round to nearest is used or not. It is based on the assumption that one of the rounding functions described in Section 4.7.1 is implemented.

```
LOGICAL  ::  left = .FALSE., right = .FALSE.
...
x = 1.
x_p1q = x + 0.25*SPACING (x)
x_p3q = x + 0.75*SPACING (x)
IF (x_p1q == x)            left  = .TRUE.
IF (x_p3q == NEAREST (x,2.))  right = .TRUE.
...
eps = EPSILON (x)
IF (left .AND. right) eps = eps/2.  ! round to nearest is implemented
```

4.8.4 Manipulation of Floating-Point Numbers

Fortran 90 contains the following intrinsic functions for the analysis of floating-point numbers:

FRACTION returns the mantissa (the rational portion),
EXPONENT returns the exponent

of a model number in $\mathbb{F}_N$. The synthesis of a floating-point number given the mantissa and the exponent can be performed using the predefined functions SCALE and SET_EXPONENT.

Example (Square Root)   The following program fragment implements an algorithm for the iterative determination of $\sqrt{x}$ within *binary* arithmetic without considering special cases (e. g., negative arguments). To achieve this the exponent of x is divided by two, and an approximation for the square root of the mantissa m is used as the trial solution in a Newton method. Using the iterative scheme

$$u_{i+1} = u_i - \frac{f(u_i)}{f'(u_i)} = u_i - \frac{u_i^2 - m}{2u_i} = u_i - \frac{u_i}{2} + \frac{m}{2u_i} = 0.5\,(u_i + m/u_i)$$

the zero of $f(u) = u^2 - m$ is determined.

```
INTEGER  ::  x_exponent
REAL     ::  x, x_mantissa, m_root
...
x_exponent = EXPONENT (x)
x_mantissa = FRACTION (x)

m_root   = 0.41732 + 0.59018*x_mantissa        ! initial approximation

IF (MOD (x_exponent,2) == 1) THEN              ! odd exponent
   m_root    = m_root*0.70710
   x_mantissa = x_mantissa*0.5
   x_exponent = x_exponent + 1
END IF

DO i = 1, (DIGITS(x)/15 + 1)                    ! Newton iteration
   m_root = 0.5*(m_root + x_mantissa/m_root)
END DO

x_root = SCALE (m_root,x_exponent/2)
```

Example (Scaling of a Vector)   The following program fragment can be used in order to scale a vector so that the magnitude of its largest component is approximately 1:

```
INTEGER                ::  exp_max
REAL                   ::  norm_max
REAL, DIMENSION (1000) ::  vector
...
norm_max = MAXVAL (ABS (vector))
IF (norm_max > 0.) THEN
   exp_max = EXPONENT (norm_max)
   vector  = SCALE (vector,-exp_max)
END IF
```

It must be noted that in this program fragment *no* precautions have been taken against exponent underflow. Such measures can be quite complicated (cf. e. g., Blue [117]).

4.8.5 Parameters in INTEGER Numbers

The Fortran 90 model of integer numbers assumes that the sign and the number itself are coded separately:

$$i = (-1)^s \sum_{j=0}^{q-1} d_j b^j,$$

where

$$\text{sign}: \qquad s \in \{0,1\}$$
$$\text{numbers}: \quad d_j \in \{0,1,\ldots,b-1\}, \quad 0 \leq j \leq q-1 .$$

The two parameters b and q characterize the set of model INTEGER numbers and can be determined by any Fortran 90 program with the help of the intrinsic functions RADIX and DIGITS respectively:

```
b = RADIX  (integer_variable)  ! base of the INTEGER number system
q = DIGITS (integer_variable)  ! maximum number of digits
```

The largest INTEGER number $b^q - 1$ of the actual computer system is determined using the function HUGE:

```
integer_max = HUGE (integer_variable).
```

Due to the symmetry of the range of *model* numbers, the smallest INTEGER number is given by

```
integer_min = -integer_max.
```

The maximum number of *decimal* places is obtained using the function RANGE.

4.8.6 Case Study: Multiple Products

During the computation of a product $a_1 a_2 a_3 \cdots a_n$, it may happen—particularly for large n—that an iterative calculation produces an exponent overflow or underflow, even if the final result is well within the range of machine numbers.

For example, a sequence $\{a_i\}$ of 1000 factors is given by

$$a_i := \begin{cases} 1.5 - \dfrac{(350-i)^2}{2 \cdot 350^2} & \text{for} \quad i = 1,2,\ldots,700, \\ \exp((700-i)/300) & \text{for} \quad i = 701,702,\ldots,1000 \end{cases} \tag{4.20}$$

(see Fig. 4.12).

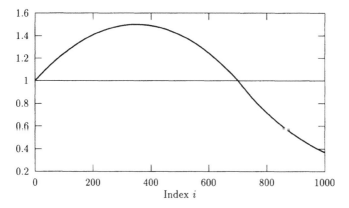

Figure 4.12: Size of the factors a_i defined in (4.20).

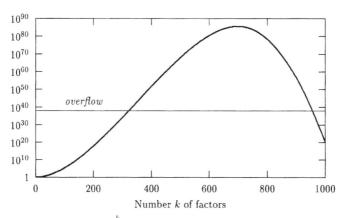

Figure 4.13: Size of $\prod_{i=1}^{k} a_i$, i.e., the intermediate results of the product.

If the program segment

```
product = 1.
DO k = 1, 1000
   product = product * factor(k)
END DO
```

is used on a computer with single-precision IEC/IEE arithmetic in order to compute the product $a_1 a_2 \cdots a_{1000}$ of the factors (4.20), then, for the index $i = 326$, the intermediate result INF (the symbol for $+\infty$) is obtained due to exponent overflow. According to the rules of the IEC/IEEE arithmetic INF $* x$ = INF for all $x > 0$ (and INF $* 0$ = NaN). Hence, all subsequent intermediate results as well as the final result are equal to INF (or NaN = *Not a Number*).

In Fortran 90 this problem can be avoided by using intrinsic functions for the manipulation of floating-point numbers.

Product without Overflow—Method 1

One possible approach is to store the exponent and the mantissa of intermediate results separately; during each iteration of the loop the mantissa of the new factor is multiplied by the intermediate result obtained thus far and the corresponding exponents are added:

```
INTEGER                :: i, n, prod_exponent
REAL                   :: prod_mantissa, product
REAL, DIMENSION (n)    :: factor
...
prod_mantissa = 1.
prod_exponent = EXPONENT (prod_mantissa)
prod_mantissa = FRACTION (prod_mantissa)

DO i = 1, n
   prod_exponent = prod_exponent + EXPONENT (factor(i))
   prod_mantissa = prod_mantissa * FRACTION (factor(i))

   prod_exponent = prod_exponent + EXPONENT (prod_mantissa)
   prod_mantissa = FRACTION (prod_mantissa)
END DO

PRINT *, "Result: "
IF ((prod_exponent <= MAXEXPONENT (product)) .AND.              &
    (prod_exponent >= MINEXPONENT (product))) THEN
   product = SET_EXPONENT (prod_mantissa,prod_exponent)
   PRINT *, product
ELSE
   PRINT *, "Outside the range of the model floating-point numbers!"
END IF
```

After the loop is terminated, the final result is calculated using the function SET_EXPONENT, provided the result is within the permissible range of floating-point numbers. In general this program does not lead to overflow or underflow since the relation

$$b^{-1} \leq \text{prod\_mantissa} < 1$$

for the normalized mantissa prod_mantissa holds. An exponent overflow[19] of the variable prod_exponent is unlikely since EXPONENT(factor(i)) and EXPONENT(prod_mantissa) always have much smaller absolute values than the largest floating-point number (i.e., $e_{\max} \ll b^{e_{\max}}(1 - b^{-p})$ for conventional floating-point number systems). In the worst case, after adding the corresponding terms to prod_exponent, the variable remains unchanged after rounding. This leads to an incorrect result, but not to an exponent overflow.[20]

Finally, the parts prod_mantissa and prod_exponent are combined and it is checked whether the result is still within the range of floating-point numbers.

---

[19]Underflow cannot occur since the exponents are integers.

[20]It is easy to see that there must be a *very* large number of factors to produce such effects, which means that prod_mantissa will also be affected by considerable errors.

Product without Overflow—Method 2

For a smaller number of factors a second approach, which is slightly faster, can also be used:

```
LOGICAL                :: complete
INTEGER                :: i, n, overall_exponent, prod_exponent
REAL                   :: prod_mantissa
REAL, DIMENSION (n)    :: factor
...
prod_exponent = 0
prod_mantissa = 1.
complete  = .TRUE.

DO i = 1, n
   IF ((prod_mantissa * FRACTION (factor(i))) < TINY (prod_mantissa)) THEN
PRINT *, "Range of model numbers is being left!"
PRINT *, "Product was calculated up to factor ", (i-1), "."
complete = .FALSE.
EXIT
   END IF
   prod_exponent = prod_exponent + EXPONENT (factor(i))
   prod_mantissa = prod_mantissa * FRACTION (factor(i))
END DO

IF (complete) THEN
   PRINT *, "Complete product calculated."
END IF

overall_exponent = prod_exponent + EXPONENT (prod_mantissa)
IF ((overall_exponent >= MINEXPONENT (prod_mantissa)) .AND.         &
    (overall_exponent <= MAXEXPONENT (prod_mantissa))) THEN
   PRINT *, "Result: "
   PRINT *, SCALE (prod_mantissa,prod_exponent)
ELSE
   PRINT *, "Result is outside the range of model numbers!"
END IF
```

In this program the exponents of the factors are added and the corresponding mantissas are multiplied by the actual intermediate result *without* scaling in between. However, since these multiplications continually cut down the variable `prod_mantissa` (the relation $b^{-1} \leq$ FRACTION $(\text{factor}(i)) < 1$ holds), an underflow can easily occur. This potential problem is avoided as a result of the check performed using the TINY function. With the above program fragment, the computation of the product is halted once the factor a_{209} is reached.

4.9 Operations with Algebraic Data

Algebraic data is a data structure whose elementary components are real numbers (cf. Section 4.2.1). If the representation of the structure is unimportant, the elements of an algebraic data structure can be simply indexed in which case the n-tuples $(x_1, x_2, \ldots, x_n) \in \mathbb{R}^n$ are the elementary data of an algebraic data structure X. Algebraic data like real arrays are handled as single entities by the computer, and the components are usually stored in contiguous memory locations.

For instance, a binary operation $Z = X \Diamond Y$ whose operands X and Y are an n-tuple and an m-tuple respectively

$$X := (x_1, \ldots, x_n) \in \mathbb{R}^n, \qquad Y := (y_1, \ldots, y_m) \in \mathbb{R}^m$$

yields a k-tuple

$$Z := (z_1, \ldots, z_k) \in \mathbb{R}^k$$

as a result. In order to specify the operation $\Diamond$, it is necessary to specify k arithmetic operations $\Diamond_1, \Diamond_2, \ldots, \Diamond_k$ which create the elements of Z given the elements x_i and y_j of X and Y respectively.

Example (Complex Numbers)   Complex numbers are treated as one-dimensional real arrays (vectors) of length two:
$$m = n = k = 2.$$

They can be represented by their real and imaginary part:

$$X = x_1 + i x_2, \qquad Y = y_1 + i y_2, \qquad Z = z_1 + i z_2.$$

The addition of complex numbers $X + Y = (x_1 + y_1) + i(x_2 + y_2) = Z$ requires the two arithmetic operations $\oplus_1$ and $\oplus_2$:

$$\oplus_1(x_1, x_2; y_1, y_2) \quad := \quad x_1 + y_1 = z_1,$$
$$\oplus_2(x_1, x_2; y_1, y_2) \quad := \quad x_2 + y_2 = z_2.$$

Multiplication $XY = (x_1 y_1 - x_2 y_2) + i(x_1 y_2 + x_2 y_1) = Z$ is defined by the two arithmetic operations $\otimes_1$ and $\otimes_2$:

$$\otimes_1(x_1, x_2; y_1, y_2) \quad := \quad x_1 y_1 - x_2 y_2 = z_1,$$
$$\otimes_2(x_1, x_2; y_1, y_2) \quad := \quad x_1 y_2 + x_2 y_1 = z_2.$$

The calculation of the absolute value of a complex number $|X| = \sqrt{x_1^2 + x_2^2}$ is a unary operation $\mathbb{C} \to \mathbb{R}$ and is accordingly defined by a single arithmetic operation:

$$\text{cabs}(x_1, x_2) \quad := \quad \sqrt{x_1^2 + x_2^2} = z_1.$$

Example (Linear Algebra)   The multiplication of a matrix $A \in \mathbb{R}^{m \times n}$ by a vector $X \in \mathbb{R}^n$ results in a vector $Y = AX \in \mathbb{R}^m$. The operation $\odot_k$ for computing single components of the matrix-vector product is then defined by

$$\odot_i (a_1, \ldots, a_{mn}; x_1, \ldots, x_n) := \sum_{j=1}^{n} a_{i+(j-1)m} y_j, \quad i = 1, \ldots, m, \tag{4.21}$$

where *column major order* of the data structure of the $m \times n$ matrix A is assumed. This works in the same way as the array storage scheme in Fortran does whereby the matrix element of the ith row and the jth column is referenced by the index $i + (j-1)m$. Hence, in this storage scheme the row index (the first index) varies the fastest.

So far the elements of algebraic data structures have been assumed to be real numbers. However, since only machine numbers are available on a computer, the elements of algebraic data structures can only contain the values of a floating-point number system $\mathbb{F}(b, p, e_{\min}, e_{\max}, denorm)$.

When implementing algebraic operations care has to be taken as to how they can be executed using operands in $\mathbb{F}$. In particular, each component of the resulting data structure must again be in $\mathbb{F}$. There are basically two ways of achieving this:

1. In analogy with definition (4.8) of arithmetic operations in $\mathbb{F}$ for the operands

$$X = (x_1, \ldots, x_n) \in \mathbb{F}^n \qquad \text{and} \qquad Y = (y_1, \ldots, y_m) \in \mathbb{F}^m$$

the component operation $\diamond_k$ of an algebraic operation $X \diamond Y$ can be defined by

$$\boxed{\diamond_k}\, (x_1, \ldots, x_n; y_1, \ldots, y_m) := \square\diamond_k(x_1, \ldots, x_n; y_1, \ldots, y_m). \qquad (4.22)$$

As with arithmetic operations, the result of an algebraic operation over $\mathbb{F}$ is determined by the *componentwise* rounding of *results* from the appropriate operations over $\mathbb{R}$.

Example (Complex Numbers)   *Complex addition* over $\mathbb{F}^2$ is defined by

$$X \boxplus Y := \square(x_1 + y_1) + i\square(x_2 + y_2) = (x_1 \boxplus y_1) + i(x_2 \boxplus y_2);$$

whence complex addition over $\mathbb{F}^2$ is defined in terms of addition over $\mathbb{F}$.

For *complex multiplication* over $\mathbb{F}^2$ an analogous definition is given by

$$X \boxdot Y := \square(x_1 y_1 - x_2 y_2) + i\square(x_1 y_2 + x_2 y_1); \qquad (4.23)$$

however, it remains unclear whether or not the expression for the two components can be evaluated in $\mathbb{F}$.

When using the IEC/IEEE data format, the correct result can always be obtained by using one of the algorithms developed by Kulisch and Miranker [262] and an accumulator of a length that adequately exceeds the format width of the numbers in $\mathbb{F}$.

2. Due to the difficulties that can occur when using the rounding function as specified in (4.22)—which requires the exact evaluation of arithmetic expressions and allows for the rounding of the components only in the final stage—it may be sufficient to evaluate these arithmetic expressions in the arithmetic of $\mathbb{F}$ and to guarantee an unambiguous definition by exactly specifying the operation order.

Example (Complex Numbers)   While complex addition remains unchanged, complex multiplication

$$X \boxdot Y := ((x_1 \boxdot y_1) \boxminus (x_2 \boxdot y_2)) + i((x_1 \boxdot y_2) \boxplus (x_2 \boxdot y_1))$$

returns a result which differs from (4.23).

In the analogous definition of the matrix-vector product with componentwise operations (4.21), not only do all rational operations have to be executed in the arithmetic of $\mathbb{F}$, but the order of summation has to be specified in order to have a proper definition of the matrix-vector product over $\mathbb{F}$.

Until now there has been no binding standard for the definition of algebraic operations over a floating-point number system. For a long time there was no need for such a standard because even the simple arithmetic operations in the floating-point number systems of current computers were not standardized (cf. Section 4.7). Moreover, algebraic operations had to be coded by the user himself: In order to compute a matrix-vector product, it was necessary to code a double nested loop explicitly. Thus the responsibility for the correctness of results was placed on the user.

Contemporary programming languages increasingly support the direct use of algebraic data types and provide suitable operations. For example, Fortran 90 provides a number of intrinsic array operations (cf. Section 4.3).

Example (XSC Languages)   The programming languages PASCAL-XSC, C-XSC and FOR-TRAN-XSC developed at the University of Karlsruhe constitute special cases. The numerous algebraic operations available in these languages are implemented in such a way that they comply with the requirement in (4.22). However, this makes it necessary that a sum of products of pairs of machine numbers be evaluated in such a way that the result is equal to the rounded exact result, as can be seen in (4.21).

The computation of the scalar product (inner product) of two vectors $u, v \in \mathbb{R}^n$ is an important subtask of many algorithms in linear algebra:

$$\langle u, v \rangle = u^\top v = \sum_{i=1}^{n} u_i v_i \ \in \mathbb{R}.$$

According to definition (4.22) it is necessary for $u, v \in \mathbb{F}^n$ in order to generate the machine result

$$u \boxdot v := \Box \langle u, v \rangle \in \mathbb{F}.$$

This can be done either by using an accumulator of extraordinary length (implemented as an array in the main memory) or by a particular iterative process (Kulisch, Miranker [262]). By using this *exact scalar product* the most common algebraic operations can be implemented in compliance with (4.22) as required in XSC languages.

4.10 Operations with Arrays

Because of the important role of linear algebra algorithms in the field of numerical data processing, the efficient storage and processing of arrays of all dimensions as well as the precision (cf. Section 4.9) is of crucial importance. The percentage of array operations in numerical data processing problems is often more than 90% of all floating-point operations.

Nearly all contemporary computers used for numerical data processing have been developed with the goal of increasing the speed of linear algebra algorithms in mind. However, they can only attain their full efficiency if accesses to the memory hierarchy and the processing of the arrays involved are well coordinated:

- The optimal exploitation of memory hierarchies requires programs which have been developed with *locality of reference* in mind. These are programs which only reference data which are stored close to one another at any given moment (see Chapter 3 and 6).

- With some vector processors the input of the pipeline must consist of floating-point numbers which are stored *contiguously* in the memory.

- On multiprocessor computers with *distributed memory* (i.e., each processor has its own main memory) data structures have to be distributed in a suitable way in the memory of all the processors involved, though it can be useful to have the same elements available on several nodes. This distribution of data onto different processors and their respective local memories is largely supported e.g., by HPF (*High Performance Fortran*), a language extension of Fortran 90 (HPFF [18], Loveman [270]).

For all these reasons, for the efficient use of high-performance computers, it is imperative that the most important basic linear algebra algorithms are optimized with respect to the particular computer architecture in use. High-level programming languages can, however, rarely be used for this purpose although optimizing compilers generate vectorized and parallelized codes which are optimized with respect to the individual features of the computer in use.

A different approach, adopted for the most widely used basic algorithms, consists of the uniform specification of relevant procedures. These procedures are subroutines implemented and optimized once on all major computer systems. Using such subroutines as opposed to coding the according parts of the algorithms by hand makes the job easier for the application programmer and leads to efficient programs on all computer systems for which such subroutines have been implemented.

4.10.1 BLAS

In 1979 a set of *Basic Linear Algebra Subroutines* (BLAS) for the computation of scalar products, vector norms, vector sums and the like was defined and implemented in Fortran (Lawson et al. [266]). This package of subroutines (now referred to as BLAS-1) soon became very popular and is now available wherever numerical data processing takes place.

The BLAS-1 package only contained programs for vector-vector operations ($O(n)$ operations) which can be implemented using a single loop, e.g.,

$$x_i := x_i + cy_i, \qquad i = 1, 2, \ldots, n.$$

However, it soon became clear that the BLAS-1 programs were not extensive enough to allow for satisfactory performance on vector computers, which were developed around the same time as the BLAS-1 subroutines. In order to overcome this shortcoming, BLAS-2 subroutines based on matrix-vector operations were developed for a range of $O(n^2)$ operations (using double nested loops). Again, these programs were not able to achieve satisfactory performance on RISC workstations and multiprocessor computers since they could not fully exploit the memory systems (in particular hierarchical and distributed memory structures). Finally then, the BLAS-3 programs (Dongarra et al. [161]) for $O(n^3)$ operations (which correspond to triple nested loops) were developed for matrix-matrix operations.

All contemporary computer systems intended for numerical data processing are expected not only to have an optimizing Fortran compiler tailor-made for the particular computer architecture, but also an efficient implementation of the BLAS-1, BLAS-2, and BLAS-3 subroutines in their runtime library. This is necessary for exploiting the potential power of advanced high-performance computer systems.

4.11 Operations with Analytic Data

Mathematical functions are the *analytic data* of numerical data processing. In this section only univariate functions are considered. Such a function maps each number x of its domain $B \subseteq \mathbb{R}$ to a unique function value $f(x) \in \mathbb{R}$:

$$f : x \mapsto f(x), \qquad x \in B \subseteq \mathbb{R}, \quad f(x) \in \mathbb{R}.$$

The function symbol, e.g., f, denotes a one-to-one *relation* $f : B \to \mathbb{R}$, while $f(x)$ denotes the *value* of the function f at the point $x \in B$.

4.11.1 Representation of Functions

If functions are used as the data of a problem, one of the main difficulties is that they have to be represented in *finite terms*. There are basically two ways of doing so: either by specifying the underlying rule of the function or by specifying the parameters of the function.

Specifying the Underlying Rule of a Function

In many cases functions can be represented by *arithmetic expressions*, i.e., a combination of rational operators, standard functions and parentheses, which determine the order of evaluation. An arithmetic expression can be specified in many programming languages by a finite chain of symbols. Moreover, this expression can be replaced by an algorithm so that for each $x \in B$ the evaluation of an arithmetic expression terminates after a finite number of steps.

Example (Air Pressure Model) For technical applications (e. g., aircraft engineering) the following relationship between the height above sea level h [km] and the atmospheric pressure p [bar] is used:

$$p = 1.0536 \left(\frac{288 - 6.5h}{288} \right)^{5.255}.$$

The function procedure listed below returns the value pressure_bar for the input height_km.

```
FUNCTION standard_pressure_bar (height_km) RESULT (pressure_bar)
   REAL, INTENT (IN)  ::  height_km          ! input value
   REAL               ::  pressure_bar       ! output value
   pressure_bar = 1.0536*((288. - 6.5*height_km)/288.)**5.255
END FUNCTION standard_pressure_bar
```

A calling of this function subprogram, e.g.,

```
pressure_lhasa_bar = standard_pressure_bar(3.7)
```

returns the calculated value 0.6663 bar, of the variable pressure_lhasa_bar, which corresponds to the atmospheric pressure at the airport of Lhasa.

Specification of Parameters

In many cases only a particular class $\mathcal{F}$ of functions is regarded, the underlying rule of which is given by a formula with a finite number of real parameters $\{c_1, \ldots, c_m\}$. Each function of that class is specified by an m-tuple of parameters.

Operations with functions specified in that way are regarded as operations whose parameters are arguments (see Section 4.11.3).

Example (Polynomials)   The class $\mathcal{F} = \mathbb{P}_d$ of all univariate polynomials of maximal degree $d \in \mathbb{N}_0$ is a finite dimensional function space. Each $(d+1)$-tuple of parameters $(c_0, c_1, \ldots, c_d)$ specifies exactly one particular polynomial in $\mathbb{P}_d$:

$$f(x) = c_0 + c_1 x + c_2 x^2 + \cdots + c_d x^d. \tag{4.24}$$

This is the representation of the polynomial with respect to the basis $\{1, x, x^2, \ldots, x^d\}$. However, each polynomial in $\mathbb{P}_d$ can also be represented, for example, as the linear combination of Chebychev polynomials $\{T_0, T_1, \ldots, T_d\}$ (cf. Chapter 9):

$$f(x) = \frac{a_0}{2} + a_1 T_1(x) + \cdots + a_d T_d(x).$$

For many numerical algorithms, this manner of specifying a polynomial may be more useful than (4.24). The $(d+1)$-tuple $(a_0, a_1, \ldots, a_d)$ is also an unambiguous characterization of a polynomial. So the *meaning* of a $(d+1)$-tuple of real parameters $(p_0, p_1, \ldots, p_d)$, i.e., the basis with respect to which it has to be interpreted, is *background information* which is not stored explicitly.

4.11.2 Implementation of Functions

For the implementation of a function to be used in numerical data processing problems, it is important to know whether just the *values* of the function are required or if operations other than function evaluation (such as the determination of derivatives, integrals, Fourier transforms etc.) have to be performed.

Computing Function Values

If only function values are required, the implementation can, for example, be accomplished as a *function procedure*, the exact form of which is adapted to the specification and type of the function.

Example (Polynomials) In order to compute the values of a polynomial $P_d \in \mathbb{P}_d$ given the coefficients $c_0, c_1, \ldots, c_d$ with respect to the basis $\{1, x, x^2, \ldots, x^d\}$, i.e.,

$$P_d(x; c_0, \ldots, c_d) = c_0 + c_1 x + c_2 x^2 + \cdots + c_d x^d,$$

it is sufficient to implement the *Horner scheme*

$$(\cdots((c_d x + c_{d-1})x + c_{d-2})x + \cdots + c_1)x + c_0.$$

Software (Polynomials) The library subroutine IMSL/MATH-LIBRARY/ppval could, for example, be used for piecewise defined polynomial functions (e.g., spline functions). The values of matrix polynomials

$$c_0 I + c_1 A + c_2 A^2 + \cdots + c_d A^d, \qquad I, A \in \mathbb{R}^{n \times n},$$

can, for example, be computed using the program IMSL/MATH-LIBRARY/polrg.

Analytic Operations (Differentiation, Integration, etc.)

If operations other than evaluation have to be performed using a function as the operand, then the second form of representation (i.e., the specification of a function by an m-tuple of parameters) can be useful, particularly if the result of such an operation can be represented in terms of the parameters $c_1, c_2, \ldots, c_m$. All operations on f can then be reduced to operations on the arithmetic data $c_1, c_2, \ldots, c_m$. In such cases the function f is handled in the same way as an algebraic data structure.

Example (Polynomials) The first derivative of a polynomial $P_d \in \mathbb{P}_d$ defined by its $d+1$ coefficients with respect to the monomial base $\{1, x, x^2, \ldots, x^d\}$ can easily be determined. The d coefficients of

$$P_d' = (c_0 + c_1 x + \cdots + c_d x^d)' = e_0 + e_1 x + \cdots + e_{d-1} x^{d-1}$$

are given by $e_i := (i+1)c_{i+1}$, $i = 0, 1, \ldots, d - 1$.

A more detailed discussion of algebraic operations on functions can be found in Section 4.11.3.

If a function is not specified by an m-tuple of parameters but rather by an arithmetic expression, it is much more difficult to reduce an operation on the function to an operation on the expression which is a string of characters. If the function is given as a *black box* procedure only, i.e., if its functionality but not its internal structure are known, then it is not possible to perform any operations on the function at all. It is necessary to restrict operations to those which compute the value returned by the function subroutine.

Example (Numerical Integration) For the purpose of numerical integration, the integrand function is often defined by the user in the form of a function subroutine. From the point of view of an integration program this subroutine behaves like a black box, returning information about f only in the form of the function values $f(x_1), f(x_2), \ldots, f(x_k)$ for the arguments $x_1, x_2, \ldots, x_k$. In order to execute the desired integration operation, the algebraic data $\{(x_i, f(x_i)) : i = 1, 2, \ldots, k\}$ has to be transformed into a function which can be integrated without difficulty, i.e., into analytic data. The function thus obtained (usually a piecewise polynomial interpolating the algebraic data) is then integrated (see Chapter 12).

4.11.3 Operations with Functions

In addition to the evaluation of a function at given points, in data processing there are other operations on functions which are also of practical importance; the arithmetic combination of several functions, substituting one function for another, differentiation, integration, and integral transforms (Fourier transforms, Laplace transforms, etc.), to name a few.

The combination of functions and the substitution of one function for another only requires the according values of all the functions involved and can therefore be done with black box representations of the functions as well.

Analytic integration, on the other hand, requires explicit knowledge of the function, even if only the value of a definite integral is desired.

Operations on Polynomials

Polynomials are the most important functions in numerical data processing because of the simplicity of all the types of operations performed on them. Linear combination, multiplication, substitution, differentiation, and integration of polynomials all result in polynomials of a known maximum degree such that the results can again be represented by a *coefficient vector*. Hence, analytic operations on polynomials are reduced to algebraic operations.

This is demonstrated by two polynomials $P \in \mathbb{P}_k, Q \in \mathbb{P}_m$ of respective maximum degrees k and m respectively:

$$P(x) = c_0 + c_1 x + c_2 x^2 + \cdots + c_k x^k, \qquad Q(x) = d_0 + d_1 x + d_2 x^2 + \cdots + d_m x^m.$$

Without loss of generality $m \leq k$ is assumed; the coefficients e_i of the resulting polynomial from operations on P and Q are:

$$\alpha P + \beta Q: \quad e_i := \begin{cases} \alpha c_i + \beta d_i, & i = 0, \ldots, m \\ \alpha c_i, & i = m+1, \ldots, k \end{cases}$$

$$PQ: \quad e_i := \sum_{l=\max(0, i-k)}^{\min(i,m)} c_{i-l} d_l, \qquad i = 0, \ldots, m+k$$

$$P(Q(x)): \quad e_i := \text{coefficient of } x^i \text{ in } \sum_{j=0}^{k} c_j \left(\sum_{l=0}^{m} d_l x^l \right)^j \qquad i = 0, \ldots, mk$$

$$P': \quad e_i := (i+1) c_{i+1}, \qquad i = 0, \ldots, k-1$$

$$\int P\, dx: \quad e_0 \text{ arbitrary}, \quad e_i := c_{i-1}/i, \qquad i = 1, \ldots, k+1.$$

Parameterized Function Sets

In addition to polynomials, other function sets $\mathcal{F}$ can also be parameterized simply and are thus well suited to the implementation of the elementary operations listed above. Examples of such function sets are

rational functions of maximum degrees k and m in the numerator and the denominator respectively:

$$f = P/Q \quad \text{where} \quad P \in \mathbb{P}_k, \quad Q \in \mathbb{P}_m,$$

trigonometric polynomials of maximum degree m:

$$f(x) = \frac{a_0}{2} + \sum_{k=1}^{m} a_k \cos kx + \sum_{k=1}^{m} b_k \sin kx,$$

and *exponential sums* of the type

$$f(x) = c_0 + \sum_{k=1}^{m} c_k \exp(d_k x);$$

the coefficients d_k of these sums can either be constants or parameters themselves.

Symbolic Processing of Analytical Data

Functions which are given by an explicit arithmetic expression can easily be combined with and substituted for each other. Differentiation can be performed manually according to well-known rules which can also be formulated as an algorithm which can be implemented on a computer. Such an implementation is no longer a numerical algorithm but rather a symbolic manipulation of the expressions. However, it is rather complicated to implement such an algorithm in an imperative programming language (such as Fortran or C) because it belongs to the field of *symbolic* data processing.

Computer algebra systems like MACSYMA, MAPLE, MATHEMATICA, AXIOM, or DERIVE were designed to manipulate mathematical formulas symbolically. These programming environments are primarily intended to be used interactively and are not particularly suitable for programming classical numerical algorithms.

For the solution of problems in *digital calculus*, it is necessary to combine numerical and symbolic algorithms into a software product with a uniform user interface and to provide the necessary programming tools; future software developments are likely to concentrate on this combination.

Automated Differentiation

The problem of determining the *value* of the derivative of an arithmetic expression for given values of the independent variable can be formulated in terms of a computational scheme which is often referred to as *automated differentiation*.

Example (Automated Differentiation) The value of an expression w_0 and the value of its derivative w_1 are combined in the pair of real numbers (w_0, w_1). Starting with the pairs

$$\begin{pmatrix} x \\ 1 \end{pmatrix} \qquad \text{for the (given) value of the free variable} \quad x \quad \text{and}$$

$$\begin{pmatrix} c \\ 0 \end{pmatrix} \qquad \text{for the (given) values of constants} \quad c,$$

the representation of the arithmetic expression only has to be processed according to the following rules.

$$\begin{pmatrix} u_0 \\ u_1 \end{pmatrix} \pm \begin{pmatrix} v_0 \\ v_1 \end{pmatrix} = \begin{pmatrix} u_0 \pm v_0 \\ u_1 \pm v_1 \end{pmatrix}$$

$$\begin{pmatrix} u_0 \\ u_1 \end{pmatrix} \begin{pmatrix} v_0 \\ v_1 \end{pmatrix} = \begin{pmatrix} u_0 v_0 \\ u_0 v_1 + u_1 v_0 \end{pmatrix} \qquad \text{product rule}$$

$$\begin{pmatrix} u_0 \\ u_1 \end{pmatrix} / \begin{pmatrix} v_0 \\ v_1 \end{pmatrix} = \begin{pmatrix} u_0/v_0 \\ (u_1 v_0 - u_0 v_1)/v_0^2 \end{pmatrix} \qquad \text{quotient rule}$$

$$\begin{pmatrix} u_0 \\ u_1 \end{pmatrix}^q = \begin{pmatrix} u_0^q \\ (q u_1 w_0)/u_0 \end{pmatrix} \qquad \text{power rule} \quad (q \in \mathbb{R})$$

$$\exp \begin{pmatrix} u_0 \\ u_1 \end{pmatrix} = \begin{pmatrix} \exp(u_0) \\ w_0 u_1 \end{pmatrix} \qquad \text{chain rule.}$$

In these equations w_0 is the value of the first component on the right-hand side. A *differentiation arithmetic* can easily be implemented by using a *preprocessor* which reformulates arithmetic expressions appropriately and, thus, performs the symbolic differentiation.

Numerical Processing of Analytic Data

For many analytical operations, algorithms cannot be formulated as easily as they can be for differentiation operations. For instance, there may be arithmetic expressions for which no closed form representation of the according integral exists. Although symbolic methods for the determination of the indefinite integral are known for large classes of integrable functions, in many cases the same approach used for black box functions has to be chosen: A set of functions $\mathcal{F}$ (e.g., piecewise polynomials) is selected for which the analytic operation can be performed with reasonable effort. Then the given function f is replaced by some $\tilde{f} \in \mathcal{F}$ such that the result of the operation applied to $\tilde{f}$ differs from that of the application of the operation on f by an acceptable amount. The choice of the auxiliary function $\tilde{f}$ is usually based only on the *values* of the originally specified function f.

Example (Numerical Integration) For the numerical determination of the integral

$$\mathrm{I}f = \int_0^1 f(x)\, dx,$$

the function f can be evaluated at the points

$$x_i = i/k, \quad i = 0, 1, \ldots, k, \qquad k \in \mathbb{N};$$

f is then replaced by the piecewise polynomial $\tilde{f}$ interpolating the points $\{(x_i, f(x_i))\}$. For $x \in [x_i, x_{i+1}]$ the relation

$$\tilde{f}(x) = [(x - x_i)f(x_{i+1}) + (x_{i+1} - x)f(x_i)]k$$

holds. A piecewise linear function $\tilde{f}$ can easily be integrated:

$$T_k f := \int_0^1 \tilde{f}(x)dx = \frac{1}{k}\left[\frac{1}{2}f(x_0) + \sum_{i=1}^{k-1} f(x_i) + \frac{1}{2}f(x_k)\right].$$

The value $T_k f$ is an approximation for $\int_0^1 f(x)dx$.

This example illustrates the fact that *no* assessment of such a numerical method can be made on the basis of only a finite set of function values *without* additional information about the function f.

Example (Numerical Integration)   The integrand could be given by the function

$$f = \tilde{f} + c|\sin(\pi kx)|, \quad c \in \mathbb{R},$$

which leads to an approximation error of

$$T_k f - If = \int_0^1 \tilde{f}(x)\,dx - \int_0^1 f(x)\,dx = -\frac{2}{\pi}c, \tag{4.25}$$

which can become arbitrarily large depending on the value of the constant c.

It follows that, for a reasonable choice of $\tilde{f}$, the function f must at least be known to belong to a particular class of functions the properties of which are well-known and can be quantified. For example, if f is known to be twice differentiable and a bound M_2 is known such that

$$|f''(x)| \leq M_2 \qquad \text{for all} \qquad x \in [0, 1],$$

then it can be shown that the error estimate

$$|T_k f - If| = \left|\int_0^1 \tilde{f}(x)\,dx - \int_0^1 f(x)\,dx\right| \leq \frac{M_2}{12k^2}$$

holds. Thus, with a known bound on the derivative M_2, the error $T_k f - If$ can be made smaller than any given (positive) error bound by choosing a sufficiently large k.

The intrinsic *finiteness* of a computer and its data affects nearly all numerical operations applied to functions. Whereas the inaccuracies discussed so far have to be expected when real numbers are replaced by floating-point numbers, the processing of functions introduces an additional, very intricate indistinctness. In all manner of applications it is important to bring these intrinsic inaccuracies under control or at least to be aware of the potential errors and their impact.

4.11.4 Functions as Results

If a function is among the *results* of a problem in numerical data processing, the way in which it is to be represented depends largely on the particular application.

If this function is required as an input parameter for another problem, then of course one of the ways described so far must be chosen. With many functions occurring in practice, a *graphical representation* on a monitor, a printer, or a plotter etc. is desired. If the resulting function is the final result of a computation then a sufficiently accurate graphical representation can even be the overall result of the problem (cf. Fig. 4.14 or Nielson, Shriver [301]).

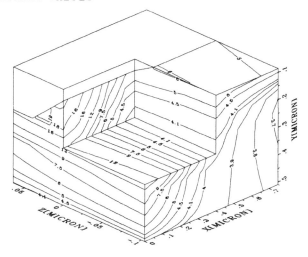

Figure 4.14: This picture illustrates the concentration function of electrons of an n-channel MOSFET transistor: The depicted cube represents a part of the transistor. The concentration of electrons is visualized with contour lines.

Chapter 5

Numerical Algorithms

Man's intellect may be praised for mastering the finesse of arithmetic, although this subtle and profound reasoning is nothing but a play of mechanism that can be performed more easily and better by clumsy machinery than by the brightest man.

DINGLER'S POLYTECHNICAL JOURNAL (1823)

In fields such as mathematics and computer science a *practical method* which implements the findings of theoretically derived solutions to a problem is expressed in the form of an algorithm.

For present purposes, an algorithm is introduced rather informally as a series of instructions which is precise enough to be processed by a human being or by a computer. Accordingly, a *numerical* algorithm is a series of instructions designed to execute numerical operations on numerical data.

This chapter firstly takes this informal understanding of an algorithm and makes it more precise. It then discusses two items of general interest for the assessment of numerical algorithms:

1. the resources (storage and computational effort) required to execute a numerical algorithm, and

2. the influence of floating-point numbers and operations on the results.

More specific forms of algorithm assessment, such as the analysis of the algorithm error, are closely connected with concrete problems and are discussed in other chapters.

5.1 The Intuitive Notion of an Algorithm

This section begins by introducing a rather informal understanding of what an algorithm is:

Definition 5.1.1 (Algorithm Concept) *An algorithm is a precise directive in the form of a finite text specifying the execution of a finite series of elementary operations; it is designed to solve problems of a particular class or type.*

The number of available elementary operations—whatever *elementary* means in a particular context—is limited, and the same is true for the execution time. The formal description of an algorithm requires a degree of preciseness which should guarantee the desired order of steps. Within a *class* of similar problems, which differ from each other only, for instance, in the values of certain parameters, a choice between different solution strategies may be necessary. In such a case *how* to choose one of the possible approaches must be defined.

Example (Bisection) From calculus it is well-known that a continuous function

$$f : [a, b] \to \mathbb{R}$$

whose values at the endpoints of the interval $[a, b]$ are of opposite sign, i.e.,

$$f(a) f(b) < 0 ,$$

has at least one *zero*: There exists a point

$$x^* \in (a, b) \quad \text{with} \quad f(x^*) = 0.$$

In order to determine an approximation of this zero, the following strategy can be used: the value $f(x_m)$ at the center of the interval $x_m = (a + b)/2$ is determined. If $f(x_m) = 0$, then a zero has been found already, namely x_m itself. Otherwise either

$$f(a) f(x_m) < 0 \quad \text{or} \quad f(x_m) f(b) < 0$$

must hold. If the first inequality holds, then the interval (a, x_m) contains at least one zero of f. If, on the other hand, the second inequality holds, then it is the interval (x_m, b) which contains at least one zero of f. Hence, an interval of length $(b - a)/2$ has been determined for further investigation.

Repeated application of this process leads either to the exact zero (when it is the center of one of the subintervals of $[a, b]$) or to a sufficiently small subinterval containing a zero of f.

This example shows a number of characteristic features of algorithmic problem solving:

1. The directive for determining smaller and smaller subintervals is *executable*. On the contrary, a mathematical statement like

   "A continuous real function $f \in C[a, b]$ whose values $f(a)$ and $f(b)$ are of opposite sign has at least one zero in (a, b)."

   is not executable and is therefore not an algorithm.

2. The execution of an algorithm proceeds *step by step*. The sequence of steps run through when an algorithm is executed is referred to as the *process* described by this algorithm.

3. The algorithm is executed by a *processor*. The processor may either be a human being or a computer.

4. Each step consists of the execution of one or several (other) sub-algorithms which are symbolized by their respective *names*. For example, "determining the function value" stands for a sub-algorithm for calculating $f(x)$ for a given argument x. The instructions of sub-algorithms are assumed to be *elementary*. Whoever or whatever executes an algorithm should know all the relevant elementary algorithms and be able to execute them.

Example (Solution of a System of Linear Equations) Once the steps required to solve a system of linear equations

$$Ax = b, \qquad A \in \mathbb{R}^{n \times n}, \quad b, x \in \mathbb{R}^n$$

are formulated as an algorithm, they do not have to be re-invented each time such a system is to be solved. The instructions used to solve such a system can be looked upon as an *elementary algorithm* which might be denoted `solve_lin_equations`.

5. Two important mechanisms for composing an algorithm can be seen in the bisection example: *choice* and *repetition*. The decision as to which execution path to follow is based on the sign of the function value at the center of the sub-interval. The bisection process is repeated until a certain condition ("zero is found" or "sufficient accuracy of the approximation is achieved") is met after which the algorithm *terminates*.

   A series of instructions can only be considered to be an algorithm if it is known, in advance, that they will terminate after a finite number of steps.

6. An algorithm must be clearly stated, i. e., its steps and the order in which they are executed can not be ambiguous. However, the specification of an algorithm should not be too detailed.

Example (Bisection) The bisection method as described earlier enables a reader with mathematical knowledge to determine a solution of the equation $xe^{-x} = 0.06$, i. e., to approximate a zero of

$$f(x) = xe^{-x} - 0.06,$$

starting, for example, with the interval $[0, 1]$ and using a pocket calculator. The adequate *number* of steps, which is only vaguely indicated in the instructions, has to be decided by the executing individual.

However, this description is not precise enough for a person without a mathematical background nor for the execution of the bisection method on a computer.

7. The language used for describing an algorithm in terms of elementary sub-algorithms (referred to as *algorithm notation*) has to be suitable for the processor. For describing instructions which are to be executed on a computer a *programming language* must be used as algorithm notation. A program is the only way to tell a computer how to execute an algorithm.

   Hence, a numerical program is a representation of a numerical algorithm suitable for execution on a computer.

5.2 Properties of Algorithms

In this section, properties already mentioned in the algorithm concept and in the bisection example on page 173 are introduced and discussed in detail.

5.2.1 Abstraction

An algorithm describes ways to solve a problem at a certain level of abstraction. The level of abstraction of the algorithm is determined by the elementary algorithms, the elementary objects and the formalisms used.

Giving a (sub-)algorithm a *name* which is used in place of a detailed description of the algorithm is one of the most important abstraction techniques.

Example (BLAS) Nowadays the BLAS-1, BLAS-2, and BLAS-3 program packages (see Section 4.10.1) are permanently installed on most computers used for numerical data processing. As a result, program developers can use prefabricated software to solve elementary linear algebra problems. The names of the BLAS routines are standardized and can be called from other programs. Their employment rids the user of a huge amount of unnecessary coding and debugging. Moreover, the BLAS routines constitute high-quality software as far as accuracy and execution time (of *machine dependently optimized* BLAS versions) are concerned.

The BLAS packages influence the program design from the very beginning as they encourage a uniform and clear programming style.

5.2.2 Generality

An algorithm is a *general* directive, which is the basis for solving not only a single instance of a problem but several (maybe all) problems of a particular class. Such a class may consist of an infinite number of problems, which differ from each other in their data (*parameters*). The parameters are used to specify one particular problem from this class.

Example (Systems of Linear Equations) Systems of linear equations

$$Ax = b, \qquad \text{where} \quad A \in \mathbb{R}^{n \times n}, \quad b, x \in \mathbb{R}^n$$

are a particular class of problems, which have $n^2 + n$ real parameters: the matrix coefficients $a_{11}, \ldots, a_{nn} \in \mathbb{R}$ and the vector components $b_1, \ldots, b_n \in \mathbb{R}$.

Example (Bisection) The parameters of the nonlinear equation to be solved by bisection are not only the numbers $a, b \in \mathbb{R}$ but also a *function* $f \in C[a, b]$. Thus, an instance of this problem type is specified by algebraic data as well as analytical data.

5.2.3 Finiteness

The description of an algorithm is necessarily of finite length (*static finiteness*). If the algorithm is supposed to return a result (see Section 5.2.4) then it is only allowed to require a finite amount of storage for intermediate results (*dynamic finiteness*).

5.2.4 Termination

An algorithm is said to be *terminating* if it stops after a finite number of steps and returns a result each time it is executed.

The termination property of an algorithm must not be confused with its finiteness. It may well be that a finite description defines a process which does *not* terminate after a finite period of time (an endless loop for example). In fact such (potentially) non-terminating algorithms may be of practical use, e. g., to control an uninterrupted process in a chemical manufacturing plant. Another example would be the operating system of a computer which is running continuously.

The considerations and examples in this book deal with terminating algorithms exclusively.

Example (Bisection)   There are different ways to terminate the bisection algorithm. Firstly, a decision has to be made as to how to characterize an approximation $\tilde{x}$ which is acceptable as a result of the algorithm: either $f(\tilde{x}) \approx 0$ (the residual criterion, cf. Fig. 14.6) or $\tilde{x} \approx x^*$ where $f(x^*) = 0$ (the error criterion, cf. Fig. 14.7) can be used. In the case of the residual criterion the process terminates as soon as $|f(\tilde{x})| \leq \tau_f$, and in the case of the error criterion as soon as the length of the current sub-interval is small enough (viz. $(b - a)/2^k < \tau_k$).

Note that in the first case it *cannot* be stated in advance as to *when* (i. e., after how many bisections) the algorithm will terminate. All that is known is that it will terminate (unless there are problems associated with machine arithmetic). The error criterion, on the other hand, makes it possible to predict the maximum number of bisections after which the process will terminate.

It may happen that an algorithm is not executable for specific data; in such a case it should terminate with an error message.

Example (Bisection)   The bisection algorithm is only applicable if the function values at the endpoints of the initial interval are of *different* sign. This prerequisite can easily be checked.

Another important prerequisite for the bisection algorithm is that f must be continuous. If this requirement is not met then the algorithm may *not* terminate. For example, for a function f defined on the interval $[0, 1]$ by

$$f = \begin{cases} -1 & \text{for} \quad x \leq 0.1 \\ +1 & \text{for} \quad x > 0.1 \end{cases}$$

the bisection algorithm does not terminate.

This example illustrates a fundamental difficulty with numerical algorithms: It is not possible to test the function f for the continuity property given a finite set of values, e. g., $f(a)$, $f(b)$, $f((a + b)/2), \ldots, f(x_{\text{stop}})$. Hence, it is not possible to create a bisection algorithm or program which automatically verifies the continuity assumption.

It is left up to the user to make sure all assumptions are met or at least to be aware of possible consequences if they are not.

The clear formulation of all prerequisites for the correct execution of the algorithm is an essential part of the *documentation* of the algorithm.

5.2.5 Deterministic Algorithms

An algorithm is said to be *deterministic* if at any point of execution there is, at most, one possible way to proceed, i.e., if the next consecutive step is uniquely determined.[1]

If there is at least one point from which there are two or more alternative ways to proceed (from which one is chosen arbitrarily) the algorithm is said to be *nondeterministic*.

A nondeterministic algorithm is said to be *stochastic* if the probabilities are known for all alternative ways of proceeding. Such algorithms are developed for problems which take too much time to be solved using a deterministic algorithm.

Example (Prime Number Test) R. Solovay and V. Strassen developed an algorithm to determine whether a given number m is prime or not. If the algorithm returns a *no* then it is certain that m is not prime. If, on the other hand, the algorithm returns a *yes*, then there is a $p \geq 0.5$ probability that m is prime. To reduce the uncertainty as to whether the affirmative result is correct, the algorithm can be re-applied to the same input value. For k repetitions the error probability $\bar{p}_k$ is reduced to 2^{-k}. Thus, any (arbitrarily small) error requirement may be met:

$\bar{p}_k$	1 %	0.1 %	0.01 %	0.001 %
$k \geq$	7	10	14	17

The remaining uncertainty of the Solovay-Strassen algorithm is compensated for by a significantly reduced computational effort. This is especially true for very large numbers as may occur in encryption. Whereas a straightforward deterministic algorithm requires $O(\sqrt{m})$ arithmetic operations (in order to check whether m is divisible by $l = 2, 3, \ldots, \lfloor \sqrt{m} \rfloor$), the effort is reduced to $O(k \log m)$ steps for the stochastic Solovay-Strassen algorithm (see Section 5.5).

5.2.6 Determinate Algorithms

An algorithm is said to be *determinate* if it returns the same result for given parameters and initial state each time it is executed.

A numerical algorithm $\boldsymbol{A}$ can be regarded as a mapping $\boldsymbol{A} : I \longrightarrow O$ of the set I of permissible input data into the set O of possible output data. In the case of a determinate algorithm this mapping is a function in the mathematical sense of the word, with exactly one output value corresponding to an input value.

The attributes *deterministic* and *determinate* must not be confused: a *deterministic* algorithm is unique in its *entirety*, whereas for a *determinate* algorithm only the result is unique.

Deterministic algorithms are determinate due to the unique way in which they are executed. However, there are nondeterministic algorithms which always produce the same result but which may proceed in different ways.

Example (Quicksort) Quicksort is an efficient sorting algorithm based on the recursive sorting of sub-arrays. Dividing the unsorted array into sub-arrays may be done in a stochastic manner. In this case the algorithm is nondeterministic. However, the result is always the same: the ordered output array. So in spite of its nondeterministic execution, the stochastic quicksort algorithm is a determinate algorithm.

---

[1] If there is no way to proceed, then it is assumed that the algorithm will terminate.

Example (Prime Number Test) The stochastic Solovay-Strassen prime number test is *nondeterministic*, i. e., it may proceed in different ways, as well as *nondeterminate*, i. e., different results are obtained when the algorithm is executed several times with the same input value.

Monte Carlo Algorithms

The name *Monte Carlo method* embraces a number of techniques for the numerical solution of mathematical problems, all of which are based on probability theory and mathematical statistics.

The following steps characterize the Monte Carlo method:

1. A stochastic model is developed for the given (determinate) problem.

2. Using this model, experiments (simulations), based on appropriately generated random numbers, are performed.

3. The results of the experiments are analyzed using statistical methods.

4. The statistical estimates thus obtained are used as (approximate) solutions to the original mathematical problem.

For example, a Monte Carlo method can be used for the numerical determination of integrals, for the solution of partial differential equations or for the solution of systems of algebraic equations. Its use is especially advantageous for multidimensional problems (Kalos, Whitlock [62]).

A *Monte Carlo algorithm* obtained by applying the Monte Carlo method to a particular problem class (e. g., numerical integration problems, see Section 12.4.4) is nondeterminate. Repeated application of a Monte Carlo algorithm with different random numbers produces different output values.

5.3 Existence of Algorithms

If an algorithm is known for a class of problems then no creative work has to be done to solve a problem belonging to this class; the instructions of the algorithm have only to be systematically followed.

For centuries mathematicians have tried to derive algorithms as general as possible; Leibniz still believed that any mathematical problem could be solved using an algorithm. However, this view has been questioned in the course of time as more and more problems have arisen for which no algorithmic solution has been found.

Example (Proofs of Formulas in Number Theory) In number theory it is not decidable whether or not a given logical formula can be proved[2]. This result (see e. g., Stegmüller [358]) is at least theoretically interesting, since the proof of a formula could be found if it was known to exist. Thus, problems like *Goldbach's conjecture* (*Every even number greater than 2 is the sum of two primes*), which have resisted all attempts at a formal proof so far, could be solved with computer aided methods. However, their practical use would be limited, because the time required would exceed all available resources.

---

[2]This holds if the calculus in use is free of formal contradictions.

If a general method has been found for solving a problem, then it can easily be checked to see if it has the properties introduced for the algorithm concept. In order to make a negative statement of the form "There is *no* algorithm for the solution of a certain problem", however, a more precise specification of this concept is required.

5.3.1 Precise Definitions of "Algorithm"

Around 1935 new concepts were proposed for making the notion of algorithm more precise. They were mainly influenced by the work of A. Church, K. Gödel, S. C. Kleene, A. M. Turing and A. A. Markov. Today several approaches are known: For example, the concepts of Turing machines, register machines, μ-recursive functions, logical formulas and so on. These concepts (and others) are equivalent as they correspond to one and the same class of computable functions, namely the class of *partially recursive functions* (see Section 5.3.2).

Church's Thesis

The equivalence of different definitions of the term *algorithm* suggests that they are appropriate for capturing the algorithm concept, i.e., that they are suitable abstractions of intuitive ideas. This proposition was first put forward in 1935 by A. Church and is nowadays referred to as *Church's thesis*: "The only effectively computable functions are those definable using Turing machines." This means that for any algorithm in an arbitrary, formal notation, a Turing machine (a particular mathematical model of data processing machines) which computes the same function can be found.

5.3.2 Computable Functions

The *theory of computability* is a branch of the theory of algorithms which deals with computable functions. A function f is said to be *computable* (in the intuitive sense) if there exists an algorithm which terminates after a finite number of steps and returns the value $f(i)$ for every input $i \in I$ that f is defined for. If $f(i)$ is not defined, then the algorithm does not terminate.

A computable function which is defined for all $i \in I$ is also said to be a *recursive function*. A computable function which is defined only on a proper subset of I is said to be a *partially recursive function*.

Testing for the existence of numerical algorithms and programs is equivalent to testing for the existence of computable functions, because any program P can be looked upon as a mapping of the set I of input values into the set O of possible output values:

$$P : I \longrightarrow O .$$

Non-Computable Functions

The following statement is merely an *existence theorem*: There are denumerably many computable functions, but the set of *non*-computable functions is not countable; thus computable functions are the exception rather than the rule. One can understand this fact using enumeration: Any algorithm is described by a finite formal text over a finite alphabet. Hence, the set of all possible formal descriptions of algorithms is countable. On the other hand, the set of functions $f : I \longrightarrow O$ for infinite I and more than one output value is not countable (Cantor's diagonal argument).

Non-computable functions *cannot* be implemented on a computer in *any* way.

Example (Termination Problem) Let $\mathcal{A}$ be the set of all algorithms and I the set of all inputs. The function

$$f_H : \mathcal{A} \times I \longrightarrow \{true, false\}$$

is defined as $f_H(\boldsymbol{A}, i) := true$ if the algorithm $\boldsymbol{A} \in \mathcal{A}$ applied to the input i terminates after a finite number of steps, and $f_H(\boldsymbol{A}, i) := false$ otherwise.

There is no algorithm which computes f_H, i.e., which takes as its input an *arbitrary* algorithm and the corresponding data and then determines whether the computation will terminate or not.

For many algorithms it is possible to decide whether or not they will terminate. The non-computability of f_H only implies that there is no *general* procedure for performing this kind of analysis for arbitrary algorithms. It thus follows that there is no automatic method which allows one to decide for an arbitrary program whether or not they contain an endless loop.

Example (Unresolved Termination Problem) For the following Fortran function subprogram, it is not known (yet) as to whether it will terminate for *all* $i \in \mathbb{N}$ or not:

```
FUNCTION unclear (i) RESULT (k)
   INTEGER, INTENT (IN)  ::  i
   INTEGER               ::  k
   k = i
   DO
      IF (k <= 1)          EXIT    ! termination
      IF (MOD(k,2) == 1) THEN
         k = 3*k + 1
      ELSE
         k = k/2
      END IF
   END DO
END FUNCTION unclear
```

As a consequence of the non-computability of f_H, it is impossible to test a program *automatically* for correctness. Any proof—or at least part of it—of correct program behavior (i.e., that the program always reacts to an input with the expected output) has to be done manually.

Example (Equivalence Problem)   The function

$$f_e : \mathcal{A} \times \mathcal{A} \longrightarrow \{true, false\}$$

which decides if two algorithms A_1, $A_2 \in \mathcal{A}$ represent the same function (i. e., if each of them returns the same output for an arbitrary input value) is also non-computable. Thus, there is no algorithm which takes two other algorithms as its input and decides automatically whether or not they have the same functionality.

5.4 Practical Solvability of Problems

The mere existence of an algorithm does not guarantee that problems of a particular class are actually solvable in practice. It may happen that there is a theoretical approach and even an algorithm available to solve a given problem, though nevertheless the problem is practically unsolvable.

Example (RSA Encryption)   The data encryption method of R. Rivest, A. Shamir, and L. Adleman—the *RSA method*—uses two different keys for encryption and decryption which are derived from two very large prime numbers p and q $(p, q > 10^{200})$. The RSA method is frequently used for *public key encryption*. In doing so the decryption key remains secret.

Encryption is based merely on the product $s_v = pq$, whereas both the factors p and q must be known for decryption.

The method could be cracked if the two prime factors of the public key s_e could be determined, as they would lead to the decryption key s_d. Thus, the theoretical solution to the problem is a "simple" prime factor decomposition. However, the problem of factorizing numbers of order 10^{200} is not solvable in practice, since, even on the fastest contemporary computers, the algorithms known today would take years to return a result (Rivest [328]).

Hence, for the time being, RSA encryption can be regarded as a safe method. The development of faster computers is said to make it even safer, because increased performance favors encryption rather than the "code crackers" as longer keys become practicable. However, the development of new algorithms might change this situation.

This example demonstrates that not only does the existence of an algorithm determine solvability, but so do the resources required for the execution of that algorithm (the complexity of the algorithm) and the degree of difficulty of the problem (the complexity of the problem).

5.5 Complexity of Algorithms

In numerical data processing, the existence of algorithms is no longer investigated, as algorithms or at least schemes for the algorithmic solution of many relevant problems already exist. It is much more important to find the *best possible* algorithms, which require as little resources as possible.

In order to determine for a particular program, e. g., a solver for linear systems, the resources needed on a particular computer, *time measurement* (either using the built-in clock of the computer or an external stop-watch) suffices. On the other hand, if the number of necessary computational steps is counted, an index is gained which is independent of the type of computer and the concrete input data of the problem. This is an abstract method of assessing the resource consumption of an algorithm.

The *complexity* of an algorithm is a measure of its execution workload. It is not only characteristic of the method in use but of the degree of difficulty (the size) of the problem, as well. Often a simple scalar index is used to characterize the degree of difficulty of a problem.

Example (Systems of Linear Equations) The workload originating from the numerical solution of a system of n linear equations with n variables is usually characterized by a parameter which depends on the dimension n. Thus, the degree of difficulty for solving a system of linear equations

$$Ax = b \quad \text{where} \quad A \in \mathbb{R}^{n \times n}, \, x, b \in \mathbb{R}^n$$

is characterized by the (size) index n.

The degree of difficulty of a problem may also be specified by several parameters.

Example (Molecular Dynamics) (Addison et al. [81]) Algorithms used in molecular dynamics simulate the motion of particles in a gravitational field. For the implementation of these algorithms on parallel computers, the investigated region is decomposed into rectangular sections each of which is assigned to one processor. Each processor computes the coordinates of all particles which are in its respective section. In order to take into account the direct interaction between all the particles, the coordinates of all the particles must be stored on each processor.

The workload is determined by the fineness of the decomposition of the investigated region as well as by the number of particles involved. Thus, the problem size may be specified by three components of a *parameter vector* (the number of sub-intervals in x- and y-directions and the number of particles).

The ordering of problems with respect to their complexity is more difficult for problems which use a parameter vector than it is for problems which are characterized by a scalar index, because an order relation for the parameter vectors must first be defined.

All statements about the execution effort of an algorithm are always made with respect to some kind of complexity index of the problem. In doing so, usually the number of *computational steps* necessary for executing the algorithm, for a certain problem size, is determined.

In spite of the abstraction (using computational steps instead of time measurements), such complexity analysis is not independent of particular hardware properties. There may be significant differences in the complexity indices obtained for the very same algorithm due to the fact that for each computer the elementary unit of work may be different. In the field of numerical data processing, it is common to regard a floating-point operation as the elementary unit of work.

Example (Gaussian Elimination) The Gaussian elimination algorithm (LU factorization) for the solution of a system of linear equations requires

$$K(n) = 2n^3/3 + 3n^2/2 - 7n/6$$

floating-point operations, where n is the number of equations. If this formula is used to characterize the computational effort of the algorithm, then computation time required for auxiliary operations such as pivoting or row interchanges is neglected.

Particularly for parallel computers, it is possible and reasonable to regard more complex operations (matrix-vector operations, matrix-matrix operations, matrix transposition, fast Fourier transforms) as elementary units of work (Hockney, Jesshope [228]). It depends very much on the computer type as to what kind and quantity of work is combined into such an elementary computational step.

5.5.1 Abstract Computer Models

Abstract computer models are used to describe the fundamental operations of a computational step and how data is referenced in a way that is independent of hardware details (Almasi, Gottlieb [86], Mayr [287], Blelloch [116]). An abstract computer model constitutes a framework for specifying which algorithms can be implemented at all. Moreover, it allows for the machine-independent complexity analysis of an algorithm.

RAM Model

The *random access machine*[3] (RAM) is the standard model of conventional computers (single-processor machines). It consists of the following components:

- a central processing unit (CPU) together with an accumulator which stores the operands before and the intermediate results after an operation is executed,

- *unlimited* memory comprising an unlimited number of memory cells which can contain numbers of any size,

- a program and

- an I/O device.

The program consists of *instructions* (concerning, for example, the addition of a number and the current value of the accumulator, the transfer of the accumulator content to the memory, loading a number from the memory to the accumulator etc.). The random access machine executes exactly one instruction per computational step. For simplicity it is assumed that all the instructions take the same time no matter what kind of operands are involved or how large they are. The program instructions are executed sequentially. This order is abandoned only for branch conditions (conditional or unconditional jumps).

Example (Summation) Summing up n numbers according to the simple pattern

$$s := 0; \quad s := s + x_1, \quad s := s + x_2, \ldots, s := s + x_n$$

clearly requires n RAM computational steps if initialization is *not* taken into account, $n+1$ steps otherwise.

---

[3]It is also called *general register machine*.

PRAM Model

The RAM model, an abstraction of a single-processor system, can be extended to the model of a multi-processor system (parallel computer) with *shared* memory. A *parallel random access machine* (PRAM) of size p is obtained by combining p RAMs and a common memory.

The most critical simplification of this model is the assumption that the time required for the execution of any instruction and particularly for any memory access operation does not depend on the number p of processors: it is assumed that each memory cell can be referenced by any processor within one time unit.

Example (Summation) In a PRAM it only takes $\lceil \log_2 n \rceil$ computational steps to add n numbers provided that $p \geq \lceil n/\log_2 n \rceil$. In any computational step, the simultaneous pairwise addition of adjacent elements (x_1, x_2), $(x_3, x_4), \ldots$ is performed. As a result, the number of terms is halved[4] repeatedly until only one term, the required sum, remains.

MPRAM Model

A *message passing random access machine* (MPRAM) is used to model parallel computers with distributed memory. Like the PRAM it consists of a number of RAMs, but in this case there are multiple memory modules, with each RAM accessing its own memory only.

It is possible for certain RAMs to communicate with one another; a connection graph specifies which RAMs actually have communication links.

VRAM Model

In addition to RAM components, a *vector random access machine* (VRAM) comprises an unlimited vector memory, vector I/O devices and specific vector instructions.

Within a time unit a VRAM can perform instructions on a fixed number of vectors from the vector memory and scalars from the scalar memory, e.g., elementwise addition of two vectors or multiplication of a vector by a scalar.

In contrast to the PRAM, which executes *sequential* RAM instructions *in parallel*, for the PRAM, extra *parallel* instructions (vector instructions), which are executed *sequentially*, are introduced.

5.5.2 Theoretical Execution Cost

Once the unit of work has been specified for a particular computer, the number of operations necessary to complete the algorithm for a problem of size n can be determined (counted). The complexity index of the algorithm, which assigns the number of required operations to each problem size, is obtained in this way.

On the basis of the abstract computer models, it is assumed that all operations take equally long regardless of the operands and the specific type of operation.

---

[4]If there is an odd number of terms, the remaining term is passed to the next step without change.

The time required for the execution of the algorithm, which depends on the problem size, can be estimated if the time required for the execution of *one* operation is known.

Thus, equally weighted operations are simply counted; it is assumed that the workload is the same for all operations. As a result of this simplification the *theoretical execution cost* can be used only for qualitative or rather crude quantitative statements.

5.5.3 Asymptotic Complexity of Algorithms

In some cases the user is interested in the qualitative behavior of an operation-count (when the problem parameters approach infinity) rather than in the exact number of operations needed for solving a concrete problem. The asymptotic growth rate of the computational effort is particularly important for the assessment of algorithms used to solve large problems.

Definition 5.5.1 (Order of Complexity)   *Depending on a problem parameter p, the complexity $C(p)$ of an algorithm is of order $f(p)$ if there are constants a and b such that*

$$C(p) \leq b\, f(p) \qquad for\ all \quad p \geq a.$$

This is expressed by the *Landau-O notation*[5]:

$$C(p) = O(f(p)). \tag{5.1}$$

Algorithms are divided into classes of complexity. In practice, the most important classes are:

Order	Class of Complexity	Example of $C(p)$
$O(1)$	constant	$c \in \mathbb{R}_+$
$O(\log p)$	logarithmic	$c \log p$
$O(p)$	linear	$c_1 p + c_0$
$O(p^2)$	quadratic	$c_2 p^2 + c_1 p + c_0$
$O(p^3)$	cubic	$c_3 p^3 + c_2 p^2 + \cdots$
$\vdots$	$\vdots$	$\vdots$
$O(p^m),\ m \in \mathbb{N}$	polynomial	$c_m p^m + c_{m-1} p^{m-1} + \cdots$
$O(c^p)$	exponential	$c^{dp} + \text{polynomial}(p)$
$O(p!)$	factorial	$cp!$

For algorithms of logarithmic complexity, the base is irrelevant since

$$\log_b(p) = \log_B(p)\, \log_b(B).$$

The items in the table above are ordered according to the rate of growth of the computational effort: for example, an algorithm of cubic complexity requires

---

[5] $O(f(p))$ is to be read: "big-O of f of p". Note that (5.1) is not an equation in the usual mathematical sense. It has to be read from left to right: $O(f(p)) = K(p)$ is meaningless.

asymptotically (for large values of the parameter p) more work than an algorithm of logarithmic complexity.

The order of complexity can be used for a rough efficiency assessment of algorithms for the solution of a particular problem. If two algorithms have the same asymptotic complexity, their computational efforts have the same growth rate for increasing problem sizes. This is true only if the two complexity orders are determined on the basis of the same problem characteristic.

Example (Sorting Algorithms)   In the case of sorting algorithms, the parameter p is usually chosen to be the number k of elements to be sorted. The amount of work is determined by the number of key comparisons required between pairs of elements. Obviously this number depends on the prior arrangement of the input data of the algorithm.

Sorting algorithms such as *bubblesort*, *insertion sort* or *quicksort* belong to the same complexity class of algorithms because, in the worst case, depending on the initial distribution, they are of quadratic complexity. On average, i.e., for random initial distributions, *quicksort* requires only an $O(k \log k)$ effort. On the other hand, the initial arrangement of the elements does not determine the complexity of *mergesort* and *heapsort*, which in any is case $O(k \log k)$. Accordingly, these methods belong to a class of asymptotically more efficient algorithms.

Example (NP-Complete Problems)   There is a class of problems (in operations research, graph theory etc.) for which all known deterministic algorithms are of exponential complexity. However, there are non-deterministic algorithms for these problems requiring polynomial computational effort. This class is called NP (which stands for *non-deterministic polynomial*). In this class there are problems known as *NP-complete*, from which all NP problems can be derived with additional effort of polynomial complexity. If just one NP-complete problems can be solved with polynomial effort then *all* problems of the NP class can be solved with polynomial effort.

A fundamental (yet unresolved) question in complexity theory is whether P = NP or P ≠ NP, i.e., if there is a deterministic method for solving all problems of the NP class with polynomial effort.

Caution is advisable when conclusions on the practical effectiveness of an algorithm are to be drawn from theoretical analyses of the asymptotic complexity. Due to the definition of the Landau-O notation, an algorithm of lower asymptotic complexity may incur a higher computational effort for parameter values p occurring in practical applications than an algorithm of higher asymptotic complexity.

Example (Comparison of Algorithms)   Two algorithms with complexity $C_1(p) = 0.67p^3$ and $C_2(p) = 260p^{2.3}$ are to be assessed with regard to computational effort. The $O(p^{2.3})$ effort of the latter algorithm means lower asymptotic complexity than the cubic behavior of the first algorithm. However, for $p \leq 5\,000$ the first algorithm requires less work than the second. For small values of p the difference is significant.

5.5.4 Problem Complexity

The complexity of an *algorithm* is a measure of the work to be done during the execution under given model assumptions.

Problem complexity, however, is not very well defined because there may exist many algorithms with different complexities to solve a problem. The notion of complexity *class* may still be transferred to problems: A complexity class of

problems comprises all problems which can be solved using algorithms belonging
to a corresponding complexity class of algorithms.

If for a given problem there exist algorithms of a minimum complexity this
suggests an intuitive *complexity of the problem*. However, in practice it is usually
impossible to determine a complexity class of problems (denoted *problem complexity* in the following). To achieve this a lower bound for the work required to
solve certain problems would have to be determined *in addition to* an optimal
algorithm whose asymptotic complexity is equal to that lower bound.

Example (Sorting)   A general sorting method, which permutes a sequence of $k \geq 2$ numbers
$a_1, a_2, \ldots, a_k$ so that for the new sequence $a_{(1)}, a_{(2)}, \ldots, a_{(k)}$ the order relation

$$a_{(1)} \leq a_{(2)} \leq \ldots \leq a_{(k)}$$

holds, and which acquires information about the ordering only by comparing two elements,
requires *at least* $O(k \log k)$ comparisons.

On the other hand, there are algorithms like *heapsort* which never require more than this
number of comparisons; hence it can be said that the complexity of the general sorting problem
is $O(k \log k)$.

Example (Systems of Linear Equations)   The Gaussian algorithm for the solution of n
linear equations is an algorithm which requires $O(n^3)$ arithmetic operations. For a long time
it had been assumed that the problem complexity of solving linear systems was also of cubic
order. In 1969 this was found not to be the case (Strassen [361]).

For the time being, the complexity of this problem can be vaguely qualified on the basis
of an explicitly known $O(n^{2.376})$ algorithm: Solving linear systems with direct methods clearly
belongs to the complexity class of $O(n^{2.376})$ problems. At the same time it is known that
algorithms require at least n^2 reference operations to the matrix elements. Hence, an informal
qualification of the complexity K of the problem is given by

$$O(n^2) \leq K \leq O(n^{2.736}).$$

The complexity of this problem is closely connected to that of matrix-matrix multiplication,
which is discussed in detail in the following case study (cf. also Pan [310]).

5.5.5 Case Study: Multiplication of Matrices

In 1850 the English mathematician J. Sylvester introduced the term *matrix*
for a rectangular array of numbers, which he took for the representation of
a linear mapping between two finite-dimensional vector spaces. The attempt
to formally describe the product (the composition) of two linear mappings
$L_1, L_2 : \mathbb{R}^n \longrightarrow \mathbb{R}^n$ prompted the natural definition of matrix-matrix multiplication:

$$(AB)_{ij} := \sum_{k=1}^{n} a_{ik} b_{kj}, \quad A, B \in \mathbb{R}^{n \times n} . \tag{5.2}$$

The *rows by columns algorithm* which can be derived from (5.2), requires n^3 multiplications and $n^2(n-1)$ additions. Hence, its asymptotic complexity with respect
to arithmetic operations is cubic for additions as well as for multiplications.

For more than a century, this method was the only algorithm known for the multiplication of matrices. Due to its simplicity it was considered optimal, so no further investigation into different methods was conducted.

In 1967, S. Winograd's discovery of a way to replace half of the n^3 multiplications of the conventional matrix-matrix multiplication algorithm by additions took the scientific community by surprise. He utilized the equality of certain inner products which can be reused once they are calculated. Winograd's paper attracted enormous attention because at that time computers executed floating-point *additions* two to three times as fast as floating-point *multiplications*. (On most contemporary computers both operations take the same time.)

Soon after the publication of Winograd's paper, V. Strassen developed a new method for matrix-matrix multiplication whose $O\left(n^{\log_2 7}\right)$ complexity is significantly better than the cubic complexity of the *classical* rows by columns method ($\log_2 7 \approx 2.807$). Strassen also raised the question of problem complexity, i.e., whether there is a smallest exponent ω such that each matrix-matrix multiplication can be performed with $O(n^\omega)$ operations. Obviously $\omega \geq 2$ holds, since all elements of both matrices have to be operands of at least one operation. However, in spite of intensive research the minimum (or at least the infimum) of those exponents and hence the problem complexity of matrix-matrix multiplication still remains unknown.

Meanwhile the exponent ω has been further improved, with the sophisticated use of tensors, bilinear and trilinear forms playing an important role. Pan [309] gives a survey of relevant publications. By 1990 the "record" was $\omega = 2.376$. Fig. 5.1 illustrates the evolution of $O(n^\omega)$.

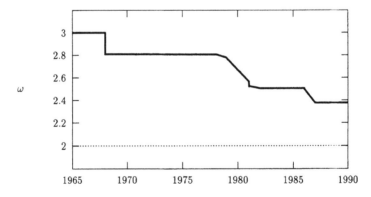

Figure 5.1: Evolution of the *complexity exponent* of matrix-matrix multiplication.

The Strassen Algorithm

The Strassen algorithm is based on the *divide and conquer* strategy. This method for algorithm development consists of two steps:

- Step 1 (*divide*): The whole problem is split into two or more equally (as equal as possible) large sub-problems which can be solved independently of one another.

- Step 2 (*conquer*): The solutions of the sub-problems are combined into a solution for the original problem.

Division steps are performed until sub-problems which are easy to solve have been extracted.

The simplest (nontrivial) subproblem of matrix-matrix multiplication is the multiplication of two 2×2 matrices $A, B \in \mathbb{R}^{2 \times 2}$

$$A = \begin{pmatrix} a_{11} & a_{12} \\ a_{21} & a_{22} \end{pmatrix}, \qquad B = \begin{pmatrix} b_{11} & b_{12} \\ b_{21} & b_{22} \end{pmatrix}.$$

With the auxiliary values

$$
\begin{aligned}
p_1 &:= (a_{11} + a_{22})(b_{11} + b_{22}) \\
p_2 &:= (a_{21} + a_{22})b_{11} \\
p_3 &:= a_{11}(b_{12} - b_{22}) \\
p_4 &:= a_{22}(b_{21} - b_{11}) \\
p_5 &:= (a_{11} + a_{12})b_{22} \\
p_6 &:= (a_{21} - a_{11})(b_{11} + b_{12}) \\
p_7 &:= (a_{12} - a_{22})(b_{21} + b_{22}),
\end{aligned}
\qquad (5.3)
$$

for the components c_{ij} of the product

$$AB =: C = \begin{pmatrix} c_{11} & c_{12} \\ c_{21} & c_{22} \end{pmatrix}$$

the following relations hold, as can be shown by straightforward calculation:

$$
\begin{aligned}
c_{11} &= p_1 + p_4 - p_5 + p_7 \\
c_{12} &= p_3 + p_5 \\
c_{21} &= p_2 + p_4 \\
c_{22} &= p_1 + p_3 - p_2 + p_6.
\end{aligned}
\qquad (5.4)
$$

This algorithm, referred to as the *Strassen algorithm* for matrix-matrix multiplication (Strassen [361]), necessitates 25 arithmetic operations (7 multiplication and 18 addition/subtraction operations), whereas the standard algorithm requires only 12 arithmetic operations (8 multiplication and 4 addition operations). In the case of 2×2 matrices the Strassen algorithm is actually inferior to the standard algorithm. However, Strassen recognized that formulas (5.3) and (5.4) remain valid if a_{ij} and b_{ij} are themselves matrices. For this generalized case the algorithms $\mathcal{A}_{m,k}$, which calculate the product of two matrices of order $m2^k$, are defined by induction on k as follows.

If n is odd, the last column of C is calculated using the standard method and the Strassen algorithm is executed for the remaining matrix of order $n - 1$.

Definition 5.5.2 *Let $\mathcal{A}_{m,0}$ be the standard algorithm for matrix-matrix multi-plication (which requires m^3 multiplication and $m^2(m-1)$ addition operations). If $\mathcal{A}_{m,k}$ is already known then $\mathcal{A}_{m,k+1}$ is recursively defined as follows: For the multiplication of matrices A and B of orders $m2^{k+1}$, the matrices A, B, and AB are divided into blocks*

$$A = \left(\begin{array}{cc} A_{11} & A_{12} \\ A_{21} & A_{22} \end{array} \right), \qquad B = \left(\begin{array}{cc} B_{11} & B_{12} \\ B_{21} & B_{22} \end{array} \right), \qquad AB = \left(\begin{array}{cc} C_{11} & C_{12} \\ C_{21} & C_{22} \end{array} \right),$$

where A_{ij}, B_{ij}, C_{ij} are matrices of order $m2^k$. Auxiliary matrices of order $m2^k$ are then calculated:

$$
\begin{aligned}
P_1 &:= (A_{11} + A_{22})(B_{11} + B_{22}) \\
P_2 &:= (A_{21} + A_{22})B_{11} \\
P_3 &:= A_{11}(B_{12} - B_{22}) \\
P_4 &:= A_{22}(B_{21} - B_{11}) \\
P_5 &:= (A_{11} + A_{12})B_{22} \\
P_6 &:= (A_{21} - A_{11})(B_{11} + B_{12}) \\
P_7 &:= (A_{12} - A_{22})(B_{21} + B_{22})
\end{aligned}
$$

$$
\begin{aligned}
C_{11} &= P_1 + P_4 - P_5 + P_7 \\
C_{12} &= P_3 + P_5 \\
C_{21} &= P_2 + P_4 \\
C_{22} &= P_1 + P_3 - P_2 + P_6,
\end{aligned}
$$

where algorithm $\mathcal{A}_{m,k}$ is used for multiplication operations and the usual (elementwise) algorithm is used for addition and subtraction operations.

The matrices P_1, P_2, ..., P_7 can all be calculated *concurrently*. This holds for the matrices C_{11}, ..., C_{22} as well. However, their computational effort is low compared to the costs of the auxiliary matrices P_1, P_2, ..., P_7, so parallel computation of the C_{ij} is usually not advantageous. In any case, the Strassen algorithm is very well suited for use on parallel computers (Bailey [98], Laderman et al. [263]).

Complexity of the Strassen Algorithm

With the Strassen algorithm for the multiplication of two matrices of order $n = 2^k$, only 7^k multiplication and less than $6 \cdot 7^k$ addition and subtraction operations are required. Specifically:

$$
\begin{aligned}
C_{\mathrm{mult}}(n) &= n^{\log_2 7} & \text{multiplications and} \\
C_{\mathrm{add}}(n) &= 6(n^{\log_2 7} - n^{\log_2 4}) & \text{additions/subtractions.}
\end{aligned}
$$

Hence, this algorithm is of algorithmic complexity $O(n^{\log_2 7})$ for addition as well as multiplication operations, which is significantly lower than the complexity of the standard algorithm.

For matrices of arbitrary order (not necessarily $n = 2^k$), the following theorem holds:

Theorem 5.5.1 *Using the Strassen algorithm, the product of two quadratic matrices of order n can be calculated with less than $28n^{\log_2 7}$ arithmetic operations.*

Proof: Strassen [361].

Software Implementing the Strassen Algorithm

It is not easy to write efficient programs implementing the Strassen algorithm. For a long time the Strassen algorithm was thus regarded as being of only theoretical interest. However, C. C. Douglas, M. Heroux, G. Slishman and R. M. Smith [171] recently published efficient implementations for use on both single-processor and parallel computers. Some software libraries already comprise implementations of the Strassen algorithm (e. g., **sgemms** and **dgemms** from the ESSL library), as well.

On most computers the break-even point $n_{\min}$, i. e., the minimum matrix order for which a Strassen implementation takes less time than the corresponding BLAS-3 subroutine, is between 32 and 256. For larger matrices a significant reduction of the required number of floating-point operations (see Fig. 5.2) and consequently a remarkable reduction of the execution time (see Fig. 5.3) is achieved.

5.5.6 Practical Effort Determination

In contrast to theoretical complexity analyses, for the empirical determination of the computational effort the *finite* rather than the asymptotic complexity of an algorithm is relevant, for example, the actual number $K(n)$ of (floating-point) operations required by the algorithm to solve a specific problem of size n.

Example (Gaussian Elimination)   The elimination algorithm (LU factorization) for solving an $n \times n$ system of linear equations requires

$$
\begin{aligned}
C_{\text{add}}(n) &= n^3/3 + n^2/2 - 5n/6 \quad \text{additions,} \\
C_{\text{mult}}(n) &= n^3/3 + n^2/2 - 5n/6 \quad \text{multiplications and} \\
C_{\text{div}}(n) &= n^2/2 + n/2 \qquad\qquad\quad\ \text{divisions}
\end{aligned}
$$

(Golub, Ortega [49]). The work required to guarantee numerical stability (pivoting, row and/or columns interchanges (Golub, Van Loan [50])) is *not* included in the complexity indices $C_{\text{add}}(n)$, $C_{\text{mult}}(n)$ and $C_{\text{div}}(n)$. This additional cost cannot easily be quantified using formulas because it depends greatly on the given data. However, its asymptotic complexity is known to be $O(n^2)$ in the worst case.

If it is not possible to determine the exact computational effort by denumerating the expected operations, a counter may be incremented for specified operations at runtime in order to determine the exact number of operations required by a particular program on a given computer system. Such experimental cost determination is made possible by *software* and *hardware monitors* (Jain [240]).

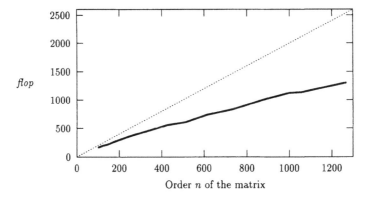

Figure 5.2: Number *flop* of floating-point operations *per element* of the resulting matrix required by the subroutines sgemmw (—) and BLAS/sgemm (·····) respectively.

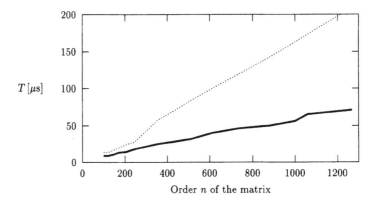

Figure 5.3: Time T required by the subroutines sgemmw (—) and BLAS/sgemm (·····) respectively on an HP workstation in microseconds *per element*.

Specification of the Term "Operation"

The practical effort determination of an algorithm requires the term *operation* to be specified more precisely than in the theoretical analyses.

In numerical data processing the computational effort required by an algorithm for solving a problem of size n is usually determined by the number of required floating-point operations. The function $K(n)$, which relates the number of floating-point operations to the problem size n, is referred to as the *computational complexity of the algorithm*. However, neither the work nor the time required by different floating-point operations are identical.

It has to be considered that not only may other operations (e.g., memory access operations) also influence the execution time, but also that different computer systems implement floating-point operations in different ways. For example, many micro-processors provide instructions for the arithmetic operations $+ - \cdot /$ as well as instructions for the computation of square roots, sine, cosine and other functions. But such instructions require much more work and hence take more time than floating-point addition and multiplication operations.

Example (Square Root) On older IBM workstations with POWER architecture the square root function was implemented as the call of a library routine which took about 50 clock cycles. The POWER2 architecture provides a square root instruction which takes only half the number of clock cycles.

Example (Multiply and Add) POWER and POWER2 processors provide an instruction to perform a combination of a multiplication and an addition operation (*floating-point multiply and add*)—the operation $ab + c$—within the same time as a single floating-point addition or multiplication. Another advantage of this composed instruction is that the intermediate result is not truncated and hence only *one* rounding error occurs (White, Dhawan [379]).

The arithmetic floating-point operations in the instruction set of a processor are *not*—in contrast to the simplifying assumptions of the RAM model— homogeneous in their expenditure of time. For example, floating-point additions and multiplications usually require significantly fewer clock cycles than floating-point divisions. This must not be neglected either for the cost assessment of algorithms or for the empirical performance assessment of computer systems (see Section 3.5).

In order to make cost assessment easier and more homogeneous, *standardized floating-point operations* are sometimes used. Each floating-point operation is counted with respect to its weight, where, for instance, the number of required clock cycles is the weight function.

The following standardization of execution costs is common[6] (Addison et al. [81]):

Floating-point operation	Weight
Addition, subtraction, multiplication	1 flop
Division, square root	10 – 30 flop
Exponential function, trigonometric functions	50 flop

The number of floating-point operations determined in this way is independent of the *types* of the floating-point operations involved, but it is not independent of the hardware features of the processor. If the instruction set contains composed instructions such as the *floating-point multiply and add* instruction mentioned above then the standardization has to take this into account.

---

[6]The weight function suitable for a particular computer is best determined by way of experiment. Flop/s indices based on anything other than addition and multiplication operations have to be handled with caution.

5.6 Representation of Algorithms

An algorithm can be represented using:

Natural Language: This representation can be adapted to any technical terminology. If desired, it can allow for vague statements though algorithms should be specified precisely enough to be understood by human beings.

Structured Representation: In algorithm and program development, graphic algorithm representations such as structured flowcharts (*Nassi-Shneiderman diagrams*) are often used.

Programming Languages: Programming languages are artificial languages used to define, in a form understandable to humans, actions required of a computer to solve a problem.

In this book two types of representation are used: program fragments in Fortran 90 and pseudocode representations.

5.6.1 Fortran 90

Program fragments written in Fortran 90 (Adams et al. [2], Ueberhuber and Meditz [78]) are used wherever the precise representation of numerical algorithms and language elements designed particularly for numerical data processing are required. Usually program *fragments* are used, for example, declarations which are not essential to the comprehension of a program are omitted. However, this is not intended to encourage the omission of explicit declarations (and thus to use the obsolete implicit type declarations of FORTRAN 77).

Example (Bisection)   The following program fragment implements (not very efficiently) the bisection algorithm:

```
...
DO iteration = 1, DIGITS(a)
   x_center = (a + b)/2.
   IF (ABS(f(x_center)) < tol_e) EXIT
   IF (f(a)*f(x_center) < 0.) THEN
      b = x_center
   ELSE
      a = x_center
   END IF
END DO
...
```

In many parts of this book (e.g., in Chapter 6) Fortran 90 program fragments are used to show how algorithms can be implemented. However, this does not mean that those program fragments are always *optimal*. As an example, all the matrix-matrix multiplication algorithms in Section 6.8 can be replaced by a single call to the Fortran 90 function MATMUL; this is recommended for small matrices, for which efficiency is not critical.

5.6.2 Pseudocode

The representation of algorithms by means of a pseudocode is even more fragmentary than the method described above. It is applied wherever non-numerical language constructs are used and/or where simplified representation facilitates understanding.

Complete listings of executable programs were deliberately *not* included in this book because there is a lot of high-quality numerical software (see Chapter 7) available today. The reader should use the packages mentioned in the book.

Language Constructs

The *assignment statement* is represented by the operator $:=$.

The following *control statements* are used:

infinite loop	**do**
	$\cdots$
	end do

---

indexed loop	**do** loop_variable $=$ initval, initval+increment, ..., endval
	$\cdots$
	end do

WHILE *loop*	**do while** condition
	$\cdots$
	end do

FORALL *loop*	**do for** variable $\in$ index_set
	$\cdots$
	end do

IF *statement*	**if** condition **then** statement

---

	if condition **then**
	$\cdots$
	end if

---

	if condition **then**
	$\cdots$
	else
	$\cdots$
	end if

The syntactic meaning of these constructs is derived from the Fortran 90 language specification (Adams et al. [2]). For instance, the infinite loop can also be left with a conditional **exit** statement in pseudocode:

> **do**
>   . . .
>   **if** condition **then exit**
>   . . .
> **end do**

In pseudocode an infinite loop *without* an exit statement may be used when the details of the loop termination are not relevant.

5.7 Influence of Rounding Errors on Numerical Algorithms

If operations on algebraic and analytical data are not executed symbolically, then they must be reduced to arithmetic operations (rational operations and evaluations of standard functions) applied to the components of structured data. Strictly speaking, each numerical algorithm consists only of arithmetic operations, aside from comparisons that are required for case and branch operations.

Numerical data processing typically employs arithmetic algorithms with an enormous number of arithmetic operations (often 10^8 and more). It is thus necessary to analyze the impact of floating-point arithmetic on a very long sequence of arithmetic operations (Higham [57]).

5.7.1 Arithmetic Algorithms

The solution of a system of 3 linear equations in 3 variables is used as an example of an arithmetic algorithm:

$$
\begin{aligned}
a_{11}x_1 + a_{12}x_2 + a_{13}x_3 &= a_{14} \\
a_{21}x_1 + a_{22}x_2 + a_{23}x_3 &= a_{24} \\
a_{31}x_1 + a_{32}x_2 + a_{33}x_3 &= a_{34}.
\end{aligned}
\tag{5.5}
$$

The *Gaussian elimination algorithm* is the standard method of solving systems of linear equations. For the linear system (5.5), the first equation is multiplied by $-a_{21}/a_{11}$ and added to the second equation. A new equation which contains only x_2 and x_3 is thus obtained. A similar equation is obtained by adding the first equation multiplied by $-a_{31}/a_{11}$ to the third equation:

$$
\begin{aligned}
\bar{a}_{22}x_2 + \bar{a}_{23}x_3 &= \bar{a}_{24} \\
\bar{a}_{32}x_2 + \bar{a}_{33}x_3 &= \bar{a}_{34}.
\end{aligned}
$$

The application of the same elimination process to the new system of equations leads to a *single* equation in the variable x_3

$$
\bar{\bar{a}}_{33}x_3 = \bar{\bar{a}}_{34},
$$

the solution of which can easily be calculated:

$$x_3 := \bar{\bar{a}}_{34}/\bar{\bar{a}}_{33}.$$

Inserting x_3 in the equation $\bar{a}_{22}x_2 + \bar{a}_{23}x_3 = \bar{a}_{24}$ results in x_2, and the last remaining component of the solution, x_1, can be obtained by substituting x_3 and x_2 in the original first equation.

Obviously, the *input data*

$$
\begin{array}{cccc}
a_{11}, & a_{12}, & a_{13}, & a_{14} \\
a_{21}, & a_{22}, & a_{23}, & a_{24} \\
a_{31}, & a_{32}, & a_{33}, & a_{34}
\end{array}
$$

are processed by the following *arithmetic operations*:

> **do** $i = 2, 3$
> $m_i := a_{i1}/a_{11}$
>   **do** $j = 2, 3, 4$
> $\bar{a}_{ij} := a_{ij} - m_i \cdot a_{1j}$ (5.6)
>   **end do**
> **end do**
> $\overline{m}_3 := \bar{a}_{32}/\bar{a}_{22}$
> **do** $j = 3, 4$
> $\bar{\bar{a}}_{3j} := \bar{a}_{3j} - \overline{m}_3 \cdot \bar{a}_{2j}$
> **end do**
> $x_3 := \bar{\bar{a}}_{34}/\bar{\bar{a}}_{33}$
> $x_2 := (\bar{a}_{24} - \bar{a}_{23} \cdot x_3)/\bar{a}_{22}$
> $x_1 := (a_{14} - a_{12} \cdot x_2 - a_{13} \cdot x_3)/a_{11}.$

Altogether, 28 arithmetic operations on 12 numerical data elements $a_{11}, \ldots, a_{34}$ and 11 intermediate results lead to 3 final results x_1, x_2, x_3.

As in this simple elimination algorithm, every arithmetic algorithm represents a directive as to how to derive results from given numerical data applying a sequence of arithmetic operations on those data and intermediate results. Such an arithmetic algorithm can always be represented in an imperative programming language such as Fortran or C.

Obviously any such algorithm describes a mathematical mapping which assigns an m-tuple of real numbers (results) to a k-tuple of real numbers (data). In the above example $k = 12$ and $m = 3$. The mapping is not necessarily defined for any k-tuple of data (in the above example it is assumed that $a_{11} \neq 0$) the same must be true for the intermediate results $\bar{a}_{22}$ and $\bar{\bar{a}}_{33}$.

Equivalent Algorithms

A mathematical representation of a mapping from the data space into the space
of possible results can be expressed independently of a concrete algorithm, in this
case by a system of linear equations

$$Ax = b, \qquad A \in \mathbb{R}^{n \times n}, \quad x, b \in \mathbb{R}^n. \qquad (5.7)$$

The mathematical problem is to find the vector x which satisfies the equation (5.7)
for a given matrix A and vector b (right-hand side of the system). There exist
various algorithms designed for the solution of this fundamental mathematical
problem, which may use different sequences of instructions and intermediate results but which lead to the same results when applied to the same data. For
instance, in the above example, x_3 could have been the first variable to be eliminated.

Definition 5.7.1 (Equivalent Arithmetic Algorithms) *Different arith-*
metic algorithms which correspond to the same mathematical mapping of given
data k-tuples onto result m-tuples are said to be equivalent.

Equivalent arithmetic algorithms return identical results when identical input
data is used. Note that exact operations (operations free of rounding errors)
on real numbers are assumed here. The influence of floating-point numbers and
floating-point arithmetic is discussed later in the book.

For the sake of simplicity, it is assumed that for two equivalent algorithms
their sets of *irregular data*, i.e. those input data for which the algorithms cannot be executed, may possibly differ. For example, if x_3 is eliminated from the
system (5.5) first, then $a_{33} \neq 0$ must hold whereas it does not matter if $a_{11} = 0$.

In the above example the set of irregular data of the linear mathematical problem (5.5) is considerably smaller than that of the standard elimination algorithm
(5.6) or of any analogous arithmetic algorithm: For the mathematical mapping
$A, b \mapsto x$, which is defined by a system of linear equations, it is sufficient that the
determinant $\det(A)$ does not vanish. A well-defined result exists for any data
set with a regular matrix A. An arithmetic algorithm whose set of irregular data
does not exceed this minimal set usually employs case distinctions by various
kinds of branching, e. g., based on a *pivot strategy* (see Chapter 13).

The use of branch conditions which depend on input data or on intermediate
results does not change the basic structure of an arithmetic algorithm: In principle, for each regular data k-tuple of real numbers, a well-defined sequence of
arithmetic operations applied to data and intermediate results is performed.

Similarly, the fact that parts of an arithmetic algorithm may be independent
of one another and that they can thus be executed *in parallel* as well, does not
change the principle of the algorithm. If it does not make any difference to the
result whether or not certain steps are executed sequentially or in parallel, then
(for the sake of simplicity) it may be assumed that operations are executed in
sequential order. Compare the first nested loop in algorithm (5.6).

Obviously, the inherent parallelizability of such an algorithm is important if the algorithm is to be executed on a parallel computer, but it is not important for the analysis of rounding errors, which is discussed later in this chapter.

5.7.2 Implementation of Arithmetic Algorithms

The implementation of an arithmetic algorithm as a program which is executed on a computer with floating-point arithmetic is usually *not* equivalent to the *exact algorithm* based on real numbers and arithmetic operations without rounding errors, which *cannot* be implemented on a computer.

Even for data in the form of machine numbers the implementation of an arithmetic algorithm is not equivalent to the exact algorithm. This is due to the fact that $\mathbb{F}$ is not complete with respect to exact arithmetic operations. Only in improbable special cases will the final results and all intermediate results belong to the underlying set $\mathbb{F}$ of floating-point numbers.

By the translation of an algorithm written in a high-level programming language into object code, arithmetic operations have to be replaced by pseudo-operations defined on the set $\mathbb{F}$ of machine numbers (Kulisch, Miranker [262]). Thus, the object code does not represent the same algorithm as it is expressed in the source code, and hence it is a different mathematical mapping from the input data into the space of results.

In numerical data processing it is fundamentally important to understand in what respects an implementation differs from the *exact* mapping, i.e., the mapping which complies with the original goals. In particular, it is necessary to understand which characteristics of an algorithm influence whether the results of the implementation (also referred to as *machine algorithm*) differ significantly or just slightly from those of the exact algorithm.

In fact, the implementations of equivalent arithmetic algorithms generally represent different machine algorithms which return different results. It may happen that results returned by implementations of equivalent algorithms differ significantly from the exact results. In such cases it does make a difference which algorithm is implemented for the solution of a given problem.

Example (Quadratic Equation)   The solutions y_1 and y_2 to the quadratic equation

$$y^2 + a_1 y + a_0 = 0$$

are calculated. With the formula

$$y_1 = \frac{\sqrt{a_1^2 - 4a_0} - a_1}{2}$$

y_1 is obtained first, and

$$y_2 = -a_1 - y_1$$

or

$$y_2 = a_0/y_1$$

determines y_2. The two arithmetic algorithms

Algorithm QE-1			Algorithm QE-2		
z_1	:=	$a_1 \cdot a_1$	z_1	:=	$a_1 \cdot a_1$
z_2	:=	$4 \cdot a_0$	z_2	:=	$4 \cdot a_0$
z_3	:=	$z_1 - z_2$	z_3	:=	$z_1 - z_2$
z_4	:=	$\sqrt{z_3}$	z_4	:=	$\sqrt{z_3}$
z_5	:=	$z_4 - a_1$	z_5	:=	$z_4 - a_1$
y_1	:=	$z_5/2$	y_1	:=	$z_5/2$
y_2	:=	$-a_1 - y_1$	y_2	:=	a_0/y_1.

differ only in the last step: Because $y_1 + y_2 = -a_1$ and $y_1 y_2 = a_0$ (according to Vietà's rule), the two algorithms for calculating y_2 are equivalent.

Now the implementation of Algorithm QE-1 and Algorithm QE-2 are to be analyzed in floating-point arithmetic based on $\mathbb{F}(10, 4, -9, 9, true)$ and round to the nearest. To start with, the data $a_1 = -.5000 \cdot 10^1$ and $a_0 = -.1000 \cdot 10^0$ belong to $\mathbb{F}$. The exact zeros of the quadratic equation are

$$y_1 = 5.01992\ldots, \qquad y_2 = -0.0199206\ldots.$$

It can easily be shown that the intermediate results $\tilde{z}_1$, $\tilde{z}_2$, and $\tilde{z}_3$ of the machine algorithm do not differ from the exact intermediate results since they belong to $\mathbb{F}$. For

$$z_4 = \sqrt{25.4} = 5.03984\ldots$$

the implementation yields

$$\tilde{z}_4 = .5040 \cdot 10^1$$

and without further rounding errors

$$\tilde{z}_5 = .1004 \cdot 10^2 \qquad \text{and} \qquad \tilde{y}_1 = .5020 \cdot 10^1.$$

The two algorithms return the following floating-point results:

$$\begin{aligned} \text{QE-1:} \quad & \tilde{y}_2 &=& \quad -.2000 \cdot 10^{-1}, \\ \text{QE-2:} \quad & \tilde{y}_2 &=& \quad -.1992 \cdot 10^{-1}. \end{aligned}$$

Although $\tilde{y}_1 = \square y_1$ and has therefore optimal accuracy, only Algorithm QE-2 returns $\tilde{y}_2 = \square y_2$. For Algorithm QE-1, however, $\tilde{y}_2$ and y_2 differ significantly (by about 0.4 %). It is shown later that this is not mere coincidence.

5.7.3 Error Propagation

In general it has to be assumed that for each step of an arithmetic algorithm the implemented operation $\boxdot$ of the machine algorithm (in object code) differs slightly from the exact operation $\circ$. Hence, an arithmetic algorithm with total N individual operations gives rise to N separate rounding errors. The relative magnitude of the rounding error occurring in a single floating-point operation (with operands in the range of normalized machine numbers) is limited by *eps*:

$$x \boxdot y = (x \circ y)(1 + \rho) \qquad \text{where} \qquad |\rho| \le eps.$$

However, there is another effect to consider: As a result of rounding errors, all intermediate results of a machine algorithm differ from the intermediate results of the corresponding exact algorithm. As intermediate results however, they are operands of further steps in the algorithm—that is why they are calculated at all—and all following operations have perturbed operands. Even if no *new* errors arise from these operations, the errors in the operands usually affect the result. One single rounding error can therefore perturb a series of intermediate results and can substantially perturb the final result. This effect is called *error propagation*.

Numerical Stability of Arithmetic Algorithms

Error analysis of an arithmetic algorithm is carried out to determine to what extent the error of an arbitrary intermediate result affects the final result. Implementations of algorithms which are *analytically* equivalent, i. e., which return identical results for identical data, may derive their results in different ways via different intermediate results. The extent of error propagation thus differs with different implementations meaning that they return results of different accuracy with identical data.

An algorithm is said to be *numerically stable* if rounding errors do not seriously influence the accuracy of the results. If, on the other hand, there is extreme error propagation, the algorithm is said to be *numerically unstable*. Of course a stable algorithm is preferred to an unstable one for the implementation.

Error Propagation Assessment

In performing an error propagation assessment it has to be taken into account that every arithmetic algorithm, as a mathematical mapping, has a characteristic sensitivity of its results to data perturbations: Any changes in the input data cause characteristic changes in the results. This sensitivity is quantified by the *condition number* and is a fundamental property of the *mathematical* relationship (see Chapter 2).

An arithmetic algorithm which represents an ill-conditioned relationship between input data and results (i. e., reacts very sensitively to data perturbations) is also likely to react sensitively to changes in the intermediate results. Thus, the sensitivity to changes in the intermediate results must be assessed in relation to the condition number of the mathematical relationship represented by the exact algorithm.

Example (Quadratic Equation) It can easily be verified that the relative condition number of the mapping which assigns the result (y_1, y_2) to the input data (a_1, a_0) is around 1. Thus the mathematical relationship between data and results is well-conditioned. The relative perturbation of the intermediate result z_4 of approximately $3 \cdot 10^{-5}$, caused by computing the square root in a four digit decimal arithmetic, is thus supposed to influence the final result proportionally. However, the perturbation $4 \cdot 10^{-3}$ of the final result, as occurs in algorithm QE-1, is larger by two orders of magnitude. Hence, the algorithm QE-1 should be considered numerically unstable for the particular data.

5.7.4 Error Propagation Analysis

If only one rounding error occurs in an algorithm (as is the case for algorithm QE-1 in the quadratic equation example), the sensitivity of the final result concerning this single perturbation can be understood as follows:

Suppose the error arises in the computation of the intermediate result z_i, i. e., $\tilde{z}_i$ is obtained instead of z_i. Then the part of the algorithm remaining after computing z_i is looked upon as an autonomous algorithm termed the "algorithm remaining after z_i". It has the same input data as the original algorithm—$a_1, a_2, \ldots, a_m$—along with the intermediate results $z_1, z_2, \ldots, z_i$ which are explicitly used in the remainder algorithm. The quantity z_i necessarily belongs to the data of the remainder algorithm, because it would not have been calculated otherwise. The *final result* of the remainder algorithm is the same as that of the original algorithm.

Condition with Regard to Intermediate Results

The conditioning of remainder algorithms can be defined in the same way as for any other algorithm: the sensitivity of the final results to changes in input data and intermediate results—particularly the perturbation of an intermediate result z_i—are considered. This sensitivity can be quantified using a *condition number* (see Chapter 2). This helps one to estimate the extent of the changes in the final result due to the transition from z_i to $\tilde{z}_i$.

Example (Quadratic Equation)   The algorithm remaining after z_4 is as follows:

$$
\begin{aligned}
\text{for QE-1:} && z_5 &:= z_4 - a_1 & \text{for QE-2:} && z_5 &:= z_4 - a_1 \\
&& y_1 &:= z_5/2 & && y_1 &:= z_5/2 \\
&& y_2 &:= -a_1 - y_1 & && y_2 &:= a_0/y_1
\end{aligned}
$$

The remainder algorithm of QE-1 has the input data a_1 and z_4, and that of QE-2 has the data $a_0, a_1,$ and z_4 .

The mathematical mapping $(a_1, z_4) \to y_2$ represented by the remainder algorithm of QE-1 is given by

$$
y_2 = -a_1 - \frac{z_4 - a_1}{2} = -\frac{1}{2}(a_1 + z_4).
$$

The relative (first order) condition number (see Chapter 2)

$$
\frac{z_4}{y_2} \frac{\partial y_2}{\partial z_4} = \frac{z_4}{-\frac{1}{2}(a_1 + z_4)} \left(-\frac{1}{2} \right) = \frac{z_4}{a_1 + z_4}
$$

characterizes the influence of changes in z_4 on the relative accuracy of the final result y_2. For $a_1 = -5$ and $z_4 \approx 5.04$ the condition number

$$
\frac{z_4}{a_1 + z_4} \approx \frac{5.04}{0.04} \approx 125
$$

is obtained, which explains why the relative error of the final result is larger by more than two orders of magnitude, as was noticed earlier.

In principle such a sensitivity analysis can be applied to any algorithm to determine condition numbers with respect to all intermediate results[7]. However,

---

[7]It is a minor problem that such an analysis assumes continuous derivatives; but for complex algorithms this kind of detailed investigation would involve far too much work.

any condition number obtained in this way takes into account only the perturbation of one intermediate result and assumes that all other parts of the algorithm are unperturbed. This assumption is unrealistic. Usually though the influence of a perturbed intermediate result on the final result is much stronger than the indirect effects of slight changes in the remainder algorithm such as the influence of rounding errors.

The analysis of error propagation aims primarily at a *qualitative* assessment as to whether or not the condition numbers referring to the intermediate results significantly exceed the condition number with respect to the input data of the overall algorithm. In practice, however, a prohibitive amount of effort is required to calculate even approximations of these condition numbers.

What is more important is to recognize situations which endanger the stability of an arithmetic algorithm. It will turn out later in this chapter that a special arithmetic constellation is responsible for most stability problems. Provided the mathematical relationship between the input data and the results is well-conditioned, this situation can often be avoided through a skillful design of the arithmetic algorithm. For that purpose the conditioning of individual arithmetic operations will be discussed in the next section.

Error Propagation in Arithmetic Operations—Cancellation

In this chapter the sensitivity of the result of one single *exact* arithmetic operation to changes in the operands will be discussed. In general the relative magnitudes of changes, perturbations and errors are regarded within the context of floating-point arithmetic: "How many digits of the mantissa are correct?" Accordingly the analysis of error propagation is usually confined to the examination of the *relative* condition and the determination of relative condition numbers (see Chapter 2). The absolute condition of particular operations does not usually play an important role in error propagation.

In a relatively ill-conditioned operation, changes in digits of low significance in the operands will affect digits of high significance in the result. This undesirable circumstance is usually due to the *cancellation of leading digits* in the course of addition (subtraction) of two nearly equal numbers with opposite (equal) sign. In such situations the leading digits, which are identical in the mantissas of both operands, cancel one another out. Thus the perturbations of digits in the back of the operand mantissas become the perturbations of the leading digits of the resulting mantissa.

Example (Catastrophic Cancellation)   The difference of two nearly equal numbers a and b is calculated:
$$a = 7.64352, \quad b = 7.64286, \quad a - b = 0.00066.$$

If a changes by 0.00007 (this is a relative modification of 10^{-5}) to $\tilde{a} = 7.64359$, then the difference changes by 0.00007, too, i.e., to $\tilde{a} - b = 0.00073$. However, the *relative* modification of the difference is significantly larger and amounts to
$$\frac{(\tilde{a} - b) - (a - b)}{a - b} = \frac{0.00073 - 0.00066}{0.00066} \approx 10^{-1}.$$

Using the floating-point number system $\mathbb{F}(10, 6, -9, 9, true)$, the following numerical results are obtained:

$$a - b = .764352 \cdot 10^1 - .764286 \cdot 10^1 = .660000 \cdot 10^{-3}$$
$$\tilde{a} - b = .764359 \cdot 10^1 - .764286 \cdot 10^1 = .730000 \cdot 10^{-3}.$$

Hence, modification of the *last* digit of the mantissa a causes a modified *first* digit of the mantissa of $a - b$!

It is interesting that there is *no* rounding error generated when cancellation occurs. The large relative error of the result is only due to data errors *prior* to the execution of the vicious operation.

Cancellation is by far the most frequent cause of instability in arithmetic algorithms. In general the local enlargement of the relative errors (perturbations) is irreversible. If an intermediate result has only one significant digit of accuracy (due to catastrophic cancellation), then the accuracy of the final result will be limited to one correct digit.

Condition of Addition and Subtraction Operations

From the formulas

$$C_{(a+b)\leftarrow a} = \left|\frac{a}{a+b}\right| \quad \text{and} \quad C_{(a+b)\leftarrow b} = \left|\frac{b}{a+b}\right|,$$

the condition numbers of addition, and the similar formulas

$$C_{(a-b)\leftarrow a} = \left|\frac{a}{a-b}\right| \quad \text{and} \quad C_{(a-b)\leftarrow b} = \left|\frac{b}{a-b}\right|,$$

the condition numbers of subtraction, the following general condition numbers for the summation of operands with opposite signs and also for the difference of operands with equal signs can be derived:

$$C_{y\leftarrow a} = \left|\frac{a}{|a|-|b|}\right| \quad \text{and} \quad C_{y\leftarrow b} = \left|\frac{b}{|a|-|b|}\right|.$$

For convenience, the absolute values of the operands a and b were used as denominators, such that the same formula can be used for both addition *and* subtraction.

If the absolute value of the difference $a - b$ is significantly smaller than both the operands a and b, then the condition numbers are significantly larger than 1. In the above example $C \approx 10^4$.

For the addition of operands with equal signs and the subtraction of operands with opposite signs, the condition numbers

$$C_{y\leftarrow a} = \left|\frac{a}{|a|+|b|}\right| < 1 \quad \text{and} \quad C_{y\leftarrow b} = \left|\frac{b}{|a|+|b|}\right| < 1$$

are obtained. If $|b| \ll |a|$ then $C_{y\leftarrow b} \ll 1$, and any changes in b have practically no effect on the sum $a + b$.

Also, for the addition of operands with *opposite* signs the case $|b| \ll |a|$ is harmless:

$$C_{y\leftarrow a} = \left|\frac{a}{|a|-|b|}\right| \approx 1 \quad \text{and} \quad C_{y\leftarrow b} = \left|\frac{b}{|a|-|b|}\right| \ll 1.$$

Condition of Multiplication and Division Operations

The multiplication operation $y = ab$ is a well-behaved operation, as is indicated by the condition numbers

$$C_{y \leftarrow a} = \left| \frac{a}{ab} b \right| = 1 \qquad \text{and} \qquad C_{y \leftarrow b} = 1,$$

which do not depend on the magnitude and the signs of the operands.

For the division operation $y = a/b$ it also holds that

$$C_{y \leftarrow a} = \left| \frac{a}{a/b} \frac{1}{b} \right| = 1 \qquad \text{and} \qquad C_{y \leftarrow b} = \left| \frac{b}{a/b} \left(-\frac{a}{b^2} \right) \right| = 1.$$

Hence—in contrast to the addition/subtraction operation—the occurrence of a small multiplication or division result is not critical with respect to error propagation; there is no danger of error enlargement.

Example (Quadratic Equation) In Section 5.7.2, Algorithm QE-2 of the quadratic equation example calculated the second solution y_2 *without* instability, i.e., through evaluation of a well-conditioned operation, namely a division.

Condition of Standard Functions

Evaluation of standard functions yields large relative error propagation (error enlargement) whenever a *small* result is returned for an argument which is *not* close to zero. This can happen in cases such as the following:

1. For logarithmic functions $y = \log a = c \ln a$ where $a \approx 1$:

$$C_{y \leftarrow a} = \left| \frac{\partial \log a}{\partial a} \frac{a}{\log a} \right| = \left| c \frac{1}{a} \frac{a}{c \ln a} \right| = \left| \frac{1}{\ln a} \right| \approx \frac{1}{|a - 1|}.$$

2. For trigonometric functions, e.g., for $y = \sin a$ where $a \neq 0$ is close to a zero: $a \approx k\pi$, $k = \ldots, -3, -2, -1, 1, 2, 3, \ldots$

$$C_{y \leftarrow a} = \left| \frac{a}{\sin a} \cos a \right| \approx \frac{|a|}{|a - k\pi|}.$$

On the other hand, the function $\sin a$ with $|a| \approx 0$ is well-behaved with respect to error propagation:

$$C_{y \leftarrow a} = \left| \frac{a}{\sin a} \cos a \right| \approx \left| \frac{a}{a} 1 \right| = 1.$$

In a sense the error enlargement of standard functions can be looked upon as a cancellation as well. Hence, cancellation is basically the only source of instability for arithmetic algorithms. When designing stable numerical methods it is necessary to make sure that no cancellation can occur in the corresponding arithmetic algorithms.

However, it may be difficult in practice to prevent cancellation, because its occurrence depends on the *values* of the variables involved: only the addition/subtraction of nearly *equal* numbers with opposite/equal signs is critical. It may therefore be necessary to monitor for the occurrence of such an eventuality at runtime (by testing and branching). By doing so instabilities caused by adverse data constellations can be avoided.

Example (Trigonometric Expression)   The evaluation of

$$y = \frac{1 - \cos x}{x^2}$$

is well-conditioned even for values $x \approx 0$. Taylor expansion shows that

$$y = \frac{1}{2} - \frac{x^2}{24} + \cdots$$

for small $|x|$, such that small relative changes in x practically leave the value of y unchanged.

However, for the *numerical* calculation with

$$
\begin{array}{llll}
\textbf{Algorithm TE-1} & z_1 & := & \cos x \\
& z_2 & := & 1 - z_1 \\
& z_3 & := & x \cdot x \\
& y & := & z_2/z_3
\end{array}
$$

a critical cancellation occurs during the calculation of z_2 because of $z_1 = \cos x \approx 1$. The rounding error, which comes from the floating-point calculation of the intermediate result z_1, perturbs the final result to a large extent.

If, on the other hand, the trigonometric identity $1 - \cos x = 2\sin^2(x/2)$ is applied in

$$
\begin{array}{llll}
\textbf{Algorithm TE-2} & z_1 & := & x/2 \\
& z_2 & := & \sin z_1 \\
& z_3 & := & z_2 \cdot z_2 \\
& z_4 & := & 2 \cdot z_3 \\
& z_5 & := & x \cdot x \\
& y & := & z_4/z_5,
\end{array}
$$

then the final result y depends on the intermediate result z_2 in a harmless way.

For $x = 0.1$ the exact result is $y \approx 0.499583$. Using $\mathbb{F}(10, 5, -9, 9, \textit{true})$ with round to nearest for the execution of TE-1, the final result $\tilde{y} = 0.50000$ is obtained via the intermediate quantity $\tilde{z}_1 = 0.99500$; the relative error is nearly $0.1\,\%$, which is quite substantial. On the other hand, with Algorithm TE-2

$$
\begin{array}{rcl}
\tilde{z}_2 & = & .49979 \cdot 10^{-1} \\
\tilde{z}_3 & = & .24979 \cdot 10^{-2} \\
\tilde{z}_4 & = & .49958 \cdot 10^{-2} \\
\tilde{y} & = & .49958 \cdot 10^0,
\end{array}
$$

is obtained, which is the exact result rounded to the nearest number in $\mathbb{F}$.

However, for x in the vicinity of $\pi/2$, Algorithm TE-1 can be used without any danger of cancellation since $\cos x \ll 1$ holds.

Overall Analysis of an Arithmetic Algorithm

For a realistic arithmetic algorithm (i.e., one that consists of *a large number* of operations) executed in a floating-point arithmetic, the situation is as follows:

1. The operations use perturbed intermediate results (from prior calculations) as their operands.

2. The results of the operations do not belong to $\mathbb{F}$; hence, they have to be mapped (by rounding into $\mathbb{F}$), which adds new errors to the existing ones.

While in Chapter 4 the origin of rounding errors was discussed, Section 5.7 so far has dealt mainly with the consequences to the final result of the errors of single operations.

The detailed analysis of the combination of all individual effects is not practical for larger algorithms, and it is usually not necessary, either: In fact, if the rounding error is sufficiently small (this is the case with a floating-point arithmetic with sufficiently long mantissa and, hence, a sufficiently small *eps*, the universal bound for relative rounding errors), then the individual errors are generally propagated *independently of one another*. The overall error of the result is then basically equal to the sum of the effects of all individual rounding errors.

Example (Trigonometric Expression) If Algorithm TE-2 (see page 206) is executed on a computer with a binary floating-point arithmetic, a rounding error will occur for z_2, z_3, z_5, and y each. According to Section 4.7 the rounded intermediate results $\tilde{z}_i$ are related to the exact results z_i by

$$\tilde{z}_i = z_i(1 + \rho_i) \qquad \text{where} \qquad |\rho_i| \le eps.$$

Hence, because $\tilde{z}_1 = z_1$,

$$
\begin{aligned}
\tilde{z}_2 &= \sin(z_1)(1 + \rho_2) \\
\tilde{z}_3 &= (\tilde{z}_2 \cdot \tilde{z}_2)(1 + \rho_3) \\
\tilde{z}_4 &= 2 \cdot \tilde{z}_3 \\
\tilde{z}_5 &= (x \cdot x)(1 + \rho_5) \\
\tilde{y} &= (\tilde{z}_4 / \tilde{z}_5)(1 + \rho_6).
\end{aligned}
$$

Substituting these terms into each other, and using the formula

$$\frac{1}{1 + \rho_5} = \sum_{k=0}^{\infty}(-\rho_5)^k$$

for geometric series, the result

$$
\begin{aligned}
\tilde{y} &= \frac{2\left(\sin(x/2)(1 + \rho_2)\right)^2 (1 + \rho_3)}{x^2(1 + \rho_5)}(1 + \rho_6) \\
&= \frac{2\sin^2(x/2)}{x^2}\frac{(1 + \rho_2)^2(1 + \rho_3)(1 + \rho_6)}{1 + \rho_5} \\
&= y(1 + 2\rho_2 + \rho_3 - \rho_5 + \rho_6 + \cdots) \\
&\approx y(1 + \rho).
\end{aligned}
\tag{5.8}
$$

Neglected terms—symbolized by "$\ldots$" in formula (5.8)—are products or powers of ρ_i, therefore of order eps^2 or higher.

The overall relative error ρ (except for neglectable terms) of the result $\tilde{y}$ is given by the sum of the individual relative errors,

$$\rho = 2\rho_2 + \rho_3 - \rho_5 + \rho_6.$$

The factor 2 in ρ_2 corresponds exactly to the condition number $C_{y \leftarrow z_2}$, which describes the dependence of the final result on the intermediate result z_2:

$$y = \frac{2z_2^2}{x^2}, \qquad \frac{\partial y}{\partial z_2} = \frac{4z_2}{x^2}, \qquad C_{y \leftarrow z_2} = \frac{z_2}{y}\frac{\partial y}{\partial z_2} = 2.$$

In the same way it can be shown that neglecting any higher-order terms (i.e., terms of orders eps^2, eps^3 etc.) yields the estimate

$$|\rho| \leq \sum_i C_{y \leftarrow z_i}|\rho_i|.$$

This formula relates the relative error ρ of the final result of an arithmetic algorithm with the errors of the intermediate results z_i provided that all remainder algorithms are twice differentiable. This means that the overall effect of rounding errors does not exceed the sum of the effects of individual rounding errors. Due to the counterbalancing of positive and negative errors, the overall error can actually be much smaller than this sum (compare, for example, the many instances in the figures on page 217 where the overall error vanishes).

The neglect of higher-order terms is generally justified as $|\rho_i\rho_j| \ll |\rho_i|$. Thus, the assumption of independent error propagation is usually a valid description of the actual (more complex) error behavior. If, however, the number of operations is very large, then even terms of higher order may play a role. For example, in

$$(1 + eps)^n = 1 + n\,eps + \frac{n(n-1)}{2}eps^2 + \cdots$$

the quadratic term must not be neglected if $n \approx 1/eps$. With single precision arithmetic this situation occurs even for problems which are not extraordinarily large. In such cases the use of double precision arithmetic can yield much higher gains in accuracy than could be expected from doubling the mantissa length.

For a *stable* algorithm the factors $C_{y \leftarrow z_i}$ do not exceed the condition number C of the mathematical problem. If the algorithm comprises N operations this gives the rough estimate

$$|\rho| \leq NCeps. \tag{5.9}$$

In practice this bound overestimates the actual error significantly. It would only hold if for every operation the error was eps with the same sign.

Stochastic considerations furnish evidence that for randomly distributed individual rounding errors it can be expected that the overall relative rounding error will be proportional to $c\sqrt{N}$ with a factor $c < 1$.

In the case of a *well-conditioned* mathematical relationship the most important precaution for guaranteeing accurate results is to avoid cancellation in the arithmetic algorithm. However, in practice *ill-conditioned* problems with very sensitive relationships between data and exact results often occur. Examples of such situations are given by all problems where results close to zero are derived from data of "normal" magnitude:

- the evaluation of a polynomial P at a point x where $P(x) \approx 0$ and $|x| \gg 0$,

- the calculation of the scalar product of two nearly orthogonal vectors,

- the computation of the residual obtained by substituting an approximate solution into a system of linear equations.

Example (Residual Computation) The accuracy of an approximate solution $\tilde{x} = (\tilde{x}_1, \ldots, \tilde{x}_n)^\top$ of a system of linear equations $Ax = b$

$$
\begin{array}{ccccc}
a_{11}x_1 + a_{12}x_2 + \cdots + a_{1n}x_n & = & b_1 \\
a_{21}x_1 + a_{22}x_2 + \cdots + a_{2n}x_n & = & b_2 \\
\vdots \quad\quad \vdots \quad\quad\quad \vdots \quad\quad \vdots \\
a_{n1}x_1 + a_{n2}x_2 + \cdots + a_{nn}x_n & = & b_n,
\end{array}
\tag{5.10}
$$

is to be investigated. The residuals

$$
r_i := \sum_{j=1}^{n} a_{ij}\tilde{x}_j - b_i, \qquad i = 1, 2, \ldots, n
$$

are computed. They are small only if the sums are close to the b_i; ideally they would be equal. Hence, in the floating-point subtraction operation most leading digits of the mantissas are canceled; but the remaining digits are precisely those which are perturbed due to rounding errors in the evaluation of $\sum a_{ij}\tilde{x}_j$. Hence, in the worst case, the floating-point computation of the residuals r_i produces not even one significant digit of accuracy—even the sign may be wrong.

It seems that such ill-conditioned problems cannot be tackled within a floating-point arithmetic without special precautions. Either higher precision (i.e., a longer mantissa) is used or tailor-made algorithms which can adapt themselves dynamically to the appropriate precision level are employed.

For the residual computation of the last example some programs calculate the inner products

$$
a_{i1}\tilde{x}_1 + a_{i2}\tilde{x}_2 + \cdots + a_{in}\tilde{x}_n, \qquad i = 1, 2, \ldots, n
$$

in double precision arithmetic. Since the product of two floating-point numbers in $\mathbb{F}_N(b, p, e_{\min}, e_{\max})$ always belongs to the set $\mathbb{F}_N(b, 2p, 2e_{\min} - 1, 2e_{\max})$, there is *no* rounding error for the multiplication operation, provided that the relation $p_2 \geq 2p_1$ holds between single precision numbers (with mantissa length p_1 and relative accuracy eps_1) and double precision numbers (with p_2 and eps_2). For the summation of those inner products, the generated relative errors are bounded by $eps_2 \leq (eps_1)^2$ of the double precision arithmetic; hence, the first p_1 mantissa digits of the sum $\sum_j a_{ij}\tilde{y}_j$ are generally correct. Digits which move up into the leading position after cancellation in the final subtraction operation are therefore correct.

There exist implementations of the scalar product in one elementary operation which satisfy:

$$
\text{calculated value of } \sum_j a_{ij}y_j = \square\Big(\sum_j a_{ij}y_j\Big).
$$

Such an implementation of the scalar product is called an *exact scalar product*.

5.8 Case Study: Floating-Point Summation

The summation of n floating-point numbers is a basic operation in many numerical algorithms: its application ranges from inner products, norms, mean values etc. to various nonlinear functions. The calculation of

$$s_n := x_1 + x_2 + \cdots + x_n$$

appears to be a trivial task. Any programming novice is familiar with the simple scheme of *recursive summation*:

$$
\begin{aligned}
&s := x_1 \\
&\textbf{do}\ \ i = 2, 3, \ldots, n \\
&\qquad s := s + x_i \\
&\textbf{end do}
\end{aligned}
\tag{5.11}
$$

An even simpler approach is to leave the summation to the system software itself as can be done through the use of the Fortran 90 function SUM:

```
REAL                  ::  s
REAL, DIMENSION (n)   ::  x
...
s  =  SUM(x)          !  summation  s = x(1) + x(2) + ... + x(n)
```

Due to peculiarities of floating-point arithmetic, such straightforward solutions to the summation problem may return unsatisfactory results. In particular the non-associativity of floating-point addition operations constitutes a fundamental difference between the floating-point arithmetic in $\mathbb{F}$ and the arithmetic in the field $\mathbb{R}$ of real numbers.

The Rounding Error of Recursive Summation

As already mentioned in Chapter 4 (in the example on page 142), even in the addition of just three floating-point numbers, the order of operations

$$(x_1 + x_2) + x_3 \qquad \text{or} \qquad x_1 + (x_2 + x_3),$$

leads to different error bounds.

In order to derive a general error estimate, it is assumed that $x_1, \ldots, x_n \in \mathbb{F}_N$, and all intermediate results belong to intervals of real numbers which are covered by $\mathbb{F}_N$:

$$x_1 + x_2,\ x_1 + x_2 + x_3,\ \ldots, x_1 + \cdots + x_n \in \mathbb{R}_N.$$

Under these assumptions the following relationships hold:

$$
\begin{aligned}
\hat{s}_1 &= s_1 &&= x_1 \\
\hat{s}_2 &= s_1 \boxplus x_2 &&= (x_1 + x_2)(1 + \rho_2) \\
\hat{s}_3 &= \hat{s}_2 \boxplus x_3 &&= (\hat{s}_2 + x_3)(1 + \rho_3) \\
&&&= (x_1 + x_2)(1 + \rho_2)(1 + \rho_3) + x_3(1 + \rho_3) \\
&\vdots \qquad \vdots \\
\hat{s}_n &= \hat{s}_{n-1} \boxplus x_n &&= (\hat{s}_{n-1} + x_n)(1 + \rho_n) \\
&&&= (x_1 + x_2)\prod_{i=2}^{n}(1 + \rho_i) + \sum_{j=3}^{n} x_j \prod_{i=j}^{n}(1 + \rho_i).
\end{aligned}
$$

Results perturbed by rounding errors are marked with a hat ($\hat{\ }$). The formulas can be simplified using the bounds

$$
|\rho_i| \leq eps, \quad i = 2, 3, \ldots, n
$$

and

$$
\prod_{i=j}^{n}(1 + \rho_i) = 1 + r_{n-j+1} \qquad \text{where} \quad r_m \leq \frac{m \ eps}{1 - m \ eps} =: R_m,
$$

as long as $m \ eps \ < \ 1$. For the approximate solution $\hat{s}_n$, the representation

$$
\hat{s}_n = (x_1 + x_2)(1 + r_{n-1}) + \sum_{i=3}^{n} x_i(1 + r_{n-i+1})
$$

and therefore the error estimate

$$
\begin{aligned}
|\hat{s}_n - s_n| &= \left| (x_1 + x_2) r_{n-1} + \sum_{i=3}^{n} x_i r_{n-i+1} \right| \\
&\leq (|x_1| + |x_2|) R_{n-1} + \sum_{i=3}^{n} |x_i| R_{n-i+1} \qquad (5.12)
\end{aligned}
$$

is obtained. This error bound is smallest when the summation is performed with terms in increasing order. This immediately suggests an alternative to the recursive summation (5.11): additional $O(n \log n)$ work is spent on sorting the terms, and then the summation is performed with respect to increasing absolute values. However, theoretical analyses as well as practical experiments (Higham [225]) show that this and similar variants, which set parentheses according to different strategies based on the values of the terms, require too much computational effort. In the following, two $O(n)$ summation algorithms, pairwise summation and compensated summation, which usually produce significantly better results than the recursive summation (5.11) are introduced.

5.8.1 Pairwise Summation

In this algorithm, neighboring elements of the sequence $(x_1, \ldots, x_n)$ are added. This leads to a new sequence

$$\left(s_1^{(1)}, s_2^{(1)}, \ldots, s_{\lceil n/2 \rceil}^{(1)}\right) \qquad \text{where} \quad s_i^{(1)} := x_{2i-1} + x_{2i}. \tag{5.13}$$

If the number n of terms is *odd*, then the following arrangement[8] is made:

$$s_{\lceil n/2 \rceil}^{(1)} := x_n.$$

Pairwise summation of neighboring elements is applied again on the newly defined sequence (5.13), which produces the sequence

$$\left(s_1^{(2)}, s_2^{(2)}, \ldots, s_{\lceil \lceil n/2 \rceil/2 \rceil}^{(2)}\right) \qquad \text{where} \quad s_i^{(2)} := s_{2i-1}^{(1)} + s_{2i}^{(1)}. \tag{5.14}$$

This process is repeated until the final result

$$s_1^{(\lceil \log_2 n \rceil)} = x_1 + \cdots + x_n$$

is returned. Fig. 5.4 depicts this process for the special case $n = 2^4$.

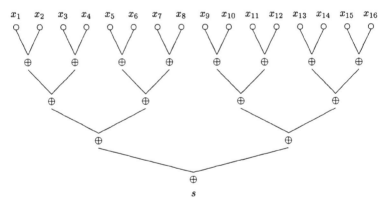

Figure 5.4: Pairwise Summation of $(x_1, x_2, \ldots, x_{16})$.

All additions in each of the $\lceil \log_2 n \rceil$ steps can be executed in parallel. This type of summation is therefore suitable for parallel computers as well.

The Rounding Error of Pairwise Summation

With similar assumptions as for recursive summation ($x_1, \ldots, x_n \in \mathbb{F}_N$ etc.) and the additional assumption $n = 2^k$, which simplifies the representation consider-

---

[8]This arrangement can be interpreted as appending an artificial term $x_{n+1} = 0$. The definition (5.14) is to be understood in that sense.

ably, the following holds:

$$
\begin{aligned}
\hat{s}_1^{(1)} &= x_1 \boxplus x_2 &&= (x_1 + x_2)(1 + \rho_1^{(1)}) \\
\hat{s}_2^{(1)} &= x_3 \boxplus x_4 &&= (x_3 + x_4)(1 + \rho_2^{(1)}) \\
&\;\;\vdots && \;\;\vdots \\
\hat{s}_{n/2}^{(1)} &= x_{n-1} \boxplus x_n &&= (x_{n-1} + x_n)(1 + \rho_{n/2}^{(1)})
\end{aligned}
$$

and furthermore

$$
\begin{aligned}
\hat{s}_1^{(2)} = \hat{s}_1^{(1)} \boxplus \hat{s}_2^{(1)} &= \left(x_1 + x_2\right)\left(1 + \rho_1^{(1)}\right)\left(1 + \rho_1^{(2)}\right) \\
&\quad + \left(x_3 + x_4\right)\left(1 + \rho_2^{(1)}\right)\left(1 + \rho_1^{(2)}\right)
\end{aligned}
$$

$$
\begin{aligned}
\;\;\vdots \qquad\qquad \;\;\vdots & \\
\hat{s}_{n/4}^{(2)} = \hat{s}_{n/2-1}^{(1)} \boxplus \hat{s}_{n/2}^{(1)} &= \left(x_{n-3} + x_{n-2}\right)\left(1 + \rho_{n/2-1}^{(1)}\right)\left(1 + \rho_{n/4}^{(2)}\right) \\
&\quad + \left(x_{n-1} + x_n\right)\left(1 + \rho_{n/2}^{(1)}\right)\left(1 + \rho_{n/4}^{(2)}\right)
\end{aligned}
$$

up to the total sum

$$
\hat{s}_1^{(k)} = \sum_{j=1}^{n} x_j \prod_{i=1}^{k} \left(1 + \rho_{\lceil j/2^i \rceil}^{(i)}\right)
$$

affected by rounding errors. With $\left|\rho_m^{(i)}\right| \leq eps$ the error estimate

$$
\left|\hat{s}_1^{(k)} - s_1^{(k)}\right| \leq R_k \sum_{i=1}^{n} |x_i| \qquad \text{where} \quad R_k := \frac{k\,eps}{1 - k\,eps} \tag{5.15}
$$

is obtained as long as $k < 1/eps$. In general this error bound is significantly smaller than the bound (5.12) for recursive summation because it is only proportional to $k = \log_2 n$ and not proportional to n as in (5.12).

5.8.2 Compensated Summation

The *compensated summation* or the *Kahan summation* uses the same strategy as recursive summation. However, for each addition operation the rounding error is estimated and compensated for with a correction term. The underlying principle of the error estimation is illustrated in Fig. 5.5, where the mantissas of the summands a and b are represented by boxes.

This principle of error estimation was introduced by W. Kahan and can be expressed by the following formula:

$$
\hat{e} = ((a \boxplus b) \boxminus a) \boxminus b = (\hat{s} \boxminus a) \boxminus b. \tag{5.16}
$$

In a binary arithmetic with rounding

$$
\hat{e} = \hat{s} - (a + b)
$$

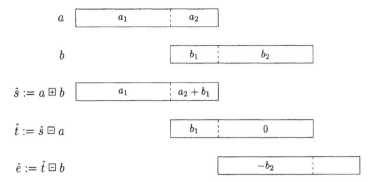

Figure 5.5: Estimation of the rounding error $\hat{s} - s = -b_2$.

holds for $a \geq b$. Thus, the rounding error is given *exactly* by (5.16) (Dekker [154]).

For compensated summation in each step the rounding error is estimated according to the Kahan principle and used for adjustment:

$$
\begin{aligned}
&s := x_1; \ e := 0 \\
&\textbf{do} \ \ i = 2, 3, \ldots, n \\
&\qquad y := x_i - e \\
&\qquad z := s + y \\
&\qquad e := (z - s) - y \\
&\qquad s := z \\
&\textbf{end do}
\end{aligned}
\tag{5.17}
$$

The Rounding Error in Compensated Summation

The following theorem holds for summation with error compensation:

Theorem 5.8.1 (Kahan Summation)   *The rounding error of the Kahan summation according to (5.17) can be estimated by*

$$
|\hat{s}_n - s_n| \leq \left(2 \, eps + O(n \, eps^2) \right) \sum_{i=1}^{n} |x_i|.
\tag{5.18}
$$

Proof: Goldberg [211].

The bound (5.18), which is, in essence, independent of the number n of summands is a significant improvement of the bounds (5.12) and (5.15) found in recursive and pairwise summation respectively, provided that n is not excessively large.

5.8.3 Comparison of Summation Methods

N. J. Higham analyzed eight different summation algorithms in a detailed study [225] and conducted practical experiments to compare them. Only the three which were introduced in the preceding sections—recursive, pairwise, and compensated summation—were of $O(n)$ complexity. One of the results of the study [225] showed that it is *not* worthwhile using algorithms of higher complexity (usually $O(n \log n)$ because of necessary sorting).

In a first experiment, 1000 independent summations of $n = 4, 8, 16, \ldots, 2048$ (each time different) random numbers uniformly distributed on $[0, 1]$ (see Chapter 17) were carried out. The results are depicted in Figures 5.6, 5.7 and 5.8. In spite of the extreme rounding errors for recursive summation being considerably lower than the theoretical error bounds, it is obvious that pairwise and compensated summation are superior to recursive summation.

The second experiment was conducted to investigate the error behavior by summing up $n = 4, 5, 6, \ldots, 2048$ random numbers uniformly distributed on $[1, 2]$ with each of the three methods. Each summation method was applied in this experiment to the first 4, then to the first 5, ... random numbers. For recursive summation this gives an overview of the trend of the rounding error during the summation of 2048 random numbers (see Fig. 5.9). Zeros of the error curve indicate that there are intermediate results (nearly) without rounding errors because of counterbalancing effects.

For pairwise summation the graphical representation cannot be interpreted in the same way, because adding a new term generally results in the grouping of all terms in the sum differently.

5.8.4 Ascending Summation

In the third experiment the advantage of ascending summation (i. e., recursive summation of terms with increasing absolute value) is demonstrated by using different summation methods for $s_n := \sum_{i=1}^{n} 1/i^2$ where $n = 4, 5, 6, \ldots, 512$. Figures 5.12, 5.13, and 5.14 depict the *absolute values* of the successively obtained relative rounding errors. In this case ascending summation is even slightly superior to pairwise summation. However, compensated summation again produced the most accurate results.

Ascending summation provides highly accurate results for situations like the previous one. This summation method is a good choice provided there is a priori information available about the magnitudes of the summands which can be utilized for ascending summation without too much additional effort. In all other cases pairwise or compensated summation should be used if small rounding errors are required.

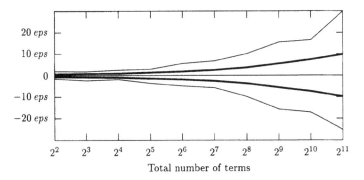

Figure 5.6: (Recursive Summation) Extrema of relative rounding errors (—) and the variance of rounding errors (—) for the recursive summation of random numbers $x_i \in [0, 1]$.

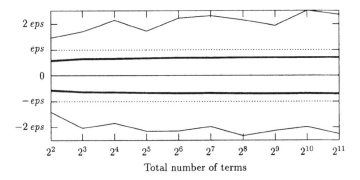

Figure 5.7: (Pairwise Summation) Extrema of relative rounding errors (—) and the variance of rounding errors (—) for the pairwise summation of random numbers $x_i \in [0, 1]$.

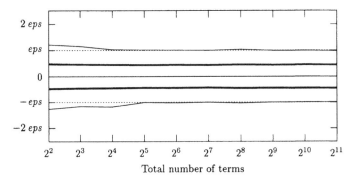

Figure 5.8: (Kahan Summation) Extrema of relative rounding errors (—) and the variance of rounding errors (—) for the Kahan summation of random numbers $x_i \in [0, 1]$.

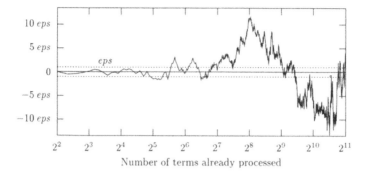

Figure 5.9: (Recursive Summation) Behavior of the relative rounding error for the recursive summation of $4, 5, 6, \ldots, 2048$ random numbers $x_i \in [1, 2]$.

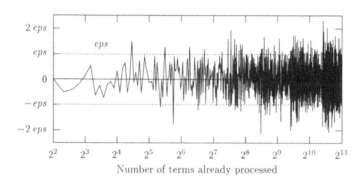

Figure 5.10: (Pairwise Summation) Behavior of the relative rounding error for the pairwise summation of $4, 5, 6, \ldots, 2048$ random numbers $x_i \in [1, 2]$.

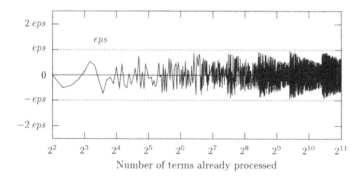

Figure 5.11: (Kahan Summation) Behavior of the relative rounding error for the compensated summation of $4, 5, 6, \ldots, 2048$ random numbers $x_i \in [1, 2]$.

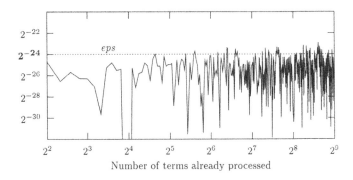

Figure 5.12: (Pairwise Summation) Absolute value of the relative rounding error for the pairwise summation of $1 + 1/4 + 1/9 + \cdots + 1/n^2$ with $n = 4, 5, 6, \ldots, 512$.

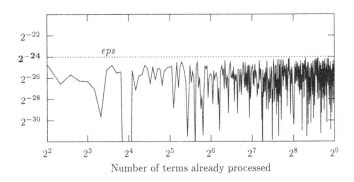

Figure 5.13: (Ascending Summation) Absolute value of the relative rounding error for the recursive summation of increasing terms $1/n^2 + \cdots + 1/9 + 1/4 + 1$ with $n = 4, 5, 6, \ldots, 512$.

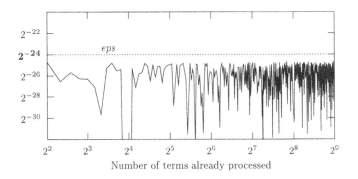

Figure 5.14: (Kahan Summation) Absolute value of the relative rounding error for the compensated summation of $1 + 1/4 + 1/9 + \cdots + 1/n^2$ with $n = 4, 5, 6, \ldots, 512$.

Chapter 6

Numerical Programs

Things are more complex than they seem to be.
Things take longer than expected.
Things cost more than expected.
If something can go wrong, it will.

MURPHY's Laws

Programming languages provide the means for a particular encoding of algorithms. A *program* is an algorithm which is formulated in a programming language and can therefore be executed by a computer.

Numerical programs are programs designed for solving numerical problems, e.g., systems of linear or nonlinear equations. They are important tools in the development of technical and scientific software.

6.1 The Quality of Numerical Programs

The attributes which determine the quality of a numerical program depend on, among other things, the particular application and the way the program is used. Because of the countless fields and modes in which programs are applied, there is no one way to measure the quality of numerical programs. However, quality assessment does, in any case, take the criteria described in the following sections into account: their individual ratings depend on the given circumstances.

6.1.1 Reliability

Reliability is a set of attributes that bear on the capability of software to maintain its performance level under stated conditions for a stated period of time. The reliability of a program is determined by the degree of certainty to which it meets the requirements specified in its documentation. Thus any reliability assessment depends on the exactness of the program specification. The more precise the specification (documentation) of the program the more meaningful the results of its reliability assessment can be.

The notion of reliability comprises the more specific features: correctness, robustness, accuracy, and consistency.

Correctness

In the field of software engineering, correctness is usually defined as follows: A program is regarded as correct if it returns *correct* results for *all* input data.

Data constellations leading to an *error case* have to be taken into account separately. However, this classical definition of correctness cannot be applied without modification to *numerical* programs since an inherent property of most numerical programs is the following: The more precisely that a numerical program (specification) is formulated, the less it meets the user's requirements and wishes.

Example (Systems of Linear Equations) If the specification of a program designed to solve systems of linear equations claims that the elimination algorithm has to be implemented in its simplest form (i.e., without pivoting or scaling), then it is exact enough to allow for even a formal proof of correctness.

   However, even if such a program is known to be a correct implementation of the elimination algorithm, it may return results which do not contain even a single digit equal to the exact solution of the problem.

   The failure of "correct" programs to produce exact solutions may be due to the influence of machine arithmetic in cases where the solution is very susceptible to data or rounding errors. Surely most users would *not* consider the program to be correct if it returned completely useless results.

This example emphasizes the need for an *application oriented specification*, which would also contain accuracy requirements. Moreover, it demonstrates that the influence of machine arithmetic must not be neglected when assessing the correctness of a numerical program.

   Apart from the machine arithmetic and the problem condition (the sensitivity of the solution to variations in the data), the discretization of the problem (i.e., the transition from a continuous problem into a discrete model which can actually be implemented on the computer) has an important influence on the results of any numerical data processing activity.

Example (Numerical Integration) All numerical integration programs are based on data sampling at a finite number of points. Either these points are fixed, i.e., independent of the integrand, or an adaptation is carried out, i.e., certain properties of the integrand are taken into account. For example, more sampling points might be used in the neighborhood of a jump discontinuity. In any case the first approximation and error estimate are based on a finite set of sampling points $\{x_1, \ldots, x_N\}$ determined by the integration algorithm (see Chapter 12). If the integration program is applied to the integrand

$$P_{2N}(x) := c \prod_{i=1}^{N} (x - x_i)^2 \qquad \text{where} \qquad c \neq 0,$$

then the polynomial P_{2N} is evaluated at the sampling points $x_1, \ldots, x_N$, i.e., only at its zeros. It follows that any algorithm will terminate independently of its accuracy requirements and return

   Integral approximation: $I_{\text{approx}} = 0$ *Error estimate:* $E_{\text{approx}} = 0$

as its result. In spite of this seemingly favorable error estimate, the actual error of the numerical result of the program can be *arbitrarily large* depending on the choice of the value c.

In this example and in similar cases—whenever inevitable discretization errors influence the results of numerical computations—the program specification must involve a restriction on as to what input data is permissible .

Example (Numerical Integration) If (depending on the integration formula) only j-times continuously differentiable functions $f \in C^j[a, b]$ are permitted as integrands and a bound M_j on the jth derivative of the integrand, i.e.,

$$|f^{(j)}(x)| \le M_j, \qquad x \in [a, b],$$

is passed as an input parameter, then it is possible to develop an integration program whose results can be guaranteed to meet an accuracy requirement

$$\left| \int_a^b f(x)\, dx - I_{\mathrm{approx}} \right| \le e_{\mathrm{abs}},$$

provided the influence of machine arithmetic is appropriately taken into account.

However, such restrictions often do not comply with the preferences and needs of the user. For example, determining bounds on the derivatives of the integrand may be harder than the original integration problem. If, on the other hand, the specification of a bound on any of the derivatives is ignored (as it is usually done in practice), then problem-related considerations concerning the correctness of the program can no longer be made. In such cases tests and heuristic considerations are the only way to assure some degree of regularity in the program's behavior.

Robustness

Robustness characterizes the degree to which a program can recognize invalid or incorrect input data, react in such a way that the user can comprehend and maintain its functionality. Maximal robustness is achieved when there is no input whatsoever that could lead to erroneous or undefined program behavior.

In the context of numerical data processing, the difficulty in applying the notion of robustness to numerical programs is determining which input data is to be considered incorrect or inadmissible.

Example (Numerical Integration) All programs for numerical integration can perform as specified in the definitions (not taking into account the influence of rounding errors) only if the integrands are polynomials which do not exceed a certain maximal degree or functions whose derivatives satisfy certain conditions. As mentioned before, the last requirement cannot usually be checked by the user. In the program documentation, this requirement is often not even mentioned and the program is praised as being "universally applicable".

This is only a documentation shortcoming since an improvement of program robustness is neither practically nor theoretically possible due to the finite nature of the integrand sampling strategy. As the program can compute only a finite number of function values, there is no algorithm which is able to check, for example, whether the derivatives are below certain bounds (which apply for *all* points of an integration region).

Example (Euclidean Vector Length) Both the subroutine BLAS/snrm2 and the program x2norm by Blue [117] for the calculation of the Euclidean length of a vector

$$\|v\|_2 = \sqrt{v_1^2 + v_2^2 + \cdots + v_n^2}$$

work correctly and without exponent overflow if the *result* is within the range of machine numbers. This property is not at all trivial. The straightforward algorithmic approach

```
sum = 0.
DO i = 1, n
    sum = sum + v(i)**2
END DO
norm_2 = SQRT(sum)
```

may result in an exponent overflow or underflow for intermediate results even if the result itself is within the range of normalized machine numbers.

A program is optimally robust if it does not simply terminate when data (vector components) produce results that do not belong to the range of machine numbers, but informs the user about this fact, for example, by using a special parameter.

An overflow usually results in a program interrupt. However, an underflow may be even worse: usually no error message is generated and the program continues with the intermediate result zero. As often, it can not be determined in advance whether such a continuation is reasonable or not, robust software should be completely free of underflows.

Accuracy

Accuracy is the program property of returning results which are equal to implicitly assumed or explicitly given reference values (the *exact* results).

Programs with floating-point number results achieve their maximal degree of accuracy when the results and the exact solutions of the problem rounded to machine numbers are equal. This ideal is usually only achieved by programs designed for particular classes of problems based on a specific machine arithmetic.

Example (ACRITH-XSC) This special arithmetic available on certain IBM computers (some IBM-compatible computers as well) allows for the development of software which returns solutions of a maximal degree of accuracy (depending on the computer). Popular arithmetics, like the IEC/IEEE-arithmetic, do *not* allow for the development of software of maximal accuracy.

Example (Elementary Functions) All programming languages suitable for numerical calculations provide subroutines for evaluating elementary functions (e. g., sin, cos, exp, log, etc.), which can be called directly by using certain keywords (e. g., SIN, COS, EXP, LOG, etc.). The algorithms implemented in such subroutines are usually hidden from the user and not mentioned in the documentation.

Numerical methods for approximating elementary functions always aim at achieving maximal accuracy. To achieve this, single precision subroutines use double precision arithmetic, and double precision subroutines use *extended precision* arithmetic. Thus, nearly maximal accuracy can be guaranteed for the most important argument ranges; however, constellations for which the accuracy of the computed function values is significantly worse are inevitable.

For example, the sine function is approximated in the neighborhood of zero by a polynomial or a rational function. Arguments of large absolute value are mapped onto the approximation interval by exploiting the periodicity of the sine function and trigonometric identities (Cody, Waite [8]). This transformation, which usually reduces the argument x by an integer multiple of $\pi/2$ such that the new argument is

$$\overline{x} \in [-\pi/4, \pi/4],$$

is responsible for inaccuracies in results for large x. This is because the leading digits are canceled when $\bar{x}$ is calculated. This is easily demonstrated on any pocket calculator: The value returned in place of the exact result

$$\sin 710_{\mathrm{rad}} = 0.0000\,60288\,70669\,15852\,65933\ldots$$

usually does not involve more than 4 or 5 correct digits in a 10 digit display.

As illustrated by the example of elementary functions, the properties of accuracy and robustness may influence one another. A subroutine which returns a result with only 4 correct digits instead of 9 or 10 without informing the user is neither accurate nor robust. The best solution is thus to have a subroutine which guarantees nearly maximal accuracy for a large scope of arguments and which generates a warning message if this condition of maximal accuracy cannot be met, eventually stopping the calculation.

Example (Systems of Linear Equations)   Once the arithmetic for the solution of a system of linear equations using Gaussian elimination has been established (by using particular hardware or through decisions made by the programmer), there is no way to improve the accuracy by algorithmic means.

The result vector may be subject to an a posteriori investigation into its accuracy, which is implemented, for example, by a number of LAPACK subroutines (see Section 13.17).

A process which converges in terms of calculus can be used to calculate an approximation of arbitrary accuracy. A solution of maximal accuracy is however not the goal of many numerical problems, because it would reduce efficiency. Instead, the numerical solution must adhere to a given accuracy limit .

Example (Archimedian Method)   Archimedes introduced a method for approximating the circumference of a circle with arbitrary accuracy by regular polygons circumscribed about or inscribed in the circle. For the length s_k of a side of a regular polygon with k sides inscribed in a circle with a diameter $d = 1$, the following recursion is obtained:

$$s_{2k} = \sqrt{2 - \sqrt{4 - s_k^2}}, \qquad \text{with initial value} \quad s_6 := \frac{1}{2}. \tag{6.1}$$

For the circumference $U_k = k\,s_k$ of the regular polygon with k sides, the following mathematical convergence statement is obtained:

$$U_k \to U \qquad \text{as} \qquad k \to \infty.$$

Hence, for each accuracy bound ε, there is a K such that the truncation error satisfies

$$|U_K - U| < \varepsilon.$$

Even the *correct* implementation of a convergent process may lead to unexpected errors in the numerical results.

Example (Archimedian Method)   A *correct* implementation of the recursion (6.1) returns the following unsatisfactory result for $k > 6 \cdot 2^9$ (a sequence ending with zero) within a floating-point arithmetic on a workstation:

k	$6 \cdot 2^1$	$6 \cdot 2^2$	$\cdots$	$6 \cdot 2^9$	$\cdots$	$6 \cdot 2^{17}$	$6 \cdot 2^{18}$	$6 \cdot 2^{19}$	$6 \cdot 2^{20}$
$k \cdot s_k$	3.105809	3.132629	$\cdots$	3.141592	$\cdots$	3.092329	3.000000	3.000000	0

In this particular case, with the solution $U = \pi$ known from the beginning, the calculation of the sequence $\{s_{12}, s_{24}, s_{48}, \ldots\}$ could be terminated with $k = 6 \cdot 2^9$.

If, however, the common termination criterion for *unknown* solutions (the constancy[1] of the sequence of results $\{\tilde{s}_k\}$) is applied, then the completely incorrect result $\tilde{U} = 0$ is returned as the *approximation* of the exact result $U = \pi$.

This example demonstrates the dangers of concentrating only on truncation errors when investigating accuracy. The situation is even more complicated if continuous information is discretized. In such cases it is impossible of obtaining reliable information regarding the accuracy of the results, either before or after. If, as is common, no further information is available restricting the set of results, then the numerical result can suffer from arbitrarily large errors, as has been demonstrated in the example concerning numerical integration.

Consistency

This property gives information as to the extent to which a program or programming system has been developed according to consistent design and implementation rules, as to whether uniform notation is used, and whether its documentation and comments are written in a homogeneous style. When a program is consistent, the internal data structures are standardized, the algorithmic approaches are similar for similar problems, and the style and format of input and output are uniform.

Example (Linear Algebra)   The software packages LAPACK (*Linear Algebra Package*) [3] and TEMPLATES [5] were developed by large teams of experts working at different universities and research institutes all over the US (and in Europe). The high degree of consistency of those programming systems is mainly due to very strict guidelines which developers have to follow.

6.1.2 Portability

Program portability is a measure of the effort necessary to transfer software from one computing environment (hardware and software) to another. If this effort is relatively small then portability is considered to be high (Cowell [144]).

The growing importance of portability is due to the fast technical progress which constantly alters the relationship between hardware and software life-cycles and—even more significantly—the relationship between hardware and software cost, which is in favor of hardware.

The true significance of portability can only be assessed after taking economic aspects into consideration. Both the development and the procurement of software constitute an investment. Investment returns depend a great deal on the degree of utilization (frequency of use, lifetime) and so all software features which prolong the life-cycle of the software help to increase profitability.

Both user preferences and the maintenance policy determine the life-cycle of a software product. High quality (e.g., high reliability) leads to more intense

---

[1] For the purpose of numerical data processing, iterations and recursions are often terminated as soon as the sequence of results $\{\tilde{s}_k\}$ has become constant, i.e., $\tilde{s}_i = \tilde{s}_{i+1} = \tilde{s}_{i+2} = \cdots$, or at least $\tilde{s}_i \approx \tilde{s}_{i+1} \approx \tilde{s}_{i+2} \approx \cdots$.

and longer use. High maintainability is actually the key to long-term use. In practice, the life-cycle of software is often *under*estimated at the time of software development or procurement.

Example (UNIX) The UNIX operating system was developed at Bell laboratories (the research institute of AT & T) in 1973 by D. M. Ritchie and K. Thompson for their own needs. Due to its portability it is commonly used on workstations and other scientific computers.

High portability decreases the effort expended on necessary adaptation work, which usually grows dramatically with longer software life-cycles. Hence, increasing portability is a way to reduce maintenance costs and increase both the software life-cycle and the profitability of the investment (labor and/or monetary).

The demand for high software portability often contradicts the demands for high efficiency and short development periods. The encapsulation of the most time-critical sub-algorithms in a (possibly small) set of computational kernels, which are implemented in a standardized higher programming language, is important to the development of portable as well as efficient software. In addition to these portable modules, machine *dependent* versions which have optimized runtime behavior, can be developed.

Example (BLAS) For a number of elementary numerical linear algebra algorithms, computational kernels were defined and put into the public domain in the form of Fortran subroutines: *Basic Linear Algebra Subroutines* (BLAS; see Section 4.10.1). There are optimized versions available for many types of computers. In order to increase efficiency, the portable Fortran version has only to be replaced by the corresponding machine-version.

The software package LAPACK (see Section 13.17) was developed based on BLAS. In order to optimize LAPACK on a new computer, only efficient implementations of BLAS routines should therefore be developed or purchased.[2]

In the field of numerical data processing, the demand for portable programs is particularly large. This is because the cost of software development is quite considerable, and the necessary software development know-how is concentrated in a small number of places (mostly in universities and research institutions). It is thus desirable to develop software which can be ported to different computing environments without wasting manpower.

Example (Sparse Systems of Linear Equations) Only a small number of universities and research institutes (e. g., Harvard or Yale) are developing high-quality software for solving sparse systems of linear equations. However, the results of such work (e. g., the *Harwell Library*; see Section 7.2.3) are used in hundreds of places all around the world. Since there is usually no support for installation and use, the degree of portability must be very high.

---

[2]The only exception would be parallel (multiprocessor) computers. A software package has been developed for such computers: PBLAS. (*Parallel* BLAS; see Section 13.21).

Device Independence

This attribute characterizes the degree of independence that software products have from specific hardware features and/or configurations.

Example (Virtual Memory) Software developed on computers with virtual memory (i.e., practically without any restrictions regarding memory capacity) may require significant computational effort to be ported to computers without virtual memory (e.g., older PCs using MS-DOS), which have limited memory resources.

Software Independence

The portability of a software product is not only determined by its device independence, but also by its software independence. This attribute is the degree to which the software product is independent of particular system software properties and of the existence of certain tools.

Example (Language Extensions) The use of programming language extensions supported by only certain manufacturers incurs additional cost when porting the software to computers produced by other manufacturers. This cost is particularly high for non-standardized programming languages which are offered by only a few manufacturers. For example, porting a PL/I program to a non-IBM environment usually requires complete recoding in a different programming language.

Measures for Improving Portability

The factors mentioned above lead naturally to possible ways of improving the portability of software systems.

Increasing device independence: All sections of the program depending on certain hardware features should be disposed of; if that cannot be done then such parts are to be concentrated at a few very thoroughly documented locations. All hardware features used must be expressed by well-documented parameters.

Increasing software independence: By using standardized, widely used system or application software—e.g., standardized programming languages (Fortran 90 [78], ANSI C [279], Pascal etc.) or standard graphics packages—a high degree of software independence can be achieved. Program sections which depend on system software should be concentrated at a few well-documented locations.

Device independent measures for increasing the efficiency of a program (see Section 6.5) may also contribute to its portability as they extend the range of applicability to less powerful computers.

6.1.3 Efficiency

The efficiency of a program characterizes the relationship between the program's performance and the amount of resources used under stated conditions (see Chapter 3). There is a distinction between execution efficiency and storage efficiency:

execution efficiency is based on the assessment of the time required for solving the problem;

storage efficiency assesses the exploitation of the memory hierarchy (registers, cache, main memory, disk storage).

Execution efficiency and storage efficiency are not independent of one another. For example, the skillful use of the memory hierarchy may improve the runtime efficiency of a program significantly.

The *compile-time efficiency* takes into account memory requirements as well as the time required for compiling the program. The higher the level of optimizing performed by the compiler, the greater the compile time requirements. Aggressive optimizing may increase the compile time by an order of magnitude (a factor 10) or more. However, this difference hardly counts if the program is used many times.

The Trade-off between Efficiency and Reliability

There are a number of arguments in favor of placing more emphasis on reliability rather than on efficiency:

- Unreliable software is of no use regardless of its efficiency.

- The subsequent cost of unreliable software may be considerably higher than the cost resulting from less efficient programs.

- There are a number of measures for improving the efficiency of inefficient programs. On the other hand it is much more difficult to increase the reliability of a program after it has been developed.

- In the context of large software systems, unreliable and/or faulty modules may increase the development cost of the overall system considerably or may completely prohibit the integration of modules into an executable program.

These are weighty arguments in favor of reliability as compared to efficiency. They are reflected in the software engineering process, where software quality control deals almost exclusively with reliability. Since there is a wealth of literature dealing with this subject, general questions of quality assurance and quality control are not discussed further in this book.

The Need for Efficiency and Its Side-Effects

The efficiency of numerical programs is an important if not crucial quality aspect of many applications. For example, for real-time systems[3] there are often time limits which have to be adhered to unconditionally in order to guarantee acceptable system behavior. Also, the context of interactive systems, runtime efficiency

---

[3] *Real-time systems* are computer systems (hardware and software) designed for the immediate control of a process. They permanently communicate with the external process.

is of huge importance, as users usually don't accept long response times. For very large scientific or engineering projects (as described, for instance, in Mesirov [291]) the required runtime for numerical computations is often the determining factor as to whether or not a project is realizable at all. For numerical program libraries efficiency, in addition to reliability, user friendliness, maintainability, and portability, is also an important attribute of quality.

Example (BLAS) Massive efficiency-improving measures (aimed at reducing runtime) are taken only at a few locations for the software package LAPACK. The BLAS (*Basic Linear Algebra Subroutines*) prove to be especially improvable, e.g., through loop unrolling. Careful coding and exhaustive documentation make sure that this does not topple reliability.

A number of real-life software projects were observed with respect to development cost and resource exploitation. It was shown that the development costs increased disproportionally with the resource exploitation targets. For example, the relatively simple Gaussian algorithm needs serious modifications if the capacity of modern vector processors and parallel computers is to be exploited to a high degree (Robert [330]). Moreover, high efficiency is not only expensive to develop but also to maintain.

Both the IMSL and NAG software libraries (see Section 7.2.3), which are of outstanding quality, combine high efficiency and reliability. The cost of developing and maintaining the libraries is shared by many users through license fees.

Adaptive Programs

Ideally, portable programs would not have machine dependent parts. On the other hand, in order to be efficient, programs have to exploit their respective hardware environments (processor characteristics, memory hierarchies etc.). Thus, it is tricky to bring together the demand for portable *and* efficient programs— aims that seem to contradict one another. A possible solution would be to enable portable programs to obtain information about their actual computing environment.[4] This information could be used to develop adaptive programs which are able to exploit the given computing environment efficiently.

This process of gathering information using portable programs which adapt themselves to the peculiarities of their respective computing environments can be performed by standardized and, therefore, portable elements of programming languages. The Fortran 90 and ANSI C standards define inquiry functions which enable a program to obtain information about the floating-point arithmetic and the floating-point representation of the respective computer system (see Section 4.8). For example, stopping criteria for iterative algorithms may be based upon such information.

The concept of making information procurement a basis for the development of portable *as well as* efficient software plays a crucial role in the development of modern computer architectures such as vector processors or parallel computers (Krommer, Ueberhuber [256]).

---

[4]This principle was first introduced by P. Naur [296] in the context of machine numbers and their arithmetic.

6.2 Reasons for Poor Efficiency

All the methods discussed in Chapter 3 aimed at increasing potential hardware performance remain a wasted effort unless efficient software which is able to exploit all those hardware characteristics is available. Above all, *optimizing* compilers are necessary in order to translate programs written in high-level programming languages such as Fortran or C into fast machine code. In fact, recent developments in compiler technology have lead to cases where programs optimized by a compiler perform better than programs written directly in an assembler language. However, even highly optimizing compilers may fall short of expectations for such simple tasks as matrix-matrix multiplication, as is demonstrated in the case study in Section 6.8.

Whenever an optimizing compiler is used, two things should be checked:

1. the efficiency of the compiled program and

2. the *accuracy* of the compiled program (as particularly aggressive optimization may result in unexpected compilation errors).

This section analyzes some of the reasons for poor program efficiency. Methods for increasing the efficiency of numerical programs are discussed in Section 6.4.

6.2.1 Lack of Parallelism

The computer architectures discussed in Chapter 3 owe their efficiency to the break with the von Neumann computer model. It is therefore the compiler's task (maybe with far-reaching support by hardware elements) to transform the source code into an executable program which actually exploits hardware potential for the parallel execution of instructions. For example, this means that on a processor utilizing the pipeline principle, the pipeline should work without stalling. With superscalar processors, all independent functional units should be provided with suitable instructions. With vector processors, the compiler should combine as many instructions as possible when creating vector instructions.

The compiler's ability to exploit parallel hardware resources can be limited by *inherently* sequential user programs, i.e., programs (or program sections) which require instructions to be executed in a particular order due to the underlying program logic. In this context a series of program instructions B is referred to as *dependent* on another series of instructions A if A has to be executed prior to B.

In the following, different types of dependencies of series of instructions are discussed in order to enable the reader to recognize and avert inherently sequential series of instructions, which may inhibit program efficiency.

Data and Control Dependencies

If the dependency between two series of instructions A and B is due to the fact that data written by A is to be read by B, this is referred to as *data dependency*.

Example (Data Dependency)   In the sequence

```
area = area + area12 - aream
errb = MAX (epsabs, epsrel*abs(area))
```

the second instruction is dependent on the first since the value of the variable `errb` cannot be calculated before the value of `area` is available.

Control dependency is caused by logical branches in control instructions as they may occur e.g., in conditional jumps or loops.

Example (Control Dependency)   In the sequence

```
IF (errsum <= errb) THEN
   result = result + area12 - aream
END IF
```

the calculation of the variable `result` depends on the Boolean expression (the condition) of the IF block since whether or not the variable `result` should be assigned to a new value depends on the result of the test `errsum <= errb`. The loop

```
DO i = 1, n
   a(i) = a(i) + b(i)
END DO
```

is another example of control dependency: the variable `a(i)` depends on the DO index i.e., the execution of the assignment depends on the control variable.

Types of Data Dependencies

In what follows w_A, w_B denote the set of variables written in the series of instructions A, B and r_A, r_B the set of variables read in A, B respectively. The series of instructions B is said to be *flow dependent* (or *properly dependent*) with respect to the variable v if $v \in w_A \cap r_B$, i.e., if v is written in A and read in B.

Example (Flow Dependency)   In the example of data dependency mentioned above

```
area = area + area12 - aream
errb = MAX (epsabs, epsrel*abs(area))
```

the second instruction is flow dependent on the first with respect to `area` since the variable `area` is written in the first and read in the second instruction.

The series of instructions B is said to be *anti-dependent* on A with respect to a variable v, if $v \in r_A \cap w_B$, i.e., if v is read in A and written in B.

Example (Anti-Dependency)   If the two instructions in the example of flow dependency are swapped, this leads to the anti-dependency

```
errb = MAX (epsabs, epsrel*abs(area))
area = area + area12 - aream
```

Although the arithmetic operations of the second instruction can be executed prior to or concurrently with the first instruction, the assignment of the right-hand side to the variable `area` must not be executed before the old value of `area` has been processed in the first instruction.

The series of instructions B is said to be *output dependent* on A with respect to the variable v if $v \in w_A \cap w_B$, i.e., v is written in both A and B.

Example (Output Dependency)   In the loop

```
DO i = 2, n-1
   a(i-1) = 2*b(i)
   a(i+1) = c(i) + 1
END DO
```

the respective iterations belonging to the indices $i = k$ and $i = k+2$ are output dependent since the variable $a(k+1)$ is written in both of them.

However, this loop cannot easily be transformed into another form which does not have output dependency. Non-trivial cases of output dependency only occur in connection with other dependencies.

If the context leaves no doubt as to which variable v a dependency is based, it is simply referred to as flow dependency, anti-dependency or output dependency[5] between A and B.

Potential Dependencies

When analyzing data dependencies the compiler has to be on the safe side and assume that data dependency actually occurs wherever it cannot be strictly ruled out. If, on the other hand, the compiler assumes that there is no dependency where there actually is and changes the sequence of instructions, this may lead to an *inaccurate* program. Hence, with respect to program efficiency, potential data dependency is as bad as actual dependency.

An obvious source of potential data dependencies is control structures and subroutine calls.

Example (Potential Data Dependencies Caused by a Control Structure)   With the instructions

```
x = a/2
IF (c <= 0) THEN
   a = d + e
ELSE
   b = d + e
END IF
```

there is (anti-)dependency between the assignment to x and the subsequent two-branch IF-block, only if the test c <= 0 holds. If the value of c is not known at the compilation time then the data dependency remains only potential.

Example (Potential Data Dependency Caused by a Subroutine Call)   With the program segment

```
CALL procedure (indicator, x)
x = y + z
```

---

[5]There is a fourth type, namely the *input dependency*, which is of no importance for numerical programs.

whether or not there is data dependency between the two instructions and—in the affirmative case—what type of data dependency there is, cannot be established as the compiler usually cannot check the way the actual parameter x is used in the subroutine procedure. For example, it may be that x is not used at all because of the value of indicator and acts just as a dummy variable. In this case there is no dependency whatsoever. If, on the other hand, x is actually referenced in procedure, it may be done either as a read or write operation. In the first case this means anti-dependency, and in the latter case output dependency of the instructions.

For array access operations, potential data dependency is often caused by index values unknown at compilation time.

Example (Potential Data Dependency with Array Access Operations)   In the loop

```
DO i = 1, n
   a(i) = b(i) + c(i)
   b(i) = a(i + k)
END DO
```

the elements of the array a are accessed in the second assignment via the *offset* k. Whether or not there is data dependency between the iterations of the loop index $i = j$ and $i = j + |k|$ and—in the affirmative case—what kind of data dependency there is, depends on the actual value of the variable k (which may be unknown at compilation time): For $k = 0$ there is no dependency, for $k < 0$ there is flow dependency and for $k > 0$ there is anti-dependency.

6.2.2 Lack of Locality of Reference

It was stated in Section 3.2 that the concept of a memory hierarchy enables the memory systems of contemporary computers to provide increasingly faster processors with data at sufficient speeds. It is a hardware and operating system task to map logical addresses to the different levels of the memory hierarchy. Ideally, the memory access operations of user programs mainly reference the fast levels of the hierarchy. This can only be the case if the programs follow the principle of locality of reference.

If the memory access operations of a user program are of insufficient locality of reference, this mechanism fails. Data then has to be fetched from slower levels of the memory hierarchy. This is said to be a *fault* or *miss*; if data cannot be found in the cache it is called a *cache miss*, if it is not stored in main memory it is called a *page fault*. The negative impact of such misses or faults on the program efficiency should not be underestimated.

Example (Lack of Locality of Reference)   Fig. 6.1 depicts the performance and efficiency of a manually optimized assembler implementation of the subroutine BLAS/sdot on an HP workstation.

The calculation of the inner product

```
sum = 0.
DO i = 1, n
   sum = sum + a(i*k)*b(i)
END DO
```

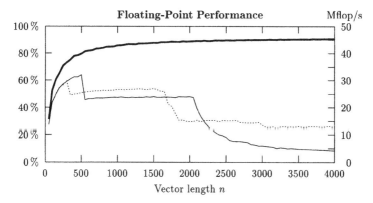

Figure 6.1: The floating-point performance of the subroutine SDOT from the vector library of an HP workstation for strides $k = 1$ (——), $k = 32$ (—), $k = 50$ (·····).

was executed for a large number of vectors with length n in the range $10 \leq n \leq 4000$ and for strides $k = 1$, $k = 32$ and $k = 50$. It is clear that for the optimal locality of reference $(k = 1)$ peak performance is approximated to a satisfactory degree at least for larger vector lengths. For other stride values, the efficiency of BLAS/sdot is generally lower and, above all, decreases with the vector length n.

The lack of locality of reference in the source code greatly deteriorates program performance because the compiler's ability to improve the locality of reference using program transformations is limited. This is because the parameters necessary to choose the best program transformations are unknown at compile time. For example, the size of the available part of the main memory might change during runtime. The same is true for important program parameters such as the dimensions of arrays etc. which may not be known before the program starts. The efficient exploitation of the memory hierarchy by the compiler is thus largely impossible. Moreover it is very complicated for the compiler to analyze reference patterns. Thus, such *automatic* determination of program transformations aimed at improving the locality of reference is not possible at present.

6.2.3 Inadequate Memory Access Patterns

The individual levels of a memory hierarchy may consist of not just one, but a number of modules of equal size. For example, interleaved memory is divided into several banks, and set-associative caches in principle consist of several directly mapped caches.[6] The mapping of logical addresses onto the different modules at a certain level of the memory hierarchy is done in a *blockwise cyclic* procedure: subsequent blocks of logical addresses are mapped onto cyclically subsequent memory modules.

---

[6]If a cache consists of Z cache lines, a directly mapped cache can also be looked upon as a 1-fold set-associative cache. Therefore, all remarks on set-associative caches are true for directly mapped caches as well.

For particular reference patterns it may be that certain memory modules are used more often than others, i.e., the majority of logical addresses are mapped onto those modules. In the case of interleaved memory (see Section 3.2.6), this means that one and the same memory module is referenced at short intervals. If the time between two access operations is shorter than the memory cycle time, then additional delays are caused by those operations. In the case of an l-fold set associative cache (see Section 3.2.4), this means that memory access operations are concentrated on m of the overall Z/l caches, which reduces the effective cache size to lm lines since $Z/l - m$ cache sets remain practically unused. Reduced cache sizes result in performance decreases, the amount of which depends on the locality of reference of the application.

Example (Inadequate Reference Patterns)   In Fig. 6.1 it can be seen that the subroutine BLAS/sdot is of *lower* efficiency for many vector lengths (in particular for $n \geq 2000$) with the smaller stride $k = 32$ than with the larger stride $k = 50$. Intuitively one would expect that a lower value of the stride would result in higher efficiency because of the principle of locality. However, the effect observed in this example is caused by the reference pattern of stride $k = 32$. The cache memory of the workstation used for the experiment is directly mapped, and its capacity is 256 KB with cache lines of 32 bytes. If only every 32nd memory word is referenced, then the effective cache size is reduced by the same factor, i.e., down to 8 KB or 2048 memory words. Thus, a significant performance decrease can be expected for vector lengths larger than $n = 2048$—as is precisely shown in Fig. 6.1. Inadequate reference patterns therefore usually lead to a poor locality of reference since, generally, only one word per fetched line is actually used.

6.2.4 Overhead

Program segments which do not directly contribute to the determination of the program's results are called *overhead*. In general there are two types of overhead:

1. more or less inevitable overhead: subroutine calls, index calculations, garbage collection, bookkeeping operations, etc. and

2. actual unnecessary operations.

The runtime increase may be small if the overhead is executed only once. However, for repetitions—especially loops—the loss of efficiency caused by overhead may be considerable.

All *subroutine calls* which make necessary memory access operations for saving registers or passing parameters belong to this category. Moreover, overhead comprises the multiple evaluation of *identical* arithmetic, logical, or other expressions. Another source of overhead (which is not often talked about) is *implicit type conversions*, which have to be executed whenever the operands of an expression are of different types, e.g., single or double precision floating-point numbers. *Logical branches* in avoidable loops influence the runtime to a great extent if their results are independent of the iteration or depend on the loop index in a predictable way.

Example (Logical Branches)   In the program segment

```
DO i = 1, n
   IF (indicator > 0) THEN
      a(i) = b(i)
   ELSE
      a(i) = 0.
   END IF
END DO
```

the result of the logical expression `indicator > 0` is the same for each loop index; hence, $n - 1$ evaluations of this expression are overhead. In

```
DO i = 1, n
   IF (MOD(i,2) == 0) THEN
      a(i) = b(i)
   ELSE
      a(i) = 0.
   END IF
END DO
```

the result of the logical expression `MOD(i,2) == 0` is not the same for each loop index, but rather it follows a very simple pattern which depends on the loop index `i`: For even indices `i`, the IF branch is executed, and for odd `i`, the ELSE branch. Hence, it constitutes considerable overhead to evaluate the IF-condition n times.

6.3 The Measurement of Performance Indices

Usually one decides to optimize self-developed numerical software when the execution of a program and the calculation of results takes too much time either from a subjective or an objective point of view. However, before modifications aimed at increasing performance can be made, the extent to which the given hardware allows for an increase in performance must be checked. It may well be that the hardware is already exploited by the software to a high or at least to a satisfactory degree. In this case a performance increase is only possible using a faster and more expensive computer system or better algorithms.

Moreover, it is necessary to check which parts of the program should be modified *before* performing any optimization operations. Numerical programs often comprise some thousand lines of code, whereas the biggest share of execution time is spent in a relatively small part of the program. The performance optimization of numerical software must concentrate on these parts of the program.

Finally, prior to optimizing, all factors which reduce the level of performance should be identified. By doing so, the large number of applicable program transformations can be reduced to a smaller number of potentially promising transformations, which also reduces the amount of labor necessary for optimization.

This section discusses the instruments available to the software developer to master that task. In addition, these instruments can also be used for the a posteriori assessment of the quality of an optimization process.

Example (Cholesky Decomposition) In the following, cholesky, a simple Fortran 77 program, is given as an example to demonstrate the use of different tools. To begin with, this program assigns values to the elements of an $n \times n$-matrix A according to

$$a_{ij} = \begin{cases} 1 & \text{for} \quad i = j \\ 1/(n|j - i|) & \text{otherwise} \end{cases}$$

and then calls the subroutine LAPACK/dottrf in order to execute the Cholesky decomposition $A = LL^\mathsf{T}$

6.3.1 Measurement of the Workload

Number of Floating-Point Operations

In order to assess the efficiency of numerical software, it is, above all, necessary to determine the number of floating-point operations actually performed. However, for most programs it is too complicated a task to find the exact number simply by counting the floating-point operations at runtime and the number of operations is dependent on the data. The only exceptions are very simple algorithms which do not depend on the actual data, such as the decomposition of dense matrices. The order of magnitude of the number of floating point operations can usually be determined simply by only taking into account those parts of the program which consume a significant amount of computing time. An important tool for this task is called *profiling* (see Section 6.3.3).

The number of floating-point operations executed at runtime may also be determined by a *hardware monitor*. However, such devices are available on very few computer systems, e.g., Cray computers.

Memory Requirements

In addition to the number of floating-point operations, memory requirements are also indicators of a program's complexity. Similar to the determination of the number of floating-point operations, the theoretical analysis of memory requirements may be quite complicated and inexact. In contrast to determining the number of floating-point operations, there are a number of tools on any UNIX computer which quantify the memory requirements. The following examples of these tools demonstrate how they are used.

The UNIX command **size** returns the *static* memory requirements determined at compile time for program instructions and data specified as the argument.

Example (size command) With a 1000×1000-matrix on a DEC workstation using OSF/1, size returns

```
> size cholesky
text    data    bss     dec     hex
32768   8192    7993216 8034176 7a9780
```

for the static size of cholesky.

In this example, the text entry specifies the memory requirements of the machine instructions; data specifies the memory requirements of the data already initialized at compile time, and `bss` the memory requirements of the data not yet initialized at compile time. The remaining entries `dec` and `hex` state the overall size of the required memory area in decimal and hexadecimal representations respectively. All values are stated in bytes.

However, the information returned by `size` often gives an inaccurate impression of the size of the memory actually used at runtime: If there are no dynamic arrays in a particular programming language (like Fortran 77) then arrays with the maximal size presumably needed are often declared. However, if only the part actually required by a particular problem is used at runtime, then `size` returns too large a value of main memory requirements. On the other hand, if there are dynamic arrays (as in Fortran 90), then a static statement about memory requirements at runtime is completely impossible. In this case `size` will return too small a value of main memory requirements.

In order to estimate the *dynamic* main memory requirements (needed at runtime), the UNIX command `ps`, which returns the *current* size of allocated main memory[7], can be used.

Example (ps Command)   If the original static allocation of a 1000×1000-matrix A is again used in `cholesky` but only a 500×500-matrix is actually stored and decomposed, then the discrepancy between static and dynamic memory requirements can be demonstrated using the `ps` command as follows:

```
> cholesky&
[1] 20423
> ps up 20423
USER      PID %CPU %MEM   VSZ  RSS TT  STAT STARTED        TIME COMMAND
ak      20423 52.0  4.4 9.80M 4.18 p3  R    13:04:00   0:02.23 cholesky
```

The figure below RSS (*Resident Set Size*) indicates the current size of the main memory (in MB) allocated to the respective process. The dynamically established value 4.18 MB is nearly half the size of the statically determined one due to the fact that only half the number of columns of A are referenced.

Attention should be paid to the fact that RSS indicates only the dynamic size of the allocated main memory, but *not* the dynamic memory requirements (i.e., the size of allocated virtual memory). Hence, the value below RSS is always smaller than the extent of the physical main memory, whereas the *virtual* memory requirements may well go beyond that.

I/O Operations

Many numerical programs read data from secondary or tertiary memory (hard disks, magnetic tapes, etc.) at the beginning of execution and write data to such storage media at the end. It may also be that data is stored there at runtime via explicit I/O operations—in particular, if the capacity of the virtual memory is exhausted. In these cases not only the number of floating-point operations and the required memory size, but also the number of I/O operations and the amount of data transferred are important characteristics used for performance assessment. Current values of these characteristics can be determined with the `time` command of the UNIX-Shell `csh` or `tcsh`.

---

[7] A sufficiently long runtime is a prerequisite for the proper use of the `ps` command.

Example (time **Command**)   If the program cholesky is modified such that the symmetric matrix A to be decomposed is read from a file and the resulting factor matrix L is written to a file as well, then the time command returns output as follows:

```
> time cholesky
82.535u 40.662s 2:08.46 95.8% 0+0k 2+18io 0pf+0w
```

The number of *blocked* I/O operations is specified in the field $x+y$io by the values x and y. Blocked I/O operations usually reference secondary memory but not, for instance, screen I/O. Thus, in the above case 2 blocks were read and 18 blocks were written.

6.3.2 Measurement of the Time

The runtime is the second crucial performance index (in addition to the work-load). Provided that the performance of the *entire* program is to be determined, it is best to use the time command described above to determine the runtime.

Example (time **Command**)   In the example discussed above,

```
> time cholesky
82.535u 40.662s 2:08.46 95.8% 0+0k 2+18io 0pf+0w
```

the entry xu contains the user CPU time x (in this case: 82.535 seconds), and ys contains the system CPU time y (in this case: 40.662 seconds) required for the execution of cholesky. The entry after the user and system CPU times specifies the response time (in this case: 128.46 seconds).

   If the coefficients of the matrix A are generated at the beginning of the program instead of being read from a file (as in the example above), and if, moreover, the output of the factor matrix L is disposed of, then the time command returns the following results:

```
> time cholesky
22.080u 0.371s 0:22.94 97.8% 0+0k 0+2io 0pf+0w
```

Thus, it is shown that nearly all the system CPU time of the program variant using secondary storage is due to the necessary I/O operations.

If, on the other hand, only certain parts of the program are to be analyzed with respect to time requirements, then the program must contain instructions for time measurement. A possible way to maintain the portability of the program is to use the C-subroutine times which is available on all UNIX systems.

Example (**Subroutine** times)   The following fragment of a C-program demonstrates how to determine user and system CPU time as well as the overall *elapsed time* using the predefined subroutine times.

```
#include <sys/times.h>
...
/* period is the granularity of the subroutine times */
period = (float) 1/sysconf(_SC_CLK_TCK);
...
start_time = times(&begin_cpu_time);
/* begin of the examined section */
...
/* end of the examined section */
end_time    = times(&end_cpu_time);
user_cpu    = period*(end_cpu_time.tms_utime - begin_cpu_time.tms_utime);
system_cpu  = period*(end_cpu_time.tms_stime - begin_cpu_time.tms_stime);
elapsed     = period*(end_time - start_time);
```

In the subroutine `times` all measured time values are stated as the multiples of a specific period of time. This period depends on the computer system and must be determined with the UNIX standard subroutine `sysconf` before `times` is used. The subroutine `times` itself must be called immediately before and after the part of the program to be measured. The argument of `times` returns the accumulated user and system CPU times, whereas the current time is returned as the function value of `times`. The difference between the respective begin and end times finally yields, together with scaling by the predetermined period of time, the actual execution times.[8]

Reliability of Time Measurement

The entirety of all the processes running on a modern computer system is of enormous complexity and cannot be understood fully. Due to multitasking and multi-user mode, the exact processing steps *cannot be reproduced*. Thus, for repeated time measurements of one and the same program or program component, there may be significant discrepancies.

These discrepancies are in opposition to the scientific principle that it should be possible to reproduce experiments and their results. If subsequent experiments return widely scattered values, it remains unclear as to which value the respective performance index should be allocated to. Since it is usually unknown as to which way the exact value has been influenced, statistical methods (such as mean value estimation), do not provide a remedy, either. It is neither possible to confirm nor to disprove statements about time related indices (e. g., conformity with a value which is expected for theoretical reasons) without in-depth scrutiny of all factors influencing the measurement.

Worse still, there is a wide range of possible manipulations of data caused by its dispersion. The selective choice of individual data may suggest that a particular theoretical consideration could be verified by conducting experiments whereas the entirety of the data (in particular their dispersion) would strictly rule out such an interpretation.

Hence, for time measurements, all factors disturbing the result have to be excluded as far as possible if the results are to be regarded as correct measurements within the meaning of experimental science. In doing so factors which lengthen (e. g., page faults) as well as those which shorten (e. g., the complete storage of the program and data in the cache) the measured computation time have to be systematically taken into account. The dispersion of measured values of the different runs is usually caused by changes in the environment. On the other hand, reproducible results are obtained if the environment of the experiment is specified precisely enough. In the following, the factors influencing the time measurement of numerical programs as well as their definition and control are discussed.

---

[8]The error caused by neglecting the time consumed by the subroutine `times` itself can be neglected on contemporary computer systems.

Cache: The cache memory (see Section 3.2.4) has a great deal of influence on the computation time. It should have a well-known, predefined status at the beginning of any time measurement. There are two extremal cases:

The cache is empty: Neither the code to be examined nor the corresponding data are in the cache. This is the default case after the program has been loaded on a multi-user and multi-tasking operating system.

The data and code are in the cache: Both the program (fragment) and the corresponding data are in the cache. If the computation time is to be measured under these assumptions, then the code should be applied to the same data twice within the same program. In this case, only the second application is used for time measurement. This strategy works assuming that the program (fragment) and its data can be held in the cache, which is usually possible for small programs and small amounts of data.

Translation Lookaside Buffer (TLB): TLB misses (see Section 3.2.5) may cause delays if *very* large amounts of data are processed. Hence for the TLB, a well-known and well-defined state is necessary as well.

Main Memory: Page faults have such an impact on the wall-clock time of a piece of code that measurements based on different numbers of page faults cannot be compared at all. Thus, in order to measure the performance of a program whose code and data can be stored in main memory, page faults should be avoided completely. One way to make it more probable that all the data is actually stored in main memory is to simply reference them before the time measurement begins.

Context Switches: A context switch (change of tasks in a multi-tasking environment) requires the execution of system routines and (if available) other pending tasks. This influences the runtime behavior significantly, e. g., because the contents of the cache are modified. If it is not the aim to determine the time requirements of a program with respect to a certain degree of system load, then the number of context switches should be reduced to a minimum, e. g., by killing all processes other than the desired one.

Interrupts: An interrupt caused by a peripheral device may seriously topple the results of the measurement and must be taken into account, provided it has been detected at all.

Optimizing Compiler: With modern RISC processors, any change in the optimization method may have a dramatic impact on the performance of the resulting object code. Hence, the optimization option used must be clearly stated in the documentation of any time measurement activity. Except for special cases (e. g., if the effect of the optimization method on the performance is being investigated) the highest level of optimization is usually the choice.

However, if the effect of different program transformations is to be investigated, then it must be noted that the optimizer might generate completely different, additional program transformations.

The programming language might influence the performance as well. For example, due to the lack of pointer arithmetic, the optimization effect of Fortran 77 compilers may greatly supersede that of C-compilers. Hence, the programming language used for the implementation of the measured program must be mentioned in the documentation.

For the assessment of time measurements, determining the CPU-time of the program (fragment) several times for the same problem with the same data and considering the variance of the results is a recommended method. Large variances are a sign of weak points in the time measurement which may be influenced by other processes active in the system. On the other hand, small variances only indicate that the environments of the different runs were essentially invariant. However, one must not jump to the conclusion that the actual situation is equal to the assumed environment (see next section, in particular Fig. 6.2 and Fig. 6.3). This fact must be confirmed by further investigations and checks.

The Measurement of Short Periods of Time

For the measurement of very short periods of time, certain difficulties may arise when the time unit of the system clock is the same order of magnitude as or even longer than the investigated times. A possible remedy would be to repeat the investigated part of the program in a loop such that the time required is long enough to be measured. In doing so, though, the time overhead for loop control must be taken into account (Addison et al. [81]).

Example (Determining the Number of Repetitions) For the matrix-matrix multiplication with the subroutine BLAS/sgemm and matrix orders of $n \leq 200$ a typical HP-workstation requires around 30 ns per floating-point operation. In order to have a runtime of at least 1 s as required for proper time measurement, the number k of necessary repetitions must be

$$k \geq \frac{1}{2n^3 30 \cdot 10^{-9}} \approx \frac{16.7 \cdot 10^6}{n^3}.$$

The factor $2n^3$ is a lower bound on the number of floating-point operations necessary for the matrix-matrix multiplication. Hence, for 100×100 matrices at least $k = 17$ repetitions are necessary in order to have a runtime of more than a second.

However, the use of repeat-loops may increase the influence of the cache significantly, as the same instructions are executed time and again, and a cache hit rate of 100 % may be achieved after the first repetition (Weicker [378]). This speedup is not desirable as it falsely suggests shorter execution times and hence better performance. Attention should be paid to the fact that this effect cannot be detected by scrutinizing the variance because the speedup effect is always the same provided that all other influencing factors remain the same. Hence, small variance values do *not* guarantee the quality of the time measurement (cf. Fig. 6.2 and Fig. 6.3).

Figure 6.2: The large and small variance (standard deviation) of measurements without systematic error (mean value = exact value).

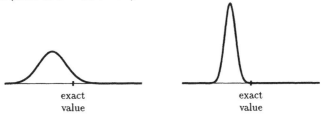

Figure 6.3: Measurements *with* systematic error (bias).

In order to obtain reliable results for time measurements of program fragments whose execution time is $1\,\text{ms} = 10^{-3}\,$s or less, an accuracy of $10\,\mu s = 10^{-5}\,$s should be achieved. Due to the reasons mentioned above (e. g., the problem with the misleading cache hits), the method of using repetitions is not useful. It is also not recommendable to increase the execution time by increasing the amount of data since a context switch might occur in a multi-tasking system.

For the measurement of short code fragments without context switches or interrupts, the *real-time* can be used. On most computers this can be measured with higher accuracy than the actual process time (determined by a call of `times(2)` in a UNIX system) whose smallest unit is usually around $10\,\text{ms} = 10^{-2}\,$s. In a UNIX system the real-time is determined by a call of `gettimeofday(2)` which returns the current time with a *granularity* of $1\,\mu s = 10^{-6}\,$s. However, this does not mean that the time is returned with the same *accuracy*. It is easy to verify the specifications by calling `gettimeofday()` repeatedly until there is a difference as compared to the previous measurement. The smallest difference δ_{min} measured in this way is a bound for the accuracy of the time measurement.

Example (Accuracy of the Time Measurement) By experiment it was determined that $\delta_{\text{min}} = 1\,\mu s$ for a PC, $24\,\mu s$ for a HP workstation, and $151\,\mu s$ for an IBM workstation. The PC value is quite satisfactory, but the relatively high value for the IBM workstation makes it virtually impossible to determine the execution time of code fragments which take less than around $2\,\text{ms}$.

The time measurement of short program fragments can be made to work as follows: For each measurement, the part of the program being observed is loaded from the disk. This guarantees that neither code nor data are in the cache at the beginning of the time measurement. This process is repeated several times.

If the operating system allows for it, then the program is given highest priority. After k runs of the program, the minimum of the measured times is used as the result: It constitutes an upper bound on the actual time requirements provided the CPU could be used exclusively for the observed part of the program.

6.3.3 Profiling

The technique of *profiling* is employed to find out which part of a program is the most time consuming and hence, the best target for performance increasing program transformations. For this purpose the program is decomposed into individual parts, and the proportion of time spent in each part is determined.

Typically, the parts into which a program is decomposed by the according tool, a *profiler*, coincide with the subroutines of the program. The advantage of such a *subroutine profiler* is that the time distribution returned can easily be identified using the source code. Moreover, the time measurement of the different parts can easily be executed.

However, if more detailed results are required, i. e., if information on the time distribution within particular subroutines is required, a *base block profiler* has to be used.

A *base block* is a sequence of instructions of maximal length in a particular part of the program which are executed in a row exactly once while the control is in this particular part. A base block profiler examines how often the base blocks are executed in a program. The best choice of base blocks to be used for the optimization of numerical programs are those of intensively used subroutines which are reused many times.

Note (Parallel Computers) On multiprocessor systems (parallel computers) it is even more complicated to determine the ratio of computation, communication and standby times than on single-processor machines. Graphical tools are frequently used to represent the runtime behavior of programs on parallel computers (Tomas, Ueberhuber [364]).

The Distribution of Execution Time Among Subroutines

Nearly all UNIX-operating systems are supplied with a subroutine profiler. In the following, two of the most widespread profilers, prof and gprof, are discussed.

With both profilers the mapping between the different subroutines and portions of the execution time is based on stochastic methods. In doing so, the program execution is interrupted at regular intervals and the subroutine which is currently being executed is determined. The *full* time span between two interrupts is allocated to this subroutine. This method is inevitably inaccurate, since it may well be that different subroutines have been executed during one interval (i. e., between two snapshots). Clearly the error becomes larger the longer the time span between two interrupts.[9]

In fact the profiling procedure itself influences the runtime behavior of the program and may garble the result ("Heisenberg uncertainty"). For example, the

---

[9]In practice, time spans of an order of magnitude of 10 ms are common.

cache contents are altered due to periodic monitoring. This may unduly increase the number of cache misses, which would not happen if the program were executed without the profiler.

Example (Profiler prof) In order to allow the subroutine profiler `prof` to determine the distribution of computation time, all subroutines must first be recompiled with the -p option.[10] The program is then executed. In order to obtain the subroutine profile, the profiler `prof` is called after the program terminates with the program name as an argument.

This procedure is demonstrated in this example using `cholesky` (with an 800×800-matrix A):

```
> f77 -O -p -o cholesky *.f
> cholesky
> prof cholesky
Profile listing generated Wed Jun  1 10:08:32 1994 with:
  prof cholesky
-------------------------------------------------------------------------
*  -p[rocedures] using pc-sampling;                                      *
*  sorted in descending order by total time spent in each procedure;    *
*  unexecuted procedures excluded                                       *
-------------------------------------------------------------------------

Each sample covers 4.00 bytes for 0.0079% of 12.3105 seconds

%time    seconds  cum %   cum sec  procedure (file)

77.7     9.5615   77.7      9.56 dgemm_ (dgemm.f)
10.1     1.2432   87.8     10.80 dtrsm_ (dtrsm.f)
 8.7     1.0703   96.5     11.88 dsyrk_ (dsyrk.f)
 3.0     0.3740   99.5     12.25 test_ (cholesky.f)
 0.4     0.0469   99.9     12.30 dgemv_ (dgemv.f)
 0.1     0.0088  100.0     12.30 ddot_ (ddot.f)
 0.0     0.0029  100.0     12.31 dscal_ (dscal.f)
 0.0     0.0020  100.0     12.31 dpotf2_ (dpotf2.f)
 0.0     0.0010  100.0     12.31 lsame_ (lsame.f)
```

`prof` lists all subroutines that have been executed at least once, at the moment of a program interrupt, in decreasing order of their share of execution time. It also lists the following data, as well:

%time the relative execution time of the subroutine [the percentage of the overall execution time],

seconds the absolute execution time of the subroutine [s],

cum % the relative accumulated execution time of all subroutines listed at that point [the percentage of the overall execution time] and

cum sec the absolute accumulated execution time of all subroutines listed at that point [s].

In this example the profiling list makes it clear that more than three quarters of the overall execution time is consumed by a single subroutine: the matrix-matrix multiplication routine **dgemm**.

The profile of the same program with a 1600×1600-matrix is as follows:

---

[10]If only selected subroutines are to be listed in the profile, it is enough to recompile those subroutines with the -p option.

```
%time     seconds   cum %    cum sec  procedure (file)

85.9      86.1553   85.9      86.16  dgemm_ (dgemm.f)
 6.3       6.2803   92.2      92.44  dtrsm_ (dtrsm.f)
 6.0       5.9844   98.1      98.42  dsyrk_ (dsyrk.f)
 1.7       1.7178   99.9     100.14  test_  (cholesky.f)
 ...
```

In this case the share of the overall execution time consumed by dgemm is even larger. Hence, the subroutine dgemm should be the basis for all measures aimed at improving the performance of the program.

In addition to the data delivered by **prof**, the profiler **gprof** returns detailed information on the *calling tree*, i.e., the way the different subroutines call one another. This kind of information can be crucial for larger program systems, where it is easy to get lost in the calling structure. With the help of this additional information, **gprof** can determine not only the accumulated runtime of a particular routine but of all the subroutines it has ever called, as well. The time-critical sections of the program system in question can thus be identified much more easily.

Example (Profiler gprof) If a subroutine profile is to be created using **gprof**, then all subroutines must first be recompiled with the option -pg.[11] One then proceeds as with **gprof**:

```
> f77 -pg -o cholesky *.f
> cholesky
> gprof cholesky
```

Due to the limited space available and the enormous length of the output, the profiling list is not given here.

The Distribution of the Computation Time Among Base Blocks

As opposed to subroutine profilers, there are no de facto standards for base block profilers. SUN systems are supplied with **tcov**, RS/6000 computers with **tprof**, computers running UNIX-System V with **lprof**, and MIPS based systems such as Silicon Graphics or DEC workstations with **pixie**. **tcov** and **lprof** are quite similar, whereas **pixie** is completely different.

Example (Profiler pixie) There is *no* need to use certain compiler options for the profiler **pixie**. Hence, the program does not have to be recompiled. Actually **pixie** is simply called with the name of the executable program as its argument, creating an executable program modified at machine-code level. The extension .pixie is attached to the name of the program. In order to create the base block profile, the modified program is executed and then **prof** is called with the program name (without any extension) as its argument. For example, for cholesky this is done as follows:

```
> pixie cholesky
> cholesky.pixie
> prof -pixie cholesky
```

---

[11] For HP computer systems the corresponding option is -G.

The profiler pixie not only performs base block profiling but also states execution times even for individual *code lines*, as can be seen in the following output segment.

```
-----------------------------------------------------------------------
*   -h[eavy] using basic-block counts;                                 *
*   sorted in descending order by the number of cycles executed in each *
*   line; unexecuted lines are excluded                                *
-----------------------------------------------------------------------


procedure (file)           line    bytes        cycles    %time    cum %

dgemm_ (dgemm.f)            284      156     688414720    61.53    61.53
dgemm_ (dgemm.f)            283       80     208158720    18.61    80.14
dsyrk_ (dsyrk.f)            249      144      78977280     7.06    87.19
dtrsm_ (dtrsm.f)            360      136      74057760     6.62    93.81
dsyrk_ (dsyrk.f)            248       84      29105280     2.60    96.42
dtrsm_ (dtrsm.f)            359       88      22438080     2.01    98.42
dgemv_ (dgemv.f)            211      132       3414488     0.31    98.73
dsyrk_ (dsyrk.f)            245       52       1878660     0.17    98.89
dgemm_ (dgemm.f)            280       44       1723946     0.15    99.05
dtrsm_ (dtrsm.f)            353       84       1411200     0.13    99.17
dgemv_ (dgemv.f)            210       80       1365350     0.12    99.30
```

So pixie returns a list of all executed program lines in the order of consumed execution time. Processor clock cycles are used as the time unit rather than seconds. The columns %time and cum % state the (accumulated) share of the execution time.

This output fragment in particular points out that 61.53 % of the total execution time is consumed by a *single* instruction, namely the one in the innermost loop of the matrix-matrix multiplication executed in dgemm.

The information given by pixie is based on the—unrealistic—assumption that *all* the data is always stored in the cache. Additional delays for memory access operations are *not* considered.

6.3.4 Spotting Performance Decreasing Factors

In most cases the analysis of program performance is quite a complicated process. Usually this is due to the *lack* of a suitable examination tool. Such tools—which are not available on all systems—make it possible to determine if, for example, there are enough program instructions independent of one another such that parallel hardware resources can be exploited to a satisfactory degree.

In a similar way, the examination of memory access delays, which may be caused by a lack of locality of reference and/or unsuitable access patterns, cannot be done properly in many cases. Even the question of if and to what degree cache misses are responsible for program delays cannot be answered with common examination tools—except for the POWER2 architecture.[12]

The UNIX command time hints at possible page faults. If the CPU utilization of the program is significantly smaller than 100 % and this phenomenon is not

---

[12]However, tools like that are likely to be available soon. A prototype of a cache profiler is e.g., CPROF (Lebeck, Wood [267]).

due to I/O operations or the existence of other active processes, then page faults are responsible for this phenomenon.

The lack of suitable tools for performance analysis seriously hinders the optimization of numerical software. In many cases the performance decreasing factors cannot be spotted precisely enough, and for this reason the causes of poor efficiency are not disposed of.

6.4 Performance Optimization

If a problem in numerical data processing cannot be solved within a given time limit on a certain computer system, and the procurement of new and more powerful hardware is not an option, the underlying method for solving the problem must be optimized. This might be done using one of the following approaches:

1. by modifying the underlying mathematical model,

2. by changing the discretization of the mathematical model,

3. by using better algorithms for the solution of the discrete problem, and

4. by implementing the algorithm more efficiently.

The task of deriving a suitable model belongs to the field of applied mathematics and will not be further discussed in this book. The task of deriving a discrete model from a continuous one as well as the solution of such a discrete problem belong to the field of numerical mathematics and are discussed in the chapters to come. This section only deals with the last approach mentioned above: the more efficient implementation of algorithms. However, this does not imply any rating (with respect to the effect on increased performance) of the different levels of optimization: More efficient coding of an existing algorithm often does not yield the same speedup as, for example, the development of a new algorithm with significantly less workload.

Moreover, the description of some program transformation techniques in the following sections does not suggest that the software developer must apply all the transformations step by step in order to achieve satisfactory program efficiency. Actually, many of the transformations are executed automatically by an optimizer. There may even be cases where hand-tuning by the programmer and automatic optimization by the compiler or by an optimizing preprocessor (KAP, VAST etc.) cancel out one another, in so far as the overall speedup effect is concerned. Inside knowledge of the computer system (especially the compiler) is necessary in order to achieve positive effects.

From the large number of possible transformations, the programmer should choose those which seem useful for improving performance and which cannot be done by the compiler with satisfactory results. Details about the transformations potentially carried out by the compiler at its different optimization levels can be found in the respective handbook. However, as to which transformations are actually carried out for the different parts of the program can usually only be discerned by analyzing the object code generated by the compiler, which requires a considerable amount of effort on the part of the user due to the poor readability of assembler code.

Thus, sometimes there is no other way to optimize the program than by trial-and-error, i. e., to try different transformations by way of experiment. Similarly, the different optimization strategies of the compiler (like the depth of loop unrolling etc.), controlled by suitable compiler directives, can be tested.

6.5 Architecture Independent Optimizations

Transformations which improve program efficiency on practically all computer systems are said to be *architecture independent*. All transformations decreasing overhead of any kind belong to this category.

In the first place, such transformations are those which *combine common terms*. By doing so, the effort for evaluating identical terms several times is disposed of. Secondly, all transformations designed to avoid *type conversions* belong to this category. For example, if variables of different types are part of an expression, then the number of necessary type conversions can be reduced to a minimum if as many variables of equal type as possible are combined in subexpressions first.

For frequently called subroutines with very short runtimes, parameter passing and the subroutine call require an unproportionally high amount of work. This can be avoided by inserting the subroutine source code instead of the subroutine call (*inlining*).[13] Most compilers provide an option for automatic inlining. However, extensive inlining may result in a decrease in performance.

Another example of architecture independent transformations are transformations which reduce the complexity of an operation without changing the result (at least algebraically). This may be as simple as replacing time consuming divisions with considerably faster multiplications by the reciprocal value or other conversions based on mathematical identities. For example, the identity

$$\exp(x_1) \exp(x_2) \cdots \exp(x_n) = \exp(x_1 + x_2 + \cdots + x_n)$$

makes it possible to replace n time consuming exponentials and $n-1$ multiplication operations with a single exponential and $n-1$ addition operations. Moreover, this transformation avoids possible underflows and overflows in the evaluation of intermediate results.

Further examples and more detailed explanations of architecture independent transformations can be found in Dowd [45].

---

[13] Inlining has to be handled cautiously if the corresponding subroutine has side-effects. Side-effects alter global or static local variables which will keep those values after the subroutine has terminated. Simple inlining usually does not reproduce these side-effects correctly. In such cases inlining may change the behavior of the program.

6.6 Loop Optimizations

The overwhelming portion of the execution time of numerical programs is spent by instructions that are located in loops. It is thus not just by chance that many optimizing program transformations consist of loop modifications. However, efficient loop transformations are heavily machine and compiler dependent. They have to be repeated when the program is ported to another computer system if the program is to yield satisfactory results there, too.

The usefulness of various loop transformations on modern uniprocessor computers (workstations etc.), on which most numerical data processing is done nowadays, will now be discussed. For vector computers completely different considerations may be relevant (see e.g., Sekera [344], Dongarra et al. [166]).

6.6.1 Loop Unrolling

Maybe the most well-known performance increasing program transformation, loop unrolling, duplicates the body of a loop several times, whereby the loop control is adapted accordingly and an appropriate startup loop is inserted.

Example (SAXPY Operation)   The subroutine BLAS/saxpy (or its double precision version BLAS/daxpy) implements the operation $y = y + ax$, where $a \in \mathbb{R}$ and $x, y \in \mathbb{R}^n$. In the simplest case—when the elements of the vectors x and y are stored in contiguous array elements—the following program segment corresponds to

```
DO i = 1, n
   y(i) = y(i) + a*x(i)
END DO
```

In the standard implementation (Dongarra et al. [11]) this loop is *three-fold*[14] unrolled, i.e., this loop is transformed into a loop with an unrolling depth of *four*:

```
nrest = MOD(n,4)
DO i = 1, nrest                  !  startup loop (non-unrolled part)
   y(i) = y(i) + a*x(i)
END DO
n1 = nrest + 1
                                 !  3-fold unrolled loop
DO i = n1, n, 4                  !  unrolling depth: 4
   y(i  ) = y(i  ) + a*x(i  )
   y(i+1) = y(i+1) + a*x(i+1)
   y(i+2) = y(i+2) + a*x(i+2)
   y(i+3) = y(i+3) + a*x(i+3)
END DO
```

Loop unrolling results in a number of performance increasing effects. Since m-fold loop unrolling decreases the number of loop iterations by a factor $m+1$, the break condition of the loop has to be checked less frequently. This reduces the loop control overhead. However, it is of much greater effect on the performance

---

[14]Some authors call this a *four-fold* unrolled loop.

increase that loop unrolling increases the degree of parallelism that the compiler can potentially employ. Without unrolling, loop control quantities must be checked between single iterations. This may create dependencies which restrict the potential number of instructions to be executed in parallel; in fact the loop iterations have to be executed strictly sequentially, because after each iteration the break condition has to be checked in order to decide whether or not to execute the next iteration. On the other hand, with m-fold unrolled loops, it is obvious to the compiler that $m+1$ subsequent iterations can be executed. In the best case—if there are no dependencies between individual iterations—the degree of software parallelism is increased by the factor $m+1$.

Example (SAXPY Operation) The SAXPY operation described above with the vector length $n = 3000$ in the original form $(m=0)$ as well as in the m-fold unrolled form is executed on a PC. The performance values are depicted in Fig. 6.4. In the best case, unrolling the SAXPY loop yields a performance increase of 50 % on the computer system which is currently used.

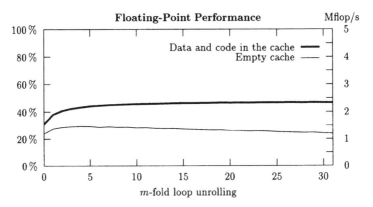

Figure 6.4: Performance increase due to the unrolling of an SAXPY loop.

Loop unrolling may also dispose of unnecessary data transfers between processor registers and the slower levels of the memory hierarchy if an array element used in an iteration is referenced again in the next iteration. The reason for this is that compilers usually do not recognize such a constellation. Rather they remove the respective variable from the register file immediately after the completion of the first iteration, so that the same value has to be fetched again from a slower level of the memory hierarchy a little later. Loop unrolling generally improves this inefficient strategy.

The extent to which loop unrolling increases the efficiency, if it does so at all, depends on a number of factors:

The size of the loop body: If the loop body already contains a sufficiently large number of instructions that can be executed in parallel, then loop unrolling (which, after all, is supposed to increase the number of parallelizable instructions) is not advantageous.

The number of iterations: If the number of iterations is smaller or just slightly larger than the unrolling factor, then a lot of time is required for the startup loop. The speedup of the main loop due to loop unrolling does not make up for this additional cost.

Loop subroutines: If the loop body contains one or more subroutine calls, then loop unrolling usually does not improve performance. Firstly, this is due to the fact that the overhead for loop control is much smaller than the overhead for the subroutine call. Moreover, subroutine calls are sources of possible data dependencies (see Section 6.2.1). In such cases loop unrolling does not increase or just marginally increases the number of potentially parallel operations, unless the subroutine source code is inlined. Finally, if the subroutine itself comprises enough instructions which can be executed in parallel, then the further increase of the software parallelism is not necessary.

Loop branches: Similar to the existence of subroutine calls, in case branch conditions appear in the middle of the loop body, the number of parallelizable statements is limited due to control dependencies.

Flow dependencies: If subsequent iterations depend on one another (flow dependency), only a limited increase in the number of parallelizable operations is possible.

Example (Inner Product) If the loop

```
sum = 0.
DO i = 1, n
   sum = sum + x(i)*y(i)
END DO
```

is 1-fold unrolled, then the following program segment is obtained (in addition to the startup loop):

```
sum = 0.
DO i = 1, n, 2
   sum = sum + x(i)  *y(i)
   sum = sum + x(i+1)*y(i+1)
END DO
```

The two instructions in the loop cannot be executed in parallel because of the dependency on the variable sum. However, if *associative* transformations are employed (cf. page 256) the number of parallelizable instructions may be increased.

One of the drawbacks of loop unrolling is that the size, the complexity, and the risk of coding errors are increased. This disadvantage can be avoided if special subroutine libraries are used as basic building blocks for the algorithm. Such optimized libraries are available on most computers, e. g., machine specific versions of BLAS routines for basic linear algebra operations,[15] the efficiency of which is only rarely achieved by user-made programs.

---

[15]The machine specific version of such a subroutine library itself is *not* portable but its use is portable due to the uniform interface definitions.

6.6.2 Unrolling Nested Loops

With nested loops the innermost one need not be the best or the only choice for loop unrolling. The same considerations and restrictions as above apply if outer loops are unrolled.

Unrolling an outer loop may be an option when one of the causes discussed above rules out unrolling the innermost loop. Another important reason for unrolling an outer loop is the existence of variables in the loop body which do not depend on the outer loop index. Unrolling outer loops may make it possible to reuse such variables several times, which reduces the number of data transfers between processor registers and the slower levels of the memory hierarchy.

Example (Unrolling Outer Loops)   In the program segment

```
DO j = 1, n
   DO i = 1, n
      a(i,j) = d*b(i,j) + c(i)
   END DO
END DO
```

c(i) does not depend on the outer loop index j. Hence, it is a logical choice of approaches to unroll the outer loop:

```
DO j = 1, n, 2
   DO i = 1, n
      a(i,j  ) = d*b(i,j  ) + c(i)
      a(i,j+1) = d*b(i,j+1) + c(i)
   END DO
END DO
```

The performance levels obtained on an HP workstation are depicted in Fig. 6.5. Loop unrolling increases the performance by 26 %.

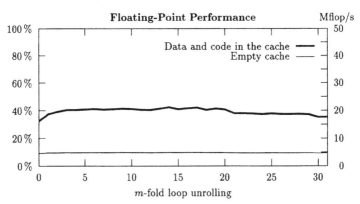

Figure 6.5: Performance increase due to unrolling the outer loop of a nested loop.

Increases in the locality of reference may be achieved through unrolling outer loops, particularly if the unrolled algorithm can be seen in terms of numerical linear algebra as a *blocked algorithm*. This is discussed in detail in Section 6.7.

6.6.3 Loop Fusion

Loop fusion is the grouping of two loop bodies within a single loop construct. The effects that can be achieved by using fused loops are to a great extent identical to those of loop unrolling: the loop control overhead is reduced and the number of parallelizable operations is increased.

Loop fusion may reduce the number of data transfers between the processor registers and the slower levels of the memory hierarchy if the different loop bodies reference the same data in corresponding iterations.

Example (Loop Fusion)   The two loops in the program segment

```
DO i = 1, n
   x = x*a(i) + b(i)
END DO
DO j = 1, n
   y = y*a(j) + c(j)
END DO
```

can be fused to the single loop

```
DO i = 1, n
   x = x*a(i) + b(i)
   y = y*a(i) + c(i)
END DO
```

The array element a(i) is thus used twice at each iteration, which reduces the number of memory access operations.

For $n = 3000$ the following performance values are obtained on an HP workstation:

	empty cache		data and code in the cache	
two individual loops	5.6 Mflop/s	(11.2 %)	15.4 Mflop/s	(30.8 %)
loop fusion	6.8 Mflop/s	(13.6 %)	30.2 Mflop/s	(60.4 %)

If data and code are held in the cache, then the floating-point performance is nearly doubled. This is possible because of the better exploitation of the floating-point pipeline and the concurrent execution of floating-point and integer operations. The resulting efficiency of 60 % peak performance is an acceptable value.

As with loop unrolling, loop fusion is applicable and reasonable only under certain circumstances. In particular, it is necessary that all loops to be fused perform the same number of iterations and that there are no dependencies between the iterations of different loops.

Moreover, it has to be taken into account that the locality of reference gets *worse* as a result of loop fusion if the loops reference different arrays!

6.6.4 The Elimination of Loop Branches

Branch conditions within the body of a loop frequently decrease performance to a considerable extent. This is mainly due to the fact that the control dependency induced by the branch reduces the number of potentially parallel executions. It is therefore obviously important to eliminate branches from loops. Whether or not this is possible depends on the way the branch condition and the iteration are related.

A branch is said to be *loop invariant* if the result of the branch condition does not depend on the loop iteration. In such a case the branch can easily be eliminated from the loop by transforming it into a branch with two loops.

Example (Loop Invariant Branch)   In the program segment

```
DO i = 1, n
   IF (indicator > 0) THEN
      y(i) = y(i) + a*x(i)
   ELSE
      y(i) = y(i) - a*x(i)
   END IF
END DO
```

the value of the logical expression indicator > 0 is independent of all loop iterations and is already available before the DO loop begins. Hence, the branch is loop invariant. If the branch is eliminated from the DO-loop, then the program segment takes the following form:

```
IF (indicator > 0) THEN
   DO i = 1, n
      y(i) = y(i) + a*x(i)
   END DO
ELSE
   DO i = 1, n
      y(i) = y(i) - a*x(i)
   END DO
END IF
```

For $n = 3000$ the following performance values are obtained on an HP workstation:

	empty cache		data and code in the cache	
IF in the loop	7.2 Mflop/s	(14.4 %)	15.0 Mflop/s	(30.0 %)
IF *outside* the loop	9.8 Mflop/s	(19.6 %)	34.8 Mflop/s	(69.6 %)

A branch is said to be *control dependent* if the result of the IF condition depends on the control variable but not on any other data. Typically this dependency follows a simple rule which makes it possible to move the branch out of the loop, splitting the original loop into one or more modified loops.

Example (Control Dependent Branch)   In the program segment

```
DO i = 1, n
   IF (MOD(i,2) == 0) THEN
      y(i) = y(i) + a*x(i)
   ELSE
      y(i) = y(i) - a*x(i)
   END IF
END DO
```

the IF branch is executed exactly for even loop indices i. The following program segment is equivalent but does not contain any IF branches:

```
DO i = 2, n, 2
   y(i) = y(i) + a*x(i)
END DO
DO i = 1, n, 2
   y(i) = y(i) - a*x(i)
END DO
```

On an HP workstation a performance value of only 5.9 Mflop/s can be achieved for a vector length $n = 3000$ and the original program segment, whereas by the program transformation, a value of 17.2 Mflop/s is achieved, a thrice better performance:

	empty cache	data and code in the cache
IF in the loop	3.2 Mflop/s (6.4 %)	5.9 Mflop/s (11.8 %)
no IF (two loops)	4.9 Mflop/s (9.8 %)	17.2 Mflop/s (34.4 %)

If the branch condition not only depends on the control variable, but also on other data (which also depends on the iteration), too, it is referred to as a *loop dependent* branch. In general, the branch *cannot* be eliminated from the loop, since the result of the IF condition depends on the respective iteration and can be expressed by a simple rule only in special cases.

Example (Loop Dependent Branch)   The program segment

```
DO i = 1, n
   IF (x(i) < 0) THEN
      y(i) = y(i) - a*x(i)
   ELSE
      y(i) = y(i) + a*x(i)
   END IF
END DO
```

basically performs a SAXPY operation with the absolute values of the components of the vector x. Hence the following program segment is equivalent:

```
DO i = 1, n
   y(i) = y(i) + a*ABS(x(i))
END DO
```

On an HP workstation a performance value of only 5.2 Mflop/s can be achieved for a vector length $n = 3000$ and the original program segment, whereas, by the program transformation, a value of 15.1 Mflop/s is obtained:

	empty cache	data and code in the cache
IF inside the loop	3.1 Mflop/s (6.2 %)	5.2 Mflop/s (10.4 %)
no IF	4.7 Mflop/s (9.4 %)	15.1 Mflop/s (30.2 %)

6.6.5 Associative Transformations

Program transformations based on the associative law

$$(x \circ y) \circ z = x \circ (y \circ z),$$

which holds for several binary operations such as real addition and multiplication or minimum and maximum, are referred to as *associative transformations*.

They have a special position within the collection of possible program transformations because the associative law does not hold for *floating-point* addition and multiplication operations (cf. Section 4.2.4)! Associative transformations alter the numerical results of floating-point operations. They are therefore *not* executed by the compiler (for which only equivalent program transformations are legal).[16] The programmer himself is responsible for such transformations, which are aimed at reducing rounding errors.

Associative transformations are applied primarily to *reduction operators*. In the context of array operations a reduction operation means that the dimension of the resulting array is smaller than the respective dimensions of the operands. For that to occur the elements of the operand arrays are usually combined componentwise and then the resulting elements are transformed to an array of lower dimension or a scalar (an array of dimension 0) by an associative operation. Suitable bracketing of the associative operation can then increase the number of potentially parallel operations.

Example (Inner Product) The simplest example of a reduction operation is the inner product, where elements of two vectors are first multiplied and then the sum of the results is calculated:

```
s = 0.
DO i = 1, n
    s = s + x(i)*y(i)
END DO
```

Since this expression can be bracketed as

$$x_1 y_1 + x_2 y_2 + \cdots + x_n y_n = (x_1 y_1 + x_3 y_3 + \cdots + x_{n-1} y_{n-1}) + (x_2 y_2 + x_4 y_4 + \cdots + x_n y_n),$$

for even n the above loop can be coded in the transformed version

```
s1 = 0.
s2 = 0.
DO i = 1, n, 2
    s1 = s1 + x(i  )*y(i  )
    s2 = s2 + x(i+1)*y(i+1)
END DO
s = s1 + s2
```

The two instructions in the loop body are—in contrast to the instructions obtained by loop unrolling—completely independent of one another. The generalization to the decomposition of a sum into m partial sums, of course increases the number independent operations.

---

[16]Some compilers have an option which explicitly allows for associative transformations.

The performances depicted in Fig. 6.6 were achieved on a workstation with vector length $n = 3000$. The performance gains for the transition from the 0-fold transformed (i. e., *not* transformed) loop to the 1-fold transformed loop are caused by the fact that a load operation and an arithmetic operation *independent* of it are executed in parallel. This parallelization is possible only through loop unrolling.

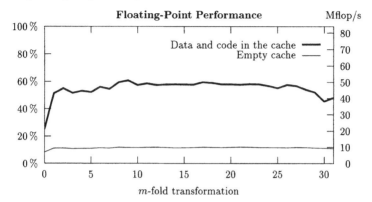

Figure 6.6: Performance gains caused by the associative transformation of the inner product.

Since floating-point addition is *not* associative, the transformed program segment is *not* equivalent to the original program segment. The programmer has to decide whether or not the changes in the influence of rounding errors are critical.

6.6.6 Loop Interchanges

When the body of a loop comprises another loop, this is referred to as loop nesting. Interchanging the order of nested loops alters the order in which the instructions in the loop body are executed and may increase the efficiency of a program. It is mainly the innermost loop which determines the order of floating-point operations and memory accesses. There are two major reasons for optimizing the structure of the innermost loop: Firstly, all the stride in which arrays are referenced should be as small as possible, as this improves the locality of reference of memory access operations. Secondly, the innermost loop should be well-suited for further optimization techniques such as loop unrolling, associative transformations and the like. These two goals may be contradictory. If this is so, optimal efficiency is possible only for a suitable permutation of the dimensions of the matrix—in the case of a matrix, this means transposing the matrix.

Example (Loop Interchanges)   In the program segment

```
REAL, DIMENSION (idim, jdim, kdim)  ::  array
...
DO i = 1, idim
   DO j = 1, jdim
      DO k = 1, kdim
         array(i,j,k) = a*array(i,j,k) + b
      END DO
   END DO
END DO
```

the three nested loops can be arranged in six different ways. The different loop orders can be named according to the scheme xyz, where x denotes the outermost loop, y the middle loop and z the innermost loop. On a DEC workstation, for the actual parameter values idim = 2, jdim = 40, and kdim = 800 the following performance values were achieved:

loop order	ijk	ikj	jik	jki	kij	kji
floating-point performance [Mflop/s]	12.3	25.7	12.3	12.6	29.1	25.4

From the point of view of locality of reference, the kji version is superior to the others as it references contiguous memory elements in subsequent iterations. However, the innermost loop comprises only two steps, which increases the loop overhead. Moreover, further optimization through loop unrolling is then redundant.[17] So the kij version achieves the best performance, even though it does not have an optimal locality of reference.

Yet another performance improvement can be achieved if the dimensions i and k of the array are permuted:

```
REAL, DIMENSION (kdim, jdim, idim)  ::  array
...
DO i = 1, idim
   DO j = 1, jdim
      DO k = 1, kdim
         array(k,j,i) = a*array(k,j,i) + b
      END DO
   END DO
END DO
```

This loop order produces an optimal locality of reference and can, moreover, be unrolled efficiently. This version achieves a performance of 43.6 Mflop/s, which is significantly better than that of the versions above.

6.7 Blocked Memory Access

The key to high-performance computing is to reuse data which has been transferred to a faster level of the memory hierarchy as often as possible before they are replaced by new data and restored at a slower level (see e. g., Lam, Rothberg, Wolf [264]). That means the ratio between floating-point operations and accesses to slower levels of the memory hierarchy should be maximal.

Blocked algorithms have proved an important tool for constructing transformations which rarely fail to achieve this goal. *Blocking* an $n \times m$ matrix A means decomposing it into pairwise disjoint submatrices A_{ij}, $i = 1, 2, \ldots, r$, $j = 1, 2, \ldots, s$. Each element of the original matrix belongs to exactly one block (submatrix).

For the sake of simplicity, it is assumed in the following that the individual blocks consist of $n_r = n/r$ consecutive rows and $m_s = m/s$ consecutive columns— the general case is discussed in Section 13.6.3. The decomposition of A then has the shape

---

[17] In this case the simple repetition of the loop body source code is a remedy.

$$A = \begin{pmatrix} A_{11} & A_{12} & \cdots & A_{1s} \\ A_{21} & A_{22} & \cdots & A_{2s} \\ \vdots & \vdots & & \vdots \\ A_{r1} & A_{r2} & \cdots & A_{rs} \end{pmatrix},$$

where

$$A_{ij} \in \mathbb{R}^{n_r \times m_s}, \qquad i = 1, 2, \ldots, r, \quad j = 1, 2, \ldots, s.$$

If an algorithm does not directly use the matrix elements but rather the submatrices A_{ij} as its operands it is referred to as a *blocked algorithm*.

Example (Blocked Matrix-Matrix Multiplication) The two matrices $A = (A_{ik}) \in \mathbb{R}^{n \times p}$ and $B = (B_{kj}) \in \mathbb{R}^{p \times m}$ are said to be *conformingly blocked* if

$$A_{ik} \in \mathbb{R}^{n_r \times p_s}, \qquad i = 1, 2, \ldots, r, \quad k = 1, 2, \ldots, s$$

and

$$B_{kj} \in \mathbb{R}^{p_s \times m_t}, \qquad k = 1, 2, \ldots, s, \quad j = 1, 2, \ldots, t$$

hold. In such a case, for the submatrices

$$[AB]_{ij}, \qquad i = 1, 2, \ldots, r, \quad j = 1, 2, \ldots, t$$

of the product matrix, the formula

$$[AB]_{ij} = \sum_{k=1}^{s} A_{ik} B_{kj}, \qquad i = 1, 2, \ldots, r, \quad j = 1, 2, \ldots, t \tag{6.2}$$

is derived. This formula already represents a blocked algorithm for computing the product AB.

Blocked algorithms are mechanically derived if linear algebra algorithms are first implemented according to their componentwise definition and one or more loops are unrolled afterwards.

Example (Blocked Matrix-Matrix-Multiplication) The nested loops

```
DO j = 1, m
   DO i = 1, n
      DO k = 1, p
         c(i,j) = c(i,j) + a(i,k)*b(k,j)
      END DO
   END DO
END DO
```

which stem from the componentwise definition of the matrix-matrix multiplication can be transformed by one-fold unrolling the two outer loops with unrolling depth 2:

```
DO j = 1, m, 2          !    unrolling depth: 2
   DO i = 1, n, 2       !    unrolling depth: 2
      DO k = 1, p, 1    !    unrolling depth: 1  (no unrolling)
         c(i  ,j  ) = c(i  ,j  ) + a(i  ,k)*b(k,j  )
         c(i+1,j  ) = c(i+1,j  ) + a(i+1,k)*b(k,j  )
         c(i  ,j+1) = c(i  ,j+1) + a(i  ,k)*b(k,j+1)
         c(i+1,j+1) = c(i+1,j+1) + a(i+1,k)*b(k,j+1)
      END DO
   END DO
END DO
```

This transformation results in four statements in the loop body which can be executed independently, implementing the matrix-matrix multiplication and addition operations

$$
\begin{pmatrix} c_{i,j} & c_{i,j+1} \\ c_{i+1,j} & c_{i+1,j+1} \end{pmatrix} = \begin{pmatrix} c_{i,j} & c_{i,j+1} \\ c_{i+1,j} & c_{i+1,j+1} \end{pmatrix} + \begin{pmatrix} a_{i,k} \\ a_{i+1,k} \end{pmatrix} \begin{pmatrix} b_{k,j} & b_{k,j+1} \end{pmatrix}.
$$

Thus, precisely the blocked algorithm (6.2) is implemented, with submatrix sizes $n_r = m_t = 2$ and $p_s = 1$.

In order to transform a matrix algorithm in such a way that a particular level of the memory hierarchy is exploited to a high degree, the algorithm is reformulated as a blocked algorithm. In order to do so, the submatrix sizes are chosen small enough such that for each partial operation of the blocked algorithm, all blocks needed as arguments can be stored in the respective level, on the one hand, and large enough so that the number of operations executed on elements of those blocks is maximal on the other hand. However, the fact that the effective size of a level of the memory hierarchy may be substantially smaller than its physical size has to be considered. For example, the operand registers of a processor not only store program data but the intermediate results of calculations as well. For cache memories the effective capacity may also be (perhaps significantly) smaller than its physical size (cf. Section 6.2.3). In order to increase the effective capacity of cache memories, the submatrices used as operands in a substep of the blocked algorithm can be copied to a contiguous memory area (referred to as *block copying*, see also Section 6.8.4), which disposes of the discrepancy between effective and physical capacity caused by the memory block placement strategy.

However, maximizing the number of executed operations is not the only key to optimal performance. At the register level, additional care should be taken that the operations executed on the elements of those blocks can be mapped on the parallel resources of the processor, which means specifically that it should be possible to execute the operations independently. For other levels of the memory hierarchy the way data is transferred (in lines or pages) may be the decisive argument for choosing a particular blocking strategy. For any line or page which has already been transferred, it is obviously reasonable to make as many operations involving elements from that line or page as possible. As a result of the complexity of all these factors, the optimal blocking strategy can often only be determined by way of experiment.

Example (Blocked Matrix-Matrix Multiplication)   The blocked algorithm (6.2) shows that the calculation of a matrix-matrix product consists of substeps of the form

$$
C_{ij} = C_{ij} + A_{ik} B_{kj},
$$

where the overall size S of the submatrices involved is given by

$$
S = n_r m_t + n_r p_s + p_s m_t
$$

and the number g of executed floating-point operations is

$$
g = 2 n_r p_s m_t.
$$

Here it is assumed that the range of the sizes n_r, p_s and m_t comprises real numbers, not only integers. The maximum of g under the constraint that S be equal to the capacity of the level of the memory hierarchy (in units of floating-point numbers) is achieved for $n_r = p_s = m_t = \sqrt{S/3}$. Thus, the operation count is maximal for quadratic or nearly quadratic matrices.

However, the block parameters $n_r = m_t = 3$ and $p_s = 1$ proved optimal in experiments aimed at the optimal exploitation of the 32 floating-point registers of an HP workstation, which does not correspond to quadratic blocking at all. The true reason for the optimal behavior of this blocking is that all its *multiply-add* operations are independent of one another and can therefore be mapped to parallel processor resources. If this block algorithm is implemented by unrolling the loops then the matrix-matrix multiplication, which fits in the cache, reaches efficiencies of 90 % and more due to the optimal exploitation of the register file.

6.7.1 Hierarchical Blocking

A blocked algorithm merely divides a task into subtasks on smaller matrices. The operations on the submatrices may themselves be described by blocked algorithms, which leads to the notion of *hierarchical blocking*.

If several levels of the memory hierarchy are to be exploited, then hierarchically blocked algorithms can be used; each blocking level exploits a corresponding level of the memory hierarchy.

Example (Hierarchically Blocked Matrix-Matrix Multiplication)   In this example the aim is not only to exploit the register file of an HP workstation, but to exploit its 256 KB cache as well. In doing so the matrix-matrix multiplication algorithm has to be blocked twice, resulting in the following program fragment:

```
DO j = 1, m, mt
  DO i = 1, n, nr
    DO k = 1, p, ps
      DO jj = j, j+mt-1, 3
        DO ii = i, i+nr-1, 3
          DO kk = k, k+ns-1
            c(ii  ,jj  ) = c(ii  ,jj  ) + a(ii  ,kk)*b(kk,jj  )
            c(ii+1,jj  ) = c(ii+1,jj  ) + a(ii+1,kk)*b(kk,jj  )
            c(ii+2,jj  ) = c(ii+2,jj  ) + a(ii+2,kk)*b(kk,jj  )
            c(ii  ,jj+1) = c(ii  ,jj+1) + a(ii  ,kk)*b(kk,jj+1)
            c(ii+1,jj+1) = c(ii+1,jj+1) + a(ii+1,kk)*b(kk,jj+1)
            c(ii+2,jj+1) = c(ii+2,jj+1) + a(ii+2,kk)*b(kk,jj+1)
            c(ii  ,jj+2) = c(ii  ,jj+2) + a(ii  ,kk)*b(kk,jj+2)
            c(ii+1,jj+2) = c(ii+1,jj+2) + a(ii+1,kk)*b(kk,jj+2)
            c(ii+2,jj+2) = c(ii+2,jj+2) + a(ii+2,kk)*b(kk,jj+2)
          END DO
        END DO
      END DO
    END DO
  END DO
END DO
```

The parameters of the inner blocks determined earlier are used explicitly. The parameters of the outer blocks can be set, for example, to mt = nr = ps = 120, giving the favorable quadratic block shape which is still small enough to store three such submatrices in the cache.

For example, for the matrix orders $m = n = p = 720$ this double blocking leads to a performance of 37.3 Mflop/s, which corresponds to 75 % efficiency. The floating-point performance of the single block variant for the same matrix order is only 14.7 Mflop/s.

The program package LAPACK, used for the solution of systems of linear equations and eigenvalue problems, comprises the blocked variants as well as the unblocked versions of many algorithms. However, blocked LAPACK algorithms employ just *one* blocking level. Thus, page faults may lead to substantial losses in performance even for the optimal block size, which ranges between 16 and 64 on the HP workstation used in the experiments (see Fig. 6.7).

A second blocking level may significantly improve the locality of reference for large matrices which cannot be held in main memory and, hence, significantly increase performance (see Fig. 6.7).

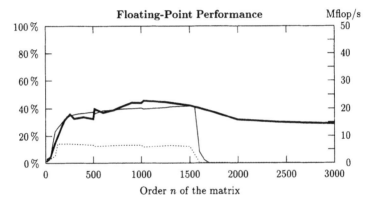

Figure 6.7: The floating-point performance of LU-factorization on an HP workstation: The routine LAPACK/sgetf2 (·····) is completely *unblocked*; LAPACK/sgetrf (——) is a *single block* routine, and slubr2 (——) is a *hierarchically double blocked* modification of the routine LAPACK/sgetrf.

6.8 Case Study: Multiplication of Matrices

The multiplication of two matrices is one of the fundamental operations in numerical linear algebra. Once an efficient implementation of this operation has been developed, other problems such as the solution of systems of linear equations (based on the LU-factorization) can be solved efficiently as well, using this fundamental software component as a building block.

The object of the discussion in this section is the *classical* algorithm for the multiplication of two quadratic[18] matrices $A, B \in \mathbb{R}^{n \times n}$:

---

[18]Assuming that the matrices are *quadratic* only helps to make the representation of the algorithms in this section more concise.

```
c = 0
DO i = 1, n, 1
   DO j = 1, n, 1
      DO k = 1, n, 1
         c(i,j) = c(i,j) + a(i,k)*b(k,j)
      END DO
   END DO
END DO
```

This algorithm requires n^3 floating-point multiplications and n^3 floating-point additions.[19] The performance measurements were executed on a widely used HP workstation. The results can be reproduced only on an identical computer system (sizes of the levels of the memory hierarchy, operating system version, compiler etc.). However, the *qualitative* results can be transferred to other computer systems, as well. Only the quantitative statements about the performance of the program variants discussed below have to be established individually for a given computing environment.

6.8.1 Loop Interchanges

Six different ways to nest the loops are obtained by permuting the three loop variables. The different variants differ in the operation executed in the innermost loop—either the scalar product of 2 vectors (SDOT) or the multiplication of a vector by a scalar and its addition to another vector (SAXPY)—and in the strides of the innermost loop for accessing the matrix elements. The following table surveys the different variants:

loop order	innermost loop	stride		
		matrix A	matrix B	matrix C
ijk	SDOT	n	1	–
jik	SDOT	n	1	–
kij	SAXPY	–	n	n
ikj	SAXPY	–	n	n
jki	SAXPY	1	–	1
kji	SAXPY	1	–	1

Since it is the innermost loop which mainly dictates the characteristic behavior of each variant, there is no significant difference between variants with identical innermost loop indices. Fig. 6.8 thus contains only three variants with different *innermost* loops.

For the ijk variant the ratio between the number of cache write operations and the number of floating-point operations is minimal (1 write in $2n$ flops), whereas the ratio is much less favorable with the other two variants (n writes in $2n$ flops). This ratio is important since write operations usually take longer than other operations (HP workstations: 1.5 clock cycles for a write compared to 1 clock cycle for a read or floating-point operation).

---

[19] Algorithms of lower complexity are discussed in Section 5.5.5.

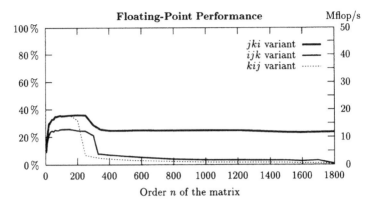

Figure 6.8: The influence of the loop order on the floating-point performance.

However, due to the high locality of reference, the jki variant is by far superior to the others. For matrix orders of 300 and more both of the other variants display a significantly decrease in performance, which is due to the higher number of cache misses. The misses are caused by the fact that frequently referenced parts of the matrices A, B, and C cannot be stored in the cache.

6.8.2 Loop Unrolling

More variants are created by loop unrolling with different unrolling depths. In the program segment

```
DO i = 1, n, 4          !  unrolling depth: 4
   DO k = 1, n, 4        !  unrolling depth: 4
      DO j = 1, n, 1     !  unrolling depth: 1   (no unrolling)
         c(i  ,j) = c(i  ,j) + a(i  ,k  )*b(k  ,j)
         c(i  ,j) = c(i  ,j) + a(i  ,k+1)*b(k+1,j)
         c(i  ,j) = c(i  ,j) + a(i  ,k+2)*b(k+2,j)
         c(i  ,j) = c(i  ,j) + a(i  ,k+3)*b(k+3,j)

         c(i+1,j) = c(i+1,j) + a(i+1,k  )*b(k  ,j)
         c(i+1,j) = c(i+1,j) + a(i+1,k+1)*b(k+1,j)
         c(i+1,j) = c(i+1,j) + a(i+1,k+2)*b(k+2,j)
         c(i+1,j) = c(i+1,j) + a(i+1,k+3)*b(k+3,j)

         c(i+2,j) = c(i+2,j) + a(i+2,k  )*b(k  ,j)
         c(i+2,j) = c(i+2,j) + a(i+2,k+1)*b(k+1,j)
         c(i+2,j) = c(i+2,j) + a(i+2,k+2)*b(k+2,j)
         c(i+2,j) = c(i+2,j) + a(i+2,k+3)*b(k+3,j)

         c(i+3,j) = c(i+3,j) + a(i+3,k  )*b(k  ,j)
         c(i+3,j) = c(i+3,j) + a(i+3,k+1)*b(k+1,j)
         c(i+3,j) = c(i+3,j) + a(i+3,k+2)*b(k+2,j)
         c(i+3,j) = c(i+3,j) + a(i+3,k+3)*b(k+3,j)
      END DO
   END DO
END DO
```

each of the outer loops are three-fold unrolled. The variant with increments 4, 4 and 1 for the i, k, and j loops is called the $i_4k_4j_1$ variant. The loop body contains 16 statements as opposed to a single statement as is the case without unrolling.

Loop unrolling not only increases the locality of reference, but also gives the Fortran compiler more opportunities to execute performance increasing program transformations.

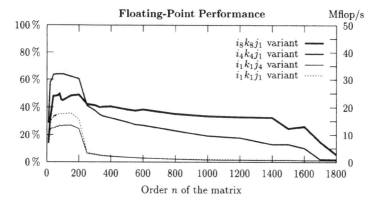

Figure 6.9: Performance comparison between different types of loop unrolling with the ikj variant.

Fig. 6.9 depicts the comparison of the floating-point performances of the ikj variants with different unrolling depths. The $i_1k_1j_1$ variant (i.e., without unrolling) generally performs the weakest, whereas the $i_8k_8j_1$ variant, which is obtained from the 7-fold unrolling of the two outer loops, achieves uniformly satisfactory performance. If, however, all the required data (in this example *all* elements of the matrix B, four rows of the matrix C and 16 elements of the matrix A) can be stored in the cache, then the $i_4k_4j_1$ variant with three-fold loop unrolling results in the best performance.

The behavior of less extensively unrolled variants for smaller matrices is due to better exploitation of the registers. Three-fold loop unrolling requires just around 30 floating-point registers for matrix elements, whereas 7-fold loop unrolling would require around 100 registers. However, as there are only 64 floating-point registers available on the HP workstation used for the case study, register reallocations adversely affect the performance. Three-fold loop unrolling is thus better for small matrices which can be stored in the cache as a whole because this variant makes better use of the registers. For larger matrices, however, this effect is outweighed by the high locality of reference and, hence, better exploitation of the cache, which is achieved by 7-fold loop unrolling and leads to higher floating point performance.

All ikj variants display a significant performance deterioration once the matrices do not fit into the cache ($n > 200$) or the main memory ($n > 1600$); (see Fig. 6.9).

Fig. 6.10 depicts the comparison between different types of loop unrolling with the jki variant. In contrast to the previous variant, this one displays little performance loss for large matrices.

Loop unrolling does not necessarily increase the performance. From Fig. 6.10 it can be seen that the $j_1 k_1 i_4$ variant performs worse than even the $j_1 k_1 i_1$ variant.

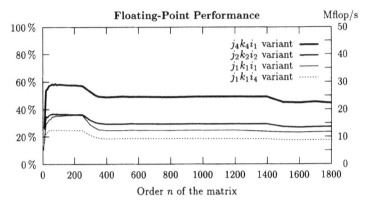

Figure 6.10: Performance comparison between different types of loop unrolling with the jki variant.

6.8.3 Blocking

In the following it is assumed, for the sake of the convenient representation of blocked matrix-matrix multiplication algorithms, that A and B are decomposed into quadratic submatrices

$$A = \begin{pmatrix} A_{11} & A_{12} & \cdots & A_{1b} \\ A_{21} & A_{22} & \cdots & A_{2b} \\ \vdots & \vdots & & \vdots \\ A_{b1} & A_{b2} & \cdots & A_{bb} \end{pmatrix} \quad \text{and} \quad B = \begin{pmatrix} B_{11} & B_{12} & \cdots & B_{1b} \\ B_{21} & B_{22} & \cdots & B_{2b} \\ \vdots & \vdots & & \vdots \\ B_{b1} & B_{b2} & \cdots & B_{bb} \end{pmatrix},$$

to which one of the six variants of matrix-matrix multiplication is applied, for example

```
DO ii = 1, n, nb
  DO kk = 1, n, nb
    DO jj = 1, n, nb
      C_{ii,jj} := C_{ii,jj} + A_{ii,kk} B_{kk,jj}
    END DO
  END DO
END DO
```

Any of the six variants may be chosen for the multiplication $A_{ii,kk} B_{kk,jj}$ whence there are 36 variants of blocked matrix-matrix multiplication.

The order of the outer loops is irrelevant for the performance of blocked matrix-matrix multiplication. It is still the innermost loop which mainly determines the level of performance. The following program segment stems from the ikj-jki variant of blocked matrix-matrix multiplication operation:

```
DO ii = 1, n, nb
   DO kk = 1, n, nb
      DO jj = 1, n, nb
         DO j = jj, MIN(n, jj+nb-1)
            DO k = kk, MIN(n, kk+nb-1)
               DO i = ii, MIN(n, ii+nb-1)
                  c(i,j) = c(i,j) + a(i,k)*b(k,j)
               END DO
            END DO
         END DO
      END DO
   END DO
END DO
```

Fig. 6.11 is based on performance data determined with the block size nb = 128, which proved optimal on the workstation used for this case study. However, this is not a very significant optimum. For other blocking factors (ranging from 80 to 256) the performance is nearly the same.

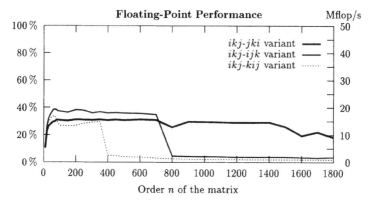

Figure 6.11: Blocked matrix-matrix multiplication operations (*without* loop unrolling).

6.8.4 Block Copying

Additional auxiliary arrays containing copies of blocks (submatrices of A, B, or C) may boost the performance of blocked multiplication routines. All computations are performed on the auxiliary arrays so that there are no cache conflicts and so that the stride for accessing the elements never exceeds the number of rows nb in the block. Without those auxiliary arrays the stride could equal n in the worst case.

Another advantage of using auxiliary arrays is that the blocks may be *transposed* while they are copied so that the innermost loops access the elements with minimal stride.

Block copying is a good option for the matrix A. The ijk variant is then also of high locality of reference, and has a favorably small number of write operations.

The following program segment with block copying and the transposition of A stems from the ikj-jki variant of the blocked matrix-matrix multiplication operations:

```
DO ii = 1, n, nb
   DO kk = 1, n, nb
      DO kb = kk, MIN(n, kk+nb-1)
         DO ib = ii, MIN(n, ii+nb-1)
            aa(kb-kk+1,ib-ii+1) = a(ib,kb)  !  copy and transpose
         END DO
      END DO
      DO jj = 1, n, nb
         DO j = jj, MIN(n, jj+nb-1)
            DO k = kk, MIN(n, kk+nb-1)
               DO i = ii, MIN(n, ii+nb-1)
                  c(i,j) = c(i,j) + aa(k-kk+1,i-ii+1)*b(k,j)
               END DO
            END DO
         END DO
      END DO
   END DO
END DO
```

There is a significant increase in performance for the ikj-ijk variant with larger matrices ($n > 700$) since at this point the elements are now referenced columnwise rather than rowwise. For other variants there are hardly any gains in performance. The performance data depicted in Fig. 6.12 were again determined using the machine-specific block size nb = 128.

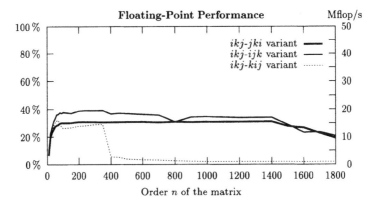

Figure 6.12: Blocked matrix-matrix multiplication operations with block copying (*without* loop unrolling).

6.8.5 Blocking, Copying and Loop Unrolling

Combining all performance increasing techniques—loop unrolling, blocking and copying—leads to special multiplication algorithms with maximal performance gains. For example, the following program segment developed by Dongarra, Mayes and Radicati [167] yields truly satisfactory performance (see Fig. 6.13):

```
DO kk = 1, n, nb
   kspan = MIN(nb, n-kk+1)
   DO ii = 1, n, nb
      ispan = MIN( nb,n-ii+1)
      ilen = 2*(ispan/2)
      DO i = ii, ii + ispan - 1
         DO k = kk, kk + kspan - 1
            ch(k-kk+1,i-ii+1) = a(i,k)
         END DO
      END DO
      DO jj = 1, n, nb
         jspan = MIN(nb, n-jj+1)
         jlen = 2*(jspan/2)
         DO j = jj, jj + jlen - 1, 2
            DO i = ii, ii + ilen - 1, 2
               t11 = 0.
               t21 = 0.
               t12 = 0.
               t22 = 0.
               DO k = kk, kk + kspan - 1
                  t11 = t11 + ch(k-kk+1,i-ii+1)*b(k,j  )
                  t21 = t21 + ch(k-kk+1,i-ii+2)*b(k,j  )
                  t12 = t12 + ch(k-kk+1,i-ii+1)*b(k,j+1)
                  t22 = t22 + ch(k-kk+1,i-ii+2)*b(k,j+1)
               END DO
               c(i  ,j  ) = c(i  ,j  ) + t11
               c(i+1,j  ) = c(i+1,j  ) + t21
               c(i  ,j+1) = c(i  ,j+1) + t12
               c(i+1,j+1) = c(i+1,j+1) + t22
            END DO
         END DO
      END DO
   END DO
END DO
```

This is a multiplication variant using blocking combined with loop unrolling. The two outer loops of the inner block (the j- and i-loops) are one-fold unrolled, and a block of the matrix A is copied and transposed.

6.8.6 Optimizing System Software

All variants of matrix-matrix multiplication operations discussed so far were manually transformed with the aim of exploiting hardware resources better and more uniformly. High reference locality has led to the better exploitation of cache

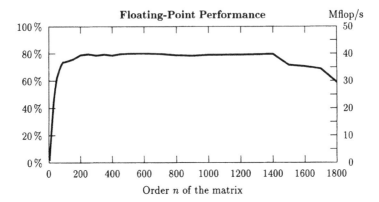

Figure 6.13: Matrix-matrix multiplication with blocking, copying and loop unrolling.

memory and registers, sequentializing the instructions in the innermost loops has improved the pipeline exploitation etc.

Many program transformations can be performed automatically by optimizing compilers. For example, HP workstations provide an optimizing Fortran pre-compiler with a special *vector library* of frequently used, simple subroutines. The subroutines of this library by and large correspond to the set of BLAS-1 routines together with some BLAS-2 and BLAS-3 routines. However, names and subroutine interfaces differ from the BLAS programs. For example, the HP Fortran pre-compiler transforms the program segment

```
DO i = 1, n, 1
   DO j = 1, n, 1
      DO k = 1, n, 1
           c(i,j) = c(i,j) + a(i,k)*b(k,j)
      END DO
   END DO
END DO
```

into the subroutine call

```
CALL blas_$sgemm ('n','n',(m),(n),(k),0.,a(1,1),lda,b(1,1),ldb,1.,c(1,1),ldc)
```

In this transformation, all three nested loops are replaced by *one* call of a subroutine from the vector library which has been optimized for a particular machine. The performance of this routine is depicted in Fig. 6.14.

Comparison of the performances of the manually tuned matrix-matrix multiplication program and the automatically optimized program clearly demonstrates the superiority of the diligent optimization carried out by experts. For large matrices, the manually optimized program never performed less than *four* times better than the automatically optimized version. The reason is the inadequate implementation of `blas_$sgemm`.

For other machines the use of automatically optimized software leads to better results; e.g., VAST/ESSC on an RS/6000 workstation nearly reaches peak performance.

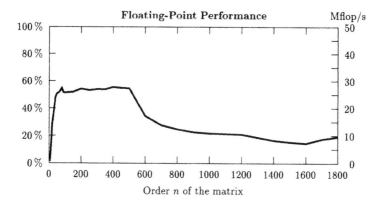

Figure 6.14: A matrix-matrix multiplication operation with pre-compiler optimization.

The choice of a suitable variant of the algorithm and the fine-tuning of programs involves considerable individual expenditure, which only pays off if performance is critical. For rarely used programs performance improvements achieved through (pre)compiler optimization are sufficient. In such cases the exquisitely concise (Fortran 90) statement `c = MATMUL(a, b)` is to be favored, leaving any optimization to the system software.

Chapter 7

Available Numerical Software

Our business is taking math and hiding it from the user.
We create tools for doing engineering and scientific work
without fooling with the math.
We focus on the applications that can be derived from the math.

JIM TUNG (Manager of The MathWorks Inc.)

Software describes all non-physical (non-hardware) parts of data processing equipment. Unlike hardware, software is not tangible. Software, although held in a physical medium, say on a disk storage unit, is composed of programs and data arranged in logical, not physical, structures.

System software comprises the operating system, language processors (compilers), and utilities (file management etc.), i. e., all programs and documentation necessary to control the system's hardware and software resources and to assist the development and execution of application programs.

Application software comprises programs written by users for a particular purpose and *packaged* programs for a wide but restricted range of purposes written to be marketed. Application software ranges from text processing and commercial problem solving to technical applications, such as the numerical simulation of crash tests in automobile development or the analysis of the dynamic behavior of semiconductor devices.

Numerical software is used to solve (mathematical) problems numerically: approximation problems, systems of linear or nonlinear algebraic or differential equations, etc. In the main, numerical software is an important tool in the development of technical and scientific application software.

7.1 The Cost of Software

Profitability studies in the field of software development are usually based on staff costs. Other cost factors (computer usage etc.) are of minor importance. The estimation of staff costs is at the heart of all calculation methods for estimating software costs.

The *lines of code*[1] (LOC) completed per person and per unit of time (generally months or years) are used as rough estimates of programmer productivity. The LOC value is determined after the completion of a software project. It includes

---

[1] Only executable lines and declaration lines are counted or estimated in imperative programming languages. Comment lines and blank lines are not considered.

all activities involved in the project (from the specification to the acceptance of the finished product).

Extensive empirical studies have shown that programmer productivity may vary by several orders of magnitude. Extreme values are about 5 and 5000 lines of code per person per month. The average value is around 250 coded lines per person per month.

Example (IMSL, NAG)   The following productivity values for numerical software systems and libraries can be used as reference points:

  EISPACK: 55 lines per person per month,

  IMSL: 160 lines per person per month,

  NAG: 260 lines per person per month.

Despite comparatively high productivity, the whole expenditure for developing numerical program libraries is enormous. More than 1000 *years* of manpower were invested in the NAG Fortran Library.

Subjective estimates of unexperienced programmers are not an appropriate basis for realistic cost studies. A programming novice writes a program with 100 or more lines within a very short time (possibly a few hours). A naive projection would thus predict a monthly production of more than 4000 lines of code. This productivity estimate is more than an order of magnitude too high! The reason is, among other things, the exponentially increasing complexity of larger software systems. Moreover, beginners do not take into account the time required for serious tests and sufficient documentation.

Productivity Factors

An empirical examination of the factors which influence programmer productivity to the greatest extent showed the following results:

The complexity of the project, requirements concerning the user interface, the practical experience of the people involved, and efficiency requirements influence the productivity of the programmer greatly. Programming methodology and the size and kind of required documentation play a key role.

These results demonstrate, for instance, that increasing user interface requirements and other critical demands lead to a *decrease* in productivity. Despite the continuously increasing quality of software products, there is no increase in productivity to be expected in the future.

The introduction of the first high-level programming languages resulted in significant productivity increases in the 1960s. Since then productivity in program development has doubled or even tripled as a result of raising the power of programming languages. Hardware performance, on the other hand, improved many orders of magnitude in the same period.

Universally Applicable Modules

Significant increases in productivity in the field of numerical data processing can be achieved by using standard software elements.

Example (LINPACK)   If somebody tried to develop a program with the functionality of the LINPACK routines sgeco, sgefa and sgesl (and the auxiliary BLAS routines saxpy, sdot, sscal, sasum and isamax), i.e., a program which solves a system of linear equations (and computes an estimate of the condition number of the system matrix) efficiently and reliably, then the following expenditure of personnel time would be required:

The 8 Fortran subroutines comprise (without blank lines and comments) about 325 lines of code. Assuming a productivity of between 150 and 450 lines per person per month, 1–2 months would be needed for the software development.

This expenditure of time is orders of magnitude larger than the comparatively negligible expense of obtaining the LINPACK routines (via the Internet) which are readily available free of charge.

This example shows that low acquisition costs (of a program package whose range of application is much greater than what is needed in this particular case) can result in significant savings in staff expenses.

Software Maintenance Costs

Software costs have to be divided into development and maintenance costs. Up until now the primary focus has been on development costs. However, maintenance costs cannot be ignored (except where *one shot* programs, which are used only once, are concerned). On very large projects, maintenance costs can amount to up to 2/3 of total costs. On the average, the cost of maintaining a numerical software product is approximately the same as developing it. Likewise, the use of standard software has significant advantages as far as maintenance costs are concerned, especially when software, such as the IMSL and the NAG libraries, is serviced professionally.

7.2 Sources of Numerical Software

If the decision has been made to utilize existing software then the next step is to choose a product appropriate to the problem. This section gives a general overview of available numerical software products. The following chapters deal with the solution of numerical problems (interpolation, integration, etc.) using numerical algorithms implemented in existing software. In these chapters, reference is made to the software products introduced in this section.

7.2.1 Numerical Application Software

Application software makes it possible to solve problems in a specific field of application. The user interface of these software products is generally designed in such a way that it is possible to define a problem within a certain scientific or technical context.

Example (VLSI Design)   The design of very large scale integrated circuits (VLSIs) is only possible if effective simulation software is used. The basic components of VLSI simulation are transistors, resistors, etc. as well as current and voltage sources. A circuit is described by systems of linear and nonlinear algebraic equations and systems of ordinary differential equations. During the simulation and analysis of a circuit, the currents and voltages in the circuit are computed using these equations. Special application software is available for VLSI design, e. g., SPICE.

Application software is composed of operative modules, control modules and interfaces. Within the scope of many operative modules and interfaces of application software, the solution of certain standard numerical data processing problems is required: systems of algebraic or differential equations, linear optimization problems, etc. are to be solved.

During the development of application software, existing software elements for the solution of these standard mathematical problems should be available. Software for standard mathematical problems would therefore be prototypical numerical software.

In the following sections various sources of ready-made numerical software are described: programs published in journals and books, numerical software libraries as well as numerical program packages.

7.2.2 Printed Programs

Programs Published in Journals

In the 1960s two periodicals—*Numerische Mathematik* and *Communications of the ACM*—started to publish numerical programs. At that time the programming language Algol 60 was predominantly used for making algorithms public. *Numerische Mathematik* stopped publishing programs a long time ago. However, many of the programs published in that journal are still used in updated versions in current program libraries or packages (e. g., in LAPACK). The ACM (*Association for Computing Machinery*) has been publishing the *Collected Algorithms of the ACM* through the journal *Transactions on Mathematical Software* (TOMS) since 1975. All the programs published in TOMS can be obtained free of charge in machine-readable form via the Internet service NETLIB (cf. Section 7.3.7).

Programs published in journals are mostly written in Fortran. The overall quality of the programs is good: All published programs are refereed by experts before they are published, as is usual in highly regarded technical journals. However, the extent and the thoroughness of these reviews do not guarantee the same level of quality as found in numerical libraries or special numerical program packages.

In addition to the programs found in *Transactions on Mathematical Software*, other journals e. g., the *Computer Journal* or the *Journal of Computational and Applied Mathematics* (JCAM) sometimes publish interesting programs and algorithms. These are not, however, available in machine readable form.

Programs Published in Books

When discussing books which deal with numerical programs, *Numerical Recipes* written by Press, Flannery, Teukolsky and Vetterling must be mentioned. Many algorithms and techniques in the most important fields of scientific computing are explained in an easily understandable manner and are illustrated in clearly structured (though not necessarily very efficient) programs.

Five versions of *Numerical Recipes*—with programs in Fortran 77 [24], Fortran 90 [25], C [26], Pascal and QuickBasic—are available; CD-ROMs with the complete source code in all the available languages in PC, Macintosh, or UNIX format are also available in book shops. Fortran and C programs can be obtained via the Internet from the Gopher server address `cfata4.harvard.edu`.

An additional volume with illustrative examples in the programming languages Fortran 77 [27], C [28] and Pascal can be obtained.

7.2.3 Numerical Software Libraries

A library is an organization where books are systematically registered, maintained, and made accessible. A software library is a similar organization in the field of computer software. The continuous work required for the running of a software library can only be accomplished with an enormous number of personnel (see Cowell [10], Chapters 10 to 14).

IMSL

IMSL (*International Mathematical and Statistical Libraries*) is the software library developed and operated by Visual Numerics Inc., 9990 Richmond Avenue, Suite 400, Houston, Texas 77042-4548, USA.

IMSL was founded in 1971 by former staff members of the IBM project *Scientific Software Package* (SSP). It is a commercial organization whose aim is to develop and distribute numerical software. The first numerical Fortran program libraries were made available to customers with IBM computers as early as 1971. In 1973, the CDC and Univac software libraries were released. Nowadays IMSL libraries can be used on all current computer systems (mainframes, workstations and PCs). Some years ago IMSL and Precision Visuals merged and became Visual Numerics Inc.

Their numerical software libraries include the IMSL Fortran Library, the IMSL C-Math Library and the IMSL C-Stat Library. These libraries are maintained and continue to be developed by a large number of leading (mainly American) scientists (see Cowell [10], Chapter 10).

NAG Software Libraries

Another numerical software library service with an international reputation comes from NAG Ltd. (*Numerical Algorithms Group*), Wilkinson House, Jordan Hill Road, Oxford OX2 8DR, England.

NAG was founded in Great Britain as the *Nottingham Algorithms Group* in 1970. Its goal was to develop numerical software for ICL computers, which were widely used at British universities at that time. Later, with the help of state subsidies, it became a not-for-profit company—independent of universities and computer manufacturers (see Cowell [10], Chapter 14).

NAG's most important numerical software library is the Fortran library; its current version (Mark 17) has 1152 Fortran 77 programs which can be called by the user, and is therefore the most extensive NAG product. In addition, there is also a C library, a Fortran 90 library and a *Parallel Library* (written in Fortran 77) for distributed-memory systems.. Users who do not need the whole Fortran library can take advantage of the *Foundation Library*.

Other Manufacturer Independent Software Libraries

In addition to the IMSL and NAG software libraries, there are other libraries which are independent of computer manufacturers:

The **Harwell Subroutine Library**—Atomic Energy Research Establishment, Computer Science and Systems Division, Harwell Laboratory, Didcot, Oxfordshire OX11 0RA, England, United Kingdom;

The **SLATEC Common Mathematical Library**—Computing Division, Los Alamos Scientific Laboratory, New Mexico 87545, USA (see Cowell [10], Chapter 11);

CMLIB —Center for Computing and Applied Mathematics, National Institute of Standards and Technology (NIST), Gaithersburg, Maryland 20899, USA;

PORT —AT&T, Bell Laboratories, Murray Hill, New Jersey 07974, USA (see Cowell [10], Chapter 13);

BOEING Math. Software Library—Boeing Computer Services Company, Tukwila, Washington 98188, USA (see Cowell [10], Chapter 12).

These libraries are, however, not as internationally well known and widely used as the IMSL and the NAG libraries.

Software Libraries Provided by Computer Companies

Most companies producing mainframes or workstations designed to process numerical data offer their customers software libraries for the optimized utilization of their machines. The programs from these libraries often (but not always) show better performance than the programs from manufacturer independent software libraries, but they are normally not portable.

Examples for such software libraries are:

ESSL: The *Engineering and Scientific Subroutine Library* (ESSL) is an IBM product which includes efficient mathematical routines. The ESSL was originally developed for IBM mainframes with *vector facilities* (3090 VF, later ES/9000). Later IBM also developed a version for RS/6000 workstations.

ESSL versions for non-IBM computers will soon be available.

ESSL contains a *vector library*, a *scalar library* and some subroutines for parallel computers. The programs in the scalar library are meant to be used on single processor computers (IBM RS/6000 workstations); the vector routines are designed to utilize the computing power provided by the *vector facility* of the IBM mainframes. ESSL routines take the memory hierarchy into account by using blocked algorithms. The data blocks are chosen in such a way that they fit the available cache memory. Regardless of this adaptation to special hardware features, efficiency can deteriorate significantly if certain loop strides are used. The auxiliary program `stride` aids the user in choosing appropriate stride values.

The vector and multi processor performance of ESSL programs for parallel computers is optimized. For example, a parallel, three-dimensional FFT program, `scft2p`, is included in the ESSL library. It works in exactly the same way as the single processor program `scft3` except that it may use more processors—if they are available—in order to speed up the computation.

The **Cray SCILIB Mathematical Library:** Cray Research Inc. offers the SCILIB library along with their computers. It includes a machine-specifically optimized collection of Fortran and C programs.

The **Convex MLIB Library:** MLIB is a library of numerical programs with outstanding floating-point performance. It includes optimized versions of BLAS-1, BLAS-2 and BLAS-3, LAPACK, MINPACK programs, etc.

The **Convex VECLIB Library:** VECLIB is a collection of optimized numerical programs which can be called by Fortran programs. It provides basic mathematical software for Convex computers and is specially designed to be compatible with the vector architecture of Convex systems.

CMSSL: The *Connection Machine Scientific Software Library* (CMSSL) is a library of numerical programs which utilize to a large extent the potential performance of Connection Machines. CMSSL routines can be called from programs written in CM Fortran, *Lisp, Fortran/Paris, C/Paris or Lisp/Paris.

7.2.4 Numerical Software Packages

A software package is a systematically developed and assembled collection of software programs designed for use in particular fields. In contrast to software libraries, most software packages lack permanent and systematic maintenance.

In 1971 the National Activity to Test Software (NATS) project was founded by the American National Science Foundation (NSF) and the Atomic Energy Commission. The objective was to produce and distribute numerical software of the highest possible quality. As prototypes, two software packages were developed:

EISPACK which solves matrix eigenproblems (Smith et al. [30]) and

FUNPACK which computes the values of special functions (Cody [138]).

They were developed, for the most part, at Argonne National Laboratory and at Stanford University. Testing was carried out at the University of Texas in Austin and at a number of other places. With the publication of the second versions of the two packages, the project was formally completed in 1976. Both packages were very successful as far as results and as far as what was learned about the organization of such projects were concerned. For the first time, a particular form of organization was chosen whereby collaborators working in different institutions in different locations were connected.

The performance and portability that EISPACK provided was a landmark in the field of numerical software. As a result, a number of software development groups use "PACK" as the suffix in their product names.

LINPACK was created to complement EISPACK for use in the field of linear equations and least squares data fitting problems (Dongarra et al. [11]).

LAPACK (*Linear Algebra Package*; see Sections 13.15 to 13.21 and 15.7) is a package of Fortran 77 subprograms designed for the direct solution of systems of linear equations and linear least squares problems (with dense or banded matrices) and for the computation of the eigenvalues and eigenvectors of matrices. LAPACK is the successor to LINPACK and EISPACK. The first version of LAPACK [3] was published in 1992. It is the best collection of linear algebra software available at the moment.

ITPACK is a software package designed for the iterative solution of large systems of linear equations with sparse matrices (especially matrices derived from the discretization of partial differential equations).

SPARSPAK is also a package created for the numerical solution of large systems of linear equations with sparse, especially positive definite matrices.

MADPACK was designed to solve systems of linear equations using the multi-grid method,

MINPACK to solve systems of nonlinear equations and optimization tasks,

TNPACK to solve large separable unconstrained minimization problems,

HOMPACK to solve systems of nonlinear equations using homotopy,

ODRPACK to solve nonlinear data fitting problems using orthogonal approximation (*orthogonal distance regression*).

PPPACK is designed to compute and manipulate *piecewise polynomials*, especially B-splines.

QUADPACK computes one-dimensional integrals and integral transformations of functions of one variable.

FFTPACK computes the fast Fourier transform of periodic data sequences.

VFFTPK is the *vectorized* version of FFTPACK. It is suitable for the simultaneous transformation of several data sequences.

ODEPACK solves initial value problems in ordinary differential equations.

ELLPACK solves elliptic partial differential equations in two dimensions on general domains or in three dimensions on cubes.

FISHPAK solves the Poisson equation in two or three dimensions.

There are a lot of excellent products among these software packages. However, the addition "PACK" (or "PAK") alone does not necessarily imply quality. Some other very good packages which do not have a "PACK" name are available. Some examples are

TOEPLITZ, which solves systems of linear equations with Toeplitz matrices,

CONFORMAL, which determines the parameters of Schwarz-Christoffel mappings;

VANHUFFEL, which solves data fitting problems in which the *orthogonal* distance between data and model is minimized;

LLSQ, which solves linear least squares problems;

PITCON which solves systems of nonlinear equations;

EDA, which is used for exploratory data analysis;

BLAS, which is used for elementary operations in linear algebra (see Section 4.10.1); and

ELEFUNT, which tests the implementation of elementary functions.

Most of these program packages can be obtained via NETLIB (see Section 7.3.7). Some packages are available from Visual Numerics Inc. or NAG Ltd. for only the cost of delivery.

7.2.5 References to Software in this Book

The extent and variety of the available numerical software is so large that a structured overview and sufficient background information is necessary in order to make the right decision in specific cases. The following chapters aim at providing the necessary overview of the available numerical software. Structured by the content of the following chapters and sections—methods, algorithms, and concepts which underlie the various programs are discussed. Advantages are emphasized and inherent weak points are worked out.

Specially marked software references, arranged according to subjects, supply the reader information about commercially available software libraries (IMSL,

NAG etc.) and about public domain numerical software (NETLIB, ELIB etc.) which can be obtained via the Internet (see Section 7.3.7).

Single programs or collections of programs are indicated using UNIX file system notation. For example,

```
IMSL/MATH-LIBRARY/qdag
```

describes the integration program qdag found in the MATH/LIBRARY section of the IMSL Fortran Library [31]. The program qdag referenced above is derived (with some internal modifications) from qag which is included in QUADPACK [22]. This is symbolized by:

```
IMSL/MATH-LIBRARY/qdag   ≈   QUADPACK/qag
```

Other software products referenced in this way are:

IMSL/MATH-LIBRARY/*	IMSL MATH/LIBRARY [31]
IMSL/STAT-LIBRARY/*	IMSL STAT/LIBRARY [32]
NAG/*	NAG library [21]
HARWELL/*	Harwell library
TOMS/ number	TOMS programs ($\to$ NETLIB)
QUADPACK/*	QUADPACK programs [22] ($\to$ NETLIB)

Normally, *Fortran 77* libraries are referenced from various Visual Numerics Inc. and NAG Ltd. products because these are the most extensive. Both software companies develop programs first in Fortran 77 and only then implement them in other programming languages and software libraries (for example, C libraries).

7.3 Software and the Internet

The conventional distribution of numerical software is done by mailing magnetic tapes, floppy discs, CD-ROMs, etc. on which the software is stored. Traditionally, technical journals and advertising brochures provide information about the availability of software products.

The importance of global computer networks as electronic media for the distribution of software has greatly increased. The essential advantage of electronic communication over conventional mailing is the reduction in the time needed to obtain the latest versions of programs, documentation, data and so on. For example, the Internet makes it possible to establish a connection between two computers anywhere in Europe, America, Asia or in other parts of the world and to transfer data within seconds. Even large software packages are very quickly transmitted in this way.

7.3.1 The Internet

The Internet evolved from the ARPANET project initiated by the Advanced
Research Projects Agency (ARPA) at the U.S. Department of Defense in the
late 1960s. The aim of this project was to develop network technologies which
make it possible to transmit data over wide distances, even if some lines or nodes
of the network are temporarily or permanently out of order.

Packet Switching Data Transmission

The conventional line switching data transmission, found in telephone networks,
was abandoned in order to avoid breakdowns. When using line switching the
message path between sender and receiver has to be determined at the beginning
of the interaction and the appropriate resources (data lines, buffer memory, etc.)
are then allocated. While the data exchange takes place the two stations are
continuously connected via the same path. If there is a breakdown in any resource
during the data transmission—for what ever reason—then the whole transmission
process has to be repeated until a correct and complete transmission has been
made.

In the ARPA network, *packet switching* data transmission is used to avoid this
problem. The sender of a message is not directly connected to the receiver (which
is why packet switching is referred to as *non-connective communication*). The
data which has to be transmitted is divided into small units of information—
packets—which are marked with a receiver identification code (destination address) and can therefore be transferred independently of one another.

In general, the path used by the packet is not determined in advance. On the
contrary, the path that each packet takes is determined by a set of computers
called *routers*: Usually each packet is sent to the nearest router and is temporarily
stored there. The router decides which other router it has to send the packet
to according to the receiver identification found in the packet. This process is
repeated until the packet has reached its destination.

If there is a disturbance between two routers during the transmission of one
packet of a message it is not necessary to repeat the transmission of the whole
message. The sender-router simply repeats the transmission of this single packet;
however, another target router may have to be determined. Therefore, packet
switching data transmission is more flexible and more suitable than line switching
for networks which are not very reliable.

Internet Protocol (IP)

In 1973 another network project was launched by ARPA. It was intended to make
the connection of different packet switching networks possible. To achieve this
objective it was necessary to define communication protocols. A *communication*
or *network protocol* is a set of formal operating rules, procedures or conventions
which govern the transmission of data over communication networks. With the
help of these protocols, two or more compatible or incompatible networks can

be interconnected in such a way that each computer perceives them as a single virtual network, the *Internet*.

Two essential problems occur when connecting different, packet switching computer networks: (i) Nodes are generally identified (addressed) differently on different networks. (ii) The routers are only able to establish connections within the subnet they belong to; they cannot connect to nodes outside the subnet they belong to.

Both problems were solved by the *Internet Protocol* (IP). The IP controls the sending and receiving of data packets by establishing a uniform identification code of the Internet nodes—the *Internet addresses*—and by making sure that the router connections are correctly established, even outside the network they belong to.

Transport Protocol (TCP)

Internet protocol allows for the exchange of packets between communication partners in different networks. It is not useful, however, for the transmission of large amounts of data for two reasons: (i) It does not guarantee that the packets of longer messages reach the receiver in the same order as they were sent. (ii) It does not guarantee that all packets are transmitted once and only once. Single packets can get lost or can be transmitted several times.

In order to guarantee the reliable transmission of packets in the right order, the *Transmission Control Protocol* (TCP) was developed and combined, along with the Internet protocol, to create the **TCP/IP** protocol. The TCP controls the handling of transmission faults, whereas the IP determines the sending and receiving of the data packets in detail via the Internet. Participation in the Internet does not depend on any specific kind of computer system because TCP/IP implementations are now available for all standard computer platforms (mainframes, workstations, PCs etc.).

Internet Development

In 1983 the ARPA network reached a point at which it was no longer in the experimental stage; it became the **ARPA Internet**. Shortly thereafter the National Science Foundation (NSF) initiated a series of projects designed to connect universities and research institutes to mainframe centers which were also financed by the NSF. As a result the NSF network evolved. The NSF network is a *backbone network*, i.e., it is only used for the connection of regional networks. Thousands of local networks in universities and research institutes are connected via the NSF network in America. In Europe a backbone network like the NSF network does not exist. Competing organizations offer partial solutions, for example the *European Backbone* (EBONE).

In the past few years, the number of computer systems connected to the Internet has increased *exponentially*—it approximately doubles every year.

The Internet and other Networks

At the beginning of the eighties, the success of the ARPA network led to the foundation of several other networks, like BITNET, CSNET, UUCP, USENET and FidoNet. These networks are generally *not* based on the Internet protocol, so they are *not* part of the Internet.

The operators of these networks decided to offer their users access to Internet services because of the outstanding importance of the Internet. For this reason they established *gateways* to the Internet. Which Internet services users can take advantage of depends on the network in question (Quarterman, Carl-Mitchell [320], Krol [19]).

7.3.2 Communication in the Internet, E-Mail

One of the most important services of the Internet or other computer networks is that which enables the communication between individual network users. Probably the best known Internet service for this task is *electronic mail (e-mail)*. The e-mail system makes the exchange of textual messages between Internet users possible, and transmission is very fast—most messages reach their receivers within a few seconds.

The e-mail system is very flexible because of its *asynchronous* communication mode: Messages can be sent whether the receiver is ready to read them or not. An incoming message is stored until the receiver reads the message, re-uses or stores it somewhere else and deletes it afterwards.

A number of programs are available for the sending, reading, editing, storing etc. of e-mail messages. The UNIX operating system, for example, provides `mail`, `mailx` and `elm`. The functionality of these programs is basically the same. All e-mail programs use the *Simple Mail Transfer Protocol* (SMTP) for the transmission of messages; whence messages can be read by programs other than the one they were sent by.

Sending e-mail messages only requires the Internet address of the receiver's computer, i. e., the *system name* and the *username*.

Example (System Name) The system name of one of the computers at the Oak Ridge National Laboratory—ornl.gov—is derived from the initials of the institution and the fact that it is a *gov*ernmental research institution. The appropriate Internet address, 128.219.128.17, reveals the identity of the system only to certain insiders.

The username is often derived from the actual name (first name, surname, initials etc.) of the user: "username@system-name" is the complete specification of the addressee.

Example (E-mail Message) The following message is written to Josef N. N. whose username is josef. He receives it on his computer which has the system name titania.tuwien.ac.at.

```
$ mail josef@titania.tuwien.ac.at
Dear Josef! Confidence is good, control is better. Yours, Wladimir Iljitsch

$
```

7.3.3 Forums for Discussion on the Internet

Mailing lists are simple extensions of the e-mail system. A mailing list includes the addresses of a group of Internet users. When a message is sent to a mailing list, it is automatically sent to all the users contained on it. In general, mailing lists are dedicated to relatively specific topics and only include Internet users who are interested in this topic. Mailing lists are often also used as forums of discussion for certain subjects. Some mailing lists are *moderated*, i. e., a moderator screens the incoming messages to determine whether or not they are appropriate for retransmission to the mailing list's subscribers.

Example (na-digest) The moderated mailing list `na-digest` contains the names of Internet users who are interested in *numerical analysis* and numerical data processing. Specific conference announcements, job offers, content lists of journals etc. can be found there. The prerequisite for being included on this mailing list is membership to the NA-NET (Numerical Analysis NETwork). Details are available at `na.help@na-net.ornl.gov`.

If the number of people interested in a certain topic exceeds a certain limit, a specific news group is established. While news groups and mailing lists serve almost the same functions, they are implemented differently. News groups use efficient ways of distributing information to extremely large groups. Certain news groups have as many as 100 000 subscribers.

Various operations in connection with news groups can be performed with special programs called **news readers**. There are many news readers, including `rn`, `xrn`, `nn`, and `tin` for Unix machines. News groups can be subscribed to, old or unwanted news groups can be canceled; contributions to news groups can be read; articles can be published etc. A graphical user interface such as the one found in `xrn`, enables the user to handle the news reader with ease (see Fig. 7.1).

7.3.4 Resource Sharing on the Internet

Internet resource sharing services provide users with the opportunity of making the resources of a single Internet node, such as special data, certain programs etc., available to other nodes. The distribution of software or information about software is part of this service.

Resource sharing often takes places using the *client server principle*. Computers which provide services to several users (computers) on a computer network, for example, a centralized data base service, are called *servers*. The computers of the users, including the software counterparts to the server software, are called *clients*. Often client software is programmed in such a way that the user does not recognize that he/she uses a remote server.

File Transfer (FTP, aFTP)

A fundamental way of using external resources is to send and receive copies of files. The program `ftp` which uses the *File Transfer Protocol* (FTP) is available for this function. `ftp` is a common tool for sharing data and programs.

Figure 7.1: xrn—A news reader based on the X-WINDOW system.

ftp makes the transfer of files either in ASCII or in binary format possible.
The ASCII mode is used when *readable* files are transferred (technical reports,
program documentation etc.). The binary mode is used to transmit executable
programs, pictures, data bases etc. ASCII is the default mode for ftp transfers.
If binary mode is desired, then a parameter has to be set.

One prerequisite for transmitting files with ftp is that the system name (or the
Internet address) of the remote computer must be known. After establishing
a computer connection using this information, the usual login procedure takes
place. The username and the password have to be typed in to prove access
authorization to the desired files. After the remote system has accepted the login
name and the password, files can be transmitted using simple commands:

get	copies a file from the remote computer;
put	transmits a file copy to the remote computer;
ls or **dir**	lists all files in the current directory;
cd	changes the directory;
binary	changes to binary mode (to transmit binary files);
ascii	changes to ASCII mode (default mode: ASCII).

Example (File Transfer Using FTP)   The following dialog protocol of an ftp transaction describes how a user establishes a connection to the Internet node uranus.tuwien.ac.at, uses the name karg to login, copies the file with the name .profile to another computer and then logs out afterwards. For emphasis, user input is printed in *italics*.

```
$ ftp uranus.tuwien.ac.at
Connected to uranus.tuwien.ac.at.
220 uranus FTP server (Version 16.2 Mon Apr 29 20:45:42 GMT 1991) ready.
Name (uranus:karg): karg
331 Password required for karg.
Password:  ···   (password input)
230 User karg logged in.
Remote system type is UNIX.
Using binary mode to transfer files.
ftp> get .profile
200 PORT command successful.
150 Opening BINARY mode data connection for .profile (1295 bytes).
226 Transfer complete.
1295 bytes received in 0.08 seconds (16.35 Kbytes/s)
ftp> bye
221 Goodbye.
$
```

If a user wants to make files accessible to other Internet users, it is necessary to give the other users his/her login name and his/her password. This kind of transmission, however endangers the safety of the computer system. This risk factor would be completely intolerable if files were made generally accessible, i.e., if all interested users had access to them.

A special service, offered by universities and other institutions, was developed to deal with this problem: **anonymous FTP or aFTP**. A user who is interested in accessing certain files on an Internet node using anonymous FTP has to identify him/herself with the username "anonymous" or "ftp". A real (secret) password is not needed in this case: normally the user's own e-mail address serves as a password. Some FTP servers actually deny access if a valid Internet address is not used as password.

There are tight restrictions on anonymous FTP users: they can normally only get file copies; they cannot install new files or modify files that already exist. There are also strict limits on the files they can copy.

Most FTP servers register the access to files and store the Internet address of the person accessing the file. In this way the user of the anonymous FTP service does *not* remain *anonymous* in the true sense of the word.

Example (File Transfer Using Anonymous FTP)   The following dialog protocol shows how a user copies the file gnuplot-3.5.tar.gz from the directory **packages/gnu** of the anonymous FTP server ftp.univie.ac.at of the University of Vienna. This file contains the GNU-PLOT visualization package.

```
$ ftp ftp.univie.ac.at
Connected to ftp.univie.ac.at.
220-     +++++++++++++++++       --------------------------------------
220-      +             +           WELCOME  to  FTP.UNIVIE.AC.AT
```

```
220-        +          +              Server for freely distributable SW
220-    +         +              -------------------------------------
...
220-
220-    There are already 5 users in your class (max 100).
220-    In case of technical problems contact manager@ftp.univie.ac.at.
220-
220 ftp FTP server (Version wu-2.4(40) Sat May 7 15:29:38 CETDST 1994) ready.
Name (ftp.univie.ac.at:username): anonymous
331 Guest login ok, send your complete e-mail address as password.
Password:  ···  (Input of one's own e-mail address)
230-
230-*************************************************************************
230-
230-                    Welcome to FTP.UNIVIE.AC.AT
230-
230-    This server is located at the Vienna University, Vienna, Austria.
...
230-*************************************************************************
230-
230-
230-Please read the file README
230-  it was last modified on Mon May  9 13:29:25 1994 - 137 days ago
230 Guest login ok, access restrictions apply.
Remote system type is UNIX.
Using binary mode to transfer files.
ftp> cd packages/gnu
250 CWD command successful.
250-Please read the file README
250-  it was last modified on Wed May 18 17:48:00 1994 - 128 days ago
250-Please read the file README-about-.gz-files
250-  it was last modified on Wed May 18 17:48:00 1994 - 128 days ago
250 CWD command successful.
ftp> dir gnuplot*
200 PORT command successful.
150 Opening ASCII mode data connection for /bin/ls.
-r--r--r--   1 gnu-adm  archive   626008 Sep 30  1993 gnuplot-3.5.tar.gz
226 Transfer complete.
ftp> binary
200 Type set to I.
ftp> get gnuplot-3.5.tar.gz
200 PORT command successful.
150 Opening BINARY mode data connection for gnuplot-3.5.tar.gz (626008 bytes).
226 Transfer complete.
626008 bytes received in 8.70 seconds (70.26 Kbytes/s)
ftp> bye
221 Goodbye.
$ ls
gnuplot-3.5.tar.gz
```

Transferring files using the Internet can take a great deal of time. Conventions and techniques which minimize transmission times have thus been developed. They make it possible to transfer several files simultaneously and guarantee that binary files are transmitted correctly. Many files available on the Internet are stored in *compressed* format, which reduces the cost of storage and transmission across the network. Large groups of logically related files can be accumulated into a single archive to make them easier to transfer.

Generally, the following conventions are used to recognize the type of the file (archive format) from the extension (suffix) of the file name.

extension	type of file	tools to be used after receipt
.Z	binary	uncompress
.arc	binary	ARChive
.shar	ASCII	SHell ARchive
.tar	binary	Tape ARchive
.uu	ASCII	uudecode
.zip	binary	unzip
.zoo	binary	zoo
.gz (or .z)	binary	GNU unzip

Transferring Files Using E-Mail

Users who have no direct FTP access may obtain data and information from FTP-by-mail servers by writing an e-mail request. The FTP-by-mail server automatically answers with a confirmation, or sends the required files.

Example (E-Mail Message to an FTP-by-Mail Server) In the answer to the e-mail message

```
$ mail parlib@hubcap.clemson.edu
Subject:
send index

$
```

information about the contents of PARLIB, a parallel computer software library at Clemson University is received.

E-mail messages are answered in the following way by (most of the) FTP-by-mail servers:

- **send index from** *library-name*
  information about the contents of *library-name* will be received,

- **send all from** *library-name*
  the whole software library will be received,

- **send** *routine-name* **from** *library-name*
  a specific program (and everything related to it) will be received,

- **send directory for** *library-name*
  sends a list of all library files.

Examples of FTP-by-mail servers are

- `netlib@ornl.gov`—a collection of numerical software,

- `statlib@temper.stat.cmu.edu`—a collection of statistical software,

- `parlib@hubcap.clemson.edu`—a collection of parallel programs for a multitude of languages and parallel processor systems and

- `tuglib@math.utah.edu`—a collection of software for TeX users.

Remote Login

Another way of sharing resources is to make *executable* programs available using *remote job control*. The simplest way to do this is to allow Internet users to log into other computers in the Internet. This makes it is possible to take advantage of remote soft- and hardware resources. In order to achieve this, the TELNET application protocol (*virtual terminal protocol*) is available on the Internet as a part of the TCP/IP protocol. `telnet` is a program which uses the TELNET protocol and, like `ftp`, a client and server system.

To log into a remote computer both computers must have TELNET software installed. Both programs co-operate; it is thus possible to work on the remote computer in the same way as using a terminal connected directly to the local computer. To use this tool it is necessary to have a user name and a password on the remote computer.

Example (Remote Login Using TELNET)   The following dialog protocol shows how to establish a connection to the Internet node `uranus.tuwien.ac.at` using `telnet`, to login using *myname*, to list all locally available files with the command *ls* and, finally, to logout using *exit*, which cancels the Internet connection.

```
$ telnet uranus.tuwien.ac.at
Trying 128.130.37.2...
Connected to uranus.tuwien.ac.at.
Escape character is '^]'.
uranus [Release A.08.00 B 9000/835]

login: myname
Password:  ···   (Input of the password)

$ ls
archive    ndv        progs      Mail       bin        hompack    pvm3
$ exit
Connection closed by foreign host.
$
```

Anonymous telneting makes it possible to log into a remote computer with a publicly known account name. These accounts often have login names like "anonymous" or "guest" or program names "archie" or "gopher". Some systems do not even require a password.

7.3.5 Finding Software in the Internet

Attempting to find software, documents etc. on a certain topic on the Internet can be impossible using previously described means because of the wide range and variety of available information and of the decentralized nature of the Internet. Several Internet search services have been developed to facilitate access to information on the Internet. These are some of the most important ones:

The **Archie system** looks for files which are available on public servers on the Internet. Archie works like an Internet "librarian" which regularly and automatically requests and registers all the files available on the Internet by anonymous FTP, and makes the index accessible in the form of a data base. Because Archie regularly polls *all* aFTP servers, this data base is always up to date.

Archie is not a single system; it consists of a group of servers. Every participating server is responsible for requesting Internet aFTP servers so that it can build up its own data base. Currently there are 14 publicly accessible Archie servers, which provide file archive directories of more than 1000 aFTP servers. The Archie data base includes more than 2.5 million files and their addresses. Only file names can be used to find files in an Archie system. The system name of the Internet nodes and the directory the file is stored in can be obtained from the data base. Handling the Archie system is easier when using the program `xarchie` because of its X-WINDOW user interface (see Fig. 7.2).

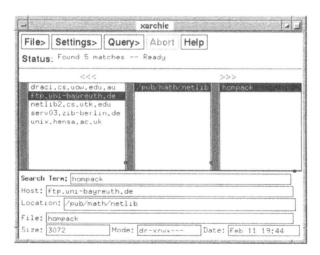

Figure 7.2: `xarchie`—An X-WINDOW user interface for the Archie system.

A distinct disadvantage of the Archie system is that the name of the file has to be known. Descriptions of the files are *not* stored. It is possible to look through a collection of files using the **WAIS system** using keys which can be defined by the user.

7.3.6 Internet Front Ends

Usually the user of Internet services quickly realizes that the simplicity of working with individual programs is reduced by the fact that different programs do not have uniform user interfaces. In order to overcome this disadvantage, programs have been developed to provide uniform access to the Internet services (FTP, TELNET, Archie and WAIS).

In the **Gopher** system, files are accessed by using a menu driven X-WINDOW interface, the `xgopher` system (see Fig. 7.3), for example.

Figure 7.3: xgopher—An X-WINDOW information service based on the Gopher system.

The most popular way to find resources is the **World-Wide Web** (WWW, or Web), a networked *hypertext* protocol and user interface. The Web makes it possible to link documents logically: To start with, a document is displayed on the screen. The user can retrieve other documents related to the displayed document by selecting specially marked keywords. The newly chosen documents can be located either on the same or on other nodes anywhere on the Internet. Each document can be referenced using the matching *uniform resource locator* (URL). Generally, an URL is just an extension of the normal file name which indicates the system name of the Internet node and an indicator as to the document's type.

Currently the Web provides the most convenient access to Internet services supported by X-WINDOW browsers (such as MOSAIC; see Fig. 7.4).

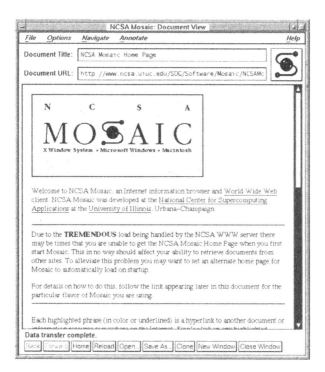

Figure 7.4: MOSAIC—An X-WINDOW Web browser.

7.3.7 NETLIB

The Internet service NETLIB (Dongarra, Grosse [165]) has become extremely important because it offers a quick, simple and efficient gateway to the procurement of *public domain* software.

NETLIB provides programs, program libraries, bibliographies, software tools etc. which are all free of charge.

Libraries: LAPACK, LINPACK, EISPACK, ITPACK, SPARSPAK, MINPACK, TN-PACK, HOMPACK, ODRPACK, PPPACK, QUADPACK, FFTPACK, VFFTPK, ODEPACK, FISHPACK, MADPACK, TOEPLITZ, CONFORMAL, VANHUFFEL, PITCON, the generally accessible parts of the PORT library etc.

TOMS Software: The programs published in *Transactions on Mathematical Software* (TOMS) can be obtained using their number.

E-Mail Access

To gain access to the NETLIB library, an e-mail message can be sent to one of the following addresses:

via Internet:	`netlib@nac.no` or
	`netlib@ornl.gov` or
	`netlib@research.att.com`
via EARN/BITNET:	`netlib@nac.norunix.bitnet`
via EUNET/uucp:	`netlib@draci.cs.uow.edu.au`

The e-mail message can have one of the following forms:

`send index`

`send index from` *library*

`send` *routines* `from` *library*

`find` *keywords*

Procuring software by e-mail is complicated and there are certain constraints (for example, only single programs can be obtained). NETLIB software can also be received by anonymous FTP using one of the addresses above.

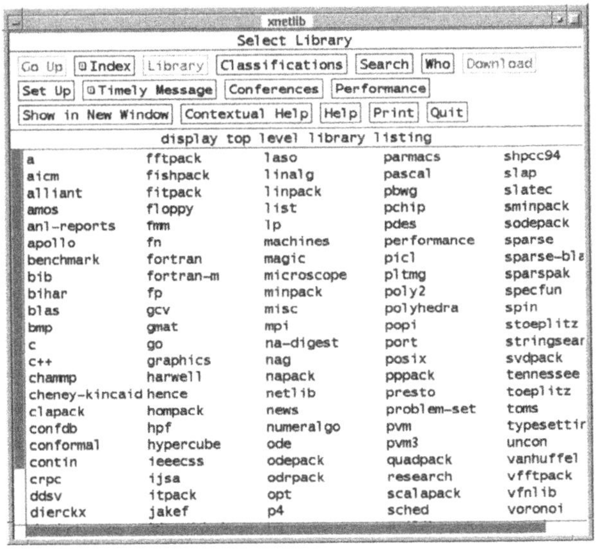

Figure 7.5: xnetlib—An X-WINDOW interface to the software library NETLIB.

WWW Access

Graphical user interfaces for NETLIB are also provided by the Web:

`ftp://netlib.att.com/netlib/master/readme.htm` and `www.netlib.org`.

However, `xnetlib`, a convenient NETLIB interface (see Fig. 7.5) which is available via NETLIB or anonymous FTP, provides even greater functionality.

The following systems are similar to NETLIB (with an FTP-by-mail access):

    `parlib@hubcap.clemson.edu`   parallel computer software;
    `statlib@temper.stat.cmu.edu` statistical software;
    `reduce-netlib@rand.org`   symbolic algebra software.

Via `ftp.dante.de` software pertaining to the "TEX world" can be obtained.

7.3.8 eLib

The Konrad Zuse Center for Information Technology in Berlin (ZIB; Heilbronner-strasse 10, D-10711 Berlin – Wilmersdorf) runs the software library ELIB. Using anonymous FTP, ELIB is reached at the Internet address `elib.zib-berlin.de` in the `pub/elib` directory. The most convenient way to access ELIB is to use the Web via the URL `http://elib.zib-berlin.de`.

7.3.9 GAMS

The *Guide to Available Mathematical Software* (GAMS) is a numerical software data base operated by the *National Institute of Standards and Technology* (NIST) in which all the subprograms of the most important mathematical libraries are documented (Boisvert et al. [120]). A short description of each function is stored there, and the subprograms are classified according to the hierarchical GAMS index. If subprograms are free of charge then they are generally directly obtainable using GAMS.

The X-WINDOW user interface **xgams** is available for the GAMS data base (see Fig. 7.6). The simplest way to gain access to these programs is to log into `gams.nist.gov` with the user name **gams** or **xgams** using TELNET. Client programs can be obtained via anonymous FTP from `enh.nist.gov:gams`.

Another way to access the GAMS data bank is through the Web using the URL `http://gams.nist.gov`.

7.4 Interactive Multifunctional Systems

If high performance methods for the solution of very large numerical problems are not required then multifunctional program systems offer a very good method of interactively solving moderately sized problems. These program systems combine the functionality of numerical, symbolic and graphical systems in comprehensive software products with a uniform user interface (Riddle [327]).

7.4.1 Exploratory Systems

The most important application fields of interactive multifunctional program systems such as MAPLE, MATHEMATICA, MATLAB, AXIOM, SENAC, MACSYMA or MATHCAD are applied in order to solve mathematical problems, analyze data, develop and analyze algorithms etc. These program systems are usually used to solve *one-of-a-kind* problems. Such systems are, therefore, not appropriate for

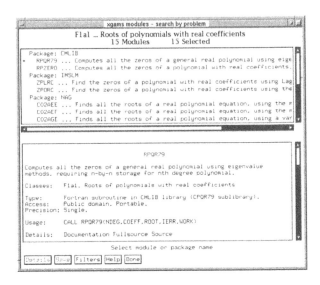

Figure 7.6: xgams—An X-WINDOW interface for the GAMS data base.

the efficient solution of large-scale problems or for recurring calculations with
changing data.

7.4.2 Numerical Systems

When the emphasis of the application lies on the numerical solution of medium-
sized problems, matrix-oriented systems like MATLAB, OCTAVE, MLAB, GAUSS
and XMATH may be considered. For the efficient solution of large-scale problems
the use of numerical software products discussed in Sections 7.2.3 and 7.2.4 is
recommended.

MATLAB

MATLAB[2] is the most famous numerically oriented multifunctional program system. Using MATLAB, equations can be defined and evaluated in a very simple
way; data and user-defined functions can be stored and re-used and the results
of calculations can be displayed graphically. The primary field of MATLAB applications is interactive numerical linear algebra: **MAT**rix **LAB**oratory (Hill,
Moler [226], Coleman, Van Loan [9]).

MATLAB covers, in addition to matrix calculations, many other fields of numerics such as the determination of the zeros of polynomials, the analysis of
data using FFTs (*fast Fourier transforms*), the numerical solution of initial value
problems in ordinary differential equations etc.

The graphical functionality makes it possible to plot two- and three-dimensional technical color graphics on a screen, printer or plotter.

---

[2]The MathWorks, Inc., 20 North Main Street, Sherborn, Massachusetts, 01770, USA.

MATLAB has an interpreter and can be programmed to extend the function-ality of the system in this way. Fortran and C programs can be called from MATLAB. As a result, the numerical solution of computationally extensive prob-lems can be sped up, and existing software can be re-used.

A variety of supplementary modules, *tool boxes*, are available for different fields of application and for special tasks: signal processing, chemometrics, opti-mization, neural networks, control systems, statistics, image processing etc.

A way to expand the MATLAB functionality in the field of symbolics is to use the *Symbolic Math Toolbox* which uses the MAPLE kernel.

OCTAVE

The public domain product OCTAVE is not only useful for learning how to use matrix oriented systems, it is also useful for solving practical numerical prob-lems. OCTAVE has the same basic designs and language constructs as MATLAB. The source code and precompiled versions for different computer systems can be obtained via anonymous FTP from `ftp.che.utexas.edu:pub/octave`.

7.4.3 Symbolic Systems

Computer algebra systems—like MAPLE, MATHEMATICA, MACSYMA, AXIOM, DERIVE etc.—make it possible to do analytic mathematics (differentiation, inte-gration etc.) with a computer.

The most important representatives of this category—MAPLE and MATHE-MATICA—are primarily computer algebra systems, but they also have a numerical and a graphical functionality.

7.4.4 Simulation Systems

Simulation uses mathematical models for experimentation instead of the real (physical or other) objects or phenomena.

Modeling: A mathematical model is established using data which describes the state of a process. The resulting model can be used to replicate real objects or processes.

Experiments: Experiments can be carried out using models with appropriate in-put data. The behavior of the model—and therefore the behavior of the real object or phenomenon—is investigated under conditions which are relevant for a specific application.

Interpretation: The results of the experiments are transfered to the real process by analogy and, thus, solve the original problem.

There are various facilities for performing simulations:

Simulation languages are special programming languages developed to simplify the formulation of simulation concepts. For example, ACSL is a simulation language used for continuous models (ordinary differential equations); SIMULA and GPSS are languages used for discrete models (queuing nets), etc.

Simulation systems are multifunctional, interactive software systems. VISSIM and ACSL GRAPHIC MODELER are two of the more important examples of simulation systems.

7.5 Problem Solving Environments

The term *problem solving environment* (PSE) stands for, roughly speaking, a software system which supports the solving of problems of a certain class using a high level user interface. PSEs are tools to solve problems which are not of a routine nature. This property distinguishes PSEs from other application software.

Example (ELLPACK) For mechanical problems which are solved using elliptic partial differential equations, there exists a large repository of application software, for example, NASTRAN, ASKA and SAP.

The problem solving environment ELLPACK, on the other hand, is not restricted to special mechanical problems: it is designed to solve general elliptic differential equations. It has an expert system front end which is responsible for the optimal adaptation of the solving algorithm to the properties of the particular problem, regardless of the specific field of application the problem stems from.

It is assumed that the user of a PSE is always a human being, i. e., not another program or computer. Thus, when designing a PSE, the user's comfort and the utility of the output (preferably in graphical form) are very important software quality criteria—as is the case in interactive multifunctional program systems (Section 7.4).

An efficient utilization of the hardware resources is, of course, important, but is generally not as important as minimizing the human effort required of PSE users (Gallopoulos, Houstis, Rice [206], Houstis, Rice, Vichnevetsky [231]).

Ideally, a PSE efficiently carries out the routine parts of a solution process without interference from the user. Furthermore, it should help the user to specify the problem, select algorithmic alternatives and determine problem oriented parameters of the algorithm.

User Interfaces

Window systems (like X-WINDOW), bit map graphics and color are important prerequisites for a PSE user interface. Plans for the use of speech input/output for future PSEs are already under way.

A difficult problem emerging in the development of PSEs is the decoding of the dialog between the user and the PSE. Because not every user's terminology is the same, a *thesaurus* (a systematically ordered collection of words from a certain field of application) has to be implemented. The thesaurus helps the PSE to reduce synonyms to uniform keywords. The user may therefore communicate with the PSE in any way he/she wishes.

Problem Solving

After the problem has been specified with sufficient accuracy, the PSE decides
which subsystem or subprogram is to be used to solve it. The selection mechanism
ranges from simple decision trees to expert systems whose knowledge base stems
from specialists in the particular field. When a knowledge base is established,
experience previously gathered in successful solution processes can be utilized by
the PSE.

The Presentation and Analysis of Results

When the internal solution mechanism has produced results, the PSE has to
put them into a form which allows the user to interpret and use them. The
visualization of numerical solutions is thus an important component of any PSE.

Information concerning the condition of a problem can be given to the user
during the presentation phase. If necessary, the user will receive a warning that
the results obtained are critically dependent on the input data, and the results
must, therefore, be interpreted very carefully.

Problem Oriented Assistance

If the user needs assistance, PSEs offer support in the form of (local or context
dependent) *help* functions. Some users are interested in knowing which solution
method was applied by the PSE and why it was chosen. Ideally, these explanations are not just a listing of facts and rules, but rather information in an
understandable form.

7.5.1 Available Problem Solving Environments

Statistical PSEs

Most of the available PSEs, such as REX, STUDENT, STATXPS, BUMP, MULTI-
STAT and GLIMPSE, are dedicated to statistical analysis.

Software systems intended for users who are not trained in mathematics and
statistics have been around for a long time. These PSEs support the statistical
examination of data, simplify complicated statistical analyses and explain program decisions to the user. There is even a variety of software systems based on
artificial intelligence methods (see, for instance Ford, Chatelin [193]) designed to
be used in statistics.

Symbolic PSEs

Symbolic mathematics is another field for which PSEs, i. e., software systems with
intensive user support, are available: Some examples are the interactive multifunctional systems MATHEMATICA, MACSYMA, MAPLE, AXIOM, DERIVE etc.
These PSEs are important in their own right and in connection with numerical

software (for example, when the derivative of a function whose extrema are to be calculated numerically is determined by symbolic differentiation).

Numerical PSEs

The first numerical PSEs were developed for the finite element method (FEM). The demand for easily usable interfaces was highest in this field. One of the first knowledge based systems was FEASA (*finite element analysis specification aid*). Besides FEASA, there are a number of other FEM expert systems; however, expert system front ends for the most important packages (e. g., NASTRAN) are still not available.

A variety of attempts have been made to provide decision trees which support the selection of appropriate algorithms or (sub)programs for given numerical problems. The documentation of most numerical software systems (the IMSL and NAG libraries for instance) include such decision aids. NITPACK (Gaffney et al. [205]) provides a decision tree for the *on-line* selection of the most appropriate software.

7.6 Case Study: Software for Elliptic PDEs

The solution of problems involving partial differential equations (PDEs) was one of the earliest computer applications. The fact that such problems still consume substantial resources in scientific computing is evidence of their importance in many areas of science and engineering.

For the solution of elliptic boundary value problems, many commercial and free (public domain) software products are available. A brief overview of the available programs is given in this section. In particular, the usefulness of the previously discussed methods of procuring numerical software is demonstrated.

The emphasis in the following software overview—as in all other sections of this book—has been placed on *public domain* products. Out of all commercial software products, only the NAG and the IMSL Fortran libraries are considered here because of their widespread use. Other commercial software packages are only dealt with if no public domain product with comparable quality is available.

There is a variety of program packages available for fields of application (mechanics, etc.) which require the solution of elliptic PDEs. In the following, software which has been developed for special applications is not considered. Only programs which generally solve elliptic boundary value problems are included in the overview.

Relevant NETLIB programs and software collected in the GAMS data base have been included in the above-mentioned framework. Only a few other programs, for instance, those which can be obtained from various institutions via anonymous FTP, are covered.

Software products designed to solve systems of linear equations which emerge from the discretization of elliptic PDEs are also available, but these are *not* considered in this section. Chapters 13 and 16 are dedicated to the solution of systems of linear equations with dense or sparse coefficient matrices.

7.6.1 Problem Classification

To start with, a few PDE terms are introduced in order to make descriptions of the functionality of the various software products possible. Exact definitions and detailed explanations can be found in standard textbooks on PDE theory (e.g., in Hackbusch [51]).

The Problem Class

For the classification of software packages, the class of problems for which they have been conceived is very important. In particular, *linear* and *nonlinear* partial differential operators are to be distinguished. However, *public domain* software deals primarily with linear problems. In this software category, programs designed to solve *second* order PDEs are predominant. For higher order differential equations, only software for the fourth order *biharmonic* equation is available.

The Partial Differential Operator

The *Laplace*, the *Poisson* and the *Helmholtz* differential equations play a fundamental role in the mathematical formulation of physical laws and so these equations are very important in many fields of scientific computation. Correspondingly, most of the programs dealing with elliptic boundary value problems solve at least one of these PDEs. Note, however, that the Helmholtz equation comprises the Poisson equation and this in turn also the Laplace equation. Thus, software which solves Helmholtz boundary value problems can also be efficiently used to solve Poisson or Laplace equations.

In addition to software packages designed to solve Helmholtz, Laplace and Poisson equations, there are also software packages which solve more general differential equations; specifically *separable* or *self-adjoint* differential equations. Programs are even available for solving *non-separable*, second order PDEs.

The Domain

The domain is an important part of the specification of any PDE boundary value problem. Two- and three-dimensional intervals (rectangles, cubes) are particularly easy domains to deal with.

Boundary Conditions

There are different types of boundary conditions: Prescribed function values along the boundary of the domain (*Dirichlet conditions*), prescribed normal derivative values (*Neumann conditions*) or a combination of both quantities prescribed on the boundary of the domain (*mixed boundary conditions*). If the solution is required to be a periodic function of one of the independent variables, then this is said to be a *periodicity condition*.

7.6.2 Software Packages for Elliptic Problems

Because of the great importance of elliptic PDEs in science and engineering, the development of *elliptic solvers* has been a central aim of many development activities from the very beginning (see Cowell [10], Chapter 9).

ELLPACK

The starting point for the development of ELLPACK in 1974 was the goal of developing a system which assessed the performance of numerical software designed to solve PDEs. It turned out that a carefully planned and organized software environment was needed to make such performance assessments possible. One requirement was a formal language for the detailed specification of the PDE problems to be solved. Another important aspect was the need to create a framework in which to include problem solving modules of different origin. After experiments with the prototype systems ELLPACK 77 and ELLPACK 78 were performed, a new PSE, ELLPACK, was developed. The book by Rice and Boisvert [29] contains a detailed description of this software system.

New versions of ELLPACK take into account the influence of modern computer architectures (vector and parallel computers), contain new methods for the numerical solution of PDEs and use expert system modules to select appropriate solution methods and suitable computer configurations.

Doubts concerning the ability of experts in the field of elliptic boundary value problems became the impetus for developing the ELLPACK project. ELLPACK was thus equipped with the ability to continuously monitor and store performance data during the process of solving PDEs. In fact, certain weak points in the knowledge of the experts could be proved using the monitored data. On the other hand, it turned out that experts knew much more than the average user (scientists, engineers etc.). This was the justification for the use of expert system technology, designed to support the numerical solution of PDEs.

ELLPACK is a software system which solves elliptic boundary value problems in two- and three-dimensional domains. ELLPACK has its own problem statement language for specifying the problem and the solution algorithm using common mathematical terminology.

Example (ELLPACK) The elliptic PDE $u_{xx} + u_{yy} + 3u_x - 4u = \exp(x + y)\sin(\pi x)$ on the two-dimensional domain $[0,1] \times [-1,2]$ with the boundary conditions

$$
\begin{array}{llll}
u & = & 0 & x = 0, \quad y \in [-1,2] \\
u & = & x & x \in [0,1], \quad y = 2 \\
u & = & y/2 & x = 1, \quad y \in [-1,2] \\
u & = & \sin(\pi x) - x/2 & x \in [0,1], \quad y = -1
\end{array}
$$

is solved by the following ELLPACK program. Difference quotients are used to approximate derivatives and Gaussian elimination is used to solve the resulting system of linear equations (which have a band matrix):

```
OPTIONS.      TIME $ MEMORY
EQUATION.     UXX + UYY + 3.*UX - 4.*U = EXP(X+Y)*SIN(PI*X)
BOUNDARY.
    U = 0.                 ON  X =  0.
    U = X                  ON  Y =  2.
    U = Y/2.               ON  X =  1.
    U = SIN(PI*X)-X/2.     ON  Y = -1.
GRID        6 X POINTS  $  12 Y POINTS
*
DISCRETIZATION.  5 POINT STAR
INDEXING.        AS IS
SOLUTION.        LINPACK BAND
*
OUTPUT.          TABLE(U)  $  PLOT(U)
END.
```

Symbolic manipulation is used in ELLPACK during the language processing of the problem specification. For example, the fact that the system to be solved is linear with constant coefficients is automatically recognized and incorporated into the solution process.

ELLIPTIC EXPERT is the expanded version of INTERACTIVE ELLPACK, which has an expert system that aids the user in selecting the best algorithm for the solution of a given elliptic boundary value problem.

PARALLEL ELLPACK is a program package with expert system support dedicated to numerically solving two- and three-dimensional elliptic boundary value problems on multiprocessor computers (Houstis, Rice, Papatheodorou [230]).

PLTMG

PLTMG (*Piecewise Linear Triangle Multi Grid*) is a software package which was developed by R. E. Bank (at the University of California in San Diego) for the solution of general, two-dimensional elliptic boundary value problems. The user must specify the domain in a triangulated form, i.e., a union of triangles.

In PLTMG the solution of a PDE problem is approximated by a continuous, piecewise linear element of a finite element space. If the initial triangulation of the domain turns out to be too coarse, i.e., the desired solution function cannot be approximated with sufficient accuracy, PLTMG carries out an *adaptive* refinement of the triangulation. During this process, only triangles whose error estimates are not small enough are divided into smaller subtriangles.

The systems of linear equations which result from discretization operations are solved using a multigrid method (cf. Section 16.8.1). The software package PLTMG also contains programs for the visualization of various data, the triangulation, the numerical solution etc.

PLTMG can be found in NETLIB in the directory pltmg, but only the codes can be obtained; the *documentation* of the software package PLTMG has been published as a book (Bank [4]) and is only available in this form.

MGGHAT

MGGHAT (*Multi Grid Galerkin Hierarchical Adaptive Triangles*), which was developed by W. Mitchell (currently at the National Institute of Standards and Technology) is a program package which was designed to numerically solve two-dimensional elliptic boundary value problems with self-adjoint differential operators. The program deals with domains whose boundaries are arbitrary polygons and with all manner of boundary conditions.

The formulation of the elliptic boundary value problem as a variational equation is discretized using the Ritz-Galerkin method with continuous piecewise polynomial functions as the basis functions of the finite element space. The user can choose between piecewise linear, quadratic or cubic polynomials as basis functions. In each case, the approximate solution MGGHAT delivers is only continuous, not differentiable.

If the initial triangulation of the domain is too coarse, then the subdivision will be adaptively refined. The systems of linear equations which result from the discretization are solved using a multigrid method (cf. Section 16.8.1). Details of the algorithms used can be found in Mitchell [292]. MGGHAT also contains programs for the visualization of the triangulation and the numerical solution of the elliptic problems.

MGGHAT can be found in NETLIB in the pdes/mgghat directory.

BIHAR

The software package BIHAR, developed by P. Bjorstad (*Stanford University*), is dedicated to the numerical solution of the biharmonic differential equation, a special elliptic PDE of fourth order which plays an important role in mechanics. This PDE can be solved in two dimensions on rectangular domains in cartesian or polar coordinates using BIHAR subprograms. Function values as well as normal derivatives on the boundary of the domain have to be specified. For the discretization of the PDE, a difference scheme is used in which the biharmonic differential operator is replaced by a 13-point difference quotient.

KASKADE

KASKADE is a modular package, developed at the Konrad Zuse Center for Information Technology in Berlin (ZIB), which makes the implementation of finite-element programs possible. KASKADE is written in C++ in order to guarantee that it can easily be reused and extended. These advantages make it easier to tolerate the fact that, when compared to Fortran programs, KASKADE is relatively inefficient.

KASKADE contains modules for the

- specification of PDE problems;
- definition, refinement etc. of triangular grids;

- discretization of PDEs using finite elements;

- control of the solution processes (grid refinement, error estimation etc.);

- solution of (sparse) systems of linear equations.

KASKADE is available via anonymous FTP from the pub/kaskade directory of elib.zib-berlin.de. Application software based on KASKADE, ELLKASK 2D and ELLKASK 3D, which solve elliptic problems on two or three dimensional domains; and KASTIO and KARDOS, which deal with *parabolic* differential equations, can also be found there.

7.6.3 Numerical Program Library Sections

Programs for the solution of elliptic PDEs are also included in the IMSL and NAG Fortran libraries. Unlike special software packages (dealt with in Section 7.6.2) for solving elliptic boundary value problems, these programs require more basic knowledge on the part of the user and much more personal effort. No user assistance is offered during the problem specification procedure and during the selection of a solution method (compare to the ELLPACK example on page 302).

IMSL Fortran Library

The program IMSL/MATH-LIBRARY/fps2h is suitable for the numerical solution of the Poisson and Helmholtz equations on two-dimensional rectangular domains. The differential operator is discretized by a finite difference scheme on an (in each dimension) equidistant grid of rectangles. The user can also specify different types of boundary conditions—Dirichlet, Neumann and periodic—and the order of the discretization method.

IMSL/MATH-LIBRARY/fps3h serves the same purpose for three-dimensional cube-shaped domains.

NAG Fortran Library

NAG/d03eaf solves the two-dimensional Laplace equation on arbitrary domains with a boundary formed by one or more closed curves. The user can choose among Dirichlet, Neumann or mixed boundary conditions. NAG/d03eaf is based on a contour integral formulation of the original PDE.

NAG/d03eef solves general elliptic PDEs on two-dimensional rectangular domains using a discretization by finite differences.

The program NAG/d03faf solves the Poisson or Helmholtz differential equations on a three-dimensional cube-shaped domain using a finite difference scheme on an equidistant mesh. The user can specify different types of boundary conditions—Dirichlet, Neumann and periodic boundary conditions—but cannot specify the order of the discretization. (An unchangeable seven point central difference scheme is used for the approximation of the Laplace operator.) During the implementation process particular attention was paid to achieve a vectorizable code.

NAG/d03maf serves as an auxiliary routine during the solution of elliptic PDEs on general two-dimensional domains. It produces a triangular grid which approximates arbitrarily shaped, two-dimensional domains. This is achieved with the help of a user defined subprogram which decides whether a point (x, y), passed as an input parameter, is inside or outside the given domain.

7.6.4 TOMS Programs

Again and again, programs dedicated to the numerical solution of elliptic PDEs are published in the journal *Transactions on Mathematical Software* (TOMS). All these programs can be obtained free of charge via the Internet service NETLIB (see Section 7.3.7) in machine readable form.

NETLIB/TOMS/541 contains a variety of subprograms for the solution of the two-dimensional Helmholtz equation in cartesian, polar, cylindrical or spherical coordinates. The domain is always rectangular (in the respective coordinate system) and various boundary conditions can be specified. A finite difference scheme based on an equidistant rectangular grid is used for the discretization of the PDE.

NETLIB/TOMS/541 has been included in the numerical program libraries CMLIB and SLATEC in an extended form known as FISHPACK. The French word for fish—*poisson*—indicates the type of equation that this package is intended for. A specially adapted version, CRAYFISHPACK, for Cray computers, is available as a commercial product.

NETLIB/TOMS/543 contains the subprogram fft9 which solves the Helmholtz boundary value problem with Dirichlet boundary conditions on rectangles. Generally, a finite difference approximation of fourth order on an equidistant mesh is used. For the Poisson equation, a sixth order method is implemented.

NETLIB/TOMS/572 solves the Helmholtz equation with Dirichlet boundary condition on arbitrarily shaped domains. The numerical solution is based on the Shortley-Weller approximation of the PDE. Details can be found in O'Leary, Widlund [303].

A similar method is also used in NETLIB/TOMS/593 for the solution of the Helmholtz equation on generally bounded, two-dimensional domains with Dirichlet or Neumann boundary conditions.

NETLIB/TOMS/651 is a set of subprograms for solving the Helmholtz boundary value problem on two- and three-dimensional rectangular or cube-shaped domains. The subprograms are quite similar to IMSL/MATH-LIBRARY/fps*h as far as the parameters and the functionality are concerned. Details of the algorithms used in these subprograms can be found in Boisvert [119].

NETLIB/TOMS/629 solves a contour integral formulation of the Laplace equation on three-dimensional domains with Dirichlet boundary conditions. The integral equation in this problem is discretized using the Galerkin method. The algorithm used in this program is described in detail in Atkinson [92].

NETLIB/TOMS/637 contains the program gencol for the numerical solution of general elliptic boundary value problems on arbitrarily shaped, connected, bounded domains. A collocation method is used for the discretization and the numerical solution of a problem.

This method is also used in NETLIB/TOMS/638 which is designed to solve general, elliptic boundary value problems on two-dimensional rectangular domains.

NETLIB/TOMS/685 contains programs designed to solve two-dimensional, separable, elliptic boundary value problems on rectangular domains using the Ritz-Galerkin discretization. Tensor product B-spline functions are used as the basis functions for the finite element space (see Section 9.9.1).

NETLIB/TOMS/732 contains programs for solving boundary value problems with self-adjoint, elliptic differential operators on general, polygonally bounded domains. The Helmholtz and Poisson equations (with Dirichlet boundary conditions) are solved very efficiently using these programs.

Chapter 8

Using Approximation in Mathematical Model Building

Soweit die Mathematik exakt ist, beschreibt sie nicht die Wirklichkeit, und soweit sie die Wirklichkeit beschreibt, ist sie nicht exakt.[1]

ALBERT EINSTEIN

In this chapter many important aspects of mathematical modeling are discussed by means of a special type of problem which is the focus of interest in the chapters to follow: An observable and continuous phenomenon occurring in reality (science, engineering etc.) is to be described using continuous mathematical models, i. e., functions. The aim of such mathematical modeling and approximation processes is to obtain *analytic data* (cf. Chapter 4) which deviates from the original phenomenon only within given bounds.

Generally speaking, approximation[2] means the representation of arbitrary objects or phenomena by other simpler objects, in such a way that the main characteristics of the original object are preserved. In this book approximation is defined to be the representation of continuous phenomena by analytic data.

Mathematical models in the form of differential equations, which require not only information on the function values but which also incorporate ordinary and/or partial derivatives, are *not* discussed in the following chapters. The same applies to difference equations as they either do not describe continuous phenomena or represent a discrete form of a differential equation.

8.1 Analytic Models

In this section no structural analogy between the original object and the model is required. In fact only *functional models* are discussed. They are assumed

1. to be usable for an adequate representation of the given object or particular information about it. Criteria for adequacy are discussed in Section 8.6.

2. to be capable of being described and processed as *analytic data* (see Chapter 4). In other words, computers must be able to store and algorithms must be able to process this data.

---

[1] As an exact science mathematics does not depict reality. If it approaches reality, it stops being exact.

[2] *approximare* (lat.) means to bring something closer.

The goal of analytic modeling is to find a function:

$$f : B \subset \mathbb{R}^n \longrightarrow \mathbb{R},$$

whose domain B is a connected subset[3] of $\mathbb{R}^n$, e.g., a finite or infinite interval. Depending on the the dimension n of the domain B there are two categories of models:

univariate models with one independent variable ($n = 1$) and

multivariate models with several independent variables ($n \geq 2$).

In addition to functions, which map each element x of their domain B to *exactly one* element of their range, space curves and surfaces in $\mathbb{R}^n$ are also suitable for model building; but such mathematical objects are always represented on a computer by model *functions*.

Example (Space Curve) If a curve in $\mathbb{R}^3$ which passes through the points

$$P_i = (u_i, v_i, w_i)^\top \in \mathbb{R}^3, \qquad i = 1, 2, \ldots, k$$

is to be determined then *three* interpolating functions $f_u(s)$, $f_v(s)$ and $f_w(s)$, depending on a parameter s, are sought which comply with the following conditions (see Section 8.7.2):

$$f(s_i) = \begin{pmatrix} f_u(s_i) \\ f_v(s_i) \\ f_w(s_i) \end{pmatrix} = \begin{pmatrix} u_i \\ v_i \\ w_i \end{pmatrix} = P_i, \qquad i = 1, 2, \ldots, k$$

for some set $\{s_1, \cdots, s_k\}$ of parameter values.

8.1.1 Elementary Functions as Models

As a consequence of the demand that the model function should be processable on a computer, it is clear, that only elementary functions are acceptable as results of the modeling process.

Definition 8.1.1 (Elementary Functions) *Elementary functions can be expressed as formulas which are characterized by a finite number of parameters. They may require only a finite number of evaluations of the standard algebraic, trigonometric and logarithmic functions available on a computer. In these formulas a distinction of different cases (enabling, e. g., piecewise defined functions) and the use of conditional statements, may occur.*

Accordingly, only polynomials, piecewise polynomial functions (splines etc.), trigonometric polynomials etc. are suitable as model functions on a computer. Functions that do *not* comply with the conditions of the previous definition, for instance, because they can only be characterized by an infinite number of coefficients of their Taylor series, have to be approximated (modeled) using elementary functions before they are processed on a computer.

---

[3] *Discrete* functions whose domain is, for example, a subset of integers, are *not* used in analytic modeling.

Example (Error Function) The error function,

$$\operatorname{erf} x = \frac{2}{\sqrt{\pi}} \int_0^x e^{-t^2} \, dt, \qquad x \geq 0, \tag{8.1}$$

is a *non*-elementary function. The integral (8.1) cannot be described in terms of a closed expression but can be approximated arbitrarily closely by elementary functions. One of these model functions is

$$g(x) = 1 - (a_1 r + a_2 r^2 + a_3 r^3 + a_4 r^4 + a_5 r^5)e^{-x^2} \qquad \text{with} \qquad r = 1/(1 + a_6 x).$$

For this function an error bound can be calculated:

$$|g(x) - \operatorname{erf} x| \leq 1.5 \cdot 10^{-7} \qquad \text{for all} \quad x \geq 0.$$

The values $a_1, \ldots, a_6$ can be looked up in the reference book of Abramowitz and Stegun [1].

Elementary Functions for Analytic Operations

Model building is inevitable for many numerical algorithms which solve mathematical problems, e.g., integration, differentiation, etc. Generally, these problems can *not* be solved *directly*. Therefore, it is necessary for constructive methods to use model building and algorithmic approximation processes:

1. The original problem is replaced by a model problem.

2. A numerical solution for the substituted model problem is computed.

3. The solution of the model problem is used as an approximation for the solution of the original problem.

The choice of a suitable class of substitute functions—the creation of a structural concept (see Chapter 1)—depends heavily on the actual problem. However, there is an important general rule:

The model problem, i.e., the problem where the original function is replaced by a model function should be exactly solvable.

For this purpose a careful choice of the class of model functions is a prerequisite. Even the class of elementary functions may be too large for some problems. In this case an appropriate confinement has to be made.

Example (Numerical Integration) The *sampling functions* (sinc-*functions*)

$$\varphi_k(t) = \frac{\sin\left(\frac{2\pi}{\Delta t}(t - k\Delta t)\right)}{\frac{2\pi}{\Delta t}(t - k\Delta t)}$$

are used in signal processing (image processing, speech analysis, etc.) as basis functions of a discrete representation of a signal $f(t)$ in accordance with the *sampling theorem* (Jerri [243], Butzer, Stens [132]):

$$f(t) = \sum_{k=-\infty}^{\infty} f(k\Delta t)\varphi_k(t).$$

Sampling functions cannot, however, serve as substitutes for the numerical integral of f, because their antiderivative *cannot* be expressed in closed form. Therefore, these functions violate the rule mentioned above. Furthermore, although they are elementary functions (according to the previous definition) they *cannot* be used as model functions for numerical integration problems.

Example (Extrema of a Function)   The determination of the extreme values of elementary functions—the only analytic data on a computer—is usually *not* practical on the basis of the calculation of the zeros of the first derivative.

The function must be replaced by a model function which allows for an easy and exact determination of the extrema. If the function is univariate, quadratic polynomials are usually used as substitute (see Fig. 8.1). Analogous methods are used for the numerical computation of other characteristic quantities of a function, e. g., its zeros (see Chapter 14).

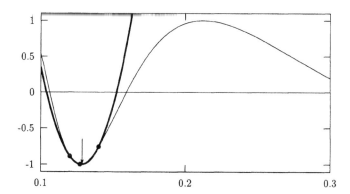

Figure 8.1: A function (—) is discretized (•); the resulting points are interpolated by a quadratic polynomial (——). The minimum of the polynomial is calculated (↓) and used as an approximation for the desired minimum of the original function.

Polynomials are the most prominent class of elementary functions used in numerical methods, because in many instances (such as integration and differentiation) exact manipulations are possible with a reasonable amount of computational work.

8.1.2 Algorithms Used as Mathematical Models

Apart from elementary functions, operator equations are also used to represent and implement models. When operator equations are used to specify a mathematical model, values of the approximating function are not obtained by evaluating a formula, rather they are obtained as a result of an iterative process. The numerical values thus depend mainly on the convergence rate of the iterative process at the point of evaluation, the termination criteria (e. g., bounds on the accuracy of the approximation) and also on the computer arithmetic.

Example (Square Root Function)   Normally the standard function $f(x) = \sqrt{x}$, the square root function, is not implemented using approximating *functions*, but by an *algorithm* describing the iterative solution of the equation $g^2 - x = 0$.

The procedure is as follows: The exponent $e(x)$ is halved, and then Newton's method is applied in computing the square root of the mantissa $m_x = M(x)$ of the floating-point number x. That is, the equation $g^2 - m_x = 0$ is solved iteratively:

$$g_i := (g_{i-1} + m_x/g_{i-1})/2, \qquad i = 1, 2, \ldots, i_{\max}.$$

The initial value g_0 and the maximum number $i_{\max}$ of iterations are chosen in consideration of the computer arithmetic (Cody, Waite [8]).

The square root function $f(x) = \sqrt{x}$ is regarded as an elementary function according to the standard IEC 559 : 1989 (*Binary Floating-Point Arithmetic for Microprocessor Systems*). The model building and the actual implementation are transparent (invisible) for the programmer.

8.2 Information and Data

After choosing the model type and the number and type of the parameters, suitable values must be found for them. To do so adequate information about the original object or phenomenon must be obtained in order to produce a sufficiently accurate model.

Terminology (Information)   The term *information* used in this book is *not* equivalent to the term found in Shannon's information theory[4]. Information in this book has the same meaning as in analytic complexity theory developed by Traub, Wozniakowski, Wasilkowski ([365], [366], [367], [373], [383] etc.). In this book, information is defined to be what is known about an object or system.

Numerical information processing is only possible on a computer if the available information can be converted into numerical data:

$$\text{discrete information} \longrightarrow \text{algebraic data}$$
$$\text{continuous information} \longrightarrow \text{analytic data}$$

8.2.1 Obtaining Algebraic Data from Discrete Information

In most cases it is not difficult to obtain algebraic data from discrete information. There are often, however, errors in the data caused, for example, by measuring a quantity imprecisely.

Example (SO_2 Prediction)   In developing a model to predict the SO_2 concentration in the atmosphere, SO_2 values measured in the immediate past can be used as information. These values are usually the average concentrations in intervals of half an hour and are therefore available in discrete form. From this starting point it is straightforward to obtain algebraic data by creating vectors of measured values. For the processing of this information, however, it is very important to know, at least roughly, the size of the data error (measurement error), which in this example by far exceeds the size of rounding errors in single precision floating-point numbers.

8.2.2 Obtaining Analytic Data from Continuous Information

It is much more difficult to extract analytic data from continuous information. In this case the procedure is split up into two steps: discretization and homogenization.

---

[4]*Information theory* deals with the mathematical description and analysis of telecommunication systems. It is based on papers by C. E. Shannon from 1947 and 1948.

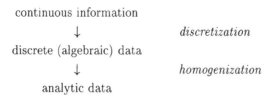

In contrast to discretization—which is easily specified using simple rules—the subsequent homogenization is problematic because its input data is strongly affected by the unavoidable loss of information resulting from the preceding discretization.

Example (Medical Data) Automatic analyses are to be performed of electrocardiograms (ECGs) which obtain information by measuring differences between the electric potential at k different tapping points on the human chest. The analog signals $U_1(t), U_2(t), \ldots, U_k(t)$ obtained in this way are time-dependent functions which have to be discretized and quantized by an analog-to-digital converter. The result of this process is a set of points on a grid (depicted in Fig. 8.2 with exaggerated coarseness).

In Fig. 8.2 it can be seen very clearly that information—e.g., the two peaks that lie between the sampling points—is lost because of the discretization. The interpolating function, which homogenizes the discrete data does not show any peak.

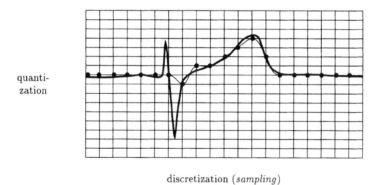

Figure 8.2: A section of an ECG (——), demonstrating discretization and quantization (•) and homogenization (interpolation) of the obtained discrete data (—).

Information lost in this way cannot be recovered by using algorithmic tricks. Information which is not contained in the discrete data can be roughly quantified, at best, using additional data (e. g., bounds for the derivatives).

Only with band-limited functions, which are discretized according to Shannon's sampling theorem with a high enough sampling rate, is *no* information lost. Signals whose Fourier spectrum vanishes outside a finite interval are reconstructible after their discretization provided the sample values are obtained with sufficiently small step sizes (Jerri [243]).

The Stochastic Perturbation of Data

Information can be lost not only during discretization, but also due to the systematic or stochastic perturbation of data. The elimination of stochastic noise from perturbed data, i.e., the reconstruction of the non-perturbed signal, can be achieved to a certain degree by approximation techniques (cf. Chapter 10). The *interpolation* of discrete data (cf. Chapter 9) only makes sense if the data is *not* stochastically perturbed.

8.2.3 The Discretization of Continuous Information

In many situations of practical relevance continuous information about an object or phenomenon (electrical voltage, visual or acoustic information etc.) for which a model is to be constructed, has to be transferred to the computer. This requires a link between the world of *real* phenomena, where the variables are generally continuous analog quantities, and the *engineer designed* world of digital information processing, where the variables are discrete quantities. The connection between the external information and the programs implemented on a computer is provided by special hard- and software used to acquire information: measuring devices, sensors, analog-to-digital converters etc., which produce discrete (algebraic) data that can be stored and processed on a computer.

Example (Process Control) In process control systems the computer is responsible for monitoring and partially or completely controlling mechanical devices and physical processes. Before being processed, continuous signals (control variables) have to be converted into another form (e.g., into pulse signals) and encoded afterwards. Only then can the signals be processed by a computer system (e.g., the autopilot of an aircraft).

Sampling and Quantization

There are two different ways to discretize continuous information:

temporal/spatial discretization, i.e., *sampling* of an independent variable and

amplitude discretization, i.e., *quantization* of the function values.

Example (Image Processing) An image is a function $a(\lambda; u, v)$ which gives the intensity (*brightness*) of the electromagnetic radiation at a given wavelength (*color*) λ, and at each point $(u, v) \in B \subset \mathbb{R}^2$. Each image must be discretized to be processed on a computer.

Usually spatial discretization is carried out on grids with $2^n \times 2^n$ points such that the result is a quadratic image *matrix*. In most cases the image matrices have the dimensions 512×512, 1024×1024 or 2048×2048 ($n = 9, 10$ or 11).

Generally, for amplitude (intensity) discretization 1 byte (8 bits) is sufficiently accurate for most purposes. The wavelengths can be taken into consideration by separately coding three frequency domains—red, green and blue—and using three different image matrices, one for each color (RGB coding).

Example (Photo CD) A special technique developed by two companies (Kodak and Philips) makes it possible to store photographs on a CD-ROM/XY. Each negative or transparency is sampled with a resolution of 2048×3072 pixels. The quantization is done using a special method invented by Kodak called YCC, which differs from the RGB coding, but also uses 3 bytes (24 bits) per pixel. Each photograph is stored in five different resolutions on the CD:

$$128 \times 192, \quad 256 \times 384, \quad 512 \times 768, \quad 1024 \times 1536, \quad \text{and} \quad 2048 \times 3072 \text{ pixels.}$$

The different degrees of spatial discretization serve different purposes, ranging from summarizing presentations to high-quality printing.

Information in the Form of Computer Programs

Continuous information is not always obtained by measuring real phenomena, but may also appear in the form of computer programs. Numerical algorithms often obtain information via parameters of function subprograms. For the calling program, for example, a program for numerical integration, this appears as a black box coding of the continuous information, e. g., the integrand function.

Example (Numerical Integration)   Most methods used for numerical integration are based on an adaptive or nonadaptive piecewise polynomial interpolation of the spatially discretized integrand function. The numerical result is obtained by integrating the interpolating polynomials (see Chapter 12). For the numerical calculation of

$$\int\limits_{0}^{\pi/2} \frac{1}{\sqrt{1 - 0.001 \sin x}}\, dx,$$

the integrand $f(x) = 1/\sqrt{1 - 0.001 \sin x}$ is coded, for instance, in the following subprogram:

```
FUNCTION  integrand (x) RESULT (f)
   REAL, INTENT (IN)  ::  x
   REAL               ::  f
   f = 1./SQRT(1. - 0.001*SIN(x))
END FUNCTION integrand
```

To make numerical integration of the function f possible, the subprogram integrand is called at suitably chosen points $x \in [0, \pi/2]$ by an integration program, with which a spatial discretization of the integrand f is carried out.

Although the function $f \in \mathcal{F}$ is *completely* specified by using formula expressions, only *incomplete information* (a *finite* number of functionals $l : \mathcal{F} \to \mathbb{R}$, i. e., function values) can be obtained from subprograms for further algorithmic processing. Even in this case the continuous information must first be discretized and then the discrete data obtained must be approximated using analytic data (elementary functions); only after these two steps have been taken can the actual processing using analytic operations take place.

Differentiation is a special operation because it is possible to differentiate functions given in the form of Fortran or C programs, i. e., a symbolic differentiation can be carried out *without* discretization and homogenization (Juedes [246]).

8.2.4 The Homogenization of Discrete Data

The discrete algebraic data obtained using discretization is, in most cases, a finite set of points situated on a grid. The discrete data only allows for the application of special methods of processing (searching, sorting etc.) because all analytical as well as many graphical methods require *continuously* defined functions, i.e., these methods can only process analytic data (which is defined for *all* the points in a set $B \subset \mathbb{R}^n$). For example, to square an area is only possible if there is actually a *surface*; discrete points cannot form the boundary of a surface—for this purpose a *continuum* of points is required (see Fig. 8.3).

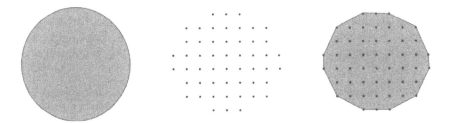

Figure 8.3: Continuum and discrete points.

In such a case continuous (analytic) data must be generated from the existing discrete (algebraic) data. Analytic model building—homogenization—is used to provide a *continuous* mathematical model, with adequate accuracy, for the discrete data.

Example (Computer Graphics)   For the representation of pointwise given relationships (see e.g., Earnshaw [177]) a multi-stage approximation method is often applied:

1.   The data points are interpolated using an *optically smooth* function, which can be a polynomial or a piecewise defined function consisting of polynomials which are *smoothly* assembled, e.g., a spline function (see Section 9.4);

2.   The continuous approximating function is discretized according to the accuracy requirements of the visualization device (plotter, rasterized screen, laser printer etc.);

3.   The discrete data obtained in the previous stage is interpolated by a polygonal arc, which describes e.g., the movements of the plotter pen (see Fig. 8.4).

Example (CAD/CAM)   The definition of curves or surfaces (e.g., the surface of a turbine blade) using discrete points provides the basis for hydrodynamic and strength calculations, graphic design (e.g., the creation of workshop drawings) and also for controlling machine tools.

Example (Simulation)   If continuous processes or objects are given only as a finite set of points (e.g., if the course of the rails in the simulation of a rail vehicle is given only pointwise) then they have to be approximated by analytic data.

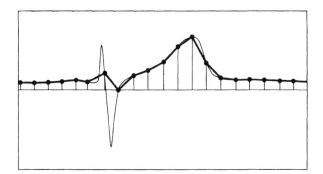

Figure 8.4: ECG discretization: Outside the interval of the two peaks, the discretization is fine enough for the interpolating polygonal arc (—) to approximate the original signal (—) quite well. Near the peaks the discretization must be made with smaller stepsizes.

8.3 Discrete Approximation

The two most common ways of constructing analytic models are using discrete approximation and using function approximation. The use of *discrete approximation (data approximation)* methods requires that discrete data from the continuous phenomenon to be modeled must already have been obtained. Therefore, the central problem is the homogenization of the discrete data. The use of *function approximation* methods requires discretization as well as homogenization.

Terminology (Data Approximation) In many books (Hayes [221]) as well as in the following sections, the term *data approximation* is often used as a synonym for discrete approximation.

If a function is to be approximated the term *function approximation* is used, although it is also a type of data approximation (see Section 8.4), because the original function f belongs to the analytic *data* of the problem.

8.3.1 Discrete Approximation Data

In discrete approximation methods data is a given finite set of points:

$$(x_i, y_i) \in \mathbb{R}^n \times \mathbb{R}, \quad i = 1, 2, \ldots, k. \tag{8.2}$$

The task is to find a function

$$g^* : B \subset \mathbb{R}^n \to \mathbb{R}$$

from a certain class of functions

$$\mathcal{G}_N = \{g(x; c_1, c_2, \ldots, c_N) : c_1, c_2, \ldots, c_N \in \mathbb{R}\}$$

whose distance (which is defined later) from the given data (8.2) is minimal.

The functions from the class $\mathcal{G}_N$ are characterized by N parameters. For these parameters to be determined uniquely by the data points (8.2), it is necessary that there be a greater number of data points than parameters; i. e., $k \geq N$ must hold.

The principle of distance minimization, on which the calculation of the function $g^* \in \mathcal{G}_N$ is based, is referred to as *best approximation*. Finding the best approximating function g^* requires the solution of a minimization problem for determining suitable values of the parameters $c_1, c_2, \ldots, c_N$. As a rule, the norm (see Section 8.6) of the *vector of residuals* $\Delta_k g - y$ of $g \in \mathcal{G}_N$ serves as a distance function D:

$$D(\Delta_k g, y) := \|\Delta_k g - y\| = \left\| \begin{pmatrix} g(x_1; c_1, c_2, \ldots, c_N) \\ g(x_2; c_1, c_2, \ldots, c_N) \\ \vdots \\ g(x_k; c_1, c_2, \ldots, c_N) \end{pmatrix} - \begin{pmatrix} y_1 \\ y_2 \\ \vdots \\ y_k \end{pmatrix} \right\|.$$

In this context $\Delta_k g = (g(x_1), g(x_2), \ldots, g(x_k))^\top$ denotes the discretized function g at points $x_1, x_2, \ldots, x_k$. The desired approximating function g^* or, strictly speaking, its parameter vector can be obtained as a solution of the minimization problem:

$$D(\Delta_k g^*, y) = \min \{ D(\Delta_k g, y) : g \in \mathcal{G}_N \}. \tag{8.3}$$

In the context of discrete approximation, two cases, with regard to the *accuracy of the data*, are distinguished: exact data and data superimposed with errors.

Error Affected Data

In the case of data superimposed with errors, the determination of an approximating function often means separating the systematic and stochastic components of the data (see Fig. 8.5). The approximating function must therefore be made independent of redundant and stochastic information. The accuracy of the approximation relating to the modeling of a continuous phenomenon is not too important as compared with the analysis and smoothing aspects.

Figure 8.5: Separation of the systematic and stochastic components of data.

Exact Data

If the data coincides with the exact values or if its deviation from the original values is negligible, the aim of constructing models is often homogenization, i. e., the generation of analytic data from discrete data. In such cases $D(\Delta_k g^*, y)$ should be as small as possible (see Section 8.3.2). However, when an efficiently processable function is required the aim can also be to yield a compact representation which is not of the utmost accuracy. In these cases the values $g^*(x_i)$ of the model function may coincide with the given data y_i only to a few decimal places.

8.3.2 Interpolation

The choice of the number N of parameters depends on the accuracy requirements of the numerical approximation problem. Generally, one starts with a low number N_{min} of parameters, which is increased step-by-step in order to reach the required level of approximation accuracy.

A special case of discrete approximation is *interpolation*. In interpolation the number N of parameters is equal to the number k of data points. The family $\mathcal{Y}_N$ of interpolating functions has to be chosen suitably (e.g., $\mathcal{Y}_N$ may contain all polynomials with degree $d \leq N$). The result of an interpolation problem, i.e., the interpolating function g^*, approximates the given discrete data (8.2)—but not necessarily the *continuous* phenomenon to be modeled (see Fig. 8.2)—as close as possible. The values of g^* at points $x_1, x_2, \ldots, x_k$ actually coincide with the given data $y_1, y_2, \ldots, y_k$:

$$
\begin{pmatrix}
g^*(x_1; c_1, c_2, \ldots, c_k) \\
g^*(x_2; c_1, c_2, \ldots, c_k) \\
\vdots \\
g^*(x_k; c_1, c_2, \ldots, c_k)
\end{pmatrix}
=
\begin{pmatrix}
y_1 \\
y_2 \\
\vdots \\
y_k
\end{pmatrix}, \quad \text{i.e.,} \quad D(\Delta_k g^*, y) = \|\Delta_k g - y\| = 0.
$$

In interpolation, solving the minimization problem (8.3) involves solving a system of $N = k$ linear or nonlinear equations to obtain the unknown parameters $c_1, c_2, \ldots, c_k$ (see Chapter 9).

8.3.3 Case Study: Water Resistance of Vessels

The *National Physical Laboratory* (Teddington, England) (Hayes [220]) was asked to find the optimal form of fishing vessels. They were given data on 94 already existing fishing vessels. For each boat there were six parameters to work with: length/width, width/draft, max. area of cross section/(width × draft) etc. The water resistance coefficient C_R for each of the boats was obtained by experiment. The size of the boats varied from 1400 to 3400 tons of displacement.

The goal was to determine the optimal form (determined by the six form parameters) for each boat size (characterized by the displacement value) which had the smallest water resistance coefficient C_R.

A possible strategy for finding a solution to this problem would have been to use the available data of the optimal vessel in certain classes of displacement, e.g., 1400 – 1600, 1600 – 1800,..., 3200 – 3400 tons. The water resistance of newly built boats would not, however, be improved using this method. A different method was thus chosen in order to utilize the available knowledge (in the form of the 94 data sets) optimally:

1. A mathematical model for the relationship between the form of the vessel and its water resistance (in fact a model where C_R is represented as a function of the parameters $p_1, \ldots, p_6$) was developed.
2. The model was used to determine the optimum parameter values for a given displacement, i.e., the function $C_R(p_1, \ldots, p_6)$ was minimized.

The ship-building engineers neither restricted the mathematical model nor made any additional requirements; whence the mathematicians had to choose a *suitable* model function on their own. During the model building process considerable difficulties arose due to the extremely low number of observations, in view of the six-dimensional parameter space. Note how rapidly the number of coefficients of a polynomial model

$$P(x_1, x_2, x_3, x_4, x_5, x_6) := \sum_{i=0}^{d} \cdots \sum_{n=0}^{d} a_{ijklmn} x_1^i x_2^j x_3^k x_4^l x_5^m x_6^n \qquad (8.4)$$

increases as the maximal exponent d increases:

maximal exponent d	1	2	3	4	5	6	...
number of coefficients	64	729	4 096	15 625	46 656	117 649	...

The number of coefficients also increases very rapidly if the restriction $i + j + k + l + m + n \leq d$ is made, which is usual in the case of multivariate polynomials $P \in \mathbb{P}_d : \mathbb{R}^n \to \mathbb{R}$ of degree d:

degree d	1	2	3	4	5	6	...
number of coefficients	7	28	84	210	462	924	...

The number of coefficients of univariate polynomials of degree d $P \in \mathbb{P}_d : \mathbb{R} \to \mathbb{R}$ is only $d + 1$.

Polynomials can only be adjusted to given data points in a *reasonable* way if there are at least as many data points as coefficients. In this case only a multivariate polynomial with $d \leq 3$ could be fitted to the 94 data points.

Thus it was necessary to obtain more information from the ship-building engineers in order to decide on the best type of model. From the practical experience of the ship-building engineers, they observed that, as a function of the bow angle α, the water resistance of the vessel had a minimum. It followed that the model function had to be at least quadratic in α because a linear function does not have a minimum (only extreme values at the boundaries of a finite domain). A multivariate polynomial model (8.4) of degree $d = 2$ (in which not all terms appeared) was thus chosen. After many experiments with various polynomials, 32 of the possible 729 coefficients of the function (8.4) were chosen, and their values were calculated using the available data. The remaining coefficients a_{ijklmn} were deliberately assumed to be zero. With the aid of this model, vessels with a water resistance coefficient C_R which was clearly lower than that of already existing vessels (in Fig. 8.6 the values of the new and old vessels are separated by a line) were built.

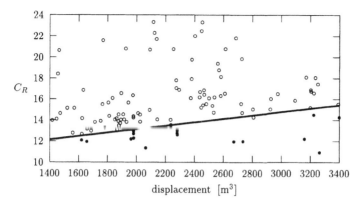

Figure 8.6: Water resistance coefficients C_R of vessels on which the model was based (○). "Optimal" vessels which were built using the model have lower C_R values (•).

Despite the usefulness of this model, it is most likely *not* an optimal model because the number of six-variate functions (8.4) with 32 non-vanishing coefficients is enormous:

$$\binom{729}{32} \approx 7.7 \cdot 10^{55}.$$

The difficulty of this problem is therefore not in numerically determining the 32 coefficients, but in choosing the right type of model, i.e., in choosing a suitable class of model functions from a function space which is too large to be manageable. Such problems are often unsolvable without the help of people working in the field.

8.4 Function Approximation

For a given function $f : B \subset \mathbb{R}^n \to \mathbb{R}$, *function approximation* is finding a model function g that is an element of a certain class of functions $\mathcal{G}_N$ whose deviation with respect to a distance D (see Section 8.6) does not exceed a given tolerance τ on the domain B:

$$D(f, g) \leq \tau.$$

In this book it is assumed (independent of the representation or availability of f) that concerning *data accuracy* the values of the function are *exact* or at most are affected by errors that are negligible in this context (e. g., rounding errors).

As already mentioned in Section 8.2, the acquisition of analytic data from continuous information can take into account only *finite* information, e. g., a finite number of function values of f.

Note that in discrete approximation, the number and the position of the data points (8.2) are given and cannot be changed. The only thing that can be adjusted is the choice of $\mathcal{G}_N$. In function approximation, however, the number and type of the defining quantities can also be chosen.

Information about the function to be approximated, for example, in the form of k scalar data values or k data vectors, is acquired by *discretizing*, i.e., by applying k *information operators* (*information functionals*)

$$l_1(f), l_2(f), \ldots, l_k(f) \in \mathbb{R}$$

(Traub, Wozniakowski [367]). This discretization can be done adaptively or non-adaptively.

8.4.1 Nonadaptive Discretization

During nonadaptive discretization of a function f, k pieces of information are acquired by applying information operations

$$N^{\mathrm{na}}(f) = \big(l_1(f), l_2(f), \ldots, l_k(f)\big),$$

that are independent of one other and whose number is fixed in advance. The acquisition of nonadaptive information can be done very efficiently on multiprocessor computers:

$$\begin{aligned}
\textit{determination of } l_1(f) \quad &\rightarrow \quad \text{processor } 1 \\
\textit{determination of } l_2(f) \quad &\rightarrow \quad \text{processor } 2 \\
\vdots \qquad\qquad &\qquad \vdots \\
\textit{determination of } l_k(f) \quad &\rightarrow \quad \text{processor } k.
\end{aligned}$$

Correspondingly, nonadaptive information is referred to as *parallel information*. The overall time needed for acquiring $N^{\mathrm{na}}(f)$ is determined by the maximum processing time used by any of the processors.

8.4.2 Adaptive Discretization

The term *adaptive* is derived from technical automation and control theory. Control systems which automatically adjust themselves to unpredictably changing conditions (e.g., changing input signals) with respect to a given effectiveness criterion, are called *adaptive systems*.

During adaptive (dynamic) discretization, the course of the sampling process depends at any time on the information already available. In this process, information about f is acquired step-by-step:

$$z_1 = l_1(f), \; z_2 = l_2(f; z_1), \ldots, \; z_i = l_i(f; z_1, z_2, \ldots, z_{i-1}), \ldots \,.$$

The ith (scalar or vector) value z_i of the information operation depends on the values $z_1, \ldots, z_{i-1}$ obtained previously:

$$N^a(f) = \big(l_1(f), l_2(f; z_1), \ldots, l_{k(f)}(f; z_1, z_2, \ldots, z_{k(f)-1})\big).$$

Therefore, a parallel evaluation of the information operations is *not* possible—adaptive information is necessarily determined in a sequential manner, which is why it is referred to as *sequential information*.

The number $k(f)$ of data points is determined dynamically, i.e., during the computation process: After determining each value z_i a stopping condition $stop_i(z_1, z_2, \ldots, z_i)$ is examined, which determines, on the basis of existing information, whether the process is terminated or allowed to proceed. After terminating during the ith step,

$$N^a(f) = \big(l_1(f), l_2(f), \ldots, l_i(f)\big)$$

represents the adaptively acquired information about f.

Example (Bisection)   Zeros of a continuous function $f : [0, 1] \to \mathbb{R}$, whose values $f(0)$ and $f(1)$ are of opposite sign, can be calculated numerically using adaptively acquired information. At first, $f(0.5)$ is calculated. Then, either $f(0.5) = 0$, i.e., 0.5 is identified as a zero, or the function values are of opposite sign at the endpoints of one of the intervals $[0, 0.5]$ or $[0.5, 1]$. Next, f is evaluated in the center of the corresponding interval. The information $(f(0), f(1), f(0.5), \ldots)$ is clearly adaptive because it is impossible to predict the next point of evaluation without knowing the previously obtained function values. For this reason a parallelization of the bisection method is impossible.

Example (Numerical Integration)   The key component of many integration programs is a *quadrature module*, which calculates an approximate value

$$q \approx \int\limits_a^b f(x)\, dx =: \mathrm{I}(f; a, b)$$

for a user-specified function procedure f and a given interval $[a, b]$ and also calculates an estimate

$$e \approx |q - \mathrm{I}(f; a, b)|$$

of the absolute value of the algorithm error. The calculation is terminated if e meets an accuracy requirement. Otherwise, the interval $[a, b]$ is bisected and the quadrature module is applied to each half. If the sum of the error estimates does not meet the accuracy requirement, the subinterval with the greater error estimate is bisected and both halves are processed in the same way as the original interval $[a, b]$. This method is continued until the sum of the error estimates is small enough. This method, the *global subdivision strategy* (see Chapter 12), is based on adaptive acquisition of information about the integrand and is thus quite similar to the bisection method. Globally adaptive integration methods can therefore only be parallelized after undergoing some modification.

8.4.3 Homogenization Algorithms

After the discretization of the function f information is available in the form of a finite set of numerical values (function values, values of derivatives, integrals etc.), viz.

$$l_1(f) = f(x_1), \ l_2(f) = f(x_2), \ldots, \ l_k(f) = f(x_k),$$

which, together with the adaptively or nonadaptively chosen abscissas (sample points) $\{x_1, x_2, \ldots, x_k\} \subset B$, represent a discrete set of points:

$$(x_1, f(x_1)), \ldots, (x_k, f(x_k)) \qquad \text{with} \quad (x_i, f(x_i)) \in B \times \mathbb{R}, \quad i = 1, 2, \ldots, k.$$

Continuous information (analytic data) is therefore reduced to *discrete* information (algebraic data). Accordingly, methods of discrete approximation, e.g., the principle of interpolation (see Chapter 9), can also be used in function approximations to determine the parameters $c_1, c_2, \ldots, c_N$ of the approximating function $g(x; c_1, c_2, \ldots, c_N)$.

After choosing the class of functions $\mathcal{G}$ (which the approximating function should be an element of) the *number* N of parameters and their concrete *values* $c_1, c_2, \ldots, c_N$ must be determined. One strategy for doing this would be an *algorithm scheme*, in which a sequence of subproblems with an increasing number of parameters is solved. For each number of parameters

$$N := N_{\min}, \; N_{\min} + \text{increment}, \; N_{\min} + 2 \cdot \text{increment}, \; \ldots$$

a model function g_N satisfying the approximation criteria (best approximation or interpolation, adaptive or nonadaptive discretization, the distribution of the interpolation nodes etc.) is determined.

$N := N_{\min}$;
error_estimate := greatest_machine_number;

do while error_estimate $> \tau$
    calculate the function $g_N = g(x; c_1, c_2, \ldots, c_N)$,
        satisfying the approximation criteria;
    calculate a new value of the error_estimate;
    $N := N + \text{increment}$
end do

To assure the termination of the algorithm, the *convergence* of the sequence $\{g_N\}$ to the approximating f must be guaranteed. The *efficiency* of approximation algorithms depends mainly on the *rate* of convergence. Therefore, in Chapter 9 and 10 convergence characterizations and the rate of convergence of sequences of approximating functions are examined.

The implementation of a specific approximation algorithm using the previous meta-algorithm must take into account the effect of rounding errors. If, for example, the tolerance τ is too small with respect to the actual level of rounding errors, the algorithm will *not* terminate without additional inquiries.

The choice of the sampling points can also influence the result of homogenization.

Example (A Function with Three Peaks)   The function

$$f(x) = \frac{0.9}{\cosh^2(10(x - 0.2))} + \frac{0.8}{\cosh^4(10^2(x - 0.4))} + \frac{0.9}{\cosh^6(10^3(x - 0.54))} \qquad (8.5)$$

is differentiable an arbitrary number of times on $[0, 1]$. Therefore, the sequence of the interpolating polynomials P_d using Chebyshev nodes converges to f as $d \to \infty$ (cf. Section 9.3.7). It is thus guaranteed that for each tolerance τ there exists a polynomial P_d which meets the accuracy requirement.

The function (8.5) is—with the exception of the small interval $[0.537, 0.543]$—*numerically* identical on a computer, with single precision IEC/IEEE-arithmetic, to the function

$$\bar{f}(x) = \frac{0.9}{\cosh^2(10(x - 0.2))} + \frac{0.8}{\cosh^4(10^2(x - 0.4))},$$

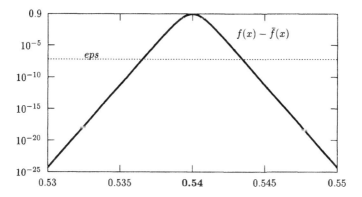

Figure 8.7: The size of the third term in the function (8.5), *logarithmically* scaled.

that consists only of the first two terms of f (which has *no* peak at $x = 0.54$), i.e., $\Box f = \Box \bar{f}$ (cf. Fig. 8.7).

Algorithms designed to compute an approximating function can only handle a finite number of function values $f(x_1), f(x_2), \ldots, f(x_k)$. Algorithms which choose interpolation nodes with

$$x_i \notin [0.537, 0.543], \quad i = 1, 2, \ldots, k,$$

e. g., equidistant interpolation nodes $x_i := 0.005 + 0.01(i-1)$

$$x_1 = 0.005, \ x_2 = 0.015, \ \ldots, \ x_{54} = 0.535, \ x_{55} = 0.545, \ \ldots, \ x_{100} = 0.995$$

cannot distinguish between f and $\bar{f}$ and yield approximating functions whose maximal deviation from f is considerably greater than the desired accuracy.

This situation is characteristic for all problems of function approximation: *No approximation algorithm which exclusively uses information in the form of (a finite number of) function values can produce an approximating function g whose distance from f is guaranteed to be less than a given tolerance.* It follows that *all* such algorithms can yield arbitrarily bad approximating functions (in spite of arbitrarily small error estimates).

8.4.4 Additional Information

For the approximation of mathematically defined functions there is often additional information, e. g., periodicity, monotonicity, zeros, minima, maxima etc. which can be used to improve the reliability and efficiency approximation algorithms (Cody, Waite [8]).

Example (Sine Function)   All implementations of a sine function procedure take into account that

$$\begin{aligned}
\sin(x + k\pi) &= (-1)^k \sin(x) \\
\sin(-x) &= -\sin(x) \\
\sin(x + \pi/2) &= \cos(x),
\end{aligned}$$

in order to simplify the approximation problem so that the interval in which an approximating function

$$g(x) \approx \sin(x), \qquad x \in [a, b],$$

is to be found, is as small as possible.

8.5 Choosing a Model Function

There is an *infinite* set G of functions, candidates for approximating a given function or a discrete set of points, e. g., the set of *all* real functions, which coincide at certain points with the data or which are a *small* distance—which has to be defined more precisely—from the data points (see Section 8.6).

In general, the set G contains both an infinite subset of *good* as well as an infinite subset of *bad* approximating functions. The question arises, as to what criteria a *good as possible* approximating function in G should satisfy.

In the following sections aspects important for the selection of approximating functions (cf. also the survey of Brodlie [127]) are considered. The *evaluation* of the properties of different model functions and the actual selection can only be done taking into account the application at hand.

8.5.1 Uniform or Piecewise Approximation

There are two ways to define an approximating function:

1. One global, uniformly defined function (e. g., *one* polynomial) is used for the whole domain B
2. A function is assembled piecewise from *several* functions, for instance, polynomials (e. g., a polygonal arc composed of linear functions).

If the aim of the approximation is to obtain a model function which is as easy to compute and manage as possible, an obvious choice of the approximating function is a single (globally defined) polynomial. According to the Weierstrass theorem (Theorem 9.3.4) for each accuracy requirement (based on maximal deviation) there exists a polynomial which meets this requirement, provided the function to be approximated is continuous on B. Although this statement is mathematically correct, it is still not possible to solve all practical problems in this way (see Chapter 9).

In many cases, for instance with strongly oscillating functions or a *non-uniform* function behavior (functions possessing very smooth parts as well as points with extremely high curvature, e. g., *peaks, edges* etc.), the approximation with *one* polynomial is problematic. The approximating polynomial, whose existence is guaranteed by the Weierstrass theorem, can possibly have—even with a modest accuracy requirement—an extremely high degree. As to *what* degree is to be regarded as *high* (or as *too high*) depends on the specific features of a particular approximation problem. This question is discussed in the following sections and chapters.

For all the reasons mentioned above it is not recommended that an approximation problem be solved with a single polynomial of high degree on the whole domain B. In most cases a more efficient solution is to divide B into disjoint subdomains $B_1, B_2, \ldots, B_m$, where separate polynomials of lower degree are determined.

8.5.2 Linear Approximation

One of the fundamental decisions in choosing a class of model functions (approximating functions)

$$\mathcal{G}_N = \{g(x; c_1, c_2, \ldots, c_N) \,|\, c_1, c_2, \ldots, c_N \in \mathbb{R},\ g : B \subset \mathbb{R}^n \to \mathbb{R}\}$$

concerns linearity. It is important to distinguish between:

1. linearity/nonlinearity with respect to *independent variables* x and
2. linearity/nonlinearity with respect to *parameters* $c_1, c_2, \ldots, c_N$.

Linearity with Respect to Independent Variables

Most processes and phenomena in the "real world" are *nonlinear*. Nevertheless models which are *linear* in x may have a satisfying informational value, and are sometimes of remarkable relevance in practical applications.

Example (Ohm's Law) Ohm's law $I = GU$ provides a relation between the current I and the voltage U. In many applications the conductance G, the factor of proportionality in Ohm's law, can be assumed to be constant. Ohm's law is then a linear model.

The electrical resistance $R = 1/G$ can be influenced by different factors. It can, for example, be affected by the temperature. On the other hand, conducting materials heat up if an electric current flows through them.

current flows $\longrightarrow$ conductor heats up
$\downarrow$
$\underbrace{\text{resistance changes nonlinearly } (\sim \text{quadratically})}$
linear model is only an *approximation*

Although Ohm's law provides only an *approximate* description of the relation between the voltage and the current, as explained above, it can still be used in many applications with advantage, because nonlinear effects are often negligible.

A good deal of mathematical methods are only applicable to linear problems. This circumstance leads to the application of linear models, even in cases where there is reason to suppose that the true relationship differs essentially from the linear one.

The errors made by replacing a nonlinear relationship with a linear one can be either quantitative or qualitative. In the case of quantitative errors, the solution of the mathematical problem describes the properties of the real process quite correctly, although there are concrete numerical results which vary considerably from the correct values. In the case of qualitative errors, some of the phenomena

which are observable in reality are not represented in the linear model. Such phenomena are said to be *essentially nonlinear* and, as a matter of principle, linear
models are unsuitable for their description. Typical examples of essentially nonlinear phenomena are saturation, bifurcation, chaos etc. In such cases nonlinear
models *must* be used.

Linearity with Respect to Parameters

Whether an approximation method is linear or nonlinear depends *not* on the
dependency of the approximating function g on its independent variable x, but
on the dependency of g on its *parameters* $c = (c_1, c_2, \ldots, c_N) \in \mathbb{R}^N$.

Definition 8.5.1 (Linear Approximation) *An approximating function g is
said to be linear if it is additive and homogeneous with respect to all of its parameters, i. e., if it has the following properties:*

  1. $g(x; b + c) = g(x; b) + g(x; c)$ (additivity)

  2. $g(x; \alpha c) = \alpha\, g(x; c)$ (homogeneity).

In this sense the use of polynomials

$$P_d(x; c_0, c_1, \ldots, c_d) = c_0 + c_1 x + c_2 x^2 + \cdots + c_d x^d, \qquad c \in \mathbb{R}^{d+1}$$

as modeling functions yields a *linear* approximation problem.

Linear approximating functions can always be represented as linear combination

$$g(x; c_1, c_2, \ldots, c_N) = \sum_{i=1}^{N} c_i g_i(x)$$

of a fixed set of (linearly independent) basis functions

$$g_i : B \subset \mathbb{R}^n \to \mathbb{R}, \qquad i = 1, 2, \ldots, N.$$

Example (Polynomials) Every univariate polynomial of degree $d \in \mathbb{N}_0$

$$P_d(x; c_0, c_1, \ldots, c_d) = c_0 + c_1 x + c_2 x^2 + \cdots + c_d x^d$$

is a linear combination of the basis functions $\{1, x, x^2, \ldots, x^d\}$. Therefore, every polynomial
depends *linearly* on its parameters $c_0, c_1, \ldots, c_d$, although polynomials (of degree $d \geq 2$) are
nonlinear functions of their independent variable x.

Example (Trigonometric Polynomials) Trigonometric sums of order d

$$S_d(x; a_0, a_1, \ldots, a_d, b_1, \ldots, b_d) = \frac{1}{2} a_0 + \sum_{i=1}^{d}(a_i \cos ix + b_i \sin ix)$$

are linear combinations of the $2d + 1$ basis functions

$$\{1, \sin x, \sin 2x, \ldots, \sin dx, \cos x, \cos 2x, \ldots, \cos dx\}.$$

They are also *linear* approximating functions, despite the fact that trigonometric polynomials
are nonlinear functions of x.

Piecewise Linear Approximation

In this case the domain B is divided into several subdomains $B_1, \ldots, B_m$, and g is a linear approximating function in each subdomain:

$$g(x; c) = \sum_{i=1}^{k_j} c_{ij} g_{ij}(x) \qquad \text{for all} \quad x \in B_j, \ j = 1, 2, \ldots, m.$$

Example (Splines) Piecewise polynomials are linear approximating functions. With appropriate smoothness at the boundaries between two adjoining subdomains they are called *spline functions* (see Section 9.5).

Properties of Linear Approximating Functions

1. Linear approximating functions require very little computational effort: determination of their parameters leads to systems of linear equations which can be solved with less effort than nonlinear systems or data fitting problems (overdetermined systems of equations).

2. Extensions of the model function by adding extra terms are straightforward.

3. Linear approximating functions are linear: approximating the two sets of data points

$$\{(x_i, y_i), \ i \in I\} \qquad \text{and} \qquad \{(x_i, \overline{y}_i), \ i \in I\}$$

and then adding the two resulting functions yields mathematically the same approximating function as would be obtained by approximating the data points $\{(x_i, y_i + \overline{y}_i), \ i \in I\}$.

The fact that approximating functions are additive also makes it easier to carry out empirical sensitivity analyses where selected data sets are superimposed with additive perturbations and their effects on the approximating functions are studied.

8.5.3 Nonlinear Approximation

In practice, mixed linear-nonlinear model functions are most often used:

$$g(x; c_0, c_1, \ldots, c_K, d_1, d_2, \ldots, d_K) = c_0 + \sum_{i=1}^{K} c_i \, g_i(x; d_1, d_2, \ldots, d_K),$$

where $c_0, \ldots, c_K$ are the linear and $d_1, \ldots, d_K$ the nonlinear parameters. If the values of the nonlinear parameters are *known* a priori then the approximation problem is *linear*.

Example (Exponential Sums) An exponential sum

$$E_d(t; a_0, a_1, \ldots, a_d, \alpha_1, \alpha_2, \ldots, \alpha_d) = a_0 + a_1 \exp(\alpha_1 t) + \cdots + a_d \exp(\alpha_d t) \qquad (8.6)$$

is *linear* with respect to the $d+1$ (proportional) parameters $a_0, a_1, \ldots, a_d$ and *nonlinear* with respect to the d time constants $\alpha_1, \alpha_2, \ldots, \alpha_d$ which determine the decreasing or increasing behavior of $E_d(t)$. If the time constants are unknown parameters, which have to be determined by the data, then the problem is nonlinear. If they are known a priori, then the problem is linear.

Properties of Nonlinear Approximating Functions

1. It may happen—especially in the case of interpolation—that the approximation problem has no solution at all: The *existence* of an approximating function has to be examined in each individual case (whereas the existence and uniqueness of linear approximating functions are guaranteed under very general conditions—see Chapters 9 and 10).

2. It is possible that there are several (even infinitely many) solutions of an approximation problem. If the *uniqueness* of the approximating function is not guaranteed, i.e., different parameters are obtained using the same data, then the interpretation of these parameter sets is problematic.

3. For determining the coefficients of nonlinear approximating functions, *iterative* methods are necessary. They not only require a great deal of computational effort, but it is also possible that they will *not* converge even if a unique solution exists.

If the properties of linear and nonlinear approximating functions are compared, it becomes clear that, from a mathematical-algorithmic point of view, linear approximating functions are the more advantageous. There are situations, however, in which nonlinear approximating functions have to be used:

1. When the parameters of a nonlinear model have a concrete meaning, as do, for example, the proportional parameters and time constants in the exponential model (8.6); and

2. when a nonlinear model with fewer parameters yields a significantly better approximation to given discrete data.

8.5.4 Global or Local Approximation

Global approximation is characterized by the property that the value of the approximating function g at an arbitrary point $x \in B$ depends on *all* data points. Therefore *all* data points have to be known in order to determine of the parameters $c_1, c_2, \ldots, c_N$. Polynomial interpolation (with *one* uniformly defined polynomial) and spline interpolation are examples of global approximation.

With global approximation the variation (perturbation) of *one* data point has an effect on the *whole* function. The function can also change markedly at points which are very far from the point of the perturbation (see Fig. 8.8).

With *local* approximation the part of the approximating function situated between two data points depends only on these two data points or possibly on data points in the immediate neighborhood. As a result it is possible to construct parts of the approximating function without knowing all of the data points. Moreover, with local approximation the variation (perturbation) of a data point only has an effect on the data points in its immediate neighborhood (see Fig. 8.9).

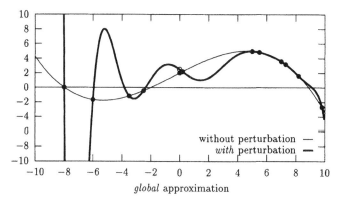

Figure 8.8: The interpolation of the data points (•) by *one* interpolating polynomial leads to extreme oscillations near the left boundary of the interpolation interval, when only a single data point is slightly changed. Here the data point at $x = 0$ is changed from $y = 2$ to $y = 2.5$ (○). All the other data points (especially the data point at $x = 0.2$ with $y = 2.2$, closest to the changed data point) remain unchanged.

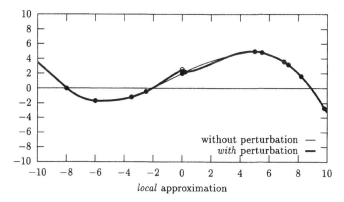

Figure 8.9: The interpolation of the same data points (•) as are found in Fig. 8.8 by a *piecewise* defined Akima function shows almost no oscillations when one data point (○) is perturbed.

Example (Polygonal Arc) The simplest univariate, local approximating function is the interpolating polygonal arc. The data points

$$(x_i, y_i) \in \mathbb{R}^2, \qquad i = 1, 2, \ldots, k$$

are approximated by a piecewise linear function

$$g(x) = y_i + (x - x_i)\frac{y_{i+1} - y_i}{x_{i+1} - x_i}, \qquad x \in [x_i, x_{i+1}], \qquad i = 1, 2, \ldots, k-1.$$

A change $y_i \to \bar{y}_i$ only has an effect on g in the interval $[x_{i-1}, x_{i+1}]$.

Example (Piecewise Cubic Functions) Piecewise interpolating functions, composed of cubic polynomials according to the principle of local approximation, are well suited for graphical function representations. Useful interpolation methods of this category were introduced by Akima [83] and by Ellis and McLain [180].

An algorithmic advantage of local approximating functions is that they require relatively little computational effort as the system of equations which define the parameters can be broken up into separate subsystems. Owing to this decoupling property local approximation methods are very suitable for parallel computers.

Local methods *cannot* be developed using *uniformly* defined approximating functions; only piecewise functions are possible candidates for local methods. Conversely, the use of piecewise functions does not necessarily lead to the creation of a local method, as the example of cubic spline functions shows (see Section 9.7).

8.5.5 Continuity and Differentiability

Continuity and differentiability are *mathematical* characteristics of the *smoothness* of a function. The more often a function is (continuously) differentiable, the smoother it is in a mathematical sense.

Example (Polygonal Arc) A polygonal arc belongs to the class C^0, i.e., to *non-smooth* functions which are continuous but *non*-differentiable. Generally, left-hand and right-hand derivatives of the polygonal arc do not coincide at the nodes.

Example (Cubic Splines) Cubic spline functions with distinct joints (see Section 9.7) belong to class C^2: The functions themselves and their first and second derivatives are continuous, but in general the third derivatives have jump discontinuities at the nodes.

Example (Polynomials) Polynomials (that are defined globally and not piecewise) belong to class C^∞, the mathematically *smoothest* functions:

$$
\begin{aligned}
P_d(x) &= c_0 + c_1 x + c_2 x^2 + \cdots + c_d x^d \\
P_d'(x) &= c_1 + 2c_2 x + \cdots + d c_d x^{d-1} \\
&\vdots \\
P_d^{(d)}(x) &= d(d-1)(d-2)\cdots 2c_d = d!c_d \\
P_d^{(d+1)}(x) &\equiv 0, \quad P_d^{(d+2)}(x) \equiv 0, \ldots \, .
\end{aligned}
$$

Polynomials, like all other functions which are differentiable an arbitrary number of times, are—mathematically speaking—*very smooth*. However, this is not always the way they look when they are used, for instance, in visualization programs. Strong oscillations can often be observed, especially for polynomials with *high* degree (this can mean—depending on the application—$d \geq 10$). They are responsible for a visually *nonsmooth* appearance (see Fig. 8.8 and Fig. 8.12).

Polygonal arcs are *not* smooth. However, if the distances between the nodes $x_{i+1} - x_i$ can be adapted to the behavior of the function f, which is to be approximated, it is always possible to find a polygonal arc g which is *optically* (within the resolution limits of the graphical device) indistinguishable from f.

Example (Robotics) The trajectory of an industrial robot can be specified as a "motion from point to point". The trajectory is defined by a sequence of discrete points in space

$$
P_i = (x_i, y_i, z_i)^\top \in \mathbb{R}^3, \qquad i = 1, 2, \ldots, k.
$$

The effector[5] of the robot runs through these points at given times $\{t_i\}$ and at certain velocities $\{v_i\}$.

From tables of trajectory points $\{P_i\}$ the trajectories as time-dependent functions $P(t)$ have to be determined by interpolation. The resulting functions can be used to control the robot. Its velocity and acceleration are given by

$$v(t) = \mathrm{d}P/\mathrm{d}t = (\dot{x}(t), \dot{y}(t), \dot{z}(t))^\mathsf{T} \quad \text{and} \quad a(t) = \mathrm{d}^2P/\mathrm{d}t^2 = (\ddot{x}(t), \ddot{y}(t), \ddot{z}(t))^\mathsf{T}.$$

The demand for smooth movement (i.e., without jolts) of the effector corresponds to the requirements for a continuous second derivative (the acceleration) of the interpolating function Connecting the points $\{P_i\}$ with a polygonal arc is thus not feasible, whereas a cubic spline function satisfies the smoothness demand.

8.5.6 Condition

With approximation problems, what matters is either the model *function* or its *parameters*. Accordingly, there exist two types of condition:

Condition of function values: The condition number k_f characterizes the sensitivity of the *values* of the approximating function with respect to changes in the data $\mathcal{D} \to \tilde{\mathcal{D}}$:

$$\|\tilde{g} - g\| \leq k_f \|\tilde{\mathcal{D}} - \mathcal{D}\|. \tag{8.7}$$

Condition of parameters: The condition number k_c characterizes the sensitivity of the *parameters* of the approximating function with respect to changes in the data $\mathcal{D} \to \tilde{\mathcal{D}}$:

$$\|\tilde{c} - c\| \leq k_c \|\tilde{\mathcal{D}} - \mathcal{D}\|. \tag{8.8}$$

The size of the condition numbers depends not only on the condition of the approximation problem, but also on the choice of norms in inequalities (8.7) and (8.8) (cf. Section 8.6).

Example (Interpolation using Polynomials) A univariate interpolation polynomial P_d is uniquely determined by the data of the interpolating problem ($d+1$ interpolation nodes and $d+1$ function values). Thus even for different representations of P_d, there exists only *one* condition number k_f for the function values, though of course this number is greatly influenced by the location of the interpolation nodes $x_1, x_2, \ldots, x_k$ (see Section 9.3.8).

The condition of the *function values* of interpolating polynomials is very unfavorable for high degrees of the polynomial (especially for equidistant nodes).

The condition number k_c of the *coefficients* strongly depends on the specific *representation* of the polynomial. For instance, the coefficients $c_0, c_1, \ldots, c_d$ of the monomial representation

$$P_d(x) = c_0 + c_1 x + c_2 x^2 + \cdots + c_d x^d$$

are generally very ill-conditioned, whereas the coefficients $a_0, a_1, \ldots, a_d$ with respect to the basis of Chebyshev polynomials

$$P_d(x) = \frac{a_0}{2} + a_1 T_1(x) + \cdots + a_d T_d(x)$$

are well conditioned.

---

[5] *Effectors* are the parts of an industrial robot used for manipulations: mechanical grabs, welding tongs, paint sprayers etc.

Although a special form of the approximating function g can sometimes be derived from theoretical analyses or numerical experience, it may be that the parameters of the approximating function can only be determined with considerable uncertainty. For instance, the determination of the *parameters* of an exponential sum is in most cases an extremely ill-conditioned problem, whereas the *function evaluation* of exponential sums is normally a well-conditioned task.

Example (Approximation using Exponential Sums)   An artificial test problem was based on the following exponential sum:

$$f(t) = 0.0951 \exp(-t) + 0.8607 \exp(-3t) + 1.5576 \exp(-5t).$$

Discretization and quantization yielded the following data points:

t_i	$\tilde{y}_i$	t_i	$\tilde{y}_i$	t_i	$\tilde{y}_i$
0.00	2.51	0.40	0.53	0.80	0.15
0.05	2.04	0.45	0.45	0.85	0.13
0.10	1.67	0.50	0.38	0.90	0.11
0.15	1.37	0.55	0.32	0.95	0.10
0.20	1.12	0.60	0.27	1.00	0.09
0.25	0.93	0.65	0.23	1.05	0.08
0.30	0.77	0.70	0.20	1.10	0.07
0.35	0.64	0.75	0.17	1.15	0.06

The quantization caused data errors in the values $\{\tilde{y}_i\}$ with a size of 0.005. The values $\{t_i\}$ are exact up to floating-point accuracy. With these values prone to errors, a reconstruction of the function f was attempted. The first attempt to approximate the data used *one* exponential function, and led to an unsatisfactory result. The second attempt led to

$$g(t) = 0.305 \exp(-1.58t) + 2.202 \exp(-4.45t).$$

This function is a satisfactory model in terms of the deviations $g(t_i) - \tilde{y}_i$ (see Fig. 8.10). g is also a very good approximation for the function f: Because of the data errors, no better behavior of the deviation $g(t) - f(t)$ can be expected.

However, concerning the approximation of the parameters of f the function g is very disappointing (compare the coefficients in f and g). The parameter approximation problem is *not* solved. Additional attempts with three exponential terms did not improve the situation—this would also contradict the principle of minimality (see Chapter 1).

8.5.7 Invariance under Scaling

Invariance under scaling transformations is important, for instance, in the field of computer graphics. Firstly, everything to be displayed graphically is defined in terms of two-dimensional Cartesian coordinates, which are called *world coordinates*. The world coordinate domain is mapped onto the normalized coordinate domain $[0, 1] \times [0, 1]$ using the *normalizing transformation*

$$\begin{aligned} \bar{x} &= c_0 + c_1 x \\ \bar{y} &= d_0 + d_1 y, \end{aligned}$$

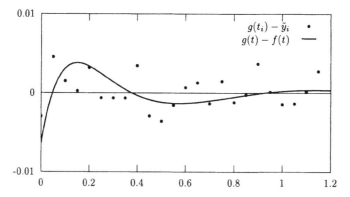

Figure 8.10: (*Visualization of the error*) Both the deviation of the approximating function g from the discrete and quantized data points (•), of the original function $f(t)$ and the deviation of the approximating function $g(t)$, from $f(t)$ (—), are around the same size as the data errors.

with suitable constants c_0, c_1, d_0, d_1. The *device transformation*

$$
\begin{aligned}
\bar{\bar{x}} &= u_0 + u_1 \bar{x} \\
\bar{\bar{y}} &= v_0 + v_1 \bar{y}
\end{aligned}
$$

maps the normalized (device independent) coordinate domain onto the domain of a concrete graphics device (Salmon, Slater [340]). An approximation technique that is invariant under scaling leads to the same picture of the function no matter whether the approximation algorithm is applied to world coordinate data or (normalized) device coordinate data. This property is important when developing black box visualization programs, whose graphical output is independent of implementation details.

8.5.8 Constraints

If the continuous phenomenon to be approximated has special features (in the univariate case e. g., monotonicity, convexity[6], a certain number of inflection points), it is desirable that the approximating function also has analogous properties.

Sometimes the approximating function does not posses the same properties as the data; this is often due to *overshooting*, i.e., the approximating function oscillates, although the data does not (see Fig. 8.11 and Fig. 8.12).

Taking into account information about the expected behavior of the solution, in the form of constraints, often considerably improves the condition of the

---

[6]A function $f : [a,b] \subset \mathbb{R} \rightarrow \mathbb{R}$ is said to be *convex*, if for all points $x_1, x_2 \in [a, b]$ the following inequality holds:

$$(f(x_1) + f(x_2))/2 \ \geq \ f((x_1 + x_2)/2).$$

For functions $f \in C^2$ convexity is characterized by $f''(x) \geq 0$.

approximation problem. This procedure is called *regularization*. So overshooting can be avoided by applying appropriate constraints (e. g., the requirement of monotonicity).

Discrete Constraints

In some cases a model function is sought where the function values or the values of one of its derivatives agree with given values at certain points. An approximating function, which is intended to be used in a standard function subprogram for the sine function, is thus expected to yield precisely the value 0 at $x = 0$ and the value 0.5 at $x = \pi/6$ etc. If the approximation problem is solved using interpolation, it suffices to add those special abscissas to the interpolation nodes. When applying the principle of best approximation, it is necessary to take these constraints into account by modifying the approximation algorithm.[7] This goal can be achieved, for instance, by choosing a special form of polynomial approximation (Hayes [221], Chapter 5):

$$\overline{P}(x) = P_R(x) + P_K(x)P(x).$$

Here P_R is a polynomial—usually with as small a degree as possible—that satisfies the constraints, and P_K is a polynomial with

$$P_K^{(k)}(x_j) = 0$$

for every point x_j where the kth derivative ($k \geq 0$) of the approximating function is involved in the constraints.

Example (Constrained Polynomial Approximation) If the approximating polynomial $P \in \mathbb{P}_d$ is required to satisfy the discrete constraints

$$\overline{P}(0) = 1, \qquad \overline{P}(1) = 0 \qquad \text{and} \quad \overline{P}'(1) = 0, \tag{8.9}$$

then it is possible to choose

$$P_R(x) = (1 - x)^2 \qquad \text{and} \qquad P_K(x) = x(1 - x)^2.$$

These two polynomials have the desired properties:

$$P_R(0) = 1, \qquad P_R(1) = 0, \qquad P_R'(1) = 0,$$
$$P_K(0) = 0, \qquad P_K(1) = 0, \qquad P_K'(1) = 0.$$

P_R is then subtracted from the given data or the given function:

$$\overline{y}_i = y_i - P_R(x_i), \qquad i = 1, 2, \ldots, k,$$

or

$$\overline{f}(x) = f(x) - P_R(x), \qquad x \in [a, b].$$

A polynomial of the form

$$P_K(x)P(x) = x(1 - x)^2[a_0 + a_1 x + a_2 x^2 + \cdots + a_{d-3}x^{d-3}]$$

is used to approximate $\{\overline{y}_i\}$ or $\overline{f}$ respectively. The polynomial $\overline{P} = P_R + P_K P$ obtained in this way satisfies the constraints (8.9).

---

[7]Note that, for instance, that there is *no* degree $d \in \mathbb{N}$ such that the best approximating polynomial P_d^* (in the sense of the maximum norm) for $\sin x$, $x \in [0, \pi/4]$, vanishes at $x = 0$.

Continuous Constraints

Often constraints do not only refer to single points but also to the whole domain B or subsets $B_r \subset B$ (which are usually intervals). Usually the constraints are lower and/or upper bounds for certain derivatives of the approximating function:

$$m(x) \le g^{(k)}(x) \le M(x), \quad x \in B_r \subset B, \quad k \in \{0, 1, 2, \ldots\}.$$

The most important cases in practice are:

Bounds on function values $(k = 0)$:

$$m \le g(x) \le M, \quad x \in B.$$

Such bounds are useful, for instance, when g is intended as an approximation of proportions or concentrations, whose values are always between $m = 0$ and $M = 1$ or between 0% and 100% respectively.

Monotonicity $(k = 1)$:

$$0 \le g'(x) \quad \text{or} \quad g'(x) \le 0, \quad x \in B.$$

This type of constraint is useful, for instance, when accumulated quantities, which cannot decrease, are to be approximated (see Fig. 8.11).

Convexity $(k = 2)$:

convex function behavior:	$0 \le m \le g''(x)$ for all	$x \in B_{\text{convex}}$
concave function behavior:	$g''(x) \le M \le 0$ for all	$x \in B_{\text{concave}}.$

The algorithmic solution of a constrained approximation problem requires the use of methods for *constrained minimization*. A summary of program packages for this type of problem can be found in the book written by Moré und Wright [20].

There are often very efficient algorithms and programs for special classes of functions: methods for *convex spline functions* can be found in de Boor [42], Shumaker [348], and Andersson and Elfving [88]; algorithms for *monotone spline functions* can be found in Fritsch and Carlson [203], Fritsch and Butland [202], Hyman [235], and Costantini [143].

8.5.9 Appearance

In addition to the properties of a function that can be formulated by mathematical means there is also a *subjective* perception of the form of the function. The visual assessment of an approximating function is mostly based on the imagination. There are always certain expectations concerning the *appropriate* approximating function that can hardly be formalized. These expectations can be visualized sometimes using freehand drawings in which information (e.g., monotonicity, convexity, non-negativity etc.) can be expressed.

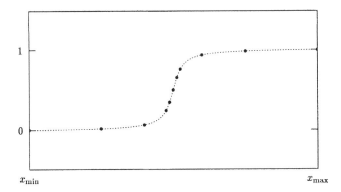

Figure 8.11: Data points (•) of a saturation function (· · ·).

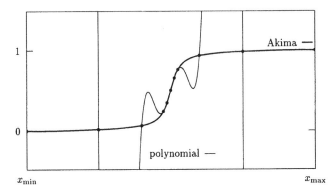

Figure 8.12: Interpolation of the data points in Fig. 8.11 using *one* interpolating polynomial of degree 10 (—) and using a piecewise cubic Akima function (—).

Some kinds of approximation correspond better to the subjectively anticipated function behavior than others. Many people, for instance, regard the Akima interpolation (Akima [83]) more similar to freehand drawing than interpolation using polynomials of high degree (the curves in Fig. 8.12 show this clearly).

Example (Saturation Function) Many economical, technical or scientific growth phenomena are characterized by the fact that there are limits which are not exceeded but which the observed values approach. Appropriate models for such phenomena are, for instance, *saturation functions* which are some form of sums (or integrals) of increments (growth functions) tending to zero. Fig. 8.11 shows data points of a symmetric saturation function[8]. It is possible, of course, to interpolate these 11 data points (•) using *one* polynomial of degree 10. In this case, however, a function (—) with a completely disappointing appearance (see Fig. 8.12) is obtained. The data points of course lie in the interval [0, 1], whereas the polynomial has values

$$P_{10}(x) \in [-4700, 4700], \quad x \in [x_{\min}, x_{\max}],$$

---

[8] A saturation function f is *symmetric*, if its growth function f' is symmetric.

that do not correspond with the expected function behavior. The only exception is a small neighborhood of the inflection point. Akima interpolation yields a function (—) with a discontinuous second derivative. The Akima function is therefore, mathematically speaking, less smooth than the polynomial—though its appearance corresponds much better with what is intuitively expected.

8.6 Choice of the Distance Function

In order to quantify the distance from the model function g to the function f being modeled, a distance function is needed. The choice of the distance function D must be adapted to the characteristics of the particular model. For data that is superimposed with stochastic perturbations, a different distance function than that used for unperturbed data is suitable.

The choice of the distance function is, however, also strongly influenced by the computational effort required in determining the parameters of an approximating function.

8.6.1 Mathematical Foundations

For the mathematical description of approximation processes, it is useful to consider functions (signals) as points or vectors in an appropriate function space (signal domain), (signal-) transformations as mappings of this space, and properties of the functions as properties of the space. Here the word *space* is used in order to give the set of functions a graphical quality.

Definition 8.6.1 (Linear Space, Vector Space)   *A set M is called a linear space or vector space if*

1. *there is an addition operation $+$ defined in M, where $(M, +)$ forms an Abelian group,*

2. *for every element $v \in M$, a multiplication operation with real (or complex) numbers α is defined, for which the following axioms hold:*

    (a) $\alpha(v_1 + v_2) = \alpha v_1 + \alpha v_2$ and
        $(\alpha_1 + \alpha_2)v = \alpha_1 v + \alpha_2 v,$ (*distributive law*)

    (b) $(\alpha_1 \alpha_2)v = \alpha_1(\alpha_2 v),$ (*associative law*)

    (c) $1 \cdot v = v.$

Examples (Linear Spaces)

- The set of n-tuples $(\alpha_1, \alpha_2, \ldots, \alpha_n)$ with real α_i and operations

$$(\alpha_1, \ldots, \alpha_n) + (\beta_1, \ldots, \beta_n) \quad := \quad (\alpha_1 + \beta_1, \ldots, \alpha_n + \beta_n)$$
$$\alpha(\alpha_1, \ldots, \alpha_n) \quad := \quad (\alpha \alpha_1, \ldots, \alpha \alpha_n)$$

forms the linear space $\mathbb{R}^n$. The common terminology used for $\mathbb{R}^n$, especially *vector* and *point*, is also used for other linear spaces.

- The set of all continuous functions defined on the interval $[a, b]$ forms the linear space denoted $C[a, b]$ (under pointwise addition and scalar multiplication)

$$
\begin{aligned}
(f + g)(x) &:= f(x) + g(x) \\
(\alpha f)(x) &:= \alpha f(x).
\end{aligned}
$$

- The set of all m-fold continuously differentiable functions on the interval $[a, b]$ forms a linear space, denoted $C^m[a, b]$.

- The set of all functions $f : \mathbb{R} \to \mathbb{R}$ for which the integral

$$
\int_a^b |f(t)|^p \, dt \qquad \text{or} \qquad \int_{-\infty}^{+\infty} |f(t)|^p \, dt \tag{8.10}
$$

exists, forms a linear space denoted $\mathcal{L}^p[a, b]$ or $\mathcal{L}^p$ respectively. In a strict sense the integrals (8.10) should be interpreted as *Lebesgue integrals*, but without great restrictions they can be thought of as "usual" integrals—*Riemann integrals*—in this chapter.

Definition 8.6.2 (Linear Combination)   *A vector*

$$
v = \alpha_1 v_1 + \alpha_2 v_2 + \cdots + \alpha_m v_m
$$

formed by the (finite) summation of vectors with scalar coefficients $\alpha_1, \alpha_2, \ldots, \alpha_m$ is said to be the linear combination of the vectors $v_1, v_2, \ldots, v_m$.

Definition 8.6.3 (Basis)   *A subset B of a vector space V is said to be a basis of V if each vector $v \in V$ can be uniquely represented as a linear combination*

$$
v = \sum_{j=1}^m \alpha_j b_j, \qquad b_j \in B, \quad m \in \mathbb{N}, \tag{8.11}
$$

of basis elements.

The scalars α_j are said to be the *coordinates* of the vector v relative to the basis B. A subset B of a vector space V is said to be a *generating system* for V if each vector $v \in V$ can be represented—*not* necessarily *uniquely*—as a linear combination of elements of B.

If a vector space V has a finite basis, then each of its bases is finite and they all have the same number of elements. This number is called the *dimension* of V. If V has no finite basis, V is *infinite-dimensional*.

Example (Euclidean Space)   The linear spaces $\mathbb{R}^n$ are finite-dimensional with dimension n. One of their bases is, for instance, the *canonical basis*:

$$
e_1 = (1, 0, \ldots, 0), \quad e_2 = (0, 1, 0, \ldots, 0), \ldots, e_n = (0, \ldots, 0, 1).
$$

Example (Continuous Functions)   The linear space of continuous functions $C[a, b]$ is an example of an *infinite*-dimensional vector space.

Example (Polynomials)   The set $\mathbb{P}_d$ of all polynomials of maximal degree d is a linear space of dimension $d+1$; however, the set $\mathbb{P}$ of *all* polynomials is an infinite-dimensional space.

Definition 8.6.4 (Normed Linear Space)   *A function $\| \, \| : V \to \mathbb{R}_+$, defined on a linear space V is called a norm if it possesses the following properties:*

1. $\|v\| = 0$ *if and only if $v = 0$*   (*definiteness*),

2. $\|\alpha v\| = |\alpha| \, \|v\|$, $\alpha \in \mathbb{R}$ *or* $\alpha \in \mathbb{C}$ (*homogeneity*), *and*

3. $\|v_1 + v_2\| \leq \|v_1\| + \|v_2\|$ (*subadditivity*).

In such cases V is said to be *normed* (under $\| \, \|$). The geometric interpretation of the norm of a vector is its length. As a result of its properties every norm defines a metric (a distance function):

$$D(u,v) := \|u - v\|.$$

Definition 8.6.5 (Metric Space)   *A set V is said to be metric space (under $D : V \times V \to \mathbb{R}_+$) if D has the following properties:*

1. $D(u,v) = 0$ *if and only if* $u = v$ (*definiteness*),

2. $D(u,v) = D(v,u)$ (*symmetry*) *und*

3. $D(u,w) \leq D(u,v) + D(v,w)$ (*subadditivity*).

The functional D is the *metric* of the space, and the number $D(u,v)$ is the *distance* between the *points* u and v.

8.6.2 Norms for Finite-Dimensional Spaces

The definition of the distance between two real numbers using the absolute value

$$D(r_1, r_2) := |r_1 - r_2|, \qquad r_1, r_2 \in \mathbb{R},$$

can be generalized to vectors u and v in $\mathbb{R}^n$ using the Euclidean distance (corresponding to the *Euclidean norm* of the difference vector):

$$D_2(u,v) := \|u - v\|_2 := \sqrt{(u_1 - v_1)^2 + (u_2 - v_2)^2 + \cdots + (u_n - v_n)^2}.$$

A further generalization can be defined using the l_p**-norm** of their difference $u - v$:

$$D_p(u,v) := \|u - v\|_p := \left(\sum_{i=1}^{n} |u_i - v_i|^p \right)^{1/p}, \qquad p \in [1, \infty). \tag{8.12}$$

In the limit $p \to \infty$ the *maximum norm*

$$D_\infty(u,v) := \|u - v\|_\infty := \max \{|u_1 - v_1|, \ldots, |u_n - v_n|\} \tag{8.13}$$

is obtained. In the case of discrete approximation (data approximation) the distance between a model function $g : \mathbb{R} \to \mathbb{R}$ and given data points

$$(x_1, y_1), (x_2, y_2), \ldots, (x_k, y_k) \in \mathbb{R}^2$$

can be defined using (8.12) and (8.13) i.e., by defining it to be the distance $D_p(\Delta_k g, y)$ between the two vectors

$$\begin{pmatrix} g(x_1) \\ g(x_2) \\ \vdots \\ g(x_k) \end{pmatrix} =: \Delta_k g \in \mathbb{R}^k \quad \text{and} \quad \begin{pmatrix} y_1 \\ y_2 \\ \vdots \\ y_k \end{pmatrix} =: y \in \mathbb{R}^k.$$

The utmost caution is advised for the interpretation of such distances!

The number $D_p(\Delta_k g, y)$ only provides information about the distance between y and the function g for a *finite* set of abscissas $\{x_1, \ldots, x_k\}$. If the vector y is the result of the discretization $\Delta_k f$ of an analytic quantity f, then $D_p(\Delta_k g, y)$ gives *no* indication as to the general qualities of approximation. The l_p-norm of the vector of deviations at discrete points

$$e_k := [e(x_1), \ldots, e(x_k)]^\top = [f(x_1) - g(x_1), \ldots, f(x_k) - g(x_k)]^\top$$

is not a norm for the *deviation function* $f - g : \mathbb{R} \to \mathbb{R}$. From the coincidence of a model function g with f at the points $x_1, \ldots, x_k$, viz.

$$\|e_k\|_p = 0$$

it *cannot* be concluded that $f - g = 0$, i.e., that $f(x) = g(x)$ for *all* $x \in \mathbb{R}$. Consequently $\|e_k\|_p$ does not define a metric on the space of real valued functions. The distance $D_p(\Delta_k g, y)$ gives no information about the behavior of the deviation function $e = f - g$ in the gaps between the points $x_1, \ldots, x_k$. The distance between the model function g and the continuous phenomenon f to be modeled *cannot* be characterized with $D_p(\Delta_k g, y)$.

l_p-Norms in Discrete Approximation

The most important l_p-norms in discrete approximation are the Euclidean norm ($p = 2$) and the 1-norm ($p = 1$). The maximum norm ($p = \infty$) plays no practical role in discrete approximation because it is not useful for separating the systematic and the stochastically perturbed parts of data.

The *Euclidean metric* (based on the l_2-norm, the Euclidean norm) is the most commonly used norm in discrete approximation (signal analysis) because

- it guarantees the existence of solutions for approximation problems,

- it enables very efficient and numerically stable algorithms,

- it has a scientific meaning (it is proportional to the energy difference of two signals), and

- it corresponds (in the sense of statistical estimation theory) with the problem of determining the distance between two signals, whose perturbations are (at least to a sufficient approximation) normally distributed.

8.6.3 Norms for Infinite-Dimensional Spaces

To define a norm which takes into account *every* value of a function (of a continuous signal), the summation in (8.12) has to be replaced by integration or (as $p \to \infty$) by the determination of extrema over the whole interval $[a, b]$. This kind of definition leads to the L_p-**norm** (defined on $\mathcal{L}^p[a, b]$) and the corresponding distance functions:

$$D_p(f, g) := \|f - g\|_p := \left(\int_a^b |f(t) - g(t)|^p dt \right)^{1/n}, \qquad p \in [1, \infty).$$

As $p \to \infty$, again, the *maximum norm*

$$D_\infty(f, g) := \|f - g\|_\infty := \max \{ |f(x) - g(x)| : x \in [a, b] \}$$

is obtained. The maximum norm is the most important norm for approximating functions. This norm provides a useful way to determine the distance between the model function $g : [a, b] \to \mathbb{R}$ and the continuous phenomenon to be modeled, the function $f : [a, b] \to \mathbb{R}$.

The importance of the L_p-norms lies particularly in theoretical analysis. The *numerical* computation of an L_p-norm would require numerical integration (or determination of extrema in the case of the maximum norm), and would thus lead to the discrete l_p-norms again.

8.6.4 Weighted Norms

In using one of the norms previously discussed, it is implicitly assumed that every value of the function $f(t)$ or every data point has the same importance, so no special weighting is necessary.

There are, however, cases where it is known a priori that some data points have been obtained by more accurate measurement than others, which are affected by higher measurement errors. In this situation the data points should be weighted: *Weights* $w_i > 0$ or *weight functions* $w(t) > 0$ are defined in such a way that more accurate values have more influence on the distance $D_{p,w}$ than others. For that purpose **weighted** l_p-**norms**

$$\|u - v\|_{p,w} := \left(\sum_{i=1}^k (w_i |u_i - v_i|)^p \right)^{1/p}, \qquad p \in [1, \infty),$$

$$\|u - v\|_{\infty,w} := \max \{ w_1 |u_1 - v_1|, \ldots, w_k |u_k - v_k| \}$$

and **weighted** L_p-**norms**

$$\|f - g\|_{p,w} := \left(\int_a^b (w(t) |f(t) - g(t)|)^p dt \right)^{1/p}, \quad p \in [1, \infty),$$

$$\|f - g\|_{\infty,w} := \max \{ w(x) |f(x) - g(x)| : x \in [a, b] \}$$

are introduced. The choice of weights can be based, for instance, on estimates of the absolute error of the y-values. In this case the weights are chosen to be inversely proportional to the error estimates.

Example (Relative Accuracy) If it is known that the *relative* error of all y-values is of the same size (i. e., if the *absolute* error of the y-values is roughly proportional to y) the weights can be defined as follows:

$$w_i := c/|y_i|, \quad \text{provided} \quad |y_i| > 0,$$

where $c > 0$ is an arbitrary constant.

8.6.5 Hamming Distance

In the space of discrete *binary* signals, the *Hamming distance* between two binary vectors $u = (u_1, \ldots, u_k)$ and $v = (v_1, \ldots, v_k)$ with $u_i, v_i \in \{0, 1\}$

$$D(u, v) = \sum_{i=1}^{k} (u_i \oplus v_i),$$

that is, the number of different components of u and v:

$$u_i \oplus v_i := (u_i + v_i) \bmod 2$$

is often used.

8.6.6 Robust Distance Functions

In statistics, methods that produce *reasonable* values, even if the properties of their input data do *not* meet theoretical assumptions on which the method has been based, upon are called *robust techniques*. Such assumptions are, for instance, certain distribution properties of the stochastic perturbations.

With discrete data obtained by measurement, there are often only a few values perturbed in an extreme way, in contrast to the other data points that are uniformly superimposed by comparatively small stochastic perturbations (*noise*). In this case the stochastic perturbation is based on a *mixed distribution*. Values that seem to be doubtful as realizations of the dominating stochastic perturbation are called *outliers*.

Example (Scanners) Multispectral scanners are used for the production of computer-generated pictures of the earth's surface. In environmental protection activities, air-borne scanners are installed in airplanes, satellites or space laboratories. In planes with propellers, as used for flying at low altitudes, electrical interference may come from the surroundings of the scanner (the ignition system of the engines, or other electrical equipment). The usual effect of such interferences are *spikes*, i.e., the resulting digital image contains pixels with a minimum or a maximum gray level.

In every sample of a population that belongs to an unbounded random variable (e. g., a normally distributed random variable) arbitrarily large and arbitrarily small values (i. e., values over or under every given boundary respectively) are to be expected with non-zero probability. The identification and removal of outliers (*suspicious values*) is therefore not that easy.

The influence of outliers on the approximating function depends strongly on the chosen distance function D_p. For larger values of p the influence of outliers is greater. Of course the greatest influence occurs when using the maximum norm. But also for $p = 2$, the Euclidean norm, this effect is often unpleasantly large. Satisfactory values are around $p \approx 1.3$ (Ekblom [178]).

Example (Robust Estimation of a Scalar) In order to minimize the influence of measurement errors δ_i in determining a scalar quantity c, one can repeat the measurement k-times. From the resulting data points

$$y_i = c + \delta_i, \quad i = 1, 2, \ldots, k$$

that value $\bar{c}$ is determined as an estimate for c that minimizes the distance

$$D(y, ce) = \|y - ce\|_p \quad \text{with} \quad y = (y_1, \ldots, y_k)^\top, \quad e = (1, \ldots, 1)^\top.$$

For the important cases of the l_1-, l_2- and l_∞-norm one gets:

$$p = 1 \quad \bar{c}_1 = \begin{cases} y_{(\frac{k+1}{2})} & k \text{ odd} \\ (y_{(\frac{k}{2})} + y_{(\frac{k}{2}+1)})/2 & k \text{ even} \end{cases} \quad \textit{sample median,}$$

$$p = 2 \quad \bar{c}_2 = \frac{1}{k} \sum_{i=1}^{k} y_i \qquad\qquad\qquad \textit{sample mean and}$$

$$p = \infty \quad \bar{c}_\infty = (y_{(1)} + y_{(k)})/2 \qquad\quad \textit{sample midrange,}$$

where $y_{(1)} \le y_{(2)} \le \ldots \le y_{(k)}$ denote the ordered components of the vector y.

For the following 10 measured values

$$99.4, \ 100.6, \ 98.4, \ 99.1, \ 101.0, \ 101.6, \ 93.8, \ 101.2, \ 99.9, \ 100$$

the estimates

$$\bar{c}_1^{10} = 100.05, \quad \bar{c}_2^{10} = 99.52 \quad \text{and} \quad \bar{c}_\infty^{10} = 97.7$$

are obtained.

Looking at the values more closely, it becomes clear that the value 93.8 is atypically small compared to the other values. If δ_i are realizations of an unbounded (e. g., normally distributed) random variable there are arbitrarily small (and arbitrarily large) values with positive probability to be expected. On the other hand, the value 93.8 is possibly the result of some kind of perturbation (difficult measuring conditions, for instance), so that it is not representative of the scalar quantity c, and can be regarded as an *outlier*.

If the value 93.8 is left out of consideration, the remaining 9 observations lead to the estimates

$$\bar{c}_1^9 = 100.20, \quad \bar{c}_2^9 = 100.16, \quad \bar{c}_\infty^9 = 100.00.$$

The change in the approximations $\bar{c}_p^9 - \bar{c}_p^{10}$,

$$\bar{c}_1^9 - \bar{c}_1^{10} = 0.15, \quad \bar{c}_2^9 - \bar{c}_2^{10} = 0.64, \quad \bar{c}_\infty^9 - \bar{c}_\infty^{10} = 2.30$$

is minimal for the robust l_1-norm and maximal for the l_∞-norm.

If the stochastic quantities $\delta_1, \ldots, \delta_k$ are the realizations of a normal distribution with zero mean, the sample mean gives an optimal estimate for the unperturbed quantity c because the sample mean has the smallest variance of all unbiased linear estimates. However, should the distribution of the perturbations deviate only slightly from the normal distribution, then this optimality is lost. That happens, for instance, if the distribution function

$$(1 - \varepsilon)\Phi + \varepsilon H, \quad \varepsilon \in [0, 1]$$

is a *mixture* of Φ, the normal distribution function, and H, an unknown distribution function. With increasing ε the variance of the sample mean $\bar{c}_2$ can increase very rapidly. In a statistical sense therefore, $\bar{c}_2$ is *not* a robust estimating function, whereas $\bar{c}_p$ with $p \in [1, 2)$ reacts much less sensitively to changes in the distribution function.

The central limit theorem states that in many situations convergence to a normal distribution is to be expected for the distribution function of a sum of independent, random quantities. This theorem is often used as a justification for using a normal distribution to describe random phenomena which result from the additive superposition of a great number of single random effects. The central limit theorem has thus made the use of the l_2-norm very popular, despite the fact that many random quantities found in practice are at best *approximately* normally distributed.

The main reason behind C. F. Gauss' use of the l_2-norm for his *method of least squares* was the fact that the resulting equations for the solution of a fitting problem were linear equations, which are easier to solve. This was no doubt a very important motivation at the beginning of the nineteenth century.

What happens if the assumptions of a normal distribution are slightly perturbed, and how robust estimating techniques can be obtained, are questions which were first discussed in the 1960s. It turned out that data fitting using the l_1-norm or l_p-norms with $p \in (1, 1.5]$ leads to more robust estimation methods, which require, however, considerably more computational effort.

The maximum norm (l_∞-norm) is of *no* practical importance for data analysis because it leads to estimates that are much more sensitive to extreme observations than the corresponding l_2-estimates.[9]

The choice between the l_2-norm and the l_p-norm with $p \in [1, 2)$ is mainly influenced by two factors:

1. Since l_p-norms with $p \in [1, 1.5]$ lead to robust techniques they are superior, in this respect, to the l_2-norm.

2. On the other hand, the least squares method requires *markedly* less computational effort than using other l_p-norms because in this case the corresponding set of equations which must be solved is *linear*.

---

[9]The maximum norm is, however, the most commonly used norm for function approximation.

For *small* data analysis problems (i. e., models with a small number of parameters), where the computational effort does not play an important role, the data fitting is preferably made—if the appropriate software is available—in the sense of an l_p-norm with $p \in [1, 1.5]$. If the computational effort is important—for instance, if the data fitting problem is very large (with hundreds or thousands of coefficients)—the least squares method is generally preferred for reasons of efficiency.

New Definitions for Distances

For the development of robust methods it is sometimes useful to define new distance functions. For this purpose one often refers to the components $u_i, v_i \in \mathbb{R}$ of the vectors $u, v \in \mathbb{R}^n$:

$$D(u, v) := \sum_{i=1}^{n} d(u_i, v_i).$$

Determining the scalar distance function $d : \mathbb{R}^2 \to \mathbb{R}$ in accordance with the maximum likelihood principle[10] leads to the following results:

for the normal distribution $\qquad\qquad\qquad d(u, v) = (u - v)^2,$

for the double exponential distribution $\quad d(u, v) = |u - v|$

and for the Cauchy distribution $\qquad\qquad d(u, v) = \log(1 + \frac{1}{2}(u - v)^2).$

The following definition of a parametric distance function (independent of any special distribution function)

$$d_a(u_i, v_i) = \begin{cases} (u_i - v_i)^2, & |u_i - v_i| < a, \\ a\left(2|u_i - v_i| - a\right), & |u_i - v_i| \geq a, \end{cases}$$

originates from Huber [233]. For $|u_i - v_i| < a$ this continuously differentiable distance function is quadratic and outside this interval it is linear. So depending on the parameter a large deviations $|u_i - v_i|$ are weighted *less* by the Huber distance than by the Euclidean distance $D_2(u, v) = \|u - v\|_2$.

In addition, the maximum likelihood principle (that leads to *M-estimates*), order statistics (*L-estimates*) and rank order tests (*R-estimates*) for defining robust distance functions (Huber [234]) are also used.

8.6.7 Orthogonal Approximation

The standard type of discrete approximation is based on the assumption that only the function values y are superimposed with an additive error δ_i:

$$y_i = g(x_i; c) + \delta_i, \quad i = 1, 2, \ldots, k.$$

---

[10]The *maximum likelihood principle* serves for obtaining statistical estimates $\hat{c}$ for a parameter vector c. For discrete probability distributions, for instance, the choice of the estimate is made in such a way that the event $\{c = \hat{c}\}$ is the most probable one.

If the errors $\delta_1, \ldots, \delta_k$ are independent random quantities that are normally distributed with a mean value $\mu = 0$ and a variance σ^2, then the maximum likelihood estimate for the parameter vector $c = (c_1, \ldots, c_N)^\top$ is the solution of the following approximation problem with respect to the l_2-norm:

$$D_2(\Delta_k g^*, y) \;=\; \min\{D_2(\Delta_k g, y) : g \in \mathcal{G}_N\}$$

$$\text{with} \quad D_2(\Delta_k g, y) := \sqrt{\sum_{i=1}^{k}(g(x_i; c) - y_i)^2}.$$

In this case the parameter vector is therefore best obtained using the *least squares method*, where the *vertical* distance between data points and approximation function is measured with Euclidean distance.

In practice, however, situations occur where the values x_i of the independent variable are superimposed with additive errors ε_i, for instance, because of measurement errors or discretization inaccuracy:

$$y_i = g(x_i + \varepsilon_i; c) + \delta_i, \quad i = 1, 2, \ldots, k.$$

Example (Air Pollutants)   There are many different pollutants in the air. Only a few of them can be measured regularly. In order to use the concentration of a certain substance (e. g., SO_2) as an *indicator* for the concentration of another pollutant (e. g., HCl), a model for the relationship between the two of them based on simultaneous measurements must therefore be developed. In such cases data points $\{(x_i, y_i)\}$ where both y- and x-values are affected by measurement errors are obtained.

In such cases it is more suitable to choose a distance definition that is based on the shortest, i. e., the orthogonal distance of each data point $(x_i, y_i), i = 1, 2, \ldots, k$, from the approximating curve in the univariate case ($x \in \mathbb{R}$) or from the approximating surface in the multivariate case ($x \in \mathbb{R}^n$, $n \geq 2$), respectively (see Fig. 8.13):

$$\{(x, g(x)) : \quad x \in B\}.$$

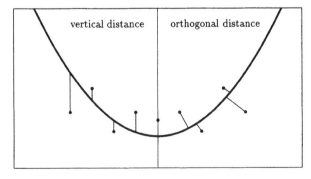

Figure 8.13: Vertical and orthogonal distances between data points and a function.

Terminology (Orthogonal Approximation) The terminology for methods based on orthogonal distance definitions is not standardized. For instance, Golub and Van Loan [50] as well as Van Huffel and Vandewalle [371] use the term *total least squares*, whereas Powell und Macdonald [316] speak of *generalized least squares*.

Boggs, Byrd and Schnabel [118] use the term *orthogonal distance regression* whereas Ammann, and Van Ness [87] use *orthogonal regression*[11].

Note: The term *orthogonal approximation* must not be confused with approximation using *orthogonal functions*. That is, functions that are orthogonal to one another with regard to an inner product (o. g., Chebyshev polynomials, trigonometric polynomials etc.).

Orthogonal approximation is not only preferred to conventional approximation techniques (considering vertical distances) when perturbed independent variables occur, it is also preferable when *visualization* is the goal of the approximation. Considering only vertical distances gives the optical effect that the steeper parts of the function are more weighted than the flat ones. Such uneven weighting does not appear when using orthogonal approximation.

Example (Orthogonal Distance) The two functions

$$f(x) = x^a \quad \text{and} \quad \overline{f}(x) = x^a + 0.1 \quad (a \geq 1)$$

are separated by a constant *vertical* distance $d_{f,\overline{f}}(x) \equiv 0.1$. The *orthogonal* distance of the two functions f and $\overline{f}$, however, depends greatly on x and b (see Fig. 8.14). This is due to the fact that the steepness

$$f'(x) = \overline{f}'(x) = ax^{a-1}$$

increases with larger values of x and $a > 1$.

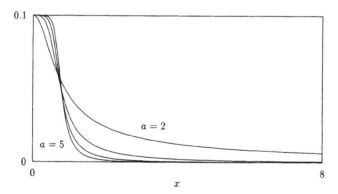

Figure 8.14: Orthogonal distances between x^a and $x^a + 0.1$.

In addition to orthogonal approximation there is another way of avoiding the optical impression of different weighting at points with larger and smaller ascent:

---

[11]The term *regression* was introduced in 1885 by F. Galton for the characterization of heredity properties. Today in mathematical statistics, regression analysis is understood as sample based analysis of stochastic dependencies and their functional description with a given sample. In numerical processing the term regression is sometimes also used for approximation.

parameterizing the data and subsequently fitting a *plane curve* (see Section 8.7.2). The advantage of this method—as compared with orthogonal approximation—is that it requires less effort; its disadvantage is that uniqueness is possibly lost, i.e., the approximating *curve* may not remain a *function*.

8.7 Transformation of the Problem

When dealing with an approximation problem, it is always necessary to decide if a change of variable could improve the quality of approximation if, for instance, unwanted oscillations of an interpolating function can be avoided. Generally speaking the most successful transformations are those that level out regionally differing behavior of the data or the function to be approximated (Hayes [221], Chapter 5). Steeply ascending behavior near the left end point of the approximation interval can, for instance, be made less critical by a logarithmic transformation of the independent variable $\bar{x} = \log(x + c)$ where c is an appropriately chosen constant.

8.7.1 Ordinate Transformation

For some problems, an ordinate transformation $T : \mathbb{R} \to \mathbb{R}$ enables a simpler and sometimes better solution. However, solving the transformed problem

$$D(\Delta_k T(g^*), T(y)) = \min\{D(\Delta_k T(g), T(y)) = \sum_{i=1}^{k} [T(g(x_i; c)) - T(y_i)]^2 : g \in \mathcal{G}_N\}$$

instead of

$$D(\Delta_k g^*, y) = \min\{D(\Delta_k g, y) = \sum_{i=1}^{k} [g(x_i; c) - y_i]^2 : g \in \mathcal{G}_N\}$$

generally yields *another* solution. An appropriate weighting can be chosen so as to force

$$\sum_{i=1}^{k} [g(x_i; c) - y_i]^2 \quad \approx \quad \sum_{i=1}^{k} w_i^2 [T(g(x_i; c)) - T(y_i)]^2 \qquad (8.14)$$

and thus approximate agreement of the respective solutions. Taylor expansion shows that

$$\sum_{i=1}^{k} [T(g(x_i; c)) - T(y_i)]^2 \quad \approx \quad \sum_{i=1}^{k} [T'(y_i)(g(x_i; c) - y_i)]^2$$

and that the relation (8.14) can be obtained by choosing $w_i = 1/T'(y_i)$.

Example (Exponential Data Fitting)   Fitting the function

$$g(x; c_0, c_1, c_2) = \exp(c_0 + c_1 x + c_2 x^2) \qquad (8.15)$$

to a set of data points is a *nonlinear* approximation problem. It is common practice to form the logarithm of both the function (8.15) and the data points $\{y_i\}$ and then to solve a *linear* approximation problem by minimizing

$$\sum_{i=1}^{k} w_i^2 \left[\ln g(x_i; c) - \ln y_i\right]^2 = \sum_{i=1}^{k} w_i^2 \left[c_0 + c_1 x_i + c_2 x_i^2 - \ln y_i\right]^2$$

to obtain values of the parameters c_0, c_1 and c_2. Because $T'(y) = 1/y$, the weights that are appropriate to obtain the desired parameter values are $w_i = y_i$.

8.7.2 Curves

All mathematical models discussed up to this point are functions in the strict mathematical sense, i.e., *single valued* mappings. In order to make use of these functions for the approximation of plane curves, or space curves, their *parametric representation*,

$$\Big(x(s), y(s)\Big), \quad s \in [a, b]$$

or

$$\Big(x(s), y(s), z(s)\Big), \quad s \in [a, b]$$

respectively, is a natural starting point. In the case of plane curves, two approximation problems of the type already discussed, i.e.,

$$x(s), \quad s \in [a, b], \qquad \text{is to be approximated by} \qquad g(s), \quad \text{and}$$
$$y(s), \quad s \in [a, b], \qquad \qquad \text{by} \qquad \qquad h(s)$$

are obtained in this way. Their combined solutions

$$\Big(g(s), h(s)\Big), \quad s \in [a, b] \tag{8.16}$$

represent the desired approximating curve. If the appearance (the *optical smoothness*) of this approximating curve is important, then caution is advised. The smoothness (moderate curvature) of the plane curve or space curve, cannot be deduced from the smoothness of g and h as functions of s.

Example (Piecewise Cubic Functions) The data points

$$(s_i, g(s_i), g'(s_i)), \quad (s_i, h(s_i), h'(s_i)), \quad i = 1, 2, \ldots, k$$

and the particular specification

$$g'(s_i) = h'(s_i) = 0, \quad i = 1, 2, \ldots, k$$

define two piecewise *differentiable* functions $g(s)$ and $h(s)$ (composed of cubic polynomials) (see Section 9.4). Nevertheless, the curve

$$\Big(g(s), h(s)\Big), \quad s \in [a, b],$$

is a plane *polygonal* arc. So the plane sharp-cornered curve is considerably less smooth than any of its constituent parametric functions.

An additional difficulty is encountered when *discrete* approximation problems involving the use of curves, are to be solved: The strictly monotone parameter values

$$s_1 < s_2 < \ldots < s_k$$

connected with the points $P_1, \ldots, P_k \in \mathbb{R}^2$ are generally not part of the given data and have to be determined in a suitable way. Only then can

$$\{(s_1, x_1), (s_2, x_2), \ldots, (s_k, x_k)\} \qquad \text{and} \qquad \{(s_1, y_1), (s_2, y_2), \ldots, (s_k, y_k)\}$$

be interpolated and can the corresponding interpolating functions $g(s)$ and $f(s)$ and, therefore, the curve (8.16) be obtained.

Since the arc length is the natural parameter of a curve, the definition

$$s_1 := 0$$
$$s_i := s_{i-1} + D(P_i, P_{i-1}), \qquad i = 2, 3, \ldots, k \qquad (8.17)$$

leads to a parametric representation, where $s_i - s_{i-1}$ is an approximation of the arc length between P_{i-1} and P_i. The Euclidean distance

$$D_2(P, Q) = \|P - Q\|_2 = \sqrt{(x_P - x_Q)^2 + (y_P - y_Q)^2} \qquad (8.18)$$

is often used as a distance function D in (8.17). An alternative is to use a constant value (that does not correspond to a metric)

$$D(P, Q) = 1,$$

i.e., the *unit parameterization* with $s_i = i - 1$. For closed curves, $P_1 = P_k$ must hold (see Fig. 8.15).

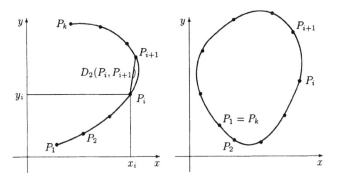

Figure 8.15: Parameterization of the points $P_1, \ldots, P_k \in \mathbb{R}^2$ for an open, and a closed plane curve. $D_2(P_i, P_{i+1})$ is the length of the segment $\overline{P_i P_{i+1}}$ connecting P_i and P_{i+1}.

The advantages and disadvantages of various distance functions in the parameterization (8.17) are discussed in detail in Epstein [181], de Boor [42] and Foley [192]. Often the Euclidean distance (8.18) is considered to be the best choice.

Chapter 9

Interpolation

Die Gleichheit, die der Mensch hier verlangen kann,
ist sicherlich der erträglichste Grad der Ungleichheit.[1]

GEORG CHRISTOPH LICHTENBERG

The k points $x_1, x_2, \ldots, x_k \in \mathbb{R}^n$ or the *interpolation nodes* and the k values $y_1, y_2, \ldots, y_k \in \mathbb{R}$ form the basis of the interpolation process. Interpolation nodes can either be given or chosen arbitrarily. Sometimes complementary information is available, from which additional demands on the interpolation function can be derived, e.g., monotonicity, convexity, or special asymptotic behavior.

The exact correspondence between the discretized model function

$$\Delta_k g := (g(x_1), g(x_2), \ldots, g(x_k))$$

and the data vector $y \in \mathbb{R}^k$ is the approximation criterion in interpolation:

$$\mathrm{dist}(\Delta_k g, y) = \Delta_k g - y = 0.$$

Example (Numerical Methods)   With many numerical methods, *non*-elementary functions (cf. Section 8.1.1) which cannot be characterized by a finite number of parameters are first mapped into finite-dimensional function spaces using interpolation (mostly using polynomial interpolation). The required operator (evaluation, differentiation, integration, etc.) is then applied to this approximation function.

Example (Signal Processing)   In signal processing a discrete signal is often interpolated first, primarily by using trigonometric interpolation functions (although other methods are occasionally used), and then filtered. For instance, high frequencies are damped by *low pass filtering*; omitting higher order terms following a Chebyshev expansion (obtained by interpolation) is a special type of low pass filtering.

Example (Real-Time Processing)   For fast function evaluation, as required, for instance, in real-time applications, the old tabular interpolation methods are sometimes used, because a *quick-and-dirty* function evaluation may be faster and more suitable for some modern computer architectures (e.g., *systolic arrays*, McKeown [288]) than the calling of a function procedure. However, evaluating a function procedure leads to a considerably higher degree of accuracy.

9.1 Interpolation Problems

In general, all *interpolation problems* consist of three subtasks:

1. Choosing a suitable class of functions.

---

[1]The equality which is required in this case is indeed the most tolerable degree of inequality.

2. Determining the parameters of an appropriate member of this function space.

3. Applying the operations required (evaluation, differentiation, integration, etc.) to the derived interpolation function.

The properties (condition number, etc.) and the algorithmic effort must be examined separately for the two computational subtasks of the interpolation (parameter determination and function evaluation).

The computational effort for the parameter determination critically depends on the function space—especially on whether the interpolation problem is linear or nonlinear with respect to the function parameters. The overall computational effort, however, not only depends on the class of the function, but also on the type of problem to be solved. The number of necessary interpolation functions (and possible dependencies among them), for example, and the number of operations which have to be executed (e.g., the number of times the interpolation function must be evaluated) play a role in the total effort.

9.1.1 Choosing a Class of Functions

Depending on the purpose of the interpolation, a suitable class $\mathcal{G}_k$ of model (approximation) functions has to be chosen. In particular, every function g of the function space $\mathcal{G}_k$ must be determined by k parameters:

$$\mathcal{G}_k = \{g(x; c_1, c_2, \ldots, c_k) \mid g : B \subset \mathbb{R}^n \to \mathbb{R}, \ c_1, c_2, \ldots, c_k \in \mathbb{R}\}.$$

The importance of choosing a suitable class of model functions is illustrated in Fig. 9.1. All four functions shown there comply with the interpolation criterion of agreeing with the given points. Nevertheless, these interpolation functions are not equally suitable for all purposes. Only by placing additional demands on certain properties of the functions of $\mathcal{G}_k$ (e.g., continuity, smoothness, monotonicity, convexity etc.) can undesirable properties (e.g., jump discontinuities, poles etc.) be avoided.

A detailed discussion of the elements of modeling necessary for choosing a suitable class of functions can be found in Chapter 8.

9.1.2 Determination of the Parameters of an Interpolation Function

Once a class $\mathcal{G}_k$ of possible approximation functions has been chosen, one particular function $g \in \mathcal{G}_k$ that agrees with the given values $y_1, y_2, \ldots, y_k$ at the k interpolation nodes has to be determined.

According to the principle of interpolation (see Section 8.3.2), the function g, which is characterized by the k parameter values $c_1, c_2, \ldots, c_k$, must exactly match (agree) with the data points $(x_1, y_1), (x_2, y_2), \ldots, (x_k, y_k)$, i.e., g has to satisfy

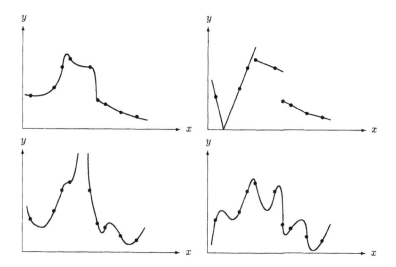

Figure 9.1: Four different functions which all interpolate the same data points (•).

the following system of k equations:

$$
\begin{aligned}
g(x_1; c_1, c_2, \ldots, c_k) &= y_1, \\
g(x_2; c_1, c_2, \ldots, c_k) &= y_2, \\
\vdots \quad\quad &\quad \vdots \\
g(x_k; c_1, c_2, \ldots, c_k) &= y_k.
\end{aligned}
\tag{9.1}
$$

These equations can either be linear or nonlinear depending on whether g is a linear or a nonlinear function of its parameters (see Chapter 8).

It is not necessary to determine the parameters $c_1, c_2, \ldots, c_k$ by solving the system of equations (9.1) if only *values* of the interpolation function g are required. This can be accomplished in a simpler way; for instance, by using the Neville algorithm (cf. Section 9.3.4).

9.1.3 Manipulation of the Interpolation Function

The aim of interpolation is always to obtain analytical data by homogenizing discrete data (cf. Chapter 4 and Chapter 8).

The operations (evaluation, integration, differentiation, root finding, or other manipulations) required for the solution of the original problem are applied to the analytical data obtained by interpolation.

Terminology (Extrapolation) Especially in the *univariate* case, the terms *interpolation* and *extrapolation* are often used to distinguish between the evaluation of g inside or outside the data interval $[x_{\min}, x_{\max}]$ respectively.

9.2 Mathematical Foundations

The general interpolation problem can be stated as: Find a function $g \in \mathcal{G}_k$ for which the k *functionals*[2]

$$\mathrm{F}_i : \mathcal{G}_k \to \mathbb{R}, \qquad i = 1, 2, \ldots, k,$$

take on the preassigned values $u_1, \ldots, u_k$, i.e.,

$$\mathrm{F}_i(g) = u_i, \quad i = 1, 2, \ldots, k.$$

The agreement of the model function $g \in \mathcal{G}_k$ with k given values $y_1, \ldots, y_k$ at k nodes $x_1, \ldots, x_k$ is a special case of the general interpolation problem.

Interpolation is often preferred to best approximation because it is less costly. For the sake of efficiency only linear interpolation problems (which are the most important problems in practical applications) are considered in the following sections: Only *linear* functionals $l_1, \ldots, l_k$ are used, and the elements $g(x; c_1, \ldots, c_k)$ of the function space $\mathcal{G}_k$ are assumed to be *linear* in their parameters $c_1, \ldots, c_k$.

9.2.1 The General Interpolation Problem

All the interpolation problems in this chapter are special cases of the following mathematical problem:

Definition 9.2.1 (General Interpolation Problem) *Within a given class $\mathcal{G}_k$ of functions that are linear in their k parameters, an element g must be determined such that the k linear functionals*

$$l_i : \mathcal{G}_k \to \mathbb{R}, \qquad i = 1, 2, \ldots, k$$

applied to g take on the prescribed values $u_1, \ldots, u_k \in \mathbb{R}$:

$$l_i\, g(x; c_1, \ldots, c_k) = u_i, \quad i = 1, 2, \ldots, k.$$

The linear functionals $l_1, \ldots, l_k$ must be defined on a class of functions that contains both $\mathcal{G}_k$ (the space of all approximating functions with k parameters) and the function space $\mathcal{F}$ of all possible data functions f.

Example (Polynomial Interpolation)   In classical, univariate, polynomial interpolation, $\mathcal{G} = \mathbb{P}$ is the space of *all* polynomials and $\mathcal{G}_{d+1} = \mathbb{P}_d$ is the space of all polynomials with maximum degree d. In this case the functionals l_i are defined by the function evaluations

$$l_i g = g(x_i), \quad i = 0, 1, \ldots, d$$

at the interpolation nodes $x_0, x_1, \ldots, x_d$. A polynomial with degree d that takes on the values

$$u_i = l_i f = f(x_i), \quad i = 0, 1, \ldots, d,$$

at the $d+1$ points $x_0, x_1, \ldots, x_d$ is thus sought. For instance, by interpolating the 9 data points

---

[2]A (linear) *functional* is a (linear) mapping from a normed space into the underlying field. For example, a functional on a space of real functions maps every function to a real number.

x_i	0.083	0.25	0.30	0.38	0.54	0.58	0.67	0.79	0.92
y_i	0.43	0.50	0.79	0.86	0.79	0.36	0.29	0.21	0.14

with a polynomial $P_8 \in \mathbb{P}_8$, the function depicted in Fig. 9.2, is obtained on the interval $[x_1, x_9]$. To concentrate on the behavior of P_8 in the range of the data points, function values above 1 and below 0 have been cut off.

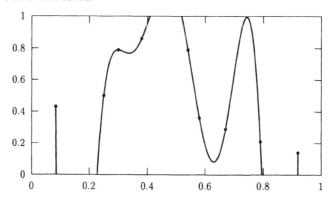

Figure 9.2: Interpolation of 9 points with a polynomial $P_8 \in \mathbb{P}_8$.

Example (Interpolation with Integration and Differentiation Functionals) On the function space $\mathcal{F} = C^1[a, b]$ three linear functionals are defined by

$$l_1 f = \int_a^b f(x)dx, \qquad l_2 f = f'(a), \qquad l_3 f = f'(b).$$

With $a = 0$, $b = \pi/2$, $f(x) = \sin x$ and $\mathcal{G}_3 = \mathbb{P}_2 \subset \mathcal{F}$, the solution of the general interpolation problem is the polynomial

$$g(x) = a_0 + a_1 x + a_2 x^2 = \frac{12 - \pi^2}{6\pi} + x - \frac{1}{\pi}x^2.$$

In $[0, \pi/2]$, this polynomial agrees with the function $\sin x$ with respect to the definite integral on the interpolation interval, and the values of the derivative at the endpoints (see Fig. 9.3).

This example demonstrates that the user may specify data other than function values $u_1 = f(x_1), \ldots, u_k = f(x_k)$ in order to determine an interpolating function.

Software (Interpolation with Non-Polynomial Functions) Interpolation with continued fractions is carried out in the subroutine NAG/e01raf. The output of this program are the coefficients of the continued fraction

$$a_0 + \cfrac{a_1(x - u_1)}{1 + \cfrac{a_2(x - u_2)}{1 + \cfrac{a_3(x - u_3)}{1 + \cdots \cfrac{}{a_d(x - u_d)}}}}.$$

The required number of these coefficients is often smaller than that required for polynomial approximation in order to comply with a given accuracy. The routine NAG/e01rbf can be used to evaluate such continued fractions.

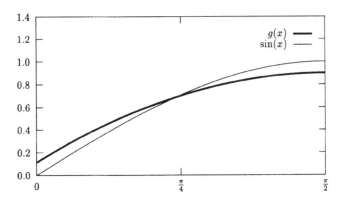

Figure 9.3: Interpolation with respect to general linear functionals.

Many questions arise in connection with the general interpolation problem: What conditions are necessary for the existence of a solution of the problem? If a solution exists, then is it unique or are there other solutions? What can be said about the accuracy $D(g, f)$ if an interpolating function g is used to approximate f, i.e., if $u_i = l_i(f)$ is assumed?

The first two questions can be answered for the general interpolation problem. For the *quantitative* assessment of the approximation accuracy, however, additional premises about $\mathcal{F}$, $\mathcal{G}$, and the functionals l_i have to be made. The approximation accuracy is discussed in Sections 9.3.5 and 9.3.6.

Unique Solvability

The unique solvability of the general interpolation problem is closely related to the linear independence of linear functionals.

Definition 9.2.2 (Linearly Independent Functionals)   *The k linear functionals $l_1, \ldots, l_k : \mathcal{G}_k \to \mathbb{R}$ are linearly independent if and only if*

$$c_1 l_1 + c_2 l_2 + \cdots + c_k l_k = 0 \tag{9.2}$$

implies $c_1 = c_2 = \cdots = c_k = 0$. The zero in (9.2) denotes the zero functional that maps every function $g \in \mathcal{G}_k$ onto 0.

It is the class $\mathcal{G}_k$, in which the interpolation function g has to be found, which is of crucial importance for the unique solvability of the general interpolation problem, and not the class $\mathcal{F}$ which contains the function f to be approximated. That is why $\mathcal{G}_k$ (and not $\mathcal{F}$) was chosen as the domain for the l_i in Definition 9.2.2.

Example (Function Value, Derivative, Integral)   For $\mathcal{G}_3 = \mathbb{P}_2$ the linear functionals

$$l_1 g = g(c), \ c \in [a, b], \qquad l_2 g = g'(c), \qquad l_3 g = \int_a^b g(x) \, dx$$

are linearly independent, but on $\mathbb{P}_1$ these three functionals are linearly *dependent*. This is due to the fact that a finite-dimensional space $\mathcal{G}_k$ and its *dual* space, the space of continuous and linear mappings $\mathcal{G}_k \to \mathbb{R}$, are of the same dimension. Hence, the space of the linear functionals $\mathbb{P}_1 \to \mathbb{R}$ has the dimension 2 and so three functionals *cannot* be linearly independent.

The unique solvability of the general interpolation problem is guaranteed by the following theorem:

Theorem 9.2.1 (Unique Solvability of General Interpolation Problems)
Let $\dim(\mathcal{G}_k) = k$; *then, the general interpolation problem*

$$l_i g = u_i, \quad i = 1, 2, \ldots, k,$$

with the given values $u_1, \ldots, u_k \in \mathbb{R}$ *possesses a unique solution if and only if the linear functionals* $l_1, \ldots, l_k : \mathcal{G}_k \to \mathbb{R}$ *are linearly independent.*

Proof: The desired element g can be represented as a unique linear combination

$$g = c_1 g_1 + c_2 g_2 + \cdots + c_k g_k$$

of basis elements $\{g_1, \ldots, g_k\}$ of the space $\mathcal{G}_k$. Therefore, the general interpolation requirement for g is:

$$l_i(c_1 g_1 + c_2 g_2 + \cdots + c_k g_k) = u_i, \quad i = 1, 2, \ldots, k.$$

Because of the linearity of the functionals l_i, the solution of the general interpolation problem is equivalent to the solution of the system of linear equations

$$\begin{pmatrix} l_1 g_1 & \cdots & l_1 g_k \\ \vdots & & \vdots \\ l_k g_1 & \cdots & l_k g_k \end{pmatrix} \begin{pmatrix} c_1 \\ \vdots \\ c_k \end{pmatrix} = \begin{pmatrix} u_1 \\ \vdots \\ u_k \end{pmatrix}. \tag{9.3}$$

This system of linear equations—and the general interpolation problem as well—possesses a unique solution if and only if $\det(l_i g_j) \neq 0$. This determinant, known as the *generalized Gram determinant*, differs from 0 if and only if the functionals $l_1, l_2, \ldots, l_k$ are linearly independent (Linz [66]). $\qquad \Box$

Example (Polynomial Interpolation) For classical interpolation by univariate polynomials $P_d \in \mathbb{P}_d$,

$$\mathcal{G}_{d+1} = \mathbb{P}_d \quad \text{and} \quad l_i P_d = P_d(x_i), \quad i = 0, 1, \ldots, d,$$

where $x_0, x_1, \ldots, x_d \in \mathbb{R}$ are the interpolation nodes. With the basis $\{1, x, x^2, \ldots, x^d\}$ of the space $\mathbb{P}_d$, the generalized Gram determinant has the following form:

$$\det(l_i x^j) = \begin{vmatrix} 1 & x_0 & x_0^2 & \cdots & x_0^d \\ 1 & x_1 & x_1^2 & \cdots & x_1^d \\ \vdots & & & & \vdots \\ 1 & x_d & x_d^2 & \cdots & x_d^d \end{vmatrix} = \prod_{j=0}^{d-1} \prod_{i=j+1}^{d} (x_i - x_j).$$

This determinant, called *Vandermonde determinant*, differs from zero if and only if all interpolation nodes $\{x_i\}$ are pairwise distinct.

The univariate polynomial interpolation problem thus always possesses a unique solution if the interpolation nodes do *not* coincide.

Example (Function Values and Derivative)   For $\mathcal{G}_3 = \mathbb{P}_2$, the following three functionals define a general interpolation problem:

$$l_1 g = g(-1), \qquad l_2 g = g(1), \qquad l_3 g = g'(0). \tag{9.4}$$

The monomial basis $\{1, x, x^2\}$ of the space $\mathbb{P}_2$ of quadratic polynomials can be used to show that

$$\det(l_i g_j) = \begin{vmatrix} 1 & -1 & 1 \\ 1 & 1 & 1 \\ 0 & 1 & 0 \end{vmatrix} = 0.$$

Hence, the general interpolation problem specified by (9.4) does *not* have a unique solution, though there may exist a whole family of solutions, depending on the l_i.

9.2.2 Interpolation of Function Values

The most important special case of the general interpolation problem is based on linear interpolating functions and the functionals

$$l_i g := g(x_i), \quad i = 1, 2, \dots, k.$$

In this case, the interpolation requirement is the matching of function values:

$$l_j \left(\sum_{i=1}^{k} a_i g_i(x) \right) = \sum_{i=1}^{k} a_i g_i(x_j) = w_j, \quad j = 1, 2, \dots, k.$$

Accordingly, the generalized Gram determinant is $\det(g_i(x_j))$.

Definition 9.2.3 (Unisolvent Functions)   *A set of functions*

$$g_i : B \subset \mathbb{R}^n \to \mathbb{R}, \quad i = 1, 2, \dots, k,$$

is said to be unisolvent on B if

$$\det(g_i(x_j)) \neq 0 \tag{9.5}$$

for arbitrary distinct points $x_1, x_2, \dots, x_k \in B$.

The requirement (9.5) is called the *Haar condition*; a unisolvent set of continuous functions is called a *Chebyshev system*. Though it is not difficult to find Chebyshev systems for univariate interpolation problems $(n = 1)$, there exist *no* Chebyshev systems at all for the interpolation of multivariate functions $(n \geq 2)$ except for the trivial case of $k = 1$.

Example (Unisolvent Functions)   The univariate monomials $\{1, x, x^2, \dots, x^d\}$ and the exponential functions $\{1, e^x, \dots, e^{dx}\}$ are unisolvent on every interval $[a, b] \subset \mathbb{R}$.

The trigonometric functions $\{1, \cos x, \sin x, \dots, \cos dx, \sin dx\}$ are unisolvent on $[-\pi, \pi)$, but *not* on the closed interval $[-\pi, \pi]$.

9.3 Univariate Polynomial Interpolation

Interpolation by univariate polynomials and the matching of function values (used as the approximation criterion) are of major importance in numerical mathematics. This type of interpolation not only serves as a basis for many numerical methods for integration, differentiation, optimization problems, differential equations etc., but also for other (piecewise) univariate and multivariate interpolation methods.

9.3.1 Univariate Polynomials

Functions defined by formulas that involve only a finite number of algebraic operations are called *elementary algebraic functions*. These functions are very well suited as approximation functions (model functions) for use on computer systems. In particular, polynomials of one independent variable—the *univariate polynomials*—are very important.

Definition 9.3.1 (Univariate Polynomial) *A function $P_d : \mathbb{R} \to \mathbb{R}$ with*

$$P_d(x; a_0, \ldots, a_d) := a_0 + a_1 x + a_2 x^2 + \cdots + a_d x^d \tag{9.6}$$

is a (real-valued) polynomial of one independent variable. $a_0, a_1, \ldots, a_d \in \mathbb{R}$ are the coefficients of the polynomial and a_0 is the constant term of the polynomial. If $a_d \neq 0$, then d is called the degree, and a_d is called the leading coefficient, of the polynomial. If $a_d = 0$ then d is called the formal degree of P_d. Polynomials with degrees $d = 1, 2, 3$ are said to be linear, quadratic and cubic respectively.

For instance, the infinite set of monomials

$$B = \{1, x, x^2, x^3, \ldots\}$$

forms a basis of the function space $\mathbb{P}$ of *all* polynomials (of any degree) so $\mathbb{P}$ is a space of infinite dimension. The space $\mathbb{P}_d$ of all polynomials with a maximal degree d has, for instance, the basis

$$B_d = \{1, x, x^2, x^3, \ldots, x^d\} \tag{9.7}$$

and is therefore a $(d+1)$-dimensional space. The monomial basis (9.7) is just one of an infinite number of bases of the vector space $\mathbb{P}_d$. The Lagrange polynomials, Bernstein polynomials, and—most importantly—the Chebyshev polynomials, are other bases of particular importance for numerical and algorithmic applications. They are discussed in the following sections.

One way to represent a polynomial is to *factor it*

$$P_d(x) = a(x - r_1)(x - r_2) \cdots (x - r_d),$$

where $r_1, r_2, \ldots, r_d$ are the zeros of the polynomial P_d. This representation is *not* in terms of any basis for $\mathbb{P}_d$.

9.3.2 Representation Forms of Univariate Polynomials

If two polynomials of degree d interpolate the same $d+1$ data points, then the two polynomials are necessarily identical according to the unique solvability of univariate polynomial interpolation problems. On the other hand, a polynomial may have *different representations* in different bases for $\mathbb{P}_d$. Two important conclusions can be drawn from these facts:

1. The order of the data has no influence on the interpolation polynomial except possibly on its representation.

2. A numerical method, founded on polynomial interpolation, can be analyzed in the most suitable representation and can be implemented in another representation (with respect to another basis).

The Monomial Representation

The mathematical standard form for the representation of a polynomial is (9.6), where a univariate polynomial is expressed as a linear combination of the monomials $1, x, x^2, \ldots, x^d$. This is called the *monomial form* of a polynomial of degree d and is mainly used for theoretical investigations. For numerical calculations, however, other representations are more useful, e.g.,

$$P_d(x; c, b_0, b_1, \ldots, b_d) = b_0 + b_1(x - c) + b_2(x - c)^2 + \cdots + b_d(x - c)^d, \quad (9.8)$$

which is obtained by expanding (9.6) around the point c. The numerical stability of (9.8) may be much better than that of (9.6) if the choice of the point c is appropriate.

Example (Evaluation of a Polynomial)   The parabola $1 + (x - 5555.5)^2$ is represented in monomial form as
$$P_2(x) = 30\,863\,580.25 - 11\,111x + x^2.$$
As a result of cancellation effects, the calculation of the values $P_2(5\,555)$ and $P_2(5\,554.5)$ with $\mathbb{F}(10, 6, -9, 9, true)$ yields the grossly erroneous results 0 and 100. On the other hand, the form (9.8),
$$\overline{P}_2(x) = 1 + (x - 5555.5)^2,$$
can be evaluated without cancellation effects and yields the exact values 1.25 and 2.

It is generally advisable to use the form (9.8) rather than form (9.6); the expansion point c should be chosen close to the center of those points at which the polynomial has to be evaluated.

The Basis of Lagrange Polynomials

The *Lagrange polynomials* $\varphi_{d,i} \in \mathbb{P}_d$ (also called Lagrange basis polynomials or Lagrange elementary polynomials) form the basis

$$B_d^L = \{\varphi_{d,0}, \varphi_{d,1}, \ldots, \varphi_{d,d}\}$$

for the vector space $\mathbb{P}_d$. For a given set $\{x_0, x_1, \ldots, x_d\}$ of $d+1$ distinct points (*interpolation nodes*), they are characterized by the following property:

$$\varphi_{d,i}(x_j) = \begin{cases} 1 & i = j \\ 0 & i \neq j. \end{cases} \tag{9.9}$$

Their existence and uniqueness are a consequence of the results in Section 9.2. They can be represented in the factored form:

$$\varphi_{d,i}(x) = \frac{x - x_0}{x_i - x_0} \cdots \frac{x - x_{i-1}}{x_i - x_{i-1}} \frac{x - x_{i+1}}{x_i - x_{i+1}} \cdots \frac{x - x_d}{x_i - x_d}.$$

Example (Lagrange Polynomials) Fig. 9.4 shows the Lagrange polynomial $\varphi_{5,3}$ for two different selections of interpolation nodes: for equidistant nodes in the interval $[-1, 1]$ and for the zeros of the Chebyshev polynomial T_5 (see Fig. 9.7).

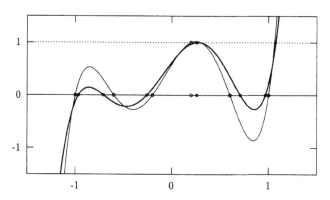

Figure 9.4: Lagrange polynomial $\varphi_{5,3}(x)$ for the Chebyshev nodes (—) and the equidistant nodes (—) on $[-1, 1]$.

Discrete Orthogonality: With respect to the inner product

$$\langle f, g \rangle_L := \sum_{k=0}^{d} f(x_k) g(x_k), \tag{9.10}$$

where $\{x_0, x_1, \ldots, x_d\}$ is a set of interpolation nodes,

$$\langle \varphi_{d,i}, \varphi_{d,j} \rangle_L = \begin{cases} 0 & i \neq j \\ 1 & i = j \end{cases}.$$

Hence, the Lagrange basis B_d^L is an orthonormal system with respect to (9.10).

The Basis of Bernstein Polynomials

The *Bernstein polynomials* $b_{d,i} \in \mathbb{P}_d$ are defined by

$$b_{d,i}(x) := \binom{d}{i} x^i (1 - x)^{d-i}, \qquad i = 0, 1, \ldots, d.$$

They constitute another basis

$$B_d^B = \{b_{d,0}, b_{d,1}, \ldots, b_{d,d}\}$$

for $\mathbb{P}_d$. Accordingly, every polynomial $P_d \in \mathbb{P}_d$ can be represented as a linear combination of the Bernstein basis B_d^B:

$$P_d = \sum_{i=0}^{d} \beta_i b_{d,i}. \qquad (9.11)$$

Usually the coefficients $\beta_0, \beta_1, \ldots, \beta_d$ are not called Bernstein coefficients but *Bézier coefficients*; (9.11) is called a *Bézier representation*.

Zeros: $b_{d,i}$ has exactly two real zeros:

$x = 0$ is a zero of order i,
$x = 1$ is a zero of order $(d - i)$.

Maxima: $b_{d,i}$ has exactly one maximum in $[0,1]$ at the point $x = i/d$.

The Basis of Chebyshev Polynomials

For numerical data processing the *Chebyshev polynomials* $\{T_0, T_1, \ldots, T_d\}$ is most important of all orthogonal polynomial systems that form a basis for $\mathbb{P}_d$:

$$
\begin{aligned}
T_0(x) &:\equiv 1 \\
T_1(x) &:= x \\
T_d(x) &:= 2x T_{d-1}(x) - T_{d-2}(x), \qquad d = 2, 3, \ldots. \qquad (9.12)
\end{aligned}
$$

Using the recursion (9.12), the polynomials $T_2, T_3, \ldots$ can be determined:

$$T_2(x) = 2x^2 - 1, \quad T_3(x) = 4x^3 - 3x, \quad T_4(x) = 8x^4 - 8x^2 + 1, \ \ldots. \qquad (9.13)$$

In (9.13) all Chebyshev polynomials T_d are expressed with respect to the monomial basis $B_d = \{1, x, x^2, \ldots, x^d\}$. Conversely, every monomial x^d can be expressed by the elements of the Chebyshev basis $B_d^T = \{T_0, T_1, \ldots, T_d\}$:

$$1 = T_0, \quad x = T_1, \quad x^2 = (T_0 + T_2)/2, \quad x^3 = (3T_1 + T_3)/4, \ \ldots.$$

Any polynomial $P_d \in \mathbb{P}_d$ can thus be written as a linear combination of Chebyshev polynomials:

$$P_d(x) = \sum_{i=0}^{d} {}' a_i T_i(x) := \frac{a_0}{2} + a_1 T_1(x) + \cdots + a_d T_d(x). \qquad (9.14)$$

Of course, the coefficients in (9.14) do not generally correspond to the ones in the monomial representation (9.6).

Notation (Primed Sum) The convention of halving the first coefficient (denoted by $\sum'$) originates from the representation of an arbitrary function as a series:

$$f(x) = \sum_{i=0}^{\infty} {}' a_i T_i(x) = \frac{a_0}{2} + a_1 T_1(x) + a_2 T_2(x) + \cdots,$$

which leads to the uniform formula (10.14) for *all* coefficients $a_0, a_1, a_2, \ldots$.

In some situations the representation

$$P_d(x) = \sum_{i=0}^{d}{}'' c_i T_i(x) := \frac{c_0}{2} + c_1 T_1(x) + \cdots + c_{d-1} T_{d-1} + \frac{c_d}{2} T_d(x) \qquad (9.15)$$

may also be useful.

Notation (Double-Primed Sum)   If the first and the last terms have been halved, then the double-primed notation $\sum''$ is used (cf. (9.15)):

$$\sum_{i=0}^{d}{}'' \alpha_i := \frac{\alpha_0}{2} + \alpha_1 + \cdots + \alpha_{d-1} + \frac{\alpha_d}{2}.$$

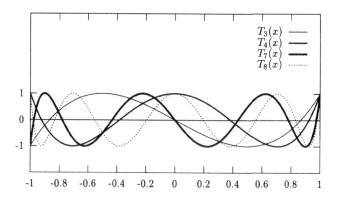

Figure 9.5: Chebyshev polynomials T_3, T_4, T_7, T_8 on $[-1, 1]$.

Zeros: $T_d(x)$ has d zeros in $[-1, 1]$ (see Fig. 9.6 and Fig. 9.7):

$$\xi_j := \xi_j^{(d)} := \cos\left(\frac{2j-1}{2d}\pi\right), \qquad j = 1, 2, \ldots, d. \qquad (9.16)$$

Extrema: In the interval $[-1, 1]$ there are $d + 1$ extrema

$$\eta_k := \eta_k^{(d)} := \cos\left(\frac{k}{d}\pi\right), \qquad k = 0, 1, \ldots, d, \qquad (9.17)$$

where T_d has a local minimum or a local maximum (Rivlin [329]). At these points

$$T_d(\eta_k) = (-1)^k, \qquad k = 0, 1, \ldots, d,$$

i.e., T_d takes the value 1 at every maximum and the value -1 at every minimum. Accordingly, $\|T_d\|_\infty = 1$ on $[-1, 1]$.

The zeros and the extrema of the Chebyshev polynomials are very important interpolation nodes.

Terminology (Chebyshev Abscissas)   When *Chebyshev abscissas* or *Chebyshev nodes* are mentioned in this book, they refer to the *zeros* of T_d. If a strict distinction is desired, it is common to use the terms *Chebyshev zeros* and *Chebyshev extrema* respectively.

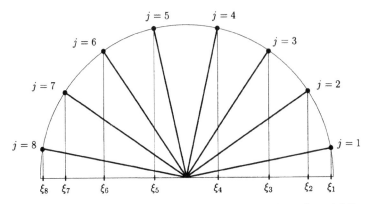

Figure 9.6: Zeros $\xi_1, \xi_2, \ldots, \xi_8$ of the Chebyshev polynomial T_8.

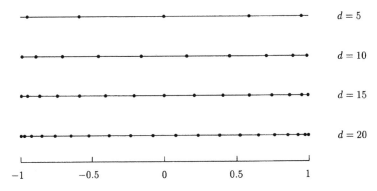

Figure 9.7: Location of the zeros of the Chebyshev polynomials T_5, T_{10}, T_{15}, T_{20}.

Discrete Orthogonality: With respect to the inner product

$$\langle f, g \rangle_T := \sum_{k=0}^{d} f(\xi_k) g(\xi_k), \tag{9.18}$$

where $\{\xi_0, \ldots, \xi_d\}$ is the set of zeros of T_{d+1}, the following property holds (Cheney [37]):

$$\langle T_i, T_j \rangle_T = \begin{cases} 0 & i \neq j \\ (d+1)/2 & i = j = 1, 2, \ldots, d \\ d+1 & i = j = 0. \end{cases} \tag{9.19}$$

With respect to the inner product

$$\langle f, g \rangle_U := \frac{1}{2} f(\eta_0) g(\eta_0) + f(\eta_1) g(\eta_1) + \cdots$$

$$\cdots + f(\eta_{d-1}) g(\eta_{d-1}) + \frac{1}{2} f(\eta_d) g(\eta_d)$$

$$= \sum_{k=0}^{d}{}'' f(\eta_k) g(\eta_k), \tag{9.20}$$

where $\{\eta_0, \ldots, \eta_d\}$ is the set of extrema of T_d, a similar property exists:

$$\langle T_i, T_j \rangle_U = \begin{cases} 0 & i \neq j \\ d/2 & i = j = 1, 2, \ldots, d \\ d & i = j = 0. \end{cases} \tag{9.21}$$

Note (Continuous Orthogonality) The *continuous* orthogonality of Chebyshev polynomials is discussed in Chapter 10 (see Formula (10.11)).

Minimax Property: The Chebyshev polynomials are distinguished by the following property:

Theorem 9.3.1 (Chebyshev) *The polynomial*

$$\frac{1}{2^{d-1}} T_d$$

has the smallest maximum norm, in the interval $[-1, 1]$, *of all polynomials* $P_d \in \mathbb{P}_d$ *with a leading coefficient* $a_d = 1$. *Because* $\|T_d\|_\infty = 1$, *the maximum norm of* P_d *is* $1/2^{d-1}$.

Proof: Rivlin [329].

The minimax property of the Chebyshev polynomials can also be formulated as follows: The Chebyshev polynomial T_d, with $a_d = 2^{d-1}$, has the *largest* leading coefficient, and the monomial x^d, with $a_d = 1$, has the smallest leading coefficient of all polynomials complying with $\|P_d\|_\infty = 1$ on $[-1, 1]$ and with $a_d \neq 0$. The advantage of an expansion using Chebyshev polynomials is therefore the rate of decrease of its coefficients. If terms with negligibly small coefficients are omitted, then expansions with rapidly decreasing coefficients lead to very concise approximations.

A detailed discussion of the properties of Chebyshev polynomials can be found in the books by Rivlin [329], Fox and Parker [194].

9.3.3 The Calculation of Coefficients

In order to approximate a given function $f : \mathbb{R} \to \mathbb{R}$ using interpolation by a polynomial $P_d \in \mathbb{P}_d$, $d + 1$ linear functionals $l_0, l_1, \ldots, l_d$ must be specified and then the polynomial which complies with the interpolation requirement

$$l_i P_d = l_i f, \quad i = 0, 1, \ldots, d$$

must be determined. In principle, the coefficients of P_d could be obtained by solving the system of equations (9.3) using the basic elements $g_i := x^i$ and the values $u_i := l_i f$. In doing so, however, a computational complexity of order $O(d^3)$ has to be accepted. In some very important special cases it is possible to express the polynomial P_d explicitly and to determine its coefficients using algorithms of considerably less complexity.

The Lagrange Basis

Assuming $l_i f = f(x_i)$, with pairwise distinct nodes $x_0, x_1, \ldots, x_d$, the assumption

$$P_d(x) = \sum_{i=0}^{d} f(x_i)\, \varphi_{d,i}(x) \tag{9.22}$$

can be made, where $\varphi_{d,0}, \varphi_{d,1}, \ldots, \varphi_{d,d}$ are the Lagrange polynomials of degree d with respect to the set of nodes $\{x_0, x_1, \ldots, x_d\}$. By definition, the polynomials $\varphi_{d,i}$ have the property (9.9) whence $P_d(x_i) = f(x_i)$, i.e., P_d is the desired interpolation polynomial. The representation (9.22) is the *Lagrange (interpolation) formula*.

The basis functions $\varphi_{d,i}$ depend only on the interpolation nodes $x_0, \ldots, x_d$. The Lagrange representation thus clearly exhibits the linear structure of the interpolation polynomial P_d with respect to the values $f(x_0), \ldots, f(x_d)$. Because of the effort needed to calculate the functions $\varphi_{d,i}$, the Lagrange formula is rarely used for algorithmic applications.

The Chebyshev Basis

A basis for $\mathbb{P}_d$ with special advantages for numerical data processing is the Chebyshev basis $\{T_0, T_1, \ldots, T_d\}$. The coefficients $c_0, c_1, \ldots, c_d$ of the representation (9.15) are determined by the system of linear equations

$$\sum_{i=0}^{d}{}'' c_i T_i(x_j) = f(x_j), \quad j = 0, 1, \ldots, d. \tag{9.23}$$

If it is possible to use the Chebyshev zeros $\{\xi_i\}$ or the Chebyshev extreme points $\{\eta_i\}$ as interpolation nodes, then the solution of the system of equations (9.23) can be obtained using very simple calculations because in such cases the discrete orthogonality of the Chebyshev polynomials can be exploited.

Using Chebyshev extreme points, both sides of the system of linear equations

$$P_d(\eta_i) = \sum_{j=0}^{d}{}'' c_j T_j(\eta_i) = f(\eta_i), \quad i = 0, 1, \ldots, d, \tag{9.24}$$

can be multiplied by the matrix

$$\begin{pmatrix} \frac{1}{2}T_0(\eta_0) & T_0(\eta_1) & \cdots & T_0(\eta_{d-1}) & \frac{1}{2}T_0(\eta_d) \\ \frac{1}{2}T_1(\eta_0) & T_1(\eta_1) & \cdots & T_1(\eta_{d-1}) & \frac{1}{2}T_1(\eta_d) \\ \vdots & \vdots & & \vdots & \vdots \\ \frac{1}{2}T_d(\eta_0) & T_d(\eta_1) & \cdots & T_d(\eta_{d-1}) & \frac{1}{2}T_d(\eta_d) \end{pmatrix}$$

on the left hand side, which yields

$$c_j = \frac{2}{d} \sum_{k=0}^{d}{}'' f(\eta_k) T_j(\eta_k), \quad j = 0, 1, \ldots, d,$$

as a result of the discrete orthogonality (9.21). As the Chebyshev polynomials can also be represented in the form

$$T_j(x) = \cos(j \arccos x),$$

the simple expression

$$c_j = \frac{2}{d} \sum_{k=0}^{d}{}'' f(\eta_k) \cos \frac{jk\pi}{d}, \qquad j = 0, 1, \ldots, d \tag{9.25}$$

can be obtained for the coefficients of (9.24) if the Chebyshev extreme points (9.17) are used as the interpolation nodes.

In exactly the same way, if the zeros (9.16) are used, then the system of linear equations

$$P_d(\xi_i) = \sum_{j=0}^{d}{}' a_j T_j(\xi_i), \qquad i = 0, 1, \ldots, d$$

is obtained for the coefficients $a_0, a_1, \ldots, a_d$ of the representation (9.14). The discrete orthogonality (9.19) of the Chebyshev polynomials implies

$$
\begin{aligned}
a_j &= \frac{2}{d+1} \sum_{k=0}^{d}{}' f(\xi_k) T_j(\xi_k) \\
&= \frac{2}{d+1} \sum_{k=0}^{d}{}' f(\xi_k) \cos\left(j \frac{2k+1}{2d+2}\pi\right), \qquad j = 0, 1, \ldots, d. \tag{9.26}
\end{aligned}
$$

Software (Determination of an Interpolating Polynomial) The two subroutines NAG/e01aef and NAG/e02aff calculate the coefficients of the Chebyshev representation of the interpolation polynomial with respect to given data points. Whereas the subroutine NAG/e02aff is only suitable to interpolation at the Chebyshev extreme points, NAG/e01aef allows for the interpolation of function values—and possibly additional derivative values—at arbitrarily distributed, pairwise distinct abscissas.

The subroutine SLATEC/polint calculates the coefficients—the *divided differences*—of the *Newton representation* of the interpolation polynomial for arbitrarily distributed data points (de Boor [42]).

Example (Complicated Approximation of a Function) The function

$$f(x) = \frac{0.9}{\cosh^2(10(x - 0.2))} + \frac{0.8}{\cosh^4(10^2(x - 0.4))} + \frac{0.9}{\cosh^6(10^3(x - 0.54))} \tag{9.27}$$

can be differentiated an infinite number of times on $[0, 1]$. Hence, the sequence of the interpolation functions at the Chebyshev nodes converges to f as $d \to \infty$ (see Section 9.3.7) whence the existence of a polynomial P_d which differs from f by no more than τ is guaranteed for every tolerance τ. However, for small values τ, polynomials of very high degrees are necessary (see Fig. 9.8, 9.9, and 9.10).

9.3.4 The Evaluation of Polynomials

One of the most important ways to manipulate interpolation polynomials is to evaluate (to determine) the functional value(s) at one or more points.

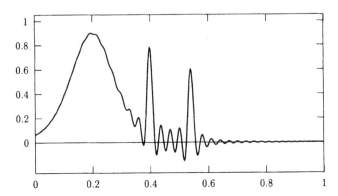

Figure 9.8: Interpolation of the function (9.27) using *one* polynomial derived from the 100 Chebyshev abscissas transformed on $[0, 1]$.

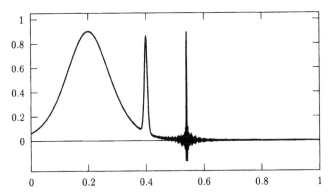

Figure 9.9: Interpolation of the function (9.27) using *one* polynomial derived from the 1000 Chebyshev abscissas transformed on $[0, 1]$.

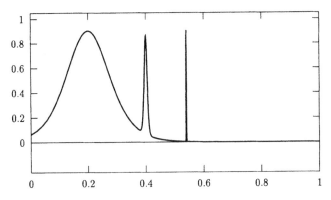

Figure 9.10: Interpolation of the function (9.27) using *one* polynomial derived from the 10 000 Chebyshev abscissas transformed on $[0, 1]$.

The Horner Method

For a polynomial in the monomial form

$$P_d(x) = c_0 + c_1 x + \cdots + c_d x^d \qquad (9.28)$$

there is a well-known recursion—the *Horner method*— used to calculate the value of P_d at the point x:

$$
\begin{aligned}
&p := c_d \\
&\textbf{do}\ \ i = 1, 2, \ldots, d \\
&\qquad p := xp + c_{d-i} \qquad\qquad\qquad\qquad (9.29)\\
&\textbf{end do}
\end{aligned}
$$

The Horner method requires d addition and d multiplication operations for the calculation of the value $P_d(x) := p$. The computational effort of this method is *optimal* for evaluating a polynomial with given coefficients with respect to the monomial basis $B_d = \{1, x, \ldots, x^d\}$. There is no other method which can accomplish this with fewer operations.

The assertion that the Horner method is optimal was originally assumed in a paper by Ostrowski [306] in 1954. The *algebraic theory of complexity* is often said to have first been developed in this paper. The proof of this assumption was provided by Pan [308] in 1966.

The Clenshaw Algorithm

If a polynomial in the Chebyshev form $P_d = a_0/2 + a_1 T_1 + \cdots + a_d T_d$ has to be evaluated at the point x, then it can be transformed into the form (9.28) and evaluated using the Horner method (9.29).

The transformation of the coefficients required to do this involves wasteful effort and actually worsens the accuracy of the result. It is more efficient to use the *Clenshaw algorithm*, which returns the desired value $P_d(x) := b_0$ for the coefficients $a_0, a_1, \ldots, a_d$ of the Chebyshev expansion:

$$
\begin{aligned}
&b_d := a_d \\
&b_{d-1} := 2xb_d + a_{d-1} \\
&\textbf{do}\ \ i = 2, 3, \ldots, d-1 \\
&\qquad b_{d-i} := 2xb_{d-i+1} - b_{d-i+2} + a_{d-i} \\
&\textbf{end do} \\
&b_0 := xb_1 - b_2 + a_0
\end{aligned}
$$

An exact derivation of the Clenshaw algorithm based on the recursion (9.12) and a proof of its numerical stability can be found in Fox and Parker [194].

Software (Evaluation of a Polynomial Represented in Chebyshev Form) A polynomial in Chebyshev representation can be evaluated at an arbitrary point using the subroutine NAG/e02aef or NAG/e02akf. NAG/e02akf has more flexibility than NAG/e02aef, but has a more complex parameter list.

Algorithms for Evaluating Derivatives and Integrals

There are also algorithms used to evaluate the derivatives or the integrals of polynomials,

$$P_d'(x), \quad P_d''(x), \quad \ldots, \int_I P_d(t)\, dt.$$

They are based on Chebyshev representations whose coefficients are adapted to calculate derivatives and integrals.

Software (Differentiation and Integration of Polynomials Represented in Chebyshev Form)   The subroutine NAG/e02ahf calculates the Chebyshev representation of the derivative of a polynomial which is given in Chebyshev form. The coefficients of the Chebyshev representation of the indefinite integral can be determined using NAG/e02ajf.

The Algorithm of de Casteljau

Based on the Bézier representation (9.11) with coefficients $\beta_0, \beta_1, \ldots, \beta_d$, a polynomial can be evaluated using the recursion

> **do** $i = 0, 1, \ldots, d$
> $b_i := \beta_i$
> **end do**
> **do** $k = 1, 2, \ldots, d-k$
> $b_i := (1 - x)b_i + x b_{i+1}$
> **end do**

which yields the result $P_d(x) := b_0$.

Evaluating Polynomials Without Calculating Coefficients

If only one value (or a small number of values) is required, it is not efficient to calculate the coefficients of either the Bézier or the Chebyshev representation of the interpolation polynomial and then to evaluate it at the desired point.

There are special algorithms for this problem which require less computational effort. These methods employ the principle of *successive linear interpolation*, which is based on the following theorem:

Theorem 9.3.2 (Aitken's Lemma)   *Let the values $f_0, f_1, \ldots, f_d$ be given for the set of nodes $\{x_0, x_1, \ldots, x_d\}$. If*

1. *the interpolation polynomial P_{d-1}^1 for the nodes $\{x_0, \ldots, x_{i-1}, x_{i+1}, \ldots, x_d\}$, and*

2. *the interpolation polynomial P_{d-1}^2 for the nodes $\{x_0, \ldots, x_{j-1}, x_{j+1}, \ldots, x_d\}$,*

are given for $i \neq j$, then the interpolation polynomial P_d for the overall set of nodes is given by the following linear combination:

$$P_d(x) := \frac{x_j - x}{x_i - x_j}\, P_{d-1}^1(x) - \frac{x_i - x}{x_i - x_j}\, P_{d-1}^2(x).$$

Proof: Henrici [56].

Based on Aitken's lemma, many iterative interpolation methods (which differ mainly in the order in which the pairs (x_i, f_i) are used) can be constructed. The most widely used methods are the Neville and the Aitken methods. In this section the iterated interpolation developed by E. H. Neville—the *Neville method*—is described. This method gradually applies the Aitken lemma to the polynomials P_{k-1}^{j-1} (with the nodes $\{x_{j-k}, x_{j-k+1}, \ldots, x_{j-1}\}$) and P_{k-1}^j (with the nodes $\{x_{j-k+1}, x_{j-k+2}, \ldots, x_j\}$); the resulting polynomial P_k^j then interpolates the original function at the nodes $\{x_{j-k}, x_{j-k+1}, \ldots, x_j\}$:

> do $j = 0, 1, \ldots, d$
> $\quad P_0^j := f_j$
> end do

> do $k = 1, 2, \ldots, d$
> $\quad$ do $j = k, k+1, \ldots, d$
> $$P_k^j(x) := \frac{x_j - x}{x_j - x_{j-k}} P_{k-1}^{j-1}(x) - \frac{x_{j-k} - x}{x_j - x_{j-k}} P_{k-1}^j(x) \qquad (9.30)$$
> $\quad$ end do
> end do

To use this recursion for the computation of the value of P_d at a given point ξ, all that is required is that the value ξ be substituted for x in the Neville scheme (9.30); in this case $P_d^d(\xi)$ is the desired result.

Software (Direct Interpolation of Single Function Values) If it is not necessary to obtain an explicit representation of the interpolating polynomial for a given set of data points, and only a single value of this function is required, then this value can be computed using the subroutine NAG/e01aaf which is based on the Aitken method using successive linear interpolation. In doing so, the data points can be distributed arbitrarily. In addition to the final result $P_d^d(x)$, all the results $P_k^j(x)$ of the partial interpolations are provided also. These values can be used to obtain an estimate of the absolute interpolation error $P_d^d(x) - f(x)$.

If equidistant data points are given, then the subroutine NAG/e01abf can be used. This subroutine calculates a single value of the interpolation polynomial using the *Everett formula* (Hildebrand [58]).

9.3.5 Error and Convergence Behavior

Functions which comply with a given error limit are determined as the solutions of numerical approximation problems. In the context of polynomial interpolation, the focus of this section, the following questions arise:

1. What can be said about the behavior of the error $f(x) - P_d(x)$ of an interpolation polynomial on the approximation interval $[a, b]$?

   (If the error function complies with the required tolerance, i. e., if $\|P_d - f\| < \tau$, then P_d is a solution of the approximation problem. Otherwise further questions arise.)

2. Is it at all possible to achieve the required accuracy by increasing the number of parameters of the approximation function? In other words:

   (a) Is there a sequence of interpolation polynomials $\{P_d\}$ which converges in norm to the function f as $d \to \infty$, on a fixed approximation interval, and which satisfies the definition of the approximation problem?

   (b) Is there a sequence of piecewise polynomials which converges to f?

   It is generally the case that both (a) and (b) can be answered in the affirmative. As far as efficiency is concerned, the following question arises according to the *principle of minimality* (see Chapter 1):

3. How can the interpolation nodes (or the subintervals in the case of piecewise interpolation) be chosen so that the least computational effort or the simplest approximation function, i.e., the function with the smallest number of parameters, is obtained?

9.3.6 Algorithm Error in Polynomial Interpolation

In this section the error of an interpolation polynomial P_d defined by

$$P_d(x_i) = f_i := f(x_i), \quad i = 0, 1, \dots, d, \tag{9.31}$$

is examined. According to this interpolation requirement, the error function

$$e_d(x) := P_d(x) - f(x)$$

has zeros at the interpolation nodes $x_0, x_1, \dots, x_d$. The value of the interpolation error e_d can be arbitrarily large between these zeros. Statements about the size of the error can be made only if additional information about f is available.

Theorem 9.3.3 (Representation of the Error)   *If a function $f \in C^{d+1}[a, b]$ is interpolated by the polynomial P_d at the pairwise distinct nodes*

$$a \le x_0, x_1, \dots, x_d \le b,$$

then, for each $x \in [a, b]$ in the smallest interval containing all the x_i and x, there exists ξ such that the error of the interpolation polynomial at the point x can be expressed as

$$e_d(x) := P_d(x) - f(x) = \frac{f^{(d+1)}(\xi)}{(d+1)!}(x - x_0)(x - x_1) \cdots (x - x_d). \tag{9.32}$$

Proof: The property that the error function vanishes at the interpolation nodes $x_0, \dots, x_d$ can be expressed as

$$e_d(x) = (x - x_0)(x - x_1) \cdots (x - x_d)\, q(x) = \omega_{d+1}(x)\, q(x) \tag{9.33}$$

with a suitable function q and with the *node polynomial*

$$\omega_{d+1}(x) := \prod_{j=0}^{d} (x - x_j).$$

Using $\omega_{d+1} \in \mathbb{P}_{d+1}$, the following auxiliary function can be defined for one specific point $\overline{x} \in [a, b]$ where $\overline{x} \notin \{x_0, x_1, \ldots, x_d\}$:

$$s(x) := P_d(x) - f(x) - \omega_{d+1}(x) \, q(\overline{x}).$$

Because

$$s(\overline{x}) = 0 \qquad \text{and} \qquad s(x_i) = 0, \quad i = 0, 1, \ldots, d,$$

this function has $d + 2$ zeros. The function s thus has $d + 2$ different zeros in the smallest interval I which contains all points $\overline{x}, x_0, x_1, \ldots, x_d$. Since sufficient differentiability is assumed, Rolle's theorem implies that s' has at least $d+1$ zeros in I. Further differentiation and further application of Rolle's theorem proves that s'' has at least d zeros in I. Accordingly, $s^{(d+1)}$ has at least one zero ξ in I.

Because $q(\overline{x})$ is a constant and

$$P_d^{(d+1)}(x) \equiv 0,$$

the equation

$$s^{(d+1)}(x) = f^{(d+1)}(x) - (d+1)! \, q(\overline{x})$$

is obtained. This equation can be solved for $q(\overline{x})$ at the zero $x = \xi$ of $s^{(d+1)}$, which yields

$$q(\overline{x}) = \frac{f^{(d+1)}(\xi)}{(d+1)!}.$$

The following representation of the error is obtained using (9.33):

$$e_d(x) = \frac{f^{(d+1)}(\xi)}{(d+1)!} \omega_{d+1}(x).$$

$\square$

Upper Bounds for the Interpolation Error

An upper bound on the absolute value of the $(d+1)$st derivative of f,

$$\max_{x \in I} |f^{(d+1)}(x)| = \|f^{(d+1)}\|_\infty \leq M_{d+1},$$

leads to an estimate of the algorithm error of the interpolation:

$$|e_d(x)| \leq \frac{M_{d+1}}{(d+1)!} |\omega_{d+1}(x)| \qquad \text{for all} \quad x \in I, \tag{9.34}$$

and for every L_p-norm with $1 \leq p \leq \infty$ the estimate[3]

$$\|P_d - f\| \leq \frac{M_{d+1}}{(d+1)!} \|\omega_{d+1}\| \tag{9.35}$$

---

[3] The estimates (9.34) and (9.35) of the interpolation error were obtained under the assumption that the function f can be differentiated $d + 1$ times. If this is not the case, a larger approximation error must be expected.

is obtained. As the formulas (9.32) and (9.35) show, the size of the interpolation error depends on the properties of the interpolated function f as well as on the distribution of the interpolation nodes $x_0, \ldots, x_d$, characterized by $\omega_{d+1}(x)$ or $\|\omega_{d+1}\|$. The most favorable distribution of the nodes with respect to the error estimate (9.35) is obtained if $\|\omega_{d+1}\|$ is as small as it can be. As a result of Theorem 9.3.1, the Chebyshev zeros are the optimal interpolation nodes for the L_∞-norm.

Example (Interpolation of the Sine Function) Let the values

$$f(x_i) = \sin(x_i), \qquad x_i = 0,\, 0.1,\, 0.2,\, 0.3,\, 0.4$$

be used to determine an interpolating polynomial $P_4 \in \mathbb{P}_4$ and let $\sin(0.14)$ be approximated by $P_4(0.14)$. With

$$M_5 = \max_{x \in [0.0.4]} |\sin^{(5)}(x)| = \max_I |\cos(x)| = 1$$

an upper bound for the error is obtained:

$$|e_4(0.14)| \leq \frac{1}{5!}|0.14|\,|0.14 - 0.1|\,|0.14 - 0.2|\,|0.14 - 0.3|\,|0.14 - 0.4| \approx 1.17 \cdot 10^{-7}.$$

Example (Linear Interpolation) The interpolation polynomial $P_1 \in \mathbb{P}_1$ which runs through the points $(x_0, f(x_0))$ and $(x_1, f(x_1))$ is given by

$$P_1(x) = f(x_0) + (x - x_0)\frac{f(x_1) - f(x_0)}{x_1 - x_0},$$

and the error is

$$e_1(x) = P_1(x) - f(x) = \frac{f''(\xi)}{2}(x - x_0)(x - x_1).$$

For $x \in [x_0, x_1]$ and

$$M_2 = \max\{|f''(x)| : x \in [x_0, x_1]\}$$

the estimate

$$|e_1(x)| \leq M_2\frac{|x - x_0||x - x_1|}{2} \leq M_2\frac{(x_1 - x_0)^2}{8}$$

is valid. If five equidistant values of $f(x) = \sin(x)$ in $[0, 0.4]$ are given as in the previous example, then the approximation error of the piecewise linear interpolation function which is determined by these five points is no larger than

$$M_2\frac{(x_{i+1} - x_i)^2}{8} = 0.4 \cdot \frac{0.1^2}{8} = 5 \cdot 10^{-4}.$$

At least 500 equidistant values are required in order to obtain the error bound $1.25 \cdot 10^{-7}$. In the above example, on the other hand, the polynomial P_4 was sufficient to obtain an error bound of the same order of magnitude.

Formula (9.34) is suitable for practical use in error estimates only if f is continuously differentiable a sufficient number of times and if the upper bounds of the derivatives of the function f are *explicitly* known on the approximation interval I. However, formulas of the type (9.34) are also important in situations where *no* concrete error bounds can be obtained because they allow for a qualitative description of the behavior of the error. If a function $f \in C^2[a, b]$ is interpolated, for instance, by a polygonal arc (i. e., by a piecewise linear interpolation function) on an equidistant node set (where the distance h between the interpolation nodes is $h = (b - a)/d$), then

$$|e_1(x)| \leq \frac{M_2}{8}h^2$$

implies that the error—depending on the step h—is $O(h^2)$.

9.3.7 The Convergence of Interpolation Polynomials

The question arises as to whether a global increase in the number of parameters of the approximation function ($P_d \in \mathbb{P}_d$ and $d \to \infty$) always leads to the desired results if f is assumed to be continuous on $[a, b]$. At first glance, the following theorem of K. Weierstrass (1885) seems to answer this question in the affirmative:

Theorem 9.3.4 (Weierstrass)   *For any function* $f \in C[a, b]$ *and any* $\varepsilon > 0$ *there exists a polynomial* $P \in \mathbb{P}$ *with*

$$|P(x) - f(x)| \leq \varepsilon \qquad for \ all \quad x \in [a, b].$$

Weierstrass himself only formulated non-constructive proofs of his theorem. The first constructive proof was provided by S. N. Bernstein (1912). He proved that the sequence of polynomials

$$P_d(x; f) := \sum_{i=0}^{d} f(i/d) b_{d,i} = \sum_{i=0}^{d} f(i/d) \binom{d}{i} x^i (1-x)^{d-i}, \quad d = 1, 2, 3, \dots, \quad (9.36)$$

constructed using the Bernstein polynomials $\{b_{d,i}\}$ converges uniformly to f on $[0, 1]$. The full exposition of this proof can be found in Davis [40].

Because of their extremely slow convergence, the polynomials defined by (9.36) are *not* suitable for the practical solution of approximation problems (see Fig. 9.11). However, the polynomials $P_d(x; f)$ reproduce the shape of f (number and position of the extreme points, inflection points etc.) remarkably well. For this reason they are used in graphical data processing.

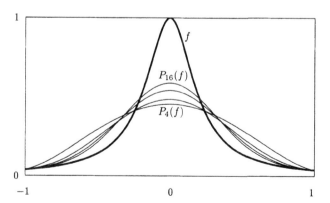

Figure 9.11: The approximation of the Runge function (—) $f(x) = (1 + 25x^2)^{-1}$ using the polynomials $P_d(x; f)$, $d = 4, 8, 12, 16$. Note the extremely slow rate of convergence.

Even the constructive proof of the Weierstrass theorem does *not* lead to an efficient method for obtaining an approximating polynomial for given f and ε.

Node Matrices

In best approximation (see Chapter 10), the algorithmic effort for determining an approximating function critically depends on the function that has to be approximated. In contrast, it is possible to simplify the algorithm using the principle of interpolation if the sequence of the interpolation nodes $\{x_{d,0}, x_{d,1}, \ldots, x_{d,d}\}$, $d = 0, 1, 2, \ldots$, is fixed in advance.

Since the resulting algorithms are non-adaptive, special computer architectures—above all parallel computers—can be exploited more efficiently by reusing already available evaluations of the function $f(x_{d,i})$ in the subsequent steps $d+1, d+2, \ldots$.

Definition 9.3.2 (Node Matrix)   *A scheme of interpolation nodes from an infinite sequence $\{P_0, P_1, P_2, \ldots\}$ of interpolation polynomials which is fixed a priori (independently of the functions that are to be approximated),*

$$
K = \begin{pmatrix}
x_{0,0} & & & \\
x_{1,0} & x_{1,1} & & \\
x_{2,0} & x_{2,1} & x_{2,2} & \\
\vdots & \vdots & \vdots & \ddots
\end{pmatrix},
$$

is called a node matrix.

The choice of a universal node matrix independent of the concrete approximation problem has great influence on the properties of a (non-adaptive) interpolation or approximation algorithm. For instance, the more extensive the set of functions for which the interpolation polynomials $P_d \to f$ converge as $d \to \infty$, the greater the applicability of the corresponding interpolation algorithm. For every desired accuracy of the approximation $\tau > 0$, this convergence guarantees that the interpolation algorithm based on a successive calculation of the polynomials with the nodes $\{x_{d,0}, x_{d,1}, \ldots, x_{d,d}\}$, $d = 0, 1, 2, \ldots$, can be successfully terminated after a finite number of steps. (At this stage, the computational effort and the influence of rounding errors are not taken into account.)

In view of the Weierstrass theorem, it is tempting to assume that the class of functions for which the convergence $P_d \to f$ can be guaranteed contains all continuous functions $f \in C[a, b]$. However, the following theorem shows that this is not true:

Theorem 9.3.5 (Faber)   *There is no universal node matrix for which the corresponding interpolation polynomials converge to f for every function $f \in C[a, b]$.*

Proof: De Vore, Lorentz [157].

Faber's theorem does not contradict the Weierstrass theorem because for *each individual* function $f \in C[a, b]$, it is possible to find a node matrix K which satisfies $P_j(f) \to f$. To that end, only the zeros in $[a, b]$ of the error function of the polynomials P_0^*, P_1^*, $P_2^*, \ldots$ (the best approximations with respect to the supremum norm) have to be combined in K.

Lebesgue Functions and Constants

If the class of functions $\mathcal{F}$ defined on $[a, b]$ is smaller than $C[a, b]$, then it is possible to find universal node matrices for which the convergence $P_d(f) \to f$ is ensured for *every* function $f \in \mathcal{F}$. In order to analyze this convergence, it is helpful to introduce the Lebesgue functions and constants.

Definition 9.3.3 (Lebesgue Function) *The sum of the absolute values of the Lagrange polynomials $\varphi_{d,i}$ with respect to the nodes $x_{d,0}, x_{d,1}, \ldots, x_{d,d}$ of the node matrix K*

$$\lambda_d(x; K) := \sum_{i=0}^{d} |\varphi_{d,i}(x)|$$

is called the Lebesgue function of K of order $d+1$ (see Fig. 9.12 and 9.14)[4].

Definition 9.3.4 (Lebesgue Constant) *The quantity*

$$\Lambda_d(K) := \max\{\lambda_d(x; K) : x \in [a, b]\}$$

is called the Lebesgue constant of K of order $d + 1$ with respect to $[a, b]$.

With the Lebesgue constant, the relationship between the error of the interpolation polynomials $\{P_d(f)\}$ with respect to the node matrix K and the optimal (i.e., the smallest possible) approximation error over *all* polynomials with degree d can be described.

The quality of a sequence of interpolating polynomials $\{P_d\}$ (which is defined by the node matrix K) with respect to the approximation of a function f can be characterized by the maximum of the interpolation error

$$E_d := E_d(f) := \|e_d\|_\infty = \max_{x \in [a,b]} |P_d(x) - f(x)|, \qquad d = 0, 1, 2, \ldots .$$

The quantities E_d become more favorable the nearer they are to

$$E_d^* := E_d^*(f) := \min\{\|P_d - f\|_\infty : P_d \in \mathbb{P}_d\} = \|P_d^* - f\|_\infty,$$

i.e., the smallest possible errors obtained using the best approximating polynomials P_d^* of degree d.

Theorem 9.3.6 *For a function $f \in C[a, b]$ and a sequence of interpolation polynomials $\{P_d(f)\}$ with respect to the node matrix K, the following inequality holds:*

$$E_d(f) \le E_d^*(f)\,(1 + \Lambda_d(K)), \qquad d = 0, 1, 2, \ldots . \tag{9.37}$$

Proof: Rivlin [329].

The Lebesgue constants Λ_d (which depend on the universal node matrix K, but not on f) are therefore a measure as to how far the interpolation error E_d can

---

[4]In Fig. 9.13 and Fig. 9.15 the Lebesgue functions of the cubic spline functions (see Section 9.7.6) are compared.

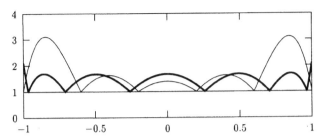

Figure 9.12: The Lebesgue function $\lambda_5(x)$ of **polynomial interpolation** at the six Chebyshev nodes (—) and the six equidistant nodes (—) on $[-1, 1]$.

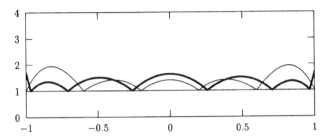

Figure 9.13: The Lebesgue function $\lambda_5(x)$ of cubic **spline interpolation** at the six Chebyshev nodes (—) and the six equidistant nodes (—) on $[-1, 1]$.

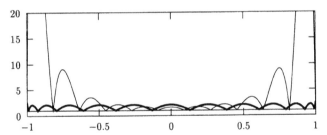

Figure 9.14: The Lebesgue function $\lambda_{11}(x)$ of **polynomial interpolation** at the 12 Chebyshev nodes (—) and the 12 equidistant nodes (—) on $[-1, 1]$.

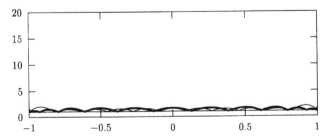

Figure 9.15: The Lebesgue function $\lambda_{11}(x)$ of cubic **spline interpolation** at the 12 Chebyshev nodes (—) and the 12 equidistant nodes (—) on $[-1, 1]$.

exceed the smallest possible error E_d^* in the worst case. The extreme interpolation error approaches the smallest possible approximation error as the Lebesgue constants of a node matrix decrease.

According to Faber's theorem, the sequence $\{\Lambda_d\}$ (which is independent of f) cannot be bounded i.e., $\Lambda_d \to \infty$. Erdös [182] proved the existence of a constant $c > 0$ for each node matrix K for which the following inequality holds:

$$\Lambda_d(K) > \frac{2}{\pi} \ln d - c, \qquad d = 1, 2, \ldots .$$

The node matrix K is crucial to non-adaptive approximation algorithms which are based on interpolation. Due to its influence on the rate of convergence it can seriously affect the efficiency of the corresponding algorithm. The algorithm based on K becomes increasingly effective as $\Lambda_d(K)$ approaches the lower bound. If $\Lambda_d(K)$ increases much more rapidly than the lower bound, then the class of functions for which the convergence $P_d \to f$ can be guaranteed may be much smaller.

The existence of the ideal node matrix, which yields the smallest possible Lebesgue constant Λ_d^* for all polynomial degrees, has been proved non-constructively by de Boor and Pinkus [152]; however, the matrix is not (yet) known in practice. The zeros (9.16) of the Chebyshev polynomials are a convenient (though still not optimal) sequence of interpolation nodes.

The node matrix $K_T = (\xi_{d,j})$ is based on the interpolation interval $[-1, 1]$. The general interval $[a, b]$ can be transformed into $[-1, 1]$ using a regular affine transformation of the independent variable:

$$\overline{x} := \frac{2x - b - a}{b - a}.$$

The following theorem shows that K_T is almost the optimal choice of universal interpolation nodes:

Theorem 9.3.7 *The Lebesgue constants of the node matrix K_T satisfy the inequality*

$$\frac{2}{\pi} \ln d + 0.9625 < \Lambda_d(K_T) \leq \frac{2}{\pi} \ln d + 1, \qquad d = 1, 2, \ldots . \tag{9.38}$$

Proof: Rivlin [329].

Even for the high polynomial degree $d = 1000$, the approximation error of an interpolation based on the Chebyshev nodes is of the same order of magnitude as for the optimal choice:

$$E_{1000}(f) \leq 6.4 \, E_{1000}^*(f),$$

i.e., even in the worst case, the approximation error is only 6.4 times as large as the error of the optimal choice of interpolation nodes.

Convergence with Respect to the Chebyshev Nodes

According to the Weierstrass theorem, $E_d^*(f) \to 0$ as $d \to \infty$ for any function in $C[a, b]$. As Erdös and Vértesi [183] have proved, for *every* node matrix (and in particular for the Chebyshev nodes), there exists a function $f \in C[a, b]$ for which the sequence of the interpolation polynomials of f diverges on the interval $[a, b]$ (except for, at most, a finite number of points). If, for example, the class of functions is restricted to *Lipschitz continuous* functions on $[-1, 1]$, i.e., functions which satisfy the inequality

$$|f(u) - f(v)| \le L|u - v|$$

for all $u, v \in [-1, 1]$ and a constant L—a *Lipschitz constant*—then the sequence of the interpolation polynomials converges in every case, which follows from the following theorem:

Theorem 9.3.8 (Jackson)   *For all functions which are Lipschitz continuous on $[-1, 1]$ with a Lipschitz constant L, the approximation error is bounded:*

$$E_d^*(f) \le \frac{L}{2d + 2}. \tag{9.39}$$

Proof: Cheney [37].

Using a combination of (9.37), (9.38), and (9.39), an upper bound (which converges to 0 as $d \to \infty$) for the approximation error of the interpolation polynomials $\{P_d\}$, for the node matrix K_T, can be obtained:

$$E_d(f) \le E_d^*(f)\left(1 + \Lambda_d\right) \le \frac{L}{2d + 2}\left(\frac{2}{\pi} \ln d + 2\right).$$

It follows that, for every Lipschitz continuous function f and any given tolerance τ, there exists a polynomial of degree d such that the polynomial $P_d(f)$, which interpolates the function with respect to the Chebyshev zeros, deviates from f by no more than τ. Hence, the set of functions for which an interpolation algorithm based on K_T converges (unless problems arise due to the computational effort and rounding errors) contains all Lipschitz continuous functions, but not all continuous functions.

Example (Piecewise Linear Function)   If the Lipschitz continuous (but not differentiable) function

$$f(x) := \begin{cases} 0 & \text{as} & x \in (-\infty, 0.4] \\ 5(x - 0.4) & \text{for} & x \in (0.4, 0.6] \\ 1 & \text{for} & x \in (0.6, \infty) \end{cases} \tag{9.40}$$

is interpolated using polynomials P_d with respect to the Chebyshev abscissas transformed on $[0, 1]$, then

$$P_d(x) \to f(x) \quad \text{as} \quad d \to \infty \quad \text{for all} \quad x \in [0, 1].$$

The interpolation function for 40 nodes is depicted in Figure 9.16.

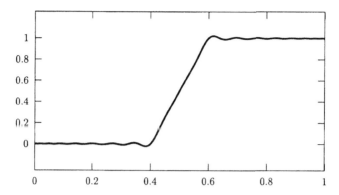

Figure 9.16: Function (9.40) interpolated by a polynomial with respect to 40 Chebyshev abscissas.

In practice, many important functions are continuously differentiable and, moreover, their derivatives of higher orders often also exist. If the Chebyshev nodes K_T are used, the convergence $E_d(f) \to 0$ is always guaranteed for these functions.

The node matrix $K_U = (\eta_{d,j})$ of the Chebyshev *extrema* (9.17) has similarly favorable properties as the node matrix of the Chebyshev zeros.

If discontinuous functions with jump discontinuities are interpolated by polynomials with respect to the Chebyshev nodes K_T, then overshooting occurs—the maximal and minimal amplitudes of which are almost independent of the polynomial degree d. This peculiarity also occurs with partial sums from the Fourier series in the neighborhood of jump discontinuities and is called the *Gibbs' phenomenon*.

Example (Function with a Jump Discontinuity) If the discontinuous function

$$f(x) = \left\{ \begin{array}{ll} 0 & \text{for } x \in (-\infty, 0.6] \\ 1 & \text{for } x \in (0.6, \infty) \end{array} \right\} \tag{9.41}$$

is interpolated by polynomials P_d with respect to the Chebyshev abscissas transformed on $[0,1]$, then $P_d \to f$ as $d \to \infty$ except at the discontinuity point $x = 0.6$ (see Fig. 9.17, 9.18 and 9.19), where the Gibbs' phenomenon can be clearly seen.

Convergence with Respect to Equidistant Nodes

In order to simplify algorithms, the *equidistant* distribution of interpolation nodes would seem to be much more attractive than the Chebyshev nodes:

$$x_{d,j} = \frac{2j}{d} - 1, \qquad j = 0, 1, \dots, d, \qquad d = 1, 2, 3, \dots.$$

In order to make a comparison with other node distributions easier, the node matrix $K_e = (x_{d,j})$ of equidistant nodes also refers to $[0, 1]$.

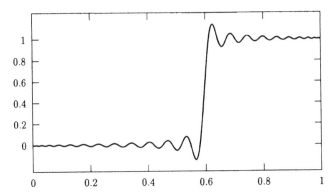

Figure 9.17: Polynomial interpolation of the function (9.41) at the 50 Chebyshev abscissas transformed to $[0, 1]$.

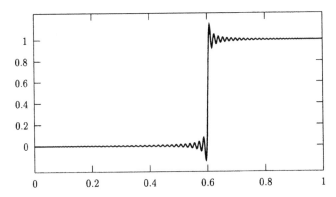

Figure 9.18: Polynomial interpolation of the function (9.41) at the 200 Chebyshev abscissas transformed to $[0, 1]$.

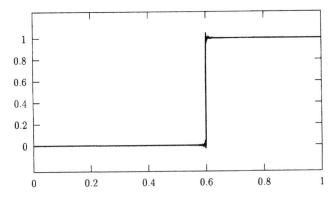

Figure 9.19: Polynomial interpolation of the function (9.41) at the 4000 Chebyshev abscissas transformed to $[0, 1]$.

Unlike K_T and K_U, where the Lebesgue constants increase *logarithmically* as $d \to \infty$, the Lebesgue constants increase *exponentially* as $d \to \infty$ if the nodes are distributed equidistantly. With equidistant nodes, even a high degree of differentiability of f is *not* sufficient to ensure the convergence $P_d \to f$ of the interpolation polynomials. Even for functions which can be differentiated an *infinite* number of times, the convergence $P_d \to f$ cannot be guaranteed.

Example (Divergence Despite Differentiability) The function (Runge [333])

$$f(x) = \frac{1}{1 + 25x^2} \qquad (\textit{Runge function})$$

can be differentiated an infinite number of times on $[-1, 1]$. However, the sequence of the interpolation polynomials with respect to equidistant nodes converges only for $|x| \leq 0.726$ being otherwise *divergent*. As a consequence, heavy oscillations, which worsen as the polynomial degree increases, occur in the neighborhood of the endpoints -1 and 1 of the interpolation interval (see Fig. 9.20 on page 385). The values of P_{10} range between -0.30 and 1.95, and those of P_{16} are between -14 and 1.35, irrespective of the original function values $f(x) \in (0.03846, 1]$.

Even on those parts of the interval $[-1, 1]$ where convergence $P_d \to f$ is guaranteed, the very low convergence rate causes extreme inefficiency in the algorithmic use of equidistant interpolation nodes.

The convergence $P_d \to f$ with respect to equidistant interpolation nodes is guaranteed only if $f : [a, b] \to \mathbb{R}$ can be extended into an entire function $\hat{f} : \mathbb{C} \to \mathbb{C}$. This is an extremely restricting prerequisite in that it requires the convergence of the power series representation of $\hat{f}$ in the whole complex plane (Hämmerlin, Hoffmann [52]).

The interpolation polynomials with respect to equidistant nodes may fail to converge even for very smooth functions. This property becomes evident even for low polynomial degrees in the form of relatively heavy oscillations of the interpolation polynomial (see Fig. 9.20).

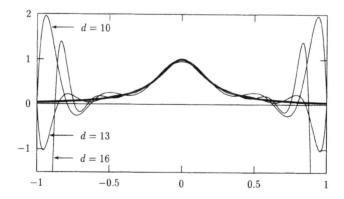

Figure 9.20: Polynomial interpolation of the Runge function (——) at equidistant nodes.

Another disadvantage of the interpolation with equidistant nodes is the large *condition number*, i.e., the sensitivity of the values $P_d(x)$ of the interpolation polynomial to changes in the data $\{f(x_0), \ldots, f(x_d)\}$ (see Section 9.3.8).

Interpolation with only *one* polynomial at equidistant nodes is thus only practically useful for rather small polynomial degrees. If, for any reason, the use of equidistant interpolation nodes is required or desired, then a *piecewise* polynomial of low degree (see Section 9.4) is recommended, especially if a large number of data points have to be interpolated.

9.3.8 The Conditioning of Polynomial Interpolation

With interpolation problems, *two* different aspects of conditioning have to be distinguished: the condition of function values and the condition of coefficients.

The Condition of Function Values

The *condition of function values* describes the sensitivity of the values $P_d(x)$ of the interpolation polynomial to perturbations of the data, i.e., the function values which define the polynomial P_d. This condition is—unlike the condition of the coefficients—*independent* of the basis used (e.g., Lagrange, Bernstein or Chebyshev basis). Accordingly, it is possible to examine how a transition from

$$\text{unperturbed data} \quad f(x_0), f(x_1), \ldots, f(x_d)$$

to

$$\text{perturbed data} \quad \tilde{f}(x_0), \tilde{f}(x_1), \ldots, \tilde{f}(x_d)$$

perturbs

$$P_d(x) = \sum_{i=0}^{d} f(x_i)\varphi_{d,i}(x) \qquad \text{into} \qquad \tilde{P}_d(x) = \sum_{i=0}^{d} \tilde{f}(x_i)\varphi_{d,i}(x).$$

The *interpolation operator*

$$L_d : \mathcal{F} \to \mathbb{P}_d$$

is the mapping that assigns a function $f \in \mathcal{F}$ the polynomial interpolating f at certain nodes $x_{d,0}, x_{d,1}, \ldots, x_{d,d}$:

$$L_d f = \sum_{i=0}^{d} f(x_{d,i})\varphi_{d,i}(x).$$

As L_d is a *linear* operator and because

$$\|L_d f - L_d \tilde{f}\| \leq \|L_d\| \, \|f - \tilde{f}\|,$$

the (absolute) condition can be characterized by $\|L_d\|$. The condition number

$$\|L_d\| := \max\{\|L_d f\|/\|f\| : f \in \mathcal{F}, \, f \neq 0\}$$

can be estimated using the Lebesgue constant. Thus,

$$
\begin{aligned}
\|L_d f\|_\infty &= \max_{x\in[a,b]} \left| \sum_{i=0}^{d} f(x_{d,i}) \varphi_{d,i}(x) \right| \\
&\le \max_{x\in[a,b]} \sum_{i=0}^{d} |f(x_{d,i})| |\varphi_{d,i}(x)| \\
&\le \max_{x\in[a,b]} |f(\tau)| \max_{x\in[a,b]} \sum_{i=0}^{d} |\varphi_{d,i}(\tau)| \\
&= \|f\|_\infty \Lambda_d(K)
\end{aligned}
$$

implies

$$
\|L_d\|_\infty = \max\{\|L_d f\|_\infty / \|f\|_\infty : f \in \mathcal{F} \setminus \{0\}\} \le \\
\le \Lambda_d(K).
$$

Table 9.1 and Fig. 9.21 contain the values of the Lebesgue constant Λ_d with respect to the interpolation interval $[-1, 1]$. These values are bounds on the maximal deviation of the interpolation polynomial of degree d if additive perturbations are applied to the data. Assuming that

$$
\begin{aligned}
P_d \quad &\text{interpolates} \quad f(x_{d,0}), f(x_{d,1}), \ldots, f(x_{d,d}) \quad \text{and} \\
\tilde{P}_d \quad &\text{interpolates} \quad \tilde{f}(x_{d,0}), \tilde{f}(x_{d,1}), \ldots, \tilde{f}(x_{d,d}),
\end{aligned}
$$

then $|\tilde{f}(x_{d,i}) - f(x_{d,i})| \le \delta$, $i = 0, 1, \ldots, d$ implies the estimate

$$
|\tilde{P}_d(x) - P_d(x)| \le \delta \Lambda_d(K) \qquad \text{for all} \quad x \in [a, b].
$$

If, for instance, $\tilde{f}(x_{d,i})$ are the values $f(x_{d,i})$ rounded to machine precision, then the use of interpolation polynomials of degree 10 may become problematic.

The exponential increase of the Lebesgue constants $\Lambda_d(K_e)$ (and with it, the increase of the condition number of polynomial interpolation) expresses the difficulties which may be encountered when working with polynomials of high degree using equidistant nodes.

Example (Temperature Data) If $f(x_{11,0}), \ldots, f(x_{11,11})$ are the values of the temperature observed over 12 monthly periods, then interpolation by a polynomial of degree 11 using the equidistant nodes 1,2,...,12 can yield undesirable results. The monthly, long-term, mean values correspond well with a sinusoidal shaped function and can thus be interpolated rather easily by a polynomial of degree 11.

The data for 1990 (cf. Fig. 9.22) is within the interval $[0.6, 21.2]$, but due to the relatively large condition number $\Lambda_{11}(K_e) = 51.2$, the variations compared with the long-term means lead to oscillations such that the values $P_d(x)$, $x \in [1, 12]$ of the interpolation polynomial are in the unrealistic interval $[-18, 22]$.

Oscillations of the interpolation polynomial, which are inconsistent with the expected course of the temperature, appear most markedly in the boundary areas.

Table 9.1, however, also shows (anticipating Section 9.7.6) the advantageous behavior of the Lebesgue constants when an increasing number of data points is interpolated using cubic spline functions. These can be used without difficulties even when the nodes are equidistant and the amount of data is huge.

Table 9.1: Lebesgue constants of the cubic spline interpolation (with *not-a-knot* boundary conditions; see Section 9.7) and interpolation by a single polynomial.

Cubic Spline Interpolation			Polynomial Interpolation (using a single polynomial)		
k	$\Lambda_k(K_T)$	$\Lambda_k(K_e)$	d	$\Lambda_d(K_T)$	$\Lambda_d(K_e)$
3	1.848	1.631	3	1.848	1.631
4	1.780	1.855	4	1.989	2.208
5	1.708	1.938	5	2.104	3.106
6	1.662	1.963	6	2.202	4.549
7	1.695	1.969	7	2.287	6.930
8	1.689	1.971	8	2.362	10.945
9	1.683	1.971	9	2.429	17.848
10	1.681	1.972	10	2.489	29.897
11	1.682	1.972	11	2.545	51.210
12	1.685	1.972	12	2.596	89.317
13	1.688	1.972	13	2.643	158.091
14	1.691	1.972	14	2.687	283.193
15	1.694	1.972	15	2.728	512.130
16	1.697	1.972	16	2.766	934.532
17	1.699	1.972	17	2.803	1716.260
18	1.701	1.972	18	2.837	3170.499
19	1.703	1.972	19	2.870	5888.154
20	1.704	1.972	20	2.901	10986.030

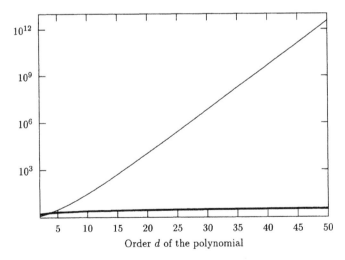

Figure 9.21: Lebesgue constants $\Lambda_d(K_e)$ and $\Lambda_d(K_T)$ for interpolation by a polynomial P_d at equidistant nodes (—) and Chebyshev nodes (—) on the interval $[-1,1]$.

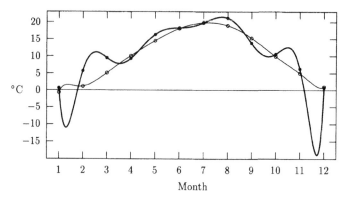

Figure 9.22: Course of the temperature in Vienna: Monthly mean values in 1990 (data points •) and long-term average temperatures from 1951 to 1990 (data points ○); each data set interpolated by a polynomial of degree 11.

The Condition of the Coefficients

The coefficients of the interpolation polynomial P_d for the data

$$(x_0, f_0), (x_1, f_1), \ldots, (x_d, f_d)$$

can, in principle, be determined by solving a system of linear equations. In determining the coefficients $a_0, \ldots, a_d$, the monomial basis $\{1, x, \ldots, x^d\}$ of the space $\mathbb{P}_d$ leads to the system of linear equations

$$\sum_{j=0}^{d} c_j x_i^j = f_i, \qquad i = 0, 1, \ldots, d, \tag{9.42}$$

whereas the basis of the Chebyshev polynomials leads to

$$\sum_{j=0}^{d} {}' a_j T_j(x_i) = f_i, \qquad i = 0, 1, \ldots, d. \tag{9.43}$$

The matrix (x_i^j) of the linear system (9.42) is a Vandermonde matrix, which is already *numerically* singular for medium-sized polynomial degrees. In this case, the linear system (9.42) becomes *practically* inconsistent. On the other hand, the matrix $(T_j(x_i))$ is an orthogonal matrix—and is thus optimally conditioned—when a special inner product is defined. Accordingly, the system of equations (9.43) is particularly easy to solve if the Chebyshev zeros (9.16) or the Chebyshev extrema (9.17) are used as interpolation nodes $\{x_i\}$ (see Section 9.3.3).

Example (Condition of the Coefficients)   For a quantitative characterization of the sensitivity of the coefficients $c_0, c_1, \ldots, c_d$ in (9.42), and $a_0, a_1, \ldots, a_d$ in (9.43), to data errors, the condition number (see Chapter 13)

$$\kappa_1(A) = \|A\|_1 \|A^{-1}\|_1$$

of the matrix $A = (x_i^j)$ and $A = (T_j(x_i))$ can be used. Fig. 9.23 shows these condition numbers for polynomials of small degrees and different choices of bases and nodes (in $[-1, 1]$):

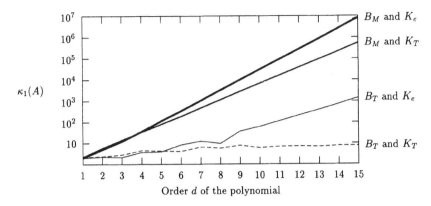

Figure 9.23: Condition numbers of coefficients depending on the polynomial degree, the basis (monomials and Chebyshev polynomials) and the order of the nodes (equidistant and Chebyshev zeros). Note the logarithmic measuring rule on the ordinate axis!

9.3.9 Choosing Interpolation Nodes

If a function f is to be approximated by only *one* interpolation polynomial on an interval $[a, b]$ and the interpolation nodes can be chosen freely, then the Chebyshev abscissas should be used. In doing so, certain advantages are obtained over equidistant nodes:

1. Because of inequality (9.37), the interpolation polynomial obtained using Chebyshev abscissas is, even for very high degrees, almost as good as the best approximating polynomial P_d^*.

2. Sensitivity to perturbations in the data (the condition of the interpolation problem) is significantly reduced, even for moderately high polynomial degrees.

3. Highly efficient computation of Chebyshev representations is made possible by evaluating (9.25) and (9.26).

If Chebyshev nodes are chosen for the approximation of a function, it is then necessary to choose the *number* of interpolation nodes and, consequently, the degree d of the polynomial. If the derivatives of f can be estimated, the error formula (9.34) can be applied. For ω_{d+1}:

$$\omega_{d+1}(x) = (x - \xi_0^{(d)})(x - \xi_1^{(d)}) \cdots (x - \xi_d^{(d)})$$

is applicable. If the $\xi_j^{(d)}$ are the zeros of the Chebyshev polynomial T_{d+1}, then ω_{d+1} is a multiple of T_{d+1}. The minimax property of the Chebyshev polynomials

(Theorem 9.3.1) then yields

$$\omega_{d+1}(x) = \frac{T_{d+1}(x)}{2^d}.$$

Moreover, the Chebyshev abscissas are the interpolation nodes in $[-1, 1]$, for which

$$\max\{|\omega_{d+1}(x)| : x \in [-1, 1]\}$$

is minimal. Again, this fact emphasizes the importance of the Chebyshev abscissas. In this special case the estimate (9.34) for the interpolation error is

$$|e_d(x)| = |P_d(x) - f(x)| \le \frac{M_{d+1}}{2^d(d+1)!} \qquad \text{for all} \quad x \in [-1, 1],$$

where M_{d+1} denotes a bound for the absolute value of $f^{(d+1)}$ on $[-1, 1]$.

Example (Approximation of the Sine Function)   In scientific programming languages, trigonometric functions are available as predefined function procedures. They are implemented by using approximating polynomials in (part of) an interval of one period. If, for instance, the sine function is to be approximated by a polynomial $P_d(x)$ interpolating $\sin x$ at the Chebyshev nodes in the interval $[-\pi/2, \pi/2]$, then

$$M_d = 1 \qquad \text{for all} \quad d = 1, 2, 3, \ldots$$

and

$$\omega_{d+1} = \left(\frac{\pi}{2}\right)^{d+1} \frac{T_{d+1}}{2^d}$$

yield the error estimate

$$|P_d(x) - \sin x| \le \frac{(\pi/2)^{d+1}}{2^d(d+1)!}.$$

The factor $(\pi/2)^{d+1}$ is a result of re-scaling $[-1, 1]$ to $[-\pi/2, \pi/2]$. The polynomial P_9 with an error bound $5 \cdot 10^{-8}$ would, for $\mathbb{F}(2, 24, -125, 128, true)$, have a satisfactory *absolute* approximation accuracy. By reducing the interpolation interval to $[0, \pi/4]$ and by simultaneously approximating $\cos(x)$, the required polynomial degree d can be further reduced.

Truncating the Chebyshev Expansion

If, as is frequently the case, there is no useful information available regarding the derivatives of f, then polynomial interpolation at the Chebyshev nodes can be used in the following way:

A degree d which is large enough to exceed the required accuracy level is chosen. The coefficients a_i of the interpolation polynomial

$$P_d(x) = \frac{a_0}{2} T_0(x) + a_1 T_1(x) + \cdots + a_d T_d(x) = \sum_{i=0}^{d}{}' a_i T_i(x) \qquad (9.44)$$

are then determined. These coefficients have the property that their absolute values $|a_0|, |a_1|, \ldots$ decrease until they reach a lower bound determined by the accuracy of the function values. If no other perturbations of the function values occur, then this lower bound equals the accuracy of the machine numbers.

Example (Approximation of the Exponential Function) If $f(x) = e^x$ is interpolated by P_{100} at the Chebyshev zeros in the interval

$$\left[-\frac{\ln 2}{2}, \frac{\ln 2}{2}\right] \approx [-0.347, \, 0.347],$$

then the absolute values $|a_0|, |a_1|, \ldots$ decrease very rapidly, as is shown in Table 9.2.

Table 9.2: Coefficients of the expansion (9.44) for the exponential function.

i	$|a_i|$	i	$|a_i|$
0	2.532	5	$5.429 \cdot 10^{-4}$
1	1.130	6	$4.498 \cdot 10^{-5}$
2	$2.715 \cdot 10^{-1}$	7	$3.049 \cdot 10^{-6}$
3	$4.434 \cdot 10^{-2}$	8	$4.528 \cdot 10^{-7}$
4	$5.474 \cdot 10^{-3}$	9	$3.294 \cdot 10^{-7}$

When using single precision arithmetic, all the values $a_{10}, a_{11}, \ldots, a_{100}$ are of magnitude $5 \cdot 10^{-7}$. Accordingly, the truncated expansion

$$\overline{P}_7(x) = \sum_{i=0}^{7} {}' a_i T_i(x)$$

with only 8 coefficients is an optimally efficient approximation. The polynomial $\overline{P}_7$ represents the best approximating polynomial with respect to the least squares error criterion (see Section 10.2).

Example (Approximation of the Logarithmic Function) If $f(x) = \ln x$ is interpolated using the 100 Chebyshev zeros in the interval $[0.71, 1.41]$, then the approximation error decreases as shown in Fig. 9.24. If the expansion is truncated after the first 10 coefficients, then the maximum absolute approximation error has already decreased to $6.41 \cdot 10^{-9}$ (see Fig. 9.25).

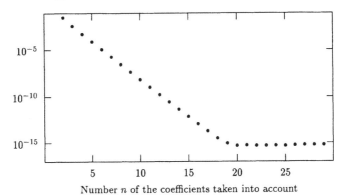

Number n of the coefficients taken into account

Figure 9.24: The maximum absolute error for truncated, 100 node Chebyshev interpolation of the logarithmic function on the interval $[1/\sqrt{2}, \sqrt{2}] \approx [0.71, 1.41]$; the first $n = 1, 2, \ldots, 30$ coefficients are used.

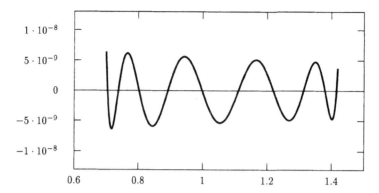

Figure 9.25: The absolute error of the Chebyshev interpolation truncated after 10 coefficients based on 100 nodes of the logarithmic function on the interval $[1/\sqrt{2}, \sqrt{2}] \approx [0.71, 1.41]$.

Successively Increasing the Degree of the Polynomial

Truncating the expansion (9.44) of an interpolation polynomial with *too high* a degree is one method to obtain an economical approximation. An alternative is successively increasing the polynomial degree until the required accuracy is reached. When using this method, not only are the zeros of the Chebyshev polynomials useful interpolation nodes, but so are the extrema. Using the extrema offers the additional advantage that—considering the transition from Q_d to Q_{2d}— all evaluations of f necessary for determining Q_d can be reused for calculating Q_{2d}. It is then possible to use

$$\max\{|Q_d(x_j^{2d}) - f(x_j^{2d})| : j = 1, 3, 5, \ldots, 2d - 1\}$$

as an error estimate, to calculate Q_{2d}, and to truncate the expansion if required.

9.3.10 Hermite Interpolation

An important method of polynomial interpolation is defined by the following $2d + 2$ linear functionals:

$$\begin{aligned} l_i f &:= f(x_i), \\ l_{d+1+i} f &:= f'(x_i), \quad i = 0, 1, \ldots, d. \end{aligned}$$

In this method of interpolation, called *Hermite interpolation* (or *osculating interpolation*), a polynomial $P_{2d+1} \in \mathbb{P}_{2d+1}$ is determined which agrees with the function f at the $d + 1$ nodes $x_0, x_1, \ldots, x_d$ with respect to function values and values of the first derivative.

If the interpolation nodes are pairwise distinct, the Hermite interpolation problem is uniquely solvable. P_{2d+1} can be expressed using basis polynomials of $\mathbb{P}_{2d+1}$ in a way similar to "ordinary" (Lagrangian) interpolation:

$$P_{2d+1}(x) = \sum_{i=0}^{d} f(x_i)\psi_{2d+1,i}(x) + \sum_{i=0}^{d} f'(x_i)\chi_{2d+1,i}(x). \tag{9.45}$$

The basis polynomials $\psi_{2d+1,i} \in \mathbb{P}_{2d+1}$

$$\psi_{2d+1,i}(x) := \left(1 - 2\varphi_{d,i}'(x_i)\,(x - x_i)\right)\varphi_{d,i}^2(x)$$

are called *elementary polynomials of the first kind.* They satisfy

$$\begin{aligned} \psi_{2d+1,i}(x_j) &= \delta_{ij} := \left\{ \begin{array}{ll} 1 & i = j \\ 0 & i \neq j \end{array} \right\} \qquad i, j = 0, 1, \dots, d. \\ \psi_{2d+1,i}'(x_j) &= 0 \end{aligned}$$

Analogously, the basis polynomials $\chi_{2d+1,i} \in \mathbb{P}_{2d+1}$

$$\chi_{2d+1,i}(x) := (x - x_i)\varphi_{d,i}^2(x)$$

are called *elementary polynomials of the second kind.* They satisfy

$$\left. \begin{array}{ll} \chi_{2d+1,i}(x_j) &= 0 \\ \chi_{2d+1,i}'(x_j) &= \delta_{ij} \end{array} \right\} \qquad i, j = 0, 1, \dots, d.$$

As the existence and the uniqueness of the elementary polynomials $\psi_{2d+1,i}$ and $\chi_{2d+1,i}$ is ensured, the formulation (9.45) leads to the desired interpolation polynomial.

One big advantage of the Hermite interpolation over conventional interpolation is the fact that undesirable oscillations are avoided by prescribing values of the first derivative.

The Hermite interpolation method can be extended by prescribing higher order derivatives f'', $f^{(3)}, \dots$ at the nodes though in practice such extensions are of little importance.

9.4 Univariate, Piecewise, Polynomial Interpolation

An alternative to using *one* uniformly defined interpolation polynomial $P_d \in \mathbb{P}_d$ is approximation using *piecewise* polynomial functions on the interval $[a, b]$:

$$g(x) := \left\{ \begin{array}{ll} P_{d_1}^1(x), & x \in [a, x_1) \\ P_{d_2}^2(x), & x \in [x_1, x_2) \\ \quad\vdots & \\ P_{d_k}^k(x), & x \in [x_{k-1}, b]. \end{array} \right. \tag{9.46}$$

As long as there are no conditions specified for the way the polynomial pieces $P_{d_1}^1, P_{d_2}^2, \dots, P_{d_k}^k$ are to be joined at the break points $x_1, x_2, \dots, x_{k-1}$—unlike in the case of spline interpolation (see Section 9.5)—, this method is a *local* interpolation method: the data points from an interval $[x_{i-1}, x_i)$ have no influence on the interpolation polynomial P^j for another interval $[x_{j-1}, x_j)$.

By choosing a suitable decomposition of the interval $[a, b]$, i. e., by choosing the number and the positions of the break points

$$a = x_0 < x_1 < \cdots < x_{k-1} < x_k = b,$$

and by choosing suitable polynomial degrees $d_1, d_2, \ldots, d_k$, the interpolation can be well adapted to local differences in the shape of the function f. The independence of the polynomial pieces $P^1, P^2, \ldots, P^k$ makes the development of highly efficient approximation algorithms possible: The nodes and the polynomial degrees can be chosen using adaptive strategies which obtain information about the function by successively increasing the number of nodes. This information can be utilized to determine the positions of the next nodes. Detailed examples of adaptive partitioning strategies are discussed in Chapter 12 in the context of numerical integration.

In this section, it is always assumed that each polynomial $P_{d_i}^i$ is defined by interpolating the function f at $d_i + 1$ nodes in the interval $[x_{i-1}, x_i]$. The behavior of g at the nodes is one reason why this kind of interpolation is not always useful. If the interpolation nodes of the polynomials $P_{d_i}^i$ are chosen in the *interior* of $[x_{i-1}, x_i]$, $i = 1, 2, \ldots, k$, then (apart from special cases) g will have jump discontinuities at the break points $x_1, x_2, \ldots, x_{k-1}$.

Software (Piecewise Quadratic Interpolation) For a given point x, the subprogram IMSL/MATH-LIBRARY/qdval returns the value of a piecewise quadratic function interpolating a supplied set of data points $(\xi_1, y_1), \ldots, (\xi_k, y_k)$. To that end, the point ξ_i nearest x is located and the quadratic interpolation polynomial, determined by the data

$$(\xi_{i-1}, y_{i-1}), \quad (\xi_i, y_i), \quad (\xi_{i+1}, y_{i+1}),$$

is computed/evaluated at x. This interpolation method corresponds, therefore, with a piecewise quadratic polynomial interpolation with break points $x_i = (\xi_i + \xi_{i+1})/2$. It should be noted that for each of these subintervals, two of the interpolation nodes are outside the subinterval and that the break points are not interpolation nodes. Therefore, in general, the interpolation function is *discontinuous* at the break points. The values of the derivative of this piecewise quadratic interpolation function can be calculated using IMSL/MATH-LIBRARY/qdder .

The inclusion of the endpoints x_{i-1} and x_i in the set of interpolation nodes for $P_{d_i}^i$, $i = 1, 2, \ldots, k$ only ensures the continuity of g. At the break points, jump discontinuities in the derivatives $g^{(l)}$, $l = 1, 2, \ldots$, still occur at least up to order l_0 after which all derivatives vanish identically:

$$P_{d_i}^{(l_0)} \equiv 0, \quad P_{d_i}^{(l_0+1)} \equiv 0, \ldots \qquad i = 1, 2, \ldots, k.$$

Whether the jump discontinuities of the function g and/or its derivatives $g', g'', \ldots$ are acceptable to the user can only be decided from case to case. Frequently, jump discontinuities are irrelevant as long as the heights of the jumps are sufficiently small. Under very general conditions it is possible to ensure that the magnitudes of the discontinuities and the whole approximation error are arbitrarily small by choosing a suitable number of suitably positioned nodes.

9.4.1 The Accuracy of the Approximation

For $d := d_1 = d_2 = \cdots = d_k$ the following theorem shows that the approximation errors

$$g(x) - f(x), \quad g'(x) - f'(x), \ \ldots$$

for $x \in [a, b]$ can—despite the jump discontinuities in the derivatives—satisfy any accuracy requirement provided the nodes are suitably chosen.

Theorem 9.4.1 *For a function $f \in C^{d+1}[a, b]$ with*

$$\|f^{(l)}\|_\infty = M_l \quad on \quad [a, b], \qquad l = 1, 2, \ldots, d+1$$

and a piecewise polynomial interpolation with step size

$$h := \max\{x_{i+1} - x_i \ : \ i = 0, 1, \ldots, k-1\},$$

the error estimate

$$|g^{(j)}(x) - f^{(j)}(x)| \leq \sum_{l=0}^{j} \frac{j!}{(d+1+j-l)!\,l!} M_{d+1+j-l} h^{d+1+l-j}, \quad j = 0, 1, \ldots, d$$

holds.

Proof: follows from the error formula (3.3.15) in Hildebrand [58].

It should be noted that in the case of piecewise polynomial interpolation, the improvement of the approximation accuracy is not accomplished by increasing the polynomial *degrees*, but by increasing the *number* of polynomial pieces. The degree d stays constant in Theorem 9.4.1 and $h \to 0$ indicates that the number of subintervals, and with it the number of polynomial pieces, is increased.

According to Theorem 9.4.1 it is possible to approximate a given function f and its derivatives $f', f'', \ldots, f^{(d)}$, as well, by using a piecewise polynomial function g—also called a *piecewise polynomial*—with arbitrary accuracy, even if g and its derivatives have jump discontinuities. There are, however, cases in which jump discontinuities are not tolerable at all.

Example (Intolerable Jump Discontinuities) If the approximation function g is used to define mechanical motion (in computer aided manufacturing, for instance, the motion of industrial robots etc.), then at the very least the continuity of g, g' and g'' is required. According to the basic law of dynamics[5] (NEWTON, 1687)

force = mass × acceleration.

The acceleration is described by the second derivative g''. The occurrence of jump discontinuities in g'' would, therefore, lead to very heavy mechanical loads and eventually to the destruction of machines, tools, or workpieces.

---

[5]For a detailed discussion of the basic equations of mechanics see, e. g., Ziegler [386].

Apart from cases such as these, piecewise polynomials with jump discontinuities are as suitable for most modeling and approximation problems as smooth polynomials, *provided* that the choice of nodes is not restricted.

If the nodes *cannot* be chosen freely, as is the case when approximating data at fixed-points, the approximating function may have to comply with certain smoothness requirements. Such requirements can be met by choosing special piecewise functions which belong to function classes of globally higher differentiability. These functions, called *spline functions*, are discussed in Section 9.5.

Note (Discontinuous Functions) When approximating discontinuous functions (with jump discontinuities) the use of piecewisely defined approximation functions which have a break point x_i at the point of discontinuity is recommended.

9.4.2 Evaluation

The evaluation of a piecewise polynomial g is carried out in two steps: For a given point x, for which the value $g(x)$ is required,

1. the index i, for which $x \in [x_{i-1}, x_i)$, is determined using, for example, a binary search (see Press et al. [24]) and

2. the ith polynomial $P_{d_i}^i$ is evaluated at point x using Horner's method.

Software (Evaluation of Piecewise Polynomials) The value at a given point of a piecewise polynomial function of arbitrary degree can be computed using the program IMSL/MATH-LIBRARY/ppval.

The program IMSL/MATH-LIBRARY/ppder determines the value of an arbitrary derivative of a piecewise polynomial function at a given point. If a whole array of these values has to be computed, then the use of IMSL/MATH-LIBRARY/pp1gd, which can do these calculations more efficiently by use of a single subroutine call, is recommended.

Software (Integration of Piecewise Polynomials) The definite integral of a piecewise polynomial function can be computed using IMSL/MATH-LIBRARY/ppitg.

9.5 Polynomial Splines

The demand for continuity and differentiability of piecewise defined functions has led to the definition of *spline functions*—functions piecewise defined on intervals whose parts join at the endpoints of the intervals in a continuous, or one or more times continuously differentiable, way.

Terminology (Spline) The English term *spline* originates from a special curve template that in the past was used, for instance, by draftsmen in ship-building. It consisted of a thin and flexible rod of wood or metal forced with mountings to connect prescribed planar points on the drawing paper. Along the rod an interpolating curve was drawn. Physically, the position assumed by the rod between the interpolation nodes is characterized by minimizing the amount of energy necessary to bend the rod, i. e.,

$$E_B(f) := \int_a^b \frac{[f''(x)]^2}{(1 + [f'(x)]^2)^3} \, dx \qquad (9.47)$$

is minimized by the interpolation function $s \in C^2[a, b]$ representing the rod.

Note (Extremal Property)   Following the minimization of the nonlinear functional (9.47) by physical splines, a mathematical spline function s can be defined by use of the *extremal property*

$$\|Ts\| = \min\{\|Tg\| : g \in \mathcal{G}, \; g(x_i) = y_i, \; i = 0, 1, \ldots, k\}.$$

If T is the differential operator $T := d^2/dx^2$, $\mathcal{G} = C^2[a, b]$, and the L_2-norm is used, then the *cubic* spline functions are obtained in this way (see Section 9.7.2).

The most important spline functions are the *polynomial* spline functions defined on the basis of piecewise polynomials.

Definition 9.5.1 (Polynomial Splines, Polynomial Spline Functions)
Polynomial splines are piecewise polynomials:

$$s(x) := \begin{cases} P_d^1(x), & x \in [x_0, x_1) \\ P_d^2(x), & x \in [x_1, x_2) \\ \vdots & \\ P_d^k(x), & x \in [x_{k-1}, x_k], \end{cases}$$

where all polynomial pieces are of the same degree d. They join in such a way that the left-hand and right-hand derivatives up to the $(d-1)$st order match with each other at the nodes $x_1, x_2, \ldots, x_{k-1}$, the inner nodes of the set $\{x_i\}$:

$$\begin{aligned} s(x_i-) &= s(x_i+), \\ s'(x_i-) &= s'(x_i+), \\ &\vdots \\ s^{(d-1)}(x_i-) &= s^{(d-1)}(x_i+), \qquad i = 1, 2, \ldots, k-1. \end{aligned} \tag{9.48}$$

For a spline function of degree d (of order $d+1$), therefore, $s \in C^{d-1}[a, b]$.

The dth derivative of a polynomial spline function is always discontinuous: Jump discontinuities in the dth order derivative *cannot* be avoided. If two polynomials have the same values and the same first d derivatives at a break point, then the function is no longer a piecewise function, but *one single* polynomial (cf. *not-a-knot* condition on page 413). The requirement (9.48) guarantees that the *smoothness* (characterized by differentiability and continuity) is *maximal* and still compatible with the definition of a piecewise polynomial.

Subsplines

If instead of (9.48) only the correspondence of the left-hand and right-hand derivatives up to the mth order, where $m < d-1$, is required, i.e., if the piecewise polynomial is an element of $C^m[a, b]$, but *not* of $C^{d-1}[a, b]$, then this piecewise polynomial is called a *subspline*. Within this class of functions, the cubic subsplines play an important role:

Definition 9.5.2 (Cubic Subsplines)   *A piecewise function s consisting of cubic polynomials which satisfies*

$$s(x_i-) = s(x_i+) \qquad \text{and} \qquad s'(x_i-) = s'(x_i+), \qquad i = 1, 2, \ldots, k-1,$$

is called a cubic subspline or a defective cubic spline.

In general, cubic subspline functions have a discontinuous second derivative, as has, for instance, the Akima function (see Section 9.8.3).

9.5.1 Oscillations and Sensitivity to Perturbations

Although the mathematical smoothness (differentiability) of a spline function increases when its order increases, the tendency to oscillate (see Fig. 9.26, 9.27 and 9.28) is a serious drawback.

The requirement of differentiability up to the $(d-1)$st degree yields systems of equations that link the polynomial pieces together. This linkage makes the spline interpolation a *global* method[6] (see Section 8.5.4), i.e., *every* data point has an influence on *every* part of the spline function.

However, in the case of spline functions of *odd* degree $d = 3, 5, 7, \ldots$, the extent of this dependency decreases if the distance grows: the influence of a change Δy_i of the data point (x_i, y_i) on $s(x)$ decreases the farther x is away from x_i.

Spline functions of *even* degree $d = 2, 4, 6, \ldots$ do not have this property of perturbation damping. Even in the case of spline functions of *odd* degree $d = 1, 3, 5, \ldots$ the damping of perturbations diminishes as d grows.

The cubic spline functions with $d = 3$ are good solutions for many interpolation problems. They are an acceptable compromise with respect to differentiability, damping of perturbations and the computational effort they require. Section 9.7 deals with cubic spline functions in detail.

Software (Testing the Properties of Splines) The routine IMSL/MATH-LIBRARY/splez serves as an interface program for many IMSL programs used for spline interpolation. Various spline functions interpolating given data points can be computed.

The interpolation method to be used can be specified by a parameter and is, therefore, easily modified. This program is therefore very useful for testing the suitability of different types of spline interpolation in concrete applications.

Software (Quadratic Splines) Quadratic spline functions that preserve the monotonicity and convexity of the given data points can be obtained using the programs from NETLIB/TOMS/574. Here, approximate values for the first order derivatives are computed. The quadratic spline function determined in this way interpolates the values of the function and the approximate values of the first derivative as well. The quadratic spline has an additional node between two break points.

Software (Quintic Splines) The software package NETLIB/TOMS/600 computes *quintic* (sub)spline functions whose break points correspond to the interpolation nodes. In the case of pairwise distinct nodes, the interpolation function is continuously differentiable four times and is therefore a spline function. However, it is possible to prescribe double or triple nodes and the corresponding values of the first and second order derivative as an interpolation constraint. In this case, the interpolation function is only three or two times continuously differentiable at these particular points and is therefore a *sub*spline. Furthermore, it is possible to deal with special cases (for instance equidistant nodes) very efficiently by using especially adapted subroutines.

---

[6] An exception are the spline functions with $d = 1$, which are broken lines.

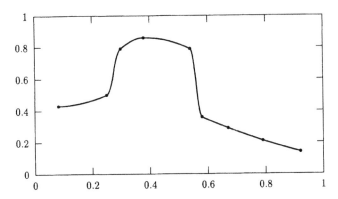

Figure 9.26: Interpolation of given data points by a *quadratic* spline.

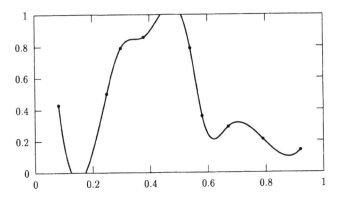

Figure 9.27: Interpolation of the same data points by a *cubic* spline.

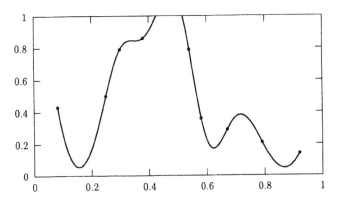

Figure 9.28: Interpolation of the above data points by a *quintic* spline.

9.5.2 The Representation of Polynomial Splines

The Hermite Representation of Cubic Splines

Cubic spline functions $s \in C^2[a, b]$ as well as cubic *subspline* functions (see Definition 9.5.2) belonging to the class of functions $C^1[a, b]$, which are less often differentiable than spline functions, can be efficiently represented using the triples

$$(r_0, f_0, s_0'), (r_1, f_1, s_1'), \quad , (r_h, f_h, s_h')$$

In every interval $[x_{i-1}, x_i]$ a cubic, Hermite interpolation polynomial is determined by the boundary conditions

$$
\begin{aligned}
P_3(x_{i-1}) &= f_{i-1} & P_3(x_i) &= f_i \\
P_3'(x_{i-1}) &= s_{i-1}' & P_3'(x_i) &= s_i'.
\end{aligned}
\tag{9.49}
$$

This interpolation polynomial in the representation

$$P_3(x) = a_0 + a_1(x - x_{i-1}) + a_2(x - x_{i-1})^2 + a_3(x - x_{i-1})^3$$

has the coefficients

$$
\begin{aligned}
a_0 &= f_{i-1} \\
a_1 &= s_{i-1}' \\
a_2 &= \frac{3}{h^2}(f_i - f_{i-1}) - \frac{1}{h}(s_i' + 2s_{i-1}') \\
a_3 &= \frac{2}{h^3}(f_i - f_{i-1}) + \frac{1}{h^2}(s_i' + s_{i-1}')
\end{aligned}
$$

where $h := x_i - x_{i-1}$. Accordingly, the interpolation of function values and values of the first derivative by a cubic spline function is called a *piecewise Hermite interpolation*. The representation of cubic spline functions by function values and values of the first derivative is also called *Hermite representation*. Since a cubic spline function in the proper meaning of the word must have continuous second derivatives, in addition to the boundary conditions (9.49), two more requirements have to be met (see Section 9.7).

Software (Calculation of Piecewise Hermite Interpolation Functions) The subroutine IMSL/MATH-LIBRARY/csher calculates the interpolatory cubic subspline function for arbitrarily given values of the function and its first derivative, and yields the coefficients of the polynomial defined for each subinterval.

Software (Calculation of Cubic Subsplines in Hermite Representation) The subroutines CMLIB/pchsp ≈ SLATEC/pchsp ≈ NETLIB/PCHIP/pchsp compute the interpolatory cubic spline function for given data points and for various boundary conditions. They yield the values of the first derivative at the nodes.

The routines NAG/e01bef ≈ CMLIB/pchim ≈ SLATEC/pchim ≈ NETLIB/PCHIP/pchim calculate approximations for the first derivative at given data points in such a way that the monotonicity of the data leads to the monotonicity of the corresponding cubic subspline as well. If the given data is only piecewise monotonic, then this is also true for the interpolation function (on the corresponding subintervals). The interpolation function has local extrema at the nodes where the monotonicity changes.

The programs CMLIB/pchic $\approx$ SLATEC/pchic $\approx$ NETLIB/PCHIP/pchic are generalizations of CMLIB/pchim. The behavior of the interpolation function in the neighborhood of the nodes at which the monotonicity changes can be controlled by a parameter. It is possible to prescribe that the interpolation function has an extremum in both of the neighboring intervals (not only at the break point). It is also possible to prescribe boundary conditions, but these may be incompatible with the monotonicity.

The subroutine NMS/pchez computes approximate values of the first derivative for given data points in such a way that the corresponding cubic spline function oscillates as little as possible and the monotonicity of the data is maintained. NMS/pchez alternatively allows for the computation of a cubic spline satisfying a *not-a-knot* boundary condition.

Software (Values, Derivatives and Integrals for the Hermite Representation) The routine NAG/e01bff evaluates a cubic subspline function in Hermite representation for a given array of abscissas. NAG/e01bgf calculates the values of the function and its first derivative using a given set of data points.

NAG/e01bhf computes the definite integral over a bounded interval of a cubic subspline function given in Hermite representation.

Truncated Power Functions

In the general case of arbitrary polynomial degrees the obvious form of representation is the one in which the spline function is represented—without especially taking continuity and differentiability into account—as a piecewise polynomial function. The spline function is represented by the $k\,(d+1)$ coefficients of all partial polynomials:

$$P_d^i = a_{i,0} + a_{i,1}(x - x_{i-1}) + \cdots + a_{i,d}(x - x_{i-1})^d, \quad x \in [x_{i-1}, x_i), \quad i = 1, 2, \ldots, k.$$

The property $s \in C^{d-1}$ can be used to represent the spline function more economically in the following way: In the first interval $[a, x_1)$ all the $d+1$ coefficients $a_{1,0}, a_{1,1}, \ldots, a_{1,d}$ are used for the characterization of the polynomial P_d^1. Due to the continuous linkage,

$$
\begin{aligned}
P_d^2(x_1) &= P_d^1(x_1) \\
(P_d^2)'(x_1) &= (P_d^1)'(x_1) \\
&\;\;\vdots \\
(P_d^2)^{(d-1)}(x_1) &= (P_d^1)^{(d-1)}(x_1)
\end{aligned}
$$

only *one* scalar quantity is necessary for the characterization of P_d^2 in the second interval $[x_1, x_2)$, namely the jump height

$$d!\,a_{2,d} = (P_d^2)^{(d)}(x_1) - (P_d^1)^{(d)}(x_1)$$

of the dth order derivative. This process can be repeated for the third, fourth,... intervals. In this way a characterization of the spline function by all the $d+1$ coefficients of the first polynomial piece and by the $k-1$ jump values of the derivatives of order d at the break points $x_1, x_2, \ldots, x_{k-1}$, is obtained. It follows that only $d+k$ coefficients are necessary for the non-redundant storage of spline functions.

As a result, the dimension of the space of the spline functions with degree d on k intervals is $d+k$.

The mathematical form of this representation of the spline function can be obtained using the *truncated power functions* of order d

$$(x - \xi)_+^d := \begin{cases} (x - \xi)^d & \text{for} \quad x \geq \xi \\ 0 & \text{for} \quad x < \xi. \end{cases}$$

The mth derivative of this function is

$$\frac{d^m (x - \xi)_+^d}{dx^m} = d(d-1)(d-2)\cdots(d-m+1)\,(x - \xi)_+^{d-m}, \quad m = 1, 2, \ldots, d.$$

In particular, the dth derivative of the truncated power function $(x - \xi)_+^d$

$$\frac{d^d (x - \xi)_+^d}{dx^d} = \begin{cases} d! & \text{for} \quad x \geq \xi \\ 0 & \text{for} \quad x < \xi, \end{cases}$$

is one of the step functions necessary for the transition from P_d^i to P_d^{i+1}. That is how the *truncated power representation*

$$s(x) = \sum_{j=0}^{d} a_{1,j}(x - x_0)^j + \sum_{i=2}^{k} a_{i,d}(x - x_{i-1})_+^d \tag{9.50}$$

is obtained. Together with the monomials, the truncated power functions form a basis

$$B_S := \{1, x, x^2, \ldots, x^d, (x - x_1)_+^d, (x - x_2)_+^d, \ldots, (x - x_{k-1})_+^d\}$$

for the space of the polynomial spline functions of degree d with inner nodes $x_1, x_2, \ldots, x_{k-1}$ (Deuflhard, Hohmann [44]).

9.6 B-Splines

For the following reasons, the basis B_S is not suitable for practical applications:

It takes too much effort: For $x \in (x_{i-1}, x_i]$, $d+i$ of the functions of the basis B_S have a non-zero value. Therefore, the evaluation of $s(x)$ for $x \in (x_{k-1}, x_k]$ might require up to $d+k$ summands in (9.50). Even if the polynomial degree d remains the same, this (maximum) computational effort increases linearly with the number of subintervals k.

This conflicts with the fact that, because s consists of polynomials of degree d, the calculation of values of the function can largely be managed with an effort of $O(d)$ of fundamental operations that is independent of the number of subintervals k. This redundant computational effort is particularly unacceptable if a very large set of data points is interpolated.

It leads to bad conditioning: If the distances $x_i - x_{i-1}$ are decreased, then the determination of the coefficients $a_{i,j}$ of the representation (9.50) turns out to be poorly conditioned, which means that they are especially sensitive to perturbations (de Boor [42]).

For these reasons, the following demands have to be made on a representation of a spline function s as a linear combination

$$s(x) = \sum_{i=1}^{k+d} c_i N_{d,i}(x) \qquad (9.51)$$

of $k+d$ linearly independent spline functions $N_{d,1}, \ldots, N_{d,k+d}$: On the one hand, only $d+1$ of the basis functions $N_{d,i}$ should take on a non-zero value in each interval $[x_{i-1}, x_i]$. In particular, the computational effort required for the evaluation of s should not depend on the number of subintervals. On the other hand, the sensitivity to perturbations of the coefficients c_i in (9.51) should, as the distances $x_i - x_{i-1}$ get smaller and smaller, increase as slowly as possible. These requirements are ideally satisfied by the *B-splines* $N_{d,1}, N_{d,2}, \ldots, N_{d,k+d}$ described in the following.

Terminology (B-Splines) On the one hand, the term *B-spline* indicates that the functions $N_{d,1}, N_{d,2}, \ldots, N_{d,k+d}$ form a *basis* for the $(k+d)$-dimensional space of spline functions. On the other hand, *B-splines* are a particular extension of the *Bernstein polynomials*, which serve, for instance, as a basis for the Bézier curves in CAD.

The definition of a B-spline basis requires the extension of the set of nodes $x_0, x_1, \ldots, x_k$ with *exterior nodes* $x_{-d}, x_{-d+1}, \ldots, x_{-1}$ and $x_{k+1}, x_{k+2}, \ldots, x_{k+d}$. Then, the normalized B-splines $N_{d,i}$ are obtained by the following recursion due to Cox [145] and de Boor [151]:

$$N_{0,i}(x) := \begin{cases} 1 & \text{for } x \in [x_{i-1}, x_i) \\ 0 & \text{otherwise} \end{cases}$$

$$N_{d,i}(x) := \frac{x - x_{i-d-1}}{x_{i-1} - x_{i-d-1}} N_{d-1,i-1}(x) \qquad (9.52)$$
$$+ \frac{x_i - x}{x_i - x_{i-d}} N_{d-1,i}(x), \quad d = 1, 2, \ldots .$$

The recursion (9.52) can also be used for the algorithmic computation of values $s(x)$ of spline functions. With suitably modified recursions it is possible to calculate values $s'(x), s''(x), \ldots$ of the derivatives in a stable way (Butterfield [131]).

The exterior nodes can be chosen in different ways, creating different systems of spline functions, which do not necessarily have to be pairwise distinct: *multiple nodes* are permitted for B-splines.[7] In most of the B-spline applications, the following equations are assumed:

$$x_{-d} = x_{-d+1} = \cdots = x_0 \qquad \text{and} \qquad x_k = x_{k+1} = \cdots = x_{k+d}.$$

In the case of a single interval ($k = 1$), the B-spline functions $N_{d,i}$ with degree $d = 0, 1, 2, 3$ are shown in Figures 9.29 to 9.32 with the above choice of exterior nodes.

---

[7]In the recursion (9.52) the convention $0/0 = 0$ is assumed in the case of coinciding nodes.

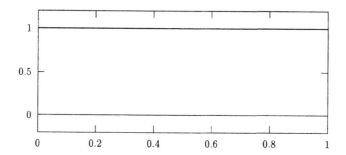

Figure 9.29: B-spline function of degree 0.

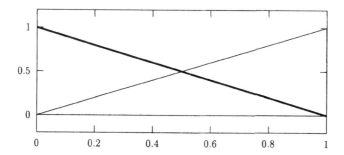

Figure 9.30: B-splines of degree 1.

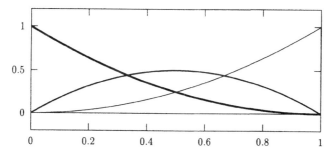

Figure 9.31: B-splines of degree 2.

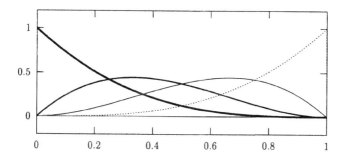

Figure 9.32: B-splines of degree 3.

Determination of Coefficients of the B-Spline Representation

From the recursion formula (9.52), the following property of the basis functions $N_{d,i}$ can be directly deduced:

$$N_{d,i} \quad \begin{cases} > 0 & \text{for} \quad x \in (x_{i-d-1}, x_i) \\ = 0 & \text{for} \quad x \notin [x_{i-d-1}, x_i). \end{cases}$$

The functions $N_{d,i}$ only take on non-zero values *locally* on $d+1$ adjoining subintervals. Thus, a function value $s(x)$ can be calculated independently of the number of subintervals k; only the $d+1$ basis functions $N_{d,i}, N_{d,i+1}, \ldots, N_{d,i+d}$ have to be taken into account for $x \in [x_{i-1}, x_i)$:

$$s(x) = \sum_{j=i}^{i+d} c_j N_{d,j}(x) \qquad \text{for} \quad x \in [x_{i-1}, x_i).$$

This representation also shows that, at most, $d+1$ unknown quantities appear in each of the linear equations that has to be solved in order to determine the coefficients and so the matrix of this linear system is a band matrix. It follows therefore that the computational effort needed to calculate the coefficients increases only linearly with the number of subintervals.

Software (Calculation of Cubic B-Splines) CMLIB/bint4 $\approx$ SLATEC/bint4 computes the coefficients of the representation of the interpolating cubic spline function in the B-spline representation for given data points. In this subroutine, the value of the first or second derivative can be alternatively prescribed as a boundary condition at each of the two interpolation endpoints. Furthermore, the exterior nodes x_{-3}, x_{-2}, x_{-1} as well as $x_{k+1}, x_{k+2}, x_{k+3}$ can be chosen arbitrarily.

9.6.1 Choice of the B-Spline Break Points

So far it has been assumed that the break points x_i of spline functions coincide with the nodes at which f has to be interpolated. The break points are the points at which a reduced smoothness of the interpolating spline occurs. The special choice mentioned above is thus useful, if, for instance, the function that has to be interpolated has an analogously reduced mathematical smoothness.

However, if the smoothness of f is *not* reduced at the interpolation nodes, it is possible—and frequently useful as well—to choose break points x_i which differ from the interpolation nodes.

In particular, the decoupling of the interpolation nodes and the break points can be used to choose the number of break points in such a way that the number of linearly independent B-splines equals the number of interpolation nodes. In doing so it is possible to drop the assumption of additional boundary conditions that would otherwise be necessary and might possibly be rather arbitrary. Interpolation nodes and break points cannot, however, be chosen completely independently; the interpolation problem is not solvable for every choice of break points. Furthermore, this choice influences the approximation quality of the interpolating spline. A detailed discussion of this topic can be found in de Boor [42].

Multiple Break Points

A further modification of the choice of the break points is to let the x_i *coincide*. If exactly m break points coincide at a position u, then u is said to be a point of *multiplicity* m. For $m = 1$ the node is called a *simple* node, for $m = 2$ the node is called a *double* node, etc. At a node with multiplicity m, the spline function s and its first $d - m$ derivatives are continuous. In the case of $m = d$, s' is discontinuous at the position u; in the case of $m = d + 1$, the spline function s has a jump discontinuity at $x = u$. No more than $d + 1$ nodes may coincide.

In particular, by choosing appropriate multiplicities, it is possible to adjust the behavior of the spline function to emulate the behavior of the function that is to be interpolated at points where it has reduced smoothness.

Software (Calculation of B-Splines of Arbitrary Degree) The subroutine CMLIB/bintk $\approx$ SLATEC/bintk computes the coefficients of the interpolating spline function with degree d using the B-splines basis, given data points and break points (*not* necessarily coinciding with the interpolation nodes).

For given values of the function and a given polynomial degree, the subroutine JCAM/spisc1 computes the corresponding spline interpolation function. The user can control the properties of the spline function by means of a number of adjustable parameters. For instance, one parameter may be used to ensure that the interpolation function preserves the monotonicity and/or convexity of the data. Furthermore, Hermite interpolation can be used, i.e., values of the derivative can be prescribed. If the user does not want to choose the polynomial degree, the routine can determine it automatically. The break points are determined automatically by the program in every case, and they do *not* normally coincide with the interpolation nodes.

9.6.2 B-Splines in Graphical Data Processing

There are a number of interesting applications for B-splines such as in the field of graphical data processing. For instance, CAD systems must enable the designer to define curves and surfaces exactly to his requirements. One common method makes it possible for the designer to prescribe a finite number of points which determine the curve or surface (Farin [185]).

In design applications it is not overly important for a curve to interpolate given points, but rather that the correlation between the positions of prescribed points and the resulting shape of the curve should be easily and intuitively understood by the designer. It turns out that the *interpolation* of given points does not comply with this requirement. The undesirable oscillations usually present in the interpolation function, and the global influence that the changing of one data point has, often leads to unexpected and unacceptable curves in certain interpolation methods.

In such a situation it is possible to employ *Schoenberg's spline approximation* For a function f defined on the interval $[a, b]$ and a sequence of nodes

$$a = x_{-d} = \cdots = x_0 \leq x_1 \leq \cdots \leq x_k = \cdots = x_{k+d} = b,$$

the Schoenberg spline function $g := g(f)$ of degree d is defined by

$$g(f) := \sum_{i=1}^{k+d} f(\xi_i) N_{d,i}, \tag{9.53}$$

where the abscissas $\xi_1 \leq \xi_2 \leq \ldots \leq \xi_{k+d}$ are determined by

$$\xi_i := (x_{i-d} + \ldots + x_{i-1})/d, \quad i = 1, 2, \ldots, k + d. \tag{9.54}$$

The function g possesses the following fundamental property:

Theorem 9.6.1 *On the interval $[a, b]$ the number of intersection points of Schoenberg's spline approximation $g(f)$ with an arbitrary line is not larger than the number of intersection points of f with the same line.*

Proof: de Boor [42].

This seemingly abstract theorem can easily be interpreted if the fact that an oscillation of f always corresponds to an inflection point of f is taken into account. On the other hand, each inflection point of f leads to an intersection point of f with a certain line. Theorem 9.6.1 therefore means that Schoenberg's spline approximation cannot oscillate more than f itself. As a result, the following special cases are obtained: All linear polynomials $f \in \mathbb{P}_2$ are reproduced exactly, $g(f) = f$, and $f \geq 0$ implies $g(f) \geq 0$. Furthermore, for a convex or concave function f, the Schoenberg spline function is also convex or concave.

Another important property can be derived directly from the definitions (9.53) and (9.54): If only *one* value $f(\xi_i)$ is changed, then this only has an effect on the $d+1$ subintervals $[x_{i-1}, x_i]$, but has no global effect on the whole function $g(f)$. Conversely, the behavior of the function $g(f)$ on the subinterval $[x_{i-1}, x_i]$ is determined only by the values $f(\xi_i), \ldots, f(\xi_{i+d})$. Therefore, Schoenberg's spline approximation is a *local* approximation method.

It should be noted that f is generally only interpolated by $g(f)$ at the endpoints a and b, but not at the interior nodes $x_1, x_2, \ldots, x_{k-1}$. The matching $g(x_i) = f(x_i)$ can, however, be forced by choosing x_i as a node with multiplicity d. Nevertheless, if the distances between the nodes $x_i - x_{i-1}$ are made sufficiently small, $g(f)$ approximates every given function f with arbitrary accuracy provided f is sufficiently smooth. The reason is that for $f \in C^r[x_0, x_k]$ and $r \geq 2$

$$\|g - f\|_\infty = O(h^2) \quad \text{with} \quad h := \max\{x_1 - x_0, \ldots, x_k - x_{k-1}\}. \tag{9.55}$$

Example (Schoenberg's Spline Approximation) For the function $f(x) = \sin(x)$ the Schoenberg spline approximation g is calculated for equidistant nodes $x_0, x_1, \ldots, x_k$ on the interval $[0, 20]$. Fig. 9.33 shows f as well as g for $d = 1$ and $k = 10$. In this case, the function g is polygonal. The graphs of the functions for the same number $k = 10$ of subintervals, but for the higher polynomial degree $d = 2$, are shown in Fig. 9.34. It should be noted that for this higher polynomial degree g approximates f *badly* at many points. By increasing the number of the subintervals to $k = 20$ (while retaining the same polynomial degree) a significantly better approximation is obtained (cf. Fig. 9.35). It is also remarkable that g approximates f especially badly in regions where f has a pronounced curvature (near the extrema).

If g is to be determined by prescribing the $k+d$ points

$$(\xi_1, y_1), \ldots, (\xi_{k+d}, y_{k+d}) \quad \text{with} \quad \xi_1 < \cdots < \xi_{k+d},$$

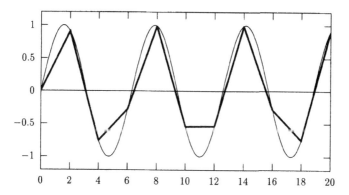

Figure 9.33: Schoenberg's spline approximation of degree 1.

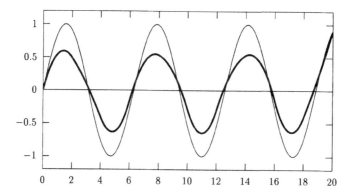

Figure 9.34: Schoenberg's spline approximation of degree 2.

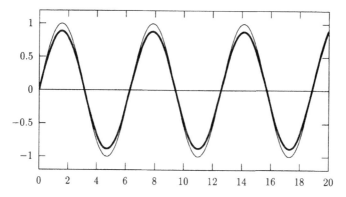

Figure 9.35: Schoenberg's spline approximation of degree 3.

then the discussion so far suggests the following method: The $k+d$ points are interpreted as values of a function f defined on the interval $[\xi_1, \xi_{k+d}]$:

$$y_i = f(\xi_i), \qquad i = 1, 2, \ldots, k+d.$$

In such cases the desired function, g, is simply Schoenberg's spline approximation $g(f)$. As a result of the approximation property (9.55) g approximates the data points at least qualitatively (provided the distances between data points are not too large). By changing *one* value y_i, the curve can be modified *locally*. It is also true that the approximating function g can never oscillate more strongly than any other interpolation function.

However, the practical implementation of this intuitive method fails in the calculation of the nodes $x_0, \ldots, x_k$ of the spline g. In general, it is *not* possible to derive a sequence of nodes $x_{-d} = \cdots = x_0 \leq x_1 \leq \ldots \leq x_k = \cdots = x_{k+d}$ for a given sequence $\xi_1 < \ldots < \xi_{k+d}$ in such a way that the ξ_i correspond with the positions defined by (9.54).

In practice this problem is solved by interpreting the given data points $(\xi_1, y_1), \ldots, (\xi_{k+d}, y_{k+d})$ as points on a plane *curve* (not necessarily of a function). The parametric form of this curve is denoted (as in Section 8.7.2) by

$$\left(f_1(\tau), \ f_2(\tau) \right), \quad \tau \in [a, b].$$

The starting point is a sequence of nodes $a = t_{-d} = \cdots = t_0 \leq t_1 \leq \ldots \leq t_k = \cdots = t_{k+d} = b$ of the *parameter interval*. One forms, according to (9.54), the sequence $\tau_1 < \ldots < \tau_{k+d}$ and views the ξ_i as being determined by the parameter value τ_i:

$$\xi_i = f_1(\tau_i), \qquad i = 1, 2, \ldots, k + d.$$

As a result, the y_i also correspond to the parameter values τ_i:

$$y_i = f_2(\tau_i), \qquad i = 1, 2, \ldots, k + d.$$

Due to the special choice of the points τ_i, the method described above can be applied to the two functions f_1 and f_2. This results in

$$g_1 := \sum_{i=1}^{k+d} \xi_i N_{d,i}(\tau) \quad \text{and} \quad g_2 := \sum_{i=1}^{k+d} y_i N_{d,i}(\tau).$$

The curve generated by the supplied points is then defined by the parameter form (g_1, g_2). Because of the monotonicity of g_1, this curve is the graph of the function $g_2(g_1^{-1})$.

The method described here can be carried out even if the assumed monotonicity of the abscissas ξ_i does not hold (the values y_i are intended to define a *unique* function). The function g_1 is no longer monotonic if the values ξ_i do not increase monotonically. The resulting curve (g_1, g_2) cannot therefore be interpreted as a function in this case. This kind of approximation can be applied, for instance, in the field of CAD.

In addition to the number of break points and the degree of the spline, the choice of parameterization—expressed in the choice of the nodes t_i—is particularly decisive in determining the shape of the curve.

Fig. 9.36 shows the curves for $k+d = 7$ and $d = 1, 2, \ldots, 6$.

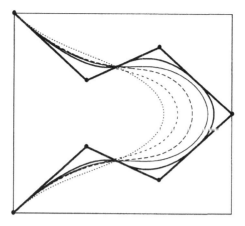

Figure 9.36: B-splines for CAD applications.

9.6.3 Software for B-Splines

A large variety of software exists for calculating and manipulating B-splines. Most of the programs rely on the algorithms and routines of de Boor [42].

Software (Nodes for the B-Spline Representation) For a polynomial degree that can be prescribed arbitrarily and for an arbitrary number of pairwise distinct interpolation nodes, IMSL/MATH-LIBRARY/bsnak as well as IMSL/MATH-LIBRARY/bsopk calculate a sequence of break points in such a way that given values of a function can later be interpolated uniquely at the nodes in the resulting space of spline functions. ·

The routine IMSL/MATH-LIBRARY/bsnak chooses the break points in such a way that—apart from the endpoints of the interpolation interval—all break points lie outside the node intervals. Therefore, the spline functions obtained are differentiable arbitrarily often at the leftmost and at the rightmost node. This is often called a *not-a-knot* choice (cf. page 413) of break points.

The subroutine IMSL/MATH-LIBRARY/bsopk chooses the break points so that the corresponding interpolation functions are optimal in the sense that they minimize the relative *worst-case*-error of the approximation with respect to the supremum norm for certain classes of functions (de Boor [42]).

Software (Solution of Interpolation Equations) The matrix of the system of linear equations (9.1) has, for every permissible choice of the break points, a shape which is referred to as *quasi-block-diagonal*. For the solution of linear systems with quasi-block-diagonal matrices (which appear in other fields of numerical data processing as well), Gaussian elimination can be specially adapted. Implementations of methods of this kind are available in NETLIB/TOMS/546 and NETLIB/TOMS/603.

Software (Coefficients of the B-Spline Representation) Using one of the three subroutines IMSL/MATH-LIBRARY/bsint, NAG/e01baf and NAG/e02baf it is possible to calculate the coefficients of the B-spline representation of a spline function that interpolates given data points. All three of the routines require a set of arbitrarily distributed data points and the desired polynomial degree as an input. In NAG/e01baf all the break points are chosen automatically, while they have to be prescribed in the two other subroutines. Favorable break points can be computed with one of the above mentioned subroutines, e. g., IMSL/MATH-LIBRARY/bsnak.

Software (Evaluation of B-Splines) The subroutine IMSL/MATH-LIBRARY/bsval evaluates a given B-spline representation at a given point. NAG/e02bbf carries out this evaluation for *cubic* spline functions. With NAG/e02bcf it is possible to simultaneously compute the value of a spline function given in B-spline representation as well as the values of the first three derivatives at a user-specified point.

The routine IMSL/MATH-LIBRARY/bsder computes the value of an arbitrary derivative (for the 0th derivative it is the value of the function itself) of a spline function, given in B-spline representation, at a given point. The subroutine IMSL/MATH-LIBRARY/bs1gd does the same but for a whole vector of abscissas simultaneously. The efficiency is thus increased if the number of evaluation points is large, because a multiple call of IMSL/MATH- LIBRARY/bsder would require greater computing time.

If the value of a definite integral of such a spline function has to be computed, then the subroutines IMSL/MATH-LIBRARY/bsitg (for general splines) and NAG/e02bdf (only for cubic splines) can be used.

The values of B-splines as well as the values of those derivatives that do not vanish at a given position x can be computed using CMLIB/bspvd $\approx$ SLATEC/bspvd. Here, the B-splines have to be specified by a sequence of break points and by their degree.

Software (Change of the Basis of a Spline Function) Using IMSL/MATH-LIBRARY/bscpp $\approx$ CMLIB/bsppp $\approx$ SLATEC/bsppp it is possible to compute the equivalent representation as a piecewise polynomial for a spline function with arbitrary degree given in B-spline representation.

9.7 Cubic Spline Interpolation

Among all polynomial splines, cubic splines play the most important role as interpolation functions because they are particularly well balanced with respect to their differentiability, to their sensitivity to perturbations and to required computational effort.

Definition 9.7.1 (Cubic Splines) *Cubic spline functions consist of polynomials of third degree for which the function value and the values of the first and second derivatives match each other at the break points of the subdivision*

$$a = x_0 < x_1 < \cdots < x_{k-1} < x_k = b \qquad (9.56)$$

of the interpolation interval $[a, b]$.

The $4k$ parameters of a cubic spline function are *not* uniquely determined by the $k + 1$ function values

$$f_0, f_1, \ldots, f_k \qquad \text{at the abscissas} \qquad x_0, x_1, \ldots, x_k \qquad (9.57)$$

and by the $3(k-1)$ requirements concerning the continuity and the differentiability at the break points $x_1, x_2, \ldots, x_{k-1}$. This non-uniqueness is due to the fact that there are only $4k-2$ defining conditions in total. For the unique determination of a cubic spline function, therefore, two additional, suitably chosen conditions— *boundary conditions*[8]—are required.

---

[8]These two conditions need not be defined at the boundaries of the interpolation interval, but for the sake of numerical stability it is most favorable to do so.

9.7.1 Boundary Conditions

The two boundary conditions additionally required for the unique definition of a cubic spline function can be specified in many different ways: natural boundary conditions, Hermite boundary conditions etc.

Natural boundary conditions

$$s''(a) = 0$$
$$s''(b) = 0. \tag{9.58}$$

This kind of boundary conditions (important only in theoretical investigations) should *not* be used for practical interpolation/approximation problems because no information concerning the second derivative, which is inherent in the data, is taken into consideration.

Hermite boundary conditions

If the values $f'(a)$, $f'(b)$ or $f''(a)$, $f''(b)$ are available, then it is possible—compared with the natural boundary conditions—to reach a significantly better approximation in the boundary patches of the interpolation interval $[a, b]$ by determining s according to one of the following two requirements:

$$\begin{array}{ll} s'(a) = f'(a) & \quad s''(a) = f''(a) \\ s'(b) = f'(b) & \text{or} \quad s''(b) = f''(b). \end{array}$$

In many practical situations, however, these values are *not* available as exact data, although estimates for the derivatives at the boundary points can be obtained using the given data points. If, for instance, a cubic polynomial $P_3(x; x_0, x_1, x_2, x_3)$ interpolates the points (x_0, f_0), (x_1, f_1), (x_2, f_2), (x_3, f_3), then $P_3'(a; x_0, x_1, x_2, x_3)$ is a useful approximation of $f'(a)$. Analogously, an approximate value of $f'(b)$ can be computed. These two approximate values are then used in the Hermite boundary conditions instead of the exact values of the derivative (Seidman, Korsan [343]):

$$\begin{array}{lll} s'(a) &=& P_3'(a; x_0, x_1, x_2, x_3) &\approx& f'(a) \\ s'(b) &=& P_3'(b; x_{k-3}, x_{k-2}, x_{k-1}, x_k) &\approx& f'(b) \end{array}$$

or

$$\begin{array}{lll} s''(a) &=& P_3''(a; x_0, x_1, x_2, x_3) &\approx& f''(a) \\ s''(b) &=& P_3''(b; x_{k-3}, x_{k-2}, x_{k-1}, x_k) &\approx& f''(b). \end{array}$$

In contrast, if only quadratic interpolation polynomials

$$\begin{array}{lll} s'(a) &=& P_2'(a; x_0, x_1, x_2) &\approx& f'(a) \\ s'(b) &=& P_2'(b; x_{k-2}, x_{k-1}, x_k) &\approx& f'(b) \end{array}$$

are used, then the approximation of the spline functions at the boundary patches is, generally, not as good as in the case of cubic interpolation polynomials.

"Not-a-knot" conditions (uniformity conditions):

$$s^{(3)}(x_1-) = s^{(3)}(x_1+) \quad \text{and} \quad s^{(3)}(x_{k-1}-) = s^{(3)}(x_{k-1}+).$$

The effect of the first condition is that the two polynomials P_3^1 and P_3^2 become one *single* cubic polynomial $P_3 := P_3^1 \equiv P_3^2$ defined uniformly on the first two intervals. Thus x_1 is not a break point (a *knot*) any longer. As a result of the second requirement, P_3^{k-1} and P_3^k become a *single* uniformly defined polynomial.

The cubic Hermite boundary conditions and the *not-a-knot* conditions are equivalent with respect to their (theoretical) approximation properties. A choice between these two boundary conditions, therefore, depends on practical testing.

Periodic boundary conditions:

$$
\begin{aligned}
s'(a) &= s'(b), \\
s''(a) &= s''(b).
\end{aligned}
\tag{9.59}
$$

Periodic boundary conditions are useful only if the underlying function f is periodic as well, i.e., if $f(x_0) = f(x_k)$, $f'(x_0) = f'(x_k), \dots$.

In the case of (9.58), $s(x)$ is called a *natural* spline function, and in the case of (9.59), $s(x)$ is said to be *periodic*.

9.7.2 Extremal Property

The cubic spline functions defined by the data (9.56) and (9.57) and by one of the boundary conditions have an important minimality property.

Theorem 9.7.1 (Extremal Property) *Let $s(x)$ be an interpolatory spline function, and let $w(x)$ be any other interpolation function that is twice continuously differentiable on $[a, b]$ satisfying either*

a) $w''(a) = w''(b) = 0$ *or*
b) $w'(a) = f'(a)$, $w'(b) = f'(b)$ *or*
c) $w(a) = w(b)$, $w'(a) = w'(b)$, $w''(a) = w''(b)$.

Then $\|s''\|_2 \leq \|w''\|_2$, i.e.,

$$
\int_a^b [s''(t)]^2 \, dt \; \leq \; \int_a^b [w''(t)]^2 \, dt.
$$

The minimality property expressed in Theorem 9.7.1 can be interpreted mathematically as follows: The curvature κ of a function w is given by

$$
\kappa(x) = \frac{w''(x)}{[1 + (w'(x))^2]^{3/2}}.
$$

For *small* values of the derivative $|w'(x)| \ll 1$, therefore, $\kappa(x) \approx w''(x)$, and thus

$$
\|w''\|_2^2 = \int_a^b [w''(t)]^2 \, dt \approx \int_a^b \kappa^2(t) \, dt.
\tag{9.60}
$$

Under the above assumption about the first derivative, (9.60) is a coarse (in many practical applications unrealistic) measure for the *total curvature* of w on $[a, b]$. The cubic spline function s is distinguished from all other interpolating functions w satisfying the boundary conditions by the fact that the integral (9.60) is minimal for $w = s$ (see also Section 9.8).

9.7.3 Error and Convergence Behavior

If $\{y_i\}$ are the values
$$y_i = f(x_i), \quad i = 0, 1, \ldots, k$$
of a function f defined on $[a, b]$, the quality of the approximation $\|s - f\|_\infty$ is of particular interest.

Theorem 9.7.2 *Let s be a cubic spline function that interpolates $f \in C^3[a, b]$ at the nodes*
$$a = x_0 < x_1 < \cdots < x_k = b.$$

With the step size
$$h := \max\{(x_{i+1} - x_i) : i = 0, 1, \ldots, k - 1\}$$

the inequality
$$\|s - f\|_\infty \le K h^j \omega \left(f^{(j)}; h \right), \quad j = 1, 2, 3 \tag{9.61}$$

holds[9], where
$$\omega \left(f^{(j)}; h \right) := \max\{|f^{(j)}(u) - f^{(j)}(v)| : u, v \in [a, b], |u - v| \le h\}$$

is the modulus of continuity of the jth derivative $f^{(j)}$.

Proof: Beatson [107].

This theorem shows—as does Theorem 9.4.1 in the case of general piecewise polynomials—that every sufficiently differentiable function can be approximated arbitrarily accurately by a cubic spline function if the step size h is chosen small enough.

The error estimate (9.61) shows that the rate of convergence is characterized by $O(h^4)$ for $f \in C^3[a, b]$, by $O(h^3)$ for $f \in C^2[a, b]$, and only by $O(h^2)$ for $f \in C[a, b]$.

9.7.4 The Calculation of Coefficients

The calculation of the coefficients of a cubic spline function requires the solution of a system of linear equations (with tridiagonal matrix) for the desired values $s_0', s_1', \ldots, s_k'$ or $s_0'', s_1'', \ldots, s_k''$. The system of equations describes the linkage between the polynomial pieces, implying that s is *twice* continuously differentiable.

---

[9] If $j = 2, 3$, then K is a constant independent of the interpolation nodes. If $j = 1$, then K depends on the arrangement of the nodes in the neighborhood of the endpoints a and b.

Software (Interpolation by Cubic Splines) The IMSL libraries provide a variety of subroutines, with which cubic spline functions interpolating given data points can be calculated:

- `IMSL/MATH-LIBRARY/csint` compute splines with *not-a-knot* boundary conditions;

- `IMSL/MATH-LIBRARY/csper` calculates periodic spline functions;

- `IMSL/MATH-LIBRARY/csdec` computes a spline function that satisfies boundary conditions prescribed by the user. It is possible to request either the *not-a-knot* condition or prescribe values for the first or second derivative at one of the endpoints. This can be done independently for the left and right boundaries (see Fig. 9.37).

`IMSL/MATH-LIBRARY/cscon` calculates a cubic spline that preserves the properties of convexity of the interpolated data and, therefore, in particular, does not oscillate excessively. The spline function obtained in this way looks optically "smooth" (see Fig. 9.38). In order to achieve this property, it is necessary to depart from the usual method of cubic spline interpolation. In `IMSL/MATH-LIBRARY/cscon` a *nonlinear* interpolation method that does *not* exactly reproduce an interpolated cubic polynomial is used (Irvine et al. [238]). In addition, further break points are inserted between the data points.

With the subroutine `IMSL/MATH-LIBRARY/csiez` an array of function values of a cubic spline which interpolates given data points can be computed. The routine internally computes the coefficients of a spline function with *not-a-knot* boundary conditions and evaluates this spline at the required points. The routine is not optimal with respect to the execution time, but it is easy to use.

Software (Discrete Cubic Splines) A special variant of cubic spline interpolation—*discrete* cubic spline interpolation—is implemented in `NETLIB/TOMS/547/dcsint`. This variant differs from conventional cubic spline interpolation mainly by replacing the matching of the first and second derivatives of neighboring polynomials at the break points with the requirement that the first and second central difference quotients match one another (Lyche [272]).

9.7.5 The Evaluation of Spline Functions

The evaluation of a cubic spline function is performed in the same way as the evaluation of any other piecewise polynomial (see Section 9.4.2); that is, the spline function is evaluated regardless of the special linkage between the individual cubic polynomials—this linkage is irrelevant in so far as the evaluation operation is concerned.

Software (Evaluation of Cubic Splines) `IMSL/MATH-LIBRARY/csval` evaluates a cubic spline function at a position arbitrarily specified by the user.

The value of an arbitrary derivative (including the derivative of order zero, which is the function itself) of a cubic spline function can be calculated using `IMSL/MATH-LIBRARY/csder`. The subroutine `IMSL/MATH-LIBRARY/cs1gd` computes the values of an arbitrary derivative of a cubic spline function (including the one of order zero) for a whole array of abscissas.

The subroutine `IMSL/MATH-LIBRARY/csitg` calculates the definite integral of a cubic spline function over a given interval $[a, b] \subset \mathbb{R}$.

9.7.6 Condition

The condition number estimate used in Section 9.3.8 to characterize the sensitivity of interpolation polynomials, with respect to changes of the interpolated function

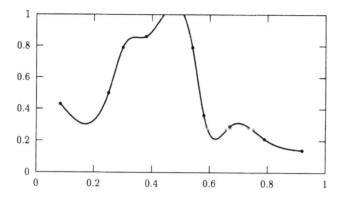

Figure 9.37: Interpolation of 9 data points by IMSL/MATH-LIBRARY/csdec.

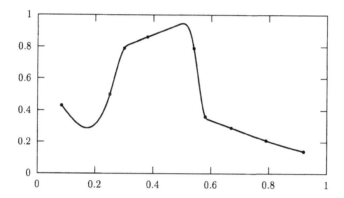

Figure 9.38: Interpolation of 9 data points by a convexity preserving spline function using the routine IMSL/MATH-LIBRARY/cscon.

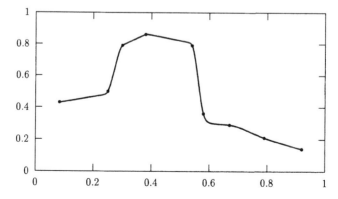

Figure 9.39: Interpolation of 9 data points by an exponential spline function (*spline under tension*; see page 421) using the routine NETLIB/TOMS/716.

values, can be applied to the cubic spline functions, as well. The estimate

$$\|L_k f\|_\infty = \max_{x \in [a,b]} \left| \sum_{i=0}^{k} f(x_{k,i}) \sigma_{k,i}(x) \right|$$

$$\leq \max_{x \in [a,b]} \sum_{i=0}^{k} |f(x_{k,i})| \, |\sigma_{k,i}(x)|$$

$$\leq \max_{x \in [a,b]} |f(x)| \max_{x \in [a,b]} \sum_{i=0}^{k} |\sigma_{k,i}(x)|$$

$$= \|f\|_\infty \Lambda_k(K)$$

implies

$$\|L_k\|_\infty = \max\{\|L_k f\|_\infty / \|f\|_\infty : f \in \mathcal{F} \setminus \{0\}, f \neq 0\} \leq \Lambda_k(K),$$

where $\sigma_{k,i}$ are the basis functions characterized by

$$\sigma_{k,i}(x_j) = \begin{cases} 1 & i = j \\ 0 & i \neq j. \end{cases}$$

Some values of the Lebesgue constants $\Lambda_k(K_e)$ and $\Lambda_k(K_T)$ are shown in Table 9.1 on page 388 and graphically in Fig. 9.40. The upper bounds of the Lebesgue constants as $k \to \infty$ (De Vore, Lorentz [157]) can be seen in this figure and in the table, as well as the nearly optimal condition of the cubic spline functions compared with interpolation by *one* uniformly defined polynomial (see Fig. 9.12 up to Fig. 9.15 on page 380).

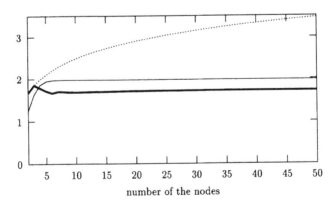

Figure 9.40: Lebesgue constants for the cubic spline interpolation (with *not-a-knot* boundary conditions) at equidistant interpolation nodes (—) and Chebyshev nodes (—) on the interval $[-1,1]$. In comparison, the logarithmically increasing Lebesgue constants for interpolation by *one* uniformly defined polynomial (·····) at the Chebyshev nodes are also shown.

Empirical Condition Analysis

If the sensitivity to data perturbations of cubic spline functions is compared in experiments with that of interpolation by *one* polynomial, then the superiority of the spline functions becomes evident. Interpolation using one uniformly defined polynomial turns out to be extremely ill-conditioned in the boundary zones of the interpolation interval and in the case of extrapolation, i.e., evaluation outside the interpolation interval.

Example (Temperature Data) If the monthly average values of the temperature in Vienna (already used in the example on page 387) are perturbed by random numbers that have a distribution corresponding to the deviation from the average over many years, then the functions shown in Fig. 9.41 and in Fig. 9.42 are obtained. The significantly worse condition of the polynomial interpolation near the boundaries and outside the interpolation interval $[1, 12]$ can be clearly seen.

In the middle region of the interpolation interval, on the other hand, both interpolation methods are roughly equivalent (in so far as the data of this example is concerned)—a result that cannot be derived from the *estimates* of the condition number.

9.8 Splines Without Undesirable Oscillations

The curve represented by a spline ruler (see page 397) is characterized by the fact that the bending energy

$$E_B := \int_a^b M^2(x)\, dx \qquad \text{with} \qquad M(x) := c\frac{g''(x)}{(1 + [g'(x)]^2)^{3/2}} \tag{9.62}$$

is minimal. Assuming that $[g'(x)]^2$ is *almost constant* or that g' is very small ($|g'(x)| \ll 1$), the minimization of the energy (9.62) can be approximately replaced by the minimization of the L_2-norm

$$\|g''\|_2 = \left(\int_a^b [g''(x)]^2\, dx\right)^{1/2}. \tag{9.63}$$

Under the previously mentioned assumption, the L_2-norm is a measure of the curvature of g over the whole interval $[a, b]$.

Among all continuously differentiable interpolation functions having a square integrable second derivative, the cubic spline functions are those for which the integral (9.63) is minimal. The cubic spline functions are the functions with minimum curvature under the above restrictions. If, however, the supposition $[g'(x)]^2 \approx \text{const}$ is violated, which happens in many practical applications, this can result in undesirable, heavy oscillations of the cubic spline functions (see Fig. 9.43).

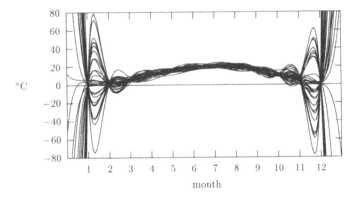

Figure 9.41: The interpolation of 30 data records (each consisting of 12 stochastically perturbed temperature means) in each case by *one* polynomial with degree $d = 11$.

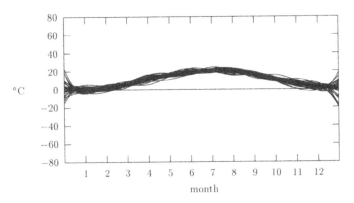

Figure 9.42: The interpolation of the same 30 data records as in Fig. 9.41 but in each case by a cubic spline function.

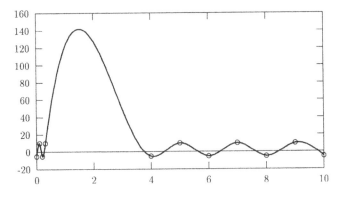

Figure 9.43: Cubic spline function, strongly oscillating in the interval $[0.3, 4]$.

Attenuation of the Oscillations by Introducing a Weight Function

An attenuation, or even a complete suppression, of undesired oscillations is possible by adding a weight function $w(x) > 0$ to the L_2-norm (9.63) (Salkauskas [339]):

$$\|g''\|_{2,w} = \left(\int_a^b w(x)[g''(x)]^2 \, dx \right)^{1/2} . \tag{9.64}$$

If the weight function w is constant between the interpolation nodes, the resulting interpolation functions which minimize (9.64) are cubic subsplines, and therefore *only once* continuously differentiable.

Note (Akima Interpolation) A possible alternative is to use the subspline functions developed by H. Akima (see Section 9.8.3). Undesirable oscillations might occur using the Akima interpolation as well, though (especially near the endpoints).

Minimizing Undesirable Oscillations

If the L_2-norm of the *first* derivative, i. e.,

$$\|g'\|_2 = \left(\int_a^b [g'(x)]^2 \, dx \right)^{1/2} , \tag{9.65}$$

is used as a measure of the oscillations, then, of all continuous interpolation functions, the polygonal arc is the optimum interpolation function which yields the smallest value for (9.65).

This result—broken lines are the optimal interpolation functions with respect to undesired oscillations—agrees completely with intuition. However, the price that has to be paid for the non-existence of oscillations is visibly reduced smoothness: every broken line has "corners" (discontinuous first derivatives) at the interpolation nodes.

9.8.1 Exponential Splines

To reduce the tendency of splines to oscillate, it it possible to enhance the mathematical model of the spline ruler by additional forces and mechanical tensions (Barsky [104]). A compromise between the cubic splines (which sometimes oscillate too much) and the cornered polygonal arc is, for instance, the interpolation function that minimizes

$$\int_a^b [g''(x)]^2 \, dx + \alpha^2 \int_a^b [g'(x)]^2 \, dx. \tag{9.66}$$

This interpolation function, introduced by Schweikert [342], is given by

$$g(x) = a_i + b_i x + c_i \sinh \alpha x + d_i \cosh \alpha x \tag{9.67}$$

on each interval $[x_i, x_{i+1})$, and is called a *spline under tension* (exponential spline). This spline has several disadvantages:

High computational effort: exponential functions are necessary to evaluate g;

There is some difficulty with stability when calculating the coefficients; special algorithmic precautions have to be taken;

Strong curvature may occur: The exponential spline (9.67) approaches a broken line as $\alpha \to \infty$. Due to the fact that g is twice continuously differentiable, strong curvature inevitably occurs near the nodes making, for example, certain visualization techniques more difficult.

Software (Splines Under Tension) An executable C-routine for the computation of exponential splines is available in NETLIB/A/TENSION. The parameter of the interpolation routine (i.e., the tension parameter α) can be set using command line parameters. The data points are read by the standard input device.

The routines in NETLIB/TOMS/716 also serve to calculate exponential splines (see Fig. 9.39 on page 417). In this case, however, the user can choose the smoothness—the degree of continuous differentiability—of the interpolation function at the interpolation nodes as well as the various boundary conditions. In fact, the different subroutines and their parameters allow for an even more extensive adaptation of the properties of the interpolation function at the request of the user. It is, for instance, possible to prescribe bounds for the interpolation function and its derivative on each subinterval or to require monotonicity or convexity.

9.8.2 ν-Splines

A class of interpolation functions consisting of piecewise cubic polynomials was introduced by G. M. Nielson [300]. In the *parametric case* (for plane curves), the same advantageous attenuation of the oscillation can be found in these functions as are found in exponential splines, but without their disadvantages. For the definition of these functions, discretization is applied on the second summand of the functional (9.66):

$$\int_a^b [g''(x)]^2 \, dx + \sum_{i=0}^{k} \nu_i [g'(x_i)]^2, \quad \nu_i \geq 0, \qquad i = 0, 1, \ldots, k. \tag{9.68}$$

In general, the functions minimizing (9.68) do *not* possess a continuous second derivative (otherwise they would be identical to the cubic spline functions); therefore, right-hand and left-hand second derivatives, $g''(x+)$ or $g''(x-)$, occur in the boundary conditions:

natural boundary conditions

$$\begin{aligned} g''(a+) &= \nu_0 g'(a) \\ g''(b-) &= \nu_k g'(b) \end{aligned}$$

periodic boundary conditions

$$\begin{aligned} g(a) &= g(b) \\ g'(a) &= g'(b) \\ g''(a+) - g''(b-) &= (\nu_0 + \nu_k) g'(a). \end{aligned}$$

It turns out that among all interpolation functions that are continuously differentiable, have a square integrable second derivative, and satisfy either natural or periodic boundary conditions, the function for which the value of the functional (9.68) is minimized has the following form:

$$s(x) = p + qx + \sum_{i=0}^{k-1} \alpha_i (x - x_i)_+^3 + \sum_{i=0}^{k-1} \beta_i (x - x_i)_+^2. \tag{9.69}$$

In the special case of $\nu_0 = \nu_1 = \cdots = \nu_k = 0$, s is the cubic spline function through the points (x_i, y_i), $i = 0, 1, \ldots, k$.

For every ν-spline function (9.69)—in the case of natural as well as of periodic boundary conditions—the following inequality

$$\sum_{i=0}^{k} \nu_i [s'(x_i)]^2 \le M_1$$

holds, where M_1 is a bound independent of the parameters ν_i, $i = 0, 1, \ldots, k$.

Theorem 9.8.1 *Let ν_k be fixed and $i \ne k$. Then $[s'(x_i)]^2$ is a decreasing function of ν_i and*

$$\lim_{\nu_i \to \infty} s'(x_i) = 0 \tag{9.70}$$

is true.

Proof: Nielson [300].

Unlike the exponential spline function, the ν-spline function interpolating the points (x_i, y_i), $i = 0, 1, \ldots, k$, does not approach a broken line for increasing parameter values

$$\nu_i \to \infty, \quad i = 0, 1, \ldots, k,$$

but instead a function with horizontal tangent lines at the nodes x_i. Such a function is therefore *not* directly suitable as a substitute for the cubic spline function. Only by changing an interpolation *function* into an interpolating *plane curve* can the oscillation attenuating property of the ν-spline functions be exploited.

ν-Spline Curves

Starting with parameterization (see Section 8.7.2), ν-spline functions $X_\nu(t)$ and $Y_\nu(t)$ are constructed to interpolate the points (t_i, x_i) and (t_i, y_i) respectively. Depending on the boundary conditions, this results either in an open or a closed curve through the data points (x_i, y_i), $i = 0, 1, \ldots, k$.

Theorem 9.8.2 *For $\dot{X}_\nu(t) \ne 0$ the derivative $d^2 Y_\nu / d X_\nu^2$ is continuous at the point t. Analogously, if $\dot{Y}_\nu(t) \ne 0$, then $d^2 X_\nu / d Y_\nu^2$ is continuous at the point t.*

Proof: Nielson [300].

This theorem directly implies the following fact that is important for the use of ν-spline functions: For $\dot{X}_\nu(t) \neq 0$ or $\dot{Y}_\nu(t) \neq 0$ the *curvature* of the curve $(X_\nu(t), Y_\nu(t))$ is *continuous* with respect to the parameter t.

The continuous curvature property is lost as $\nu_i \to \infty$. Due to (9.70) the critical case for $\nu_i \to \infty$ and $\nu_{i+1} \to \infty$ is characterized by horizontal tangents of the functions X_ν and Y_ν at the points x_i and x_{i+1}. In this case, the ν-spline curve is *linear* between (x_i, y_i) and (x_{i+1}, y_{i+1})! Therefore, the limiting process for all $\nu_i \to \infty$ yields the interpolating broken line. By choosing the ν_i sufficiently large, undesirable oscillations of the interpolation curve can be avoided.

ν-Spline Interpolation of Functions

Because of the possibility of interpreting the graph of each function as a plane curve, the method mentioned previously, which is founded on the parametric ν-spline interpolation, can even be applied if no satisfying result can be obtained using interpolation *functions*. It should be taken into consideration, though, that the interpolating ν-spline *curve*, apart from the special case of a polygonal interpolation, does not need to be a *function* $s : [a, b] \to \mathbb{R}$. Several ordinates may belong to a single abscissa.

Weighted ν-Splines

The *weighted* ν-splines are piecewise cubic interpolation functions which minimize

$$\sum_{i=0}^{k-1} \left(w_i \int_{t_i}^{t_{i+1}} [g''(t)]^2 dt \right) + \sum_{i=0}^{k} \nu_i [g'(t_i)]^2 \qquad \text{where} \quad w_i, \nu_i \geq 0 \qquad \text{for all i,}$$

instead of the functional (9.68). In addition to (9.70),

$$\lim_{w_i \to \infty} s''(t) = 0 \qquad \text{for all} \quad t \in [t_i, t_{i+1})$$

is a property of these spline functions. There are thus *two* vectors ν and ω which can be used to influence the shape of the interpolation function:

$\nu_i \geq 0$ is the *point tension* at t_i;

$w_i > 0$ is the *interval weight* of $[t_i, t_{i+1})$.

The influence of these parameters on the shape of plane ν-spline *curves* is shown in two examples illustrated in Figures 9.44 and 9.45.

Theoretical analyses and practical applications of weighted ν-spline functions in the field of CAD can be found in Foley [192].

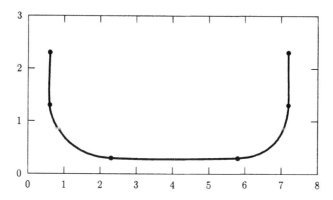

Figure 9.44: Weighted ν-spline curves with $\nu = (0,0,0,0,0,0)$, $w = (10,1,20,1,10)$.

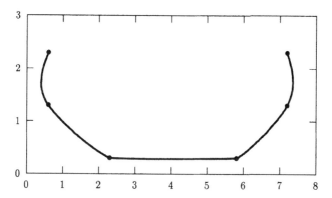

Figure 9.45: Weighted ν-spline curves with $\nu = (1,1,40,40,1,1)$, $w = (1,1,1,1,1)$.

The Calculation of ν-Splines

If the desired ν-spline function satisfies the natural boundary conditions

$$\nu_0 s'(a) - s''(a+) = 0,$$
$$\nu_k s'(b) - s''(b-) = 0,$$

then a system of linear equations with the tridiagonal matrix

$$
\begin{pmatrix}
b_0 & c_0 & & & & \mathbf{0} \\
a_1 & b_1 & c_1 & & & \\
& a_2 & \ddots & \ddots & & \\
& & \ddots & b_{k-1} & c_{k-1} \\
\mathbf{0} & & & a_k & b_k
\end{pmatrix}
\begin{pmatrix}
s_0' \\
s_1' \\
\vdots \\
s_{k-1}' \\
s_k'
\end{pmatrix}
=
\begin{pmatrix}
r_0 \\
r_1 \\
\vdots \\
r_{k-1} \\
r_k
\end{pmatrix}.
$$

is obtained for the $k+1$ unknowns $s_0', s_1', \ldots, s_k'$ (Nielson [300]). If $y_1 = y_k$, then a *periodic* ν-spline function with $s_1' = s_k'$ can be calculated. In this case, the following system of linear equations is obtained for the k unknowns $s_1', s_2', \ldots, s_k'$:

$$
\begin{pmatrix}
b_1 & c_1 & & & & a_1 \\
a_2 & b_2 & c_2 & & & \\
    & a_3 & \ddots & \ddots & & \\
    & & \ddots & b_{k-1} & c_{k-1} \\
c_k & & & & a_k & b_k
\end{pmatrix}
\begin{pmatrix}
s_1' \\ s_2' \\ \vdots \\ s_{k-1}' \\ s_k'
\end{pmatrix}
=
\begin{pmatrix}
r_1 \\ r_2 \\ \vdots \\ r_{k-1} \\ r_k
\end{pmatrix}.
$$

The two coefficient matrices for the calculation of natural or periodic ν-spline functions are positive definite and tridiagonal or cyclic tridiagonal respectively. For the solution of the systems of linear equations, therefore, Gaussian elimination modified for (cyclic) tridiagonal matrices is applicable. According to the positive definiteness of the matrices, this algorithm is numerically stable even *without pivoting*.

9.8.3 The Akima Subspline Interpolation

If the demand (9.48) for maximum differentiability of splines is reduced to

$$
s \in C^m[a,b] \quad \text{with} \quad m \in \{1,2,\ldots,d-2\},
$$

then *subsplines* (see page 398) are obtained. These functions can be chosen in a way that leads to *local* interpolation methods and to a visual appearance appropriate for many practical applications.

H. Akima [83] developed a special kind of subspline interpolation with continuously differentiable, piecewise cubic functions. Similar to the way a scientist or engineer draws a curve by hand, this interpolation method only takes the nearest data points into account when the curve is determined at a certain position. Hence, this method is a *local* interpolation scheme.

For the approximation of the first derivative of s at the point x_i, a total of five points are used: the data point in question (x_i, y_i) and two neighboring points on each side. To begin with, the four difference quotients

$$
d_j := \frac{\Delta y_j}{\Delta x_j} = \frac{y_{j+1} - y_j}{x_{j+1} - x_j}, \quad j = i-2,\, i-1,\, i,\, i+1,
$$

are determined for the four subintervals

$$
[x_{i-2}, x_{i-1}], \quad [x_{i-1}, x_i], \quad [x_i, x_{i+1}], \quad [x_{i+1}, x_{i+2}].
$$

Next, the values d_{i-2}, d_{i-1}, d_i, d_{i+1} are used for calculating the first derivative $s_i' := s'(x_i)$ of the interpolation function $s(x)$ using the following heuristic consideration: $s_i' \approx d_{i-1}$ becomes increasingly valid the smaller the difference between d_{i-2} and d_{i-1}. The same approximation is used for d_i and d_{i+1}. In order to avoid undesirable oscillations of the interpolation function, $s_i' = 0$ is enforced in the case of matching ordinates

$$
y_{i-2} = y_{i-1} = y_i, \quad \text{where} \quad d_{i-2} = d_{i-1} = 0.
$$

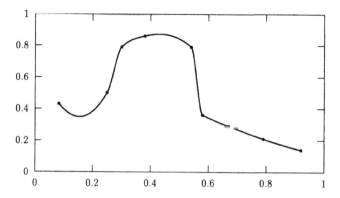

Figure 9.46: Interpolation of 9 data points by conventional Akima interpolation.

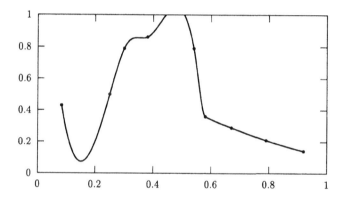

Figure 9.47: Interpolation of the same data points by the *modified* Akima interpolation using cubic polynomials (as does the conventional Akima interpolation).

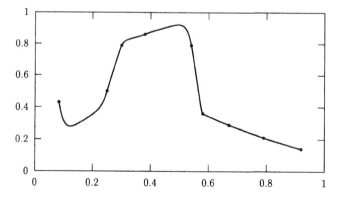

Figure 9.48: Interpolation of the data points by the *modified* Akima interpolation this time using polynomials of degree 10.

On the basis of these heuristic and intuitive considerations, s_i' is defined by

$$s_i' := \frac{w_{i-1}d_{i-1} + w_i d_i}{w_{i-1} + w_i} \tag{9.71}$$

using the weights

$$w_{i-1} := |d_{i+1} - d_i|$$
$$w_i := |d_{i-1} - d_{i-2}|.$$

Special definitions of s_i' are required if the difference quotients are not distinct:

$$s_i' := d_{i-1}, \qquad \text{if} \quad d_{i-2} = d_{i-1} \quad and \quad d_i \neq d_{i+1};$$
$$s_i' := d_i, \qquad \text{if} \quad d_i = d_{i+1} \quad and \quad d_{i-2} \neq d_{i-1};$$
$$s_i' := d_{i-1} = d_i, \qquad \text{if} \quad d_{i-1} = d_i.$$

s_i' cannot be determined using formula (9.71) if $d_{i-2} = d_{i-1} \neq d_i = d_{i+1}$. In this case, s_i' is defined by

$$s_i' := \frac{d_{i-1} + d_i}{2}.$$

If k given data points $(x_1, y_1), (x_2, y_2), \ldots, (x_k, y_k)$ are to be interpolated, it turns out that (9.71) is *not* directly applicable near the endpoints. To be able to use (9.71) at the boundary intervals, two auxiliary points outside of the interpolation interval $[x_1, x_k]$ are estimated in such a way that they enable a parabolic continuation of s outside $[x_1, x_k]$. For instance, on the right-hand side of x_k the ansatz

$$P_2(x) = a_0 + a_1(x - x_k) + a_2(x - x_k)^2 \tag{9.72}$$

is made. The coefficients a_0, a_1 and a_2 are determined using the three data points (x_{k-2}, y_{k-2}), (x_{k-1}, y_{k-1}) and (x_k, y_k). Making the assumption that

$$x_{k+2} - x_k = x_{k+1} - x_{k-1} = x_k - x_{k-2},$$

the values $y_{k+1} := P_2(x_{k+1})$ and $y_{k+2} := P_2(x_{k+2})$ can be derived from (9.72). This yields

$$d_{k+1} - d_k = d_k - d_{k-1} = d_{k-1} - d_{k-2}$$

for the difference quotients and therefore

$$d_k := 2d_{k-1} - d_{k-2}$$
$$d_{k+1} := 2d_k - d_{k-1}$$

is obtained for the right boundary area and, analogously,

$$d_0 := 2d_1 - d_2$$
$$d_{-1} := 2d_0 - d_1$$

for the left boundary area.

Software (Akima Interpolation) The subroutine IMSL/MATH-LIBRARY/csakm computes a piecewise, cubic interpolation function according to the Akima method. It yields the coefficients of each partial polynomial of the interpolation function. The function obtained can be evaluated using, for example, IMSL/MATH-LIBRARY/csval (see Fig. 9.46).

Software (Modified Akima Interpolation) A modified version of the Akima interpolation which does not attenuate undesirable oscillations as strongly as the conventional method does is implemented in the routines NETLIB/TOMS/697 (see Fig. 9.47). This modified method reproduces—in contrast to the original Akima interpolation—an interpolated cubic polynomial exactly. If desired, the modified routines also use polynomials with a degree higher than three (see Fig. 9.48).

Advantages of the Akima Interpolation

The Akima interpolation is a highly efficient method with favorable optical properties, making it suitable for various visualization applications. The efficiency of this method is outstanding because of the following reasons:

Execution time efficiency: Due to the locality of its definition, no solution of a system of equations is necessary to determine the coefficients of the interpolation function. For this reason, this algorithm is also easily parallelized.

Memory efficiency: In addition to the given data, only the auxiliary points have to be stored, i.e., the Akima interpolation works with the smallest possible memory requirement.

9.9 Multivariate Interpolation

In the previous sections, only univariate interpolation functions

$$g : B \subset \mathbb{R} \to \mathbb{R}$$

for the nodes $x_1, \ldots, x_k \in \mathbb{R}$ have been discussed. If multidimensional data

$$(x_1, y_1), \ (x_2, y_2), \ \ldots, \ (x_k, y_k) \in \mathbb{R}^n \times \mathbb{R}$$

is to be interpolated, then a *multivariate* interpolation function

$$g : B \subset \mathbb{R}^n \to \mathbb{R}$$

is required. Taking the position of the nodes $x_1, \ldots, x_k \in \mathbb{R}^n$ into account leads to the following distinction:

Data in a grid arrangement is given on a regularly (equidistantly or non-equidistantly) distributed set of nodes (see Fig. 9.49)

$$\{(x_{1j_1}, x_{2j_2}, \ldots, x_{nj_n}) : x_{ij_i} \in \{x_{i1}, \ldots, x_{ik_i}\}, \ i = 1, 2, \ldots, n\}.$$

Data which is not arranged in a grid is given on a set of nodes which systematically or unsystematically differs from the grid structure (see Fig. 9.50).

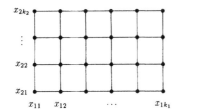
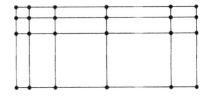

Figure 9.49: Two-dimensional equidistant and non-equidistant grids.

Figure 9.50: Points not arranged in grids: systematic and unsystematic distributions.

9.9.1 Tensor Product Interpolation

For data arranged in a grid, a multivariate interpolation function can be obtained by using an expression in the form of a product

$$g(x_1,\ldots,x_n) := g_1(x_1)\,g_2(x_2)\,\cdots\,g_n(x_n), \tag{9.73}$$

using n univariate interpolation functions

$$g_i : B_i \subseteq \mathbb{R} \to \mathbb{R}, \quad i = 1, 2, \ldots, n. \tag{9.74}$$

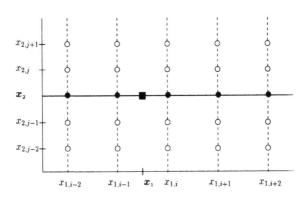

Figure 9.51: Two-dimensional tensor product interpolation at the point (x_1, x_2).

Example (Two-Dimensional Tensor Product Interpolation)   If the data is given on a two-dimensional grid, then the value of the interpolation function $g_1(x_1)\,g_2(x_2)$ at the abscissa

(x_1, x_2) can be determined by a multiple, one-dimensional interpolation. If g_2 is determined at the nodes

$$\ldots, (x_{1,i-2}, x_{2,j-2}), (x_{1,i-2}, x_{2,j-1}), \ldots, (x_{1,i-2}, x_{2,j+1}), \ldots$$
$$\ldots, (x_{1,i-1}, x_{2,j-2}), (x_{1,i-1}, x_{2,j-1}), \ldots, (x_{1,i-1}, x_{2,j+1}), \ldots$$
$$\vdots$$
$$\ldots, (x_{1,i+2}, x_{2,j-2}), (x_{1,i+2}, x_{2,j-1}), \ldots, (x_{1,i+2}, x_{2,j+1}), \ldots$$

and evaluated at the abscissa x_2, then function values at the points marked ● in Fig. 9.51 are obtained. If these values are interpolated by g_1 and evaluated at x_1, then the desired value of the bivariate interpolation function is obtained at the point (x_1, x_2) marked ■.

If the univariate interpolation functions (9.74) are uniquely determined by the one-dimensional data, then their product function (9.73) is also uniquely determined, by the multi-dimensional data. For instance, univariate polynomials or spline functions are, therefore, a suitable basis for tensor product methods.

The degrees of polynomials used for interpolation schemes on equidistant grids should not be chosen too high due to the risk of undesirable oscillations. Rather it is advisable to use piecewise interpolation.

Example (Bilinear Polynomial Interpolation)   If interpolation is carried out on a sub-rectangle $[x_{11}, x_{12}] \times [x_{21}, x_{22}]$ of a two-dimensional grid by linear polynomials first with respect to the first axial direction and then with respect to the other axial direction, then the bilinear polynomial

$$P(x_1, x_2) = a_{00} + a_{10}x_1 + a_{01}x_2 + a_{11}x_1x_2,$$

which is uniquely determined by the four values y_{11}, y_{12}, y_{21}, y_{22} at the four grid points $(x_{11}, x_{21}), \ldots, (x_{12}, x_{22})$, is obtained (see Fig. 9.52).

9.9.2 Triangulation

For data points which are not ordered in a grid, the domain is often subdivided so that the given nodes are the corner points of the subdomains.

In the case of two dimensions, neighboring nodes are often linked so that a *mesh of triangles* arises. The resulting *triangulation* is not unique—but it is possible to optimize it in accordance with certain criteria. It is, for instance, important for many applications not to let the angles of the triangles become too acute (see Fig. 9.54 on the page 432).

A separate interpolation function can be used on each triangle of the triangulation.

Example (Linear Interpolation on Triangles)   If values y_1, y_2, y_3 are given on three different points $P_1, P_2, P_3 \in \mathbb{R}^2$ (which are not situated on a line), then a plane

$$P(x_1, x_2) = a_{00} + a_{10}x_1 + a_{01}x_2$$

is uniquely determined (see Fig. 9.53). If a linear interpolation of this kind is carried out on all the triangles of a triangulation, then a bivariate continuous interpolation function on the convex hull of the nodes results.

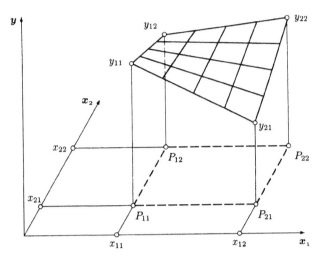

Figure 9.52: Bilinear polynomial interpolation (*hyperbolic paraboloid*).

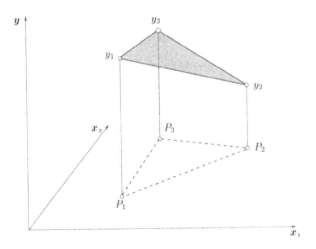

Figure 9.53: Linear interpolation (*interpolating plane*).

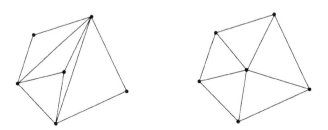

Figure 9.54: Unfavorable and favorable triangulation.

9.10 Multivariate Polynomial Interpolation

The set $\mathbb{P}^n$ of *all* multivariate polynomials in n variables is an infinite-dimensional vector space over the field of real numbers.

The simplest multivariate polynomials are the *monomials* $x_1^{d_1} x_2^{d_2} \cdots x_n^{d_n}$, which can be written in the abbreviated form x^D:

$$x^D := x_1^{d_1} x_2^{d_2} \cdots x_n^{d_n} \quad \text{with} \quad D = (d_1, d_2, \ldots, d_n) \in \mathbb{N}_0^n.$$

The set of all monomials $\{x^D\}$ forms a basis for the space $\mathbb{P}^n$, i.e., each polynomial $P \in \mathbb{P}^n$ can be uniquely represented by a finite number of monomials:

$$P = \sum_{j=1}^{J} a_j x^{D_j}, \quad a_j \neq 0, \quad j = 1, 2, \ldots, J.$$

The *degree of a monomial* x^D is defined by

$$\deg x^D := \|D\|_1 = d_1 + d_2 + \cdots + d_n,$$

and the *degree of a general multivariate polynomial*

$$P = \sum_{j=1}^{J} a_j x^{D_j}$$

is defined by

$$d := \deg P := \max\{\|D_1\|_1, \|D_2\|_1, \ldots, \|D_J\|_1\}.$$

The set

$$\mathbb{P}_d^n := \{P \in \mathbb{P}^n : \deg P \leq d\}$$

of all n-variate polynomials with maximum degree d is a linear subspace of $\mathbb{P}^n$. The space $\mathbb{P}_d^n$ is the linear closure of the monomials

$$\mathbb{P}_d^n = \text{span}\,\{x^D \;:\; \deg x^D \leq d\}$$

and is therefore a finite-dimensional space with a dimension which depends on n and d (Stroud [362]):[10]

$$dim(d, n) := \dim \mathbb{P}_d^n = |\{x^D \;:\; \deg x^D \leq d\}| = \binom{n+d}{n}.$$

The following table shows the rapid growth of $dim(d, n)$ (the number of polynomial coefficients) with increasing dimension n and degree d.

n	$d = 1$	$d = 5$	$d = 10$	$d = 15$
1	2	6	11	16
5	6	252	3 003	15 504
10	11	3 003	184 756	3 268 760
15	16	15 504	3 268 760	155 117 520

---

[10]In this case $|S|$ denotes the number of elements of the finite set S.

Polynomial Model Functions

When determining approximation functions, the decisive question arises: From which linear space of model functions $\mathcal{M}_{\mathbb{P}}$ should the approximating polynomials be chosen? To keep the algorithm error as small as possible, $\mathcal{M}_{\mathbb{P}}$ should be chosen so that all $f \in \mathcal{F}$ for a relevant class of functions $\mathcal{F}$ can be approximated by polynomials $P \in \mathcal{M}_{\mathbb{P}}$ with as high an accuracy as possible.

For $n \geq 2$, the *generalized* Weierstrass theorem (Conway [38]) says that continuous multivariate functions on compact sets can be approximated by polynomials $P \in \mathbb{P}_d^n$ with an arbitrary accuracy, provided that the polynomial degree d is chosen sufficiently high. The Taylor expansion of a (multivariate) function suggests that, if f is smooth, then a very good approximation can be reached using low polynomial degrees d. As a result of these facts, the choice of the polynomial spaces $\mathbb{P}_d^n$, $d = 1, 2, \ldots$ as model functions seems to be well founded.

The multivariate polynomials $\mathbb{P}_d^n$ can only be a suitable choice for the space $\mathcal{M}_{\mathbb{P}}$ if (for other reasons that remain to be discussed) the required dimension of $\mathcal{M}_{\mathbb{P}}$ coincides with a dimension $\dim \mathbb{P}_d^n$ of the spaces $\mathbb{P}_d^n$, $d = 1, 2, 3, \ldots$. If this is not the case, then no natural choice for the space of model functions $\mathcal{M}_{\mathbb{P}}$ exists and additional information regarding the properties of the approximation problem at hand are required.

Interpolation with Multivariate Polynomials

The multivariate polynomial interpolation problem consists of matching given function values

$$f(x_1), f(x_2), \ldots, f(x_k) \in \mathbb{R}$$

at k different points

$$x_1, x_2, \ldots, x_k \in \mathbb{R}^n, \quad n \geq 2$$

with polynomials P from a finite-dimensional linear space of polynomial model functions $\mathcal{M}_{\mathbb{P}}$. For arbitrarily given function values $f(x_1), f(x_2), \ldots, f(x_k)$ this interpolation problem can be solved if and only if the linear functionals $l_1, l_2, \ldots, l_k : \mathcal{M}_{\mathbb{P}} \to \mathbb{R}$ on $\mathcal{M}_{\mathbb{P}}$ defined by

$$l_i P := P(x_i), \quad i = 1, 2, \ldots, k,$$

are linearly independent (Davis [40]). If this condition is satisfied, then the interpolation polynomial is unique if and only if $\dim \mathcal{M}_{\mathbb{P}} = k$.

The previously mentioned facts imply that the spaces of model functions $\mathbb{P}_d^n$ are only suitable for solving the interpolation problem if the dimension $\dim \mathbb{P}_d^n$ corresponds to the number of interpolation points $x_1, x_2, \ldots, x_k$. If k is not identical to a dimension $\dim \mathbb{P}_d^n$ of the spaces $\mathbb{P}_d^n$, $d = 1, 2, 3, \ldots$, then there is no natural choice for the space of the model functions $\mathcal{M}_{\mathbb{P}}$. For this reason, the interpolation principle *cannot* be used to design approximation functions for a general number k of abscissas.

Example (Number of the Interpolation Points)   The interpolation of the data

$$(x_i, f(x_i)) \in \mathbb{R}^5 \times \mathbb{R}, \quad i = 1, 2, \ldots, k,$$

is only directly possible if the number of the coefficients $dim(d, 5)$ of the polynomials $P \in \mathbb{P}_d^5$ coincides with k

d	1	2	3	4	5	6	...
$dim(d, 5)$	6	21	56	126	252	462	...

The interpolation of $k = 32$ points requires, for instance, the definition of a special 32-dimensional subspace $\mathcal{M}_\mathbb{P}$ of $\mathbb{P}_3^5, \mathbb{P}_4^5$ or $\mathbb{P}_5^5 \ldots$. In principle, there are an infinite number of possible choices of such a subspace. Even if the principle of minimizing the computational effort is applied, there is still a very large number of possible definitions of $\mathcal{M}_\mathbb{P}$. However, none of them is distinguished in advance.

Even if the number k of interpolation points is equal to the dimension $\dim \mathbb{P}_d^n$ of one of the spaces, it is still possible that the functionals $l_1, l_2, \ldots, l_k$ are *not* linearly independent in $\mathbb{P}_d^n$. In this case, the interpolation polynomial $P \in \mathbb{P}_d^n$ is either not uniquely determined or does not exist at all. The interpolation approach *cannot* be used as a generally applicable principle for the construction of approximation functions, even if the number of abscissas k is chosen as $k := \dim \mathbb{P}_d^n$.

Example (The Non-Existence and Non-Uniqueness of Interpolating Polynomials)
A function $f \colon \mathbb{R}^2 \to \mathbb{R}$ is to be interpolated by a polynomial $P_1 \in \mathbb{P}_1^2 = \mathrm{span}\, \{1, x_1, x_2\}$ at the points

$$x_1 = (0,0), \quad x_2 = (1/2, 1/2) \quad \text{and} \quad x_3 = (1, 1).$$

Even though the number of the interpolation nodes is equal to the dimension of $\mathbb{P}_1^2$, this interpolation problem cannot generally be solved.

As the functions in $\mathbb{P}_1^2$ are affine and because all three interpolation abscissas are on one line, the function f can be interpolated at the abscissas x_1, x_2 and x_3 only if the points $(x_1, f(x_1))$, $(x_2, f(x_2))$, $(x_3, f(x_3))$ are also all on one line of $\mathbb{R}^3$. If this is the case, then there is an interpolation polynomial $P_1 \in \mathbb{P}_1^2$, but it cannot be unique because if P_1 satisfies the interpolation conditions, all polynomials of the form $P_1 + \lambda(x_1 - x_2) \in \mathbb{P}_1^2$ satisfy them as well.

9.11 Multivariate (Sub-) Spline Interpolation

In order to illustrate the effect of different node orderings, two data sets are used:

- *data arranged in a 10×10 grid* on the domain B;

- *data not arranged in a grid*: Here, 60 arbitrarily distributed nodes in the interior of B as well as 10 nodes on the boundary of B are chosen.

Example (Bivariate Interpolation Functions)   In order to assess the properties of several interpolation routines (see Fig. 9.55, Fig. 9.56, and Fig. 9.57) data points from the bivariate function

$$f(x, y) := -\exp(-(x + 1.5)^2 - (y + 1)^2)$$

on the domain $B = [-3, 3] \times [-3, 3]$ are used. The left side graph in Fig. 9.55 shows the representation of f on B using contour lines. The darker grey colors represent smaller function values; f has a global minimum at the position $(x^*, y^*) = (-1.5, -1)$. Positive values $g(x, y) > 0$ of the interpolation function are represented in white in the graphs.

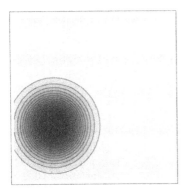 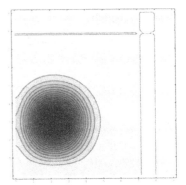

Figure 9.55: *Left*: Representation using contour lines of the function whose values are interpolated. *Right*: Interpolation of data on a 10×10 grid using NAG/e01daf.

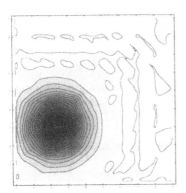 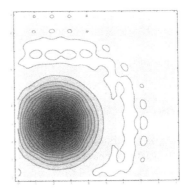

Figure 9.56: *Left*: Interpolation of data on a 10×10 grid using the program IMSL/MATH-LIBRARY/surf. *Right*: Interpolation of data on a 10×10 grid using NAG/e01sef.

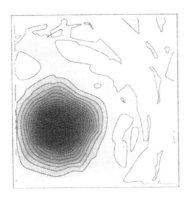 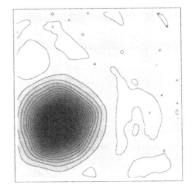

Figure 9.57: *Left*: Interpolation of 60 arbitrarily located data points using the program IMSL/MATH-LIBRARY/surf. *Right*: Interpolation of 60 arbitrarily distributed data points using NAG/e01sef.

9.11.1 Tensor Product Spline Functions

The principle of tensor product interpolation is very suitable for the multivariate spline interpolation of grid data.

The simplest tensor product spline interpolation is bilinear polynomial interpolation (see page 431). This interpolation yields a continuous interpolation function on the rectangle $[x_{11}, x_{1k_1}] \times [x_{21}, x_{2k_2}]$. However, the principle of tensor product interpolation can be applied to any spline functions (of higher order).

Software (Interpolation with Tensor Product Spline Functions) For interpolation by a two-dimensional tensor product spline function, in B-spline representation, of a set of data points on a rectangular grid, the subroutines IMSL/MATH-LIBRARY/bs2in or CMLIB/b2ink can be used. The analogous three-dimensional problem can be solved by using the programs IMSL/MATH-LIBRARY/bs3in and CMLIB/b3ink. In all these cases, the corresponding coordinates of the data points and the degree of the spline functions have to be specified for each dimension. In the case of the two CMLIB subroutines, the specification of the break points of the spline functions is optional for each dimension; in the case of the other subroutines, these points must be specified. The output data are the coefficients of the tensor product B-splines.

Using NAG/e01daf, the two-dimensional, cubic tensor product spline interpolation can be computed for data points arranged on a rectangular grid (see Fig. 9.55). The results are again the coefficients of the tensor product B-spline representation.

Software (Tensor Product Spline Evaluation) IMSL/MATH-LIBRARY/bs2vl evaluates a bivariate, tensor product spline function given in B-spline representation at a given point. The program IMSL/MATH-LIBRARY/bs3vl solves the analogous trivariate problem.

The routine NAG/e02def can compute a list of values of a two-dimensional cubic tensor product spline function given in B-spline representation. The points at which the resulting function should be evaluated can be arbitrarily distributed. For the calculation of function values at points arranged on a two-dimensional grid, it is simpler to use NAG/e02dff.

Software (Derivatives and Integrals of Tensor Product Spline Functions) With the routine IMSL/MATH-LIBRARY/bs2dr the value of an arbitrary partial derivative of a bivariate, tensor product spline function given in B-spline representation can be calculated at a prescribed point. IMSL/MATH-LIBRARY/bs2gd computes the values of an arbitrary derivative of such a spline function on a two-dimensional, rectangular grid.

The subroutine IMSL/MATH-LIBRARY/bs2ig can be used to calculate a definite integral over a rectangular integration domain which is parallel to the coordinate axes of a bivariate tensor product spline function given in B-spline representation.

There is a trivariate variant that solves the corresponding problem for three-dimensional tensor-product spline functions for each of these IMSL-subroutines. The names of these subroutines are obtained by replacing bs2 with bs3 in the names of the subroutines: IMSL/MATH-LIBRARY/bs3dr, IMSL/MATH-LIBRARY/bs3gd and IMSL/MATH-LIBRARY/bs3ig.

9.11.2 Polynomial Interpolation on Triangles

For the interpolation of irregularly arranged, two-dimensional data using piecewise polynomial functions, a triangulation is carried out first (see Section 9.9.2). The polynomials on the partial triangles are not only determined by the three function values at the corner points, but also by the conditions concerning the continuity and the differentiability of the links with the polynomials on the neighboring triangles. Some algorithms and routines do without the continuous and/or differentiable linkage between the partial polynomials in order to simplify calculations.

Software (Piecewise Polynomial Interpolation on Triangular Grids)   The subroutine IMSL/MATH-LIBRARY/surf interpolates irregularly distributed data points using a method developed by H. Akima [84], and it computes an array of function values at rectangular grid points (see Fig. 9.57). IMSL/MATH-LIBRARY/surf is based on NETLIB/TOMS/526. The continuously differentiable interpolation function is locally (on each triangle) a polynomial of fifth degree.

The subroutine NAG/e01saf computes a bivariate interpolation function for irregularly distributed nodes. This program is based on a method developed by R. J. Renka and A. K. Cline [322] and on the package NETLIB/TOMS/624. In this method the interpolation function is piecewise cubic and continuously differentiable. The function value at a given point can be computed using the program NAG/e01sbf.

NETLIB/TOMS/684 uses either a polynomial of fifth or of ninth degree (locally) on each triangle. Correspondingly, the interpolation function is then either once or twice (globally) continuously differentiable.

9.12 Related Problems and Methods

This section briefly refers to interpolation software based either on methods not mentioned so far, or on ways of looking at a problem that have not been treated. It deals with routines whose domains are not two-dimensional or are not based on a triangulation of a two-dimensional domain and subsequent piecewise polynomial interpolation, and for the underlying algorithms, bibliographic references are given.

Software (Special Multivariate Interpolation Problems and Methods)   The value of a piecewise, bivariate, quadratic polynomial that interpolates given data on a grid can be calculated at a point (x, y) using the subroutine IMSL/MATH-LIBRARY/qd2vl. For this purpose, the mesh point (x_i, y_j) closest to the given point (x, y) is determined. Furthermore, the bivariate, quadratic polynomial which interpolates the function values given at the five grid points $(x_i, y_j), (x_{i\pm1}, y_j), (x_i, y_{j\pm1})$ as well as at the closest grid points $(x_{i\pm1}, y_{j\pm1})$ is calculated. It should be noted that the interpolation nodes, with the exception of the point (x_i, y_j), lie outside the domain of definition of the bivariate, quadratic interpolation polynomial which is determined by them. In general, the interpolation function is discontinuous on the boundary of these subdomains. The subroutine IMSL/MATH-LIBRARY/qd2dr is able to compute the value of an arbitrary derivative of this function at a given point. The subroutines IMSL/MATH-LIBRARY/qd3vl and IMSL/MATH-LIBRARY/qd3dr serve similar purposes in the case of three-dimensional interpolation problems.

NETLIB/TOMS/623 is a package of programs for the calculation of a continuously differentiable function which interpolates data points on the surface of a three-dimensional sphere.

A two-dimensional interpolation function can be determined for irregularly distributed data points (see Fig. 9.57) using the routine NAG/e01sef. A modified version of the Shepard method is used (Franke, Nielson [198]). The value of this interpolating function at a given point can be determined using the program NAG/e01sff.

The packages NETLIB/TOMS/660 and NETLIB/TOMS/661 compute two- and three-dimensional interpolation functions for irregularly distributed data points. Again, a modified Shepard method is used (Renka [321]).

In NETLIB/TOMS/677 a two-dimensional, smooth interpolation function, where the data can be arbitrarily distributed, is computed. Due to the smoothness (depending on a tension parameter that can be prescribed by the user), this routine is especially suitable for the interpolation of function values that vary strongly. By choosing a suitable tension factor, the appearance of the function can be modified and undesirable oscillations in areas with a strong variation of the data can be reduced or even prevented (Bacchelli-Montefusco, Casciola [96]).

Glossary of Notation

$M_1 \setminus M_2$	Difference of two sets (M_1 minus M_2)
$M_1 \times M_2$	Cartesian product of M_1 and M_2
M^n	n-fold Cartesian product of the set M
$M^\perp$	Orthogonal complement
$\{x, y, z, \dots\}$	Set whose members are x, y, z, $\dots$
$\{x_i\}$	Sequence whose ith term is x_i, $i \in I \subseteq \mathbb{Z}$
(a, b)	Interval excluding the endpoints
$[a, b]$	Interval including the endpoints
$f : M_1 \to M_2$	Function (mapping) from M_1 to M_2
$f^{(m)}$	mth derivative of the function f
$S := A$	Definition: S is defined by A
$\approx$	Approximate equality
$\sim$	Asymptotic equivalence
$\equiv$	Identity
$\doteq$	Correspondence
$\lvert \cdot \rvert$	Absolute value
$\lVert \cdot \rVert_F$	Frobenius norm
$\lVert \cdot \rVert_2$	Euclidean norm
$\lVert \cdot \rVert_\infty$	Maximum norm (uniform norm)
$\lVert \cdot \rVert_p$	l_p- or L_p-norm
$\lVert \cdot \rVert_{p,w}$	l_p- or L_p-norm weighted with w
$\langle u, v \rangle$	Inner product (scalar product) of the vectors u and v
$\oplus$	Direct sum of two spaces
$\circ : \mathbb{R} \times \mathbb{R} \to \mathbb{R}$	Binary (arithmetic) operation
$\Box : \mathbb{R} \to \mathbb{F}$	Rounding function
$\boxplus \; \boxminus \; \boxdot \; \boxplus$	Floating-point addition, subtraction, multiplication, division
$\boxdot$	Floating-point operation, general
$\lceil x \rceil$	Smallest integer greater than or equal to x
$\lfloor x \rfloor$	Greatest integer less than or equal to x
$\Delta_k g$	Function g discretized at k sampling points
$\kappa_p(A)$	Condition number $\lVert A \rVert_p \lVert A^+ \rVert_p$ of the matrix A
$\lambda_i(x; K)$	Lebesgue function w. r. t. the ith row of node matrix K
$\lambda(A)$	Spectrum (set of eigenvalues) of the matrix A
$\rho(A)$	Spectral radius of the matrix A
ρ_A, ρ_b	Relative error of the matrix A, of the vector b
$\mu\mathrm{s}$	Microsecond (10^{-6} s)
∇	Gradient
$\varphi_{d,i}, \sigma_{d,i}$	Basis function (basis spline)
Φ	Gaussian integral (standard normal distribution function)
τ	Error tolerance of a numerical problem
$\displaystyle\int_a^b f(x)\,dx$	Definite integral of f over $[a, b]$
$\displaystyle\sum_{j=m}^{k} f(j)$	Sum $f(m) + f(m+1) + \cdots + f(k)$
$\sum{}'$	Dashed sum: Halving of the first term
$\sum{}''$	Double dashed sum: Halving of the first and the last term
ω_c	Nyquist frequency

$A^{\top}, v^{\top}$	Transpose of the matrix A, the vector v
A^{H}	Conjugate transpose of the matrix A
A^{-1}	Inverse of the matrix A
$A^{-\top}$	Inverse of the matrix $A^{\top}$
A^{+}	Generalized inverse (pseudo-inverse) of the matrix A
$b_{d,i}$	Bernstein polynomial
$\mathbb{C}$	Set of all complex numbers
$\mathbb{C}^{m \times n}$	Set of all complex $m \times n$ matrices
$C(\Omega)$, C	Set of all functions continuous on Ω or $\mathbb{R}^n$ respectively
$C^m(\Omega)$, C^m	Set of all functions having m continuous derivatives on Ω or $\mathbb{R}^n$ respectively
c^*	Parameter vector of a best approximating function
$\text{cond}_p(A)$	Condition number $\|A\|_p \|A^+\|_p$ of the matrix A
$D(\cdot, \cdot)$	Distance function
$D_{p,w}(\cdot, \cdot)$	Distance function w. r. t. $\| \cdot \|_{p,w}$
$\mathcal{D}$	Data set
$\tilde{\mathcal{D}}$	Perturbed data set
d	Degree
$\det(A)$	Determinant of the matrix A
$\text{diag}(a_{11}, \ldots, a_{nn})$	Diagonal matrix
$\text{diam}(M)$	Diameter of the set M
$\dim$	Dimension of a space
$\text{dist}(\cdot, \cdot)$	Distance function
div	Divergence of a vector field
$E_d(t; a_0, \ldots, a_d, \alpha_1, \ldots, \alpha_d)$	Exponential sum $a_0 + a_1 e^{\alpha_1 t} + \cdots + a_d e^{\alpha_d t}$
eps	Relative machine accuracy
$\mathcal{F}$	Set of functions
$\mathbb{F}$	Set of all floating-point numbers
$\mathbb{F}_D$	Set of all *denormalized* floating-point numbers
$\mathbb{F}_N$	Set of all *normalized* floating-point numbers
flop	Floating-point operation
$\mathcal{G}$, $\mathcal{G}_k$	Set of functions (with k parameters)
g^*	Best approximating function
grad	Gradient
H_f	Hessian matrix $\nabla^2 f = (\text{grad} f)'$
Hz	Hertz (cycles per second)
$I(f; a, b)$	Definite integral of $f : \mathbb{R} \to \mathbb{R}$ over $[a, b]$
$I(f; B)$	Definite integral of $f : \mathbb{R}^n \to \mathbb{R}$ over $B \subseteq \mathbb{R}^n$
KB	Kilobyte (10^3 bytes, also 2^{10} bytes)
K_c	Condition number w. r. t. coefficients
K_e	Node matrix, equidistant
K_f	Condition number w. r. t. function values
$k_{F \leftarrow x}$	Absolute (first order) condition number
$K_{F \leftarrow x}$	Relative (first order) condition number
K_T	Node matrix, Chebyshev zeros
K_U	Node matrix, Chebyshev extrema
$\mathcal{K}_i(r, B)$	Krylov space
$l_i : \mathcal{F} \to \mathbb{R}$	Linear functional
l_p	Set of all real sequences $\{a_i\}$ with convergent $\sum_{i=1}^{\infty} |a_i|^p$
$L^p[a, b]$, L^p	Set of all functions f, for which $I(|f|^p; a, b)$ or $I(|f|^p; -\infty, \infty)$ exists
$\text{Lip}(F, B)$	Lipschitz norm of F on the set B
$\ln$	Natural logarithm (base e)
$\log$, $\log_b$	Logarithm (base b)

$\max M$	Maximum value of the elements of the set M
MB	Megabyte (10^6 bytes, also 2^{20} bytes)
$\min M$	Minimum value of the elements of the set M
MK	Markowitz cost function
mod	Modulo function
ms	Millisecond (10^{-3} s)
$\mathcal{N}(F)$	Null space of the mapping F
$\mathbb{N}$	Set of all positive integers
$\mathbb{N}_0$	Set of all non-negative integers
$\mathrm{ND}(A)$	Sum of the squares of all non-diagonal elements of the matrix A
$\text{non-zero}(A)$	Number of non-zero elements of the matrix A
$\text{non-zero-block}(B)$	Number of non-zero elements of the block matrix B
ns	Nanosecond (10^{-9} s)
$O(\cdot)$	Order of convergence (Landau's asymptotic growth notation)
n	Dimension
$n_{1/2}$	Vector length which achieves $r_\infty/2$
$N^{\mathrm{a}}(\cdot)$	Adaptive information
$N^{\mathrm{na}}(\cdot)$	Non-adaptive information
$N_{d,i}$	B-spline function
$\mathbb{P}, \mathbb{P}_d$	Set of all univariate polynomials (of degree not exceeding d)
$\mathbb{P}^n, \mathbb{P}_d^n$	Set of all n-variate polynomials (of degree not exceeding d)
$P_d(x; c_0, \ldots, c_d)$	Polynomial $c_0 + c_1 x + c_2 x^2 + \cdots + c_d x^d$
P_d^*	Best approximating polynomial of degree not exceeding d
P_F	Floating-point performance [flop/s]
P_I	Instruction performance
r_∞	Asymptotic throughput
$\mathbb{R}$	Set of all real numbers
$\mathbb{R}_+$	Set of all positive real numbers
$\mathbb{R}_+^0$	Set of all non-negative real numbers
$\mathbb{R}^{m \times n}$	Set of all real $m \times n$ matrices
$\mathbb{R}_D$	Set of all real numbers covered by $\mathbb{F}_D$
$\mathbb{R}_N$	Set of all real numbers covered by $\mathbb{F}_N$
$\mathbb{R}_{\mathrm{overflow}}$	Set of all real numbers *not* covered by $\mathbb{F}$
$\mathcal{R}(F)$	Range space of the mapping F
$\mathrm{rank}(A)$	Rank of the matrix A
$S_d(x; a_0, \ldots, a_d, b_1, \ldots, b_d)$	Trigonometric polynomial $a_0/2 + \sum_{j=1}^d (a_j \cos jx + b_j \sin jx)$
span	Linear hull
$T_d \in \mathbb{P}_d$	Chebyshev polynomial of degree not exceeding d
$w(\cdot)$	Weight function
w_i	ith integration weight
$\mathbb{Z}$	Set of all integers
$zz \cdots z_2, zz \cdots z_{10}, zz \cdots z_{16}$	Number in binary, decimal- or hexadecimal representation

Bibliography

Reference Books

[1] M. Abramowitz, I. A. Stegun: *Handbook of Mathematical Functions*, 10th ed. National Bureau of Standards, Appl. Math. Ser. No. 55, U. S. Government Printing Office, 1972.

[2] J. C. Adams, W. S. Brainerd, J. T. Martin, B. T. Smith, J. L. Wagener: *Fortran 90 Handbook—Complete ANSI/ISO Reference*. McGraw-Hill, New York 1992.

[3] E. Anderson, Z. Bai, C. Bischof, J. Demmel, J. J. Dongarra, J. Du Croz, A. Greenbaum, S. Hammarling, A. McKenney, S. Ostrouchov, D. C. Sorensen: LAPACK *User's Guide*, 2nd ed. SIAM Press, Philadelphia 1995.

[4] R. E. Bank: PLTMG—*A Software Package for Solving Elliptic Partial Differential Equations—User's Guide 7.0*. SIAM Press, Philadelphia 1994.

[5] R. Barrett, M. Berry, T. Chan, J. Demmel, J. Donato, J. J. Dongarra, V. Eijkhout, R. Pozo, C. Romine, H. A. Van der Vorst: TEMPLATES *for the Solution of Linear Systems— Building Blocks for Iterative Methods*. SIAM Press, Philadelphia 1993.

[6] J. A. Brytschkow, O. I. Maritschew, A. P. Prudnikow: *Tables of Indefinite Integrals*. Gordon & Breach Science Publishers, New York 1989.

[7] J. Choi, J. J. Dongarra, D. W. Walker, SCALAPACK *Reference Manual—Parallel Factorization Routines (LU, QR, and Cholesky), and Parallel Reduction Routines (HRD, TRD, and BRD)*. Technical Report TM-12471, Mathematical Sciences Section, Oak Ridge National Laboratory, 1994.

[8] W. J. Cody, W. Waite: *Software Manual for the Elementary Functions*. Prentice-Hall, Englewood Cliffs 1981.

[9] T. F. Coleman, C. Van Loan: *Handbook for Matrix Computations*. SIAM Press, Philadelphia 1988.

[10] W. R. Cowell (Ed.): *Sources and Development of Mathematical Software*. Prentice-Hall, Englewood Cliffs 1984.

[11] J. J. Dongarra, J. R. Bunch, C. B. Moler, G. W. Stewart: LINPACK *User's Guide*. SIAM Press, Philadelphia 1979.

[12] J. J. Dongarra, R. A. Van de Geijn, R. C. Whaley: *A User's Guide to the* BLACS, Version 1.0, Technical Report, University of Tennessee, 1995.

[13] G. Engeln-Müllges, F. Uhlig: *Numerical Algorithms with C*. Springer-Verlag, Berlin Heidelberg New York Tokyo 1996.

[14] G. Engeln-Müllges, F. Uhlig: *Numerical Methods with Fortran*. Springer-Verlag, Berlin Heidelberg New York Tokyo 1996.

[15] H. Engesser, V. Claus, A. Schwill: *Encyclopedia of Information Technology*. Ellis Horwood, Chichester 1992.

[16] B. S. Garbow, J. M. Boyle, J. J. Dongarra, C. B. Moler: *Matrix Eigensystem Routines— EISPACK Guide Extension*. Lecture Notes in Computer Science Vol. 51, Springer-Verlag, Berlin Heidelberg New York Tokyo 1977.

[17] J. F. Hart et al.: *Computer Approximations*. Wiley, New York 1968.

[18] High Performance Fortran Forum (HPFF): *High Performance Fortran Language Specification*. Version 1.1, 1994.

[19] E. Krol: *The Whole Internet—User's Guide and Catalog*. O'Reilly, Sebastopol 1992.

[20] J. J. Moré, S. J. Wright: *Optimization Software Guide*. SIAM Press, Philadelphia 1993.

[21] NAG Ltd.: *NAG Fortran Library Manual—Mark 17*. Oxford 1995.

[22] R. Piessens, E. de Doncker, C. W. Ueberhuber, D. K. Kahaner: QUADPACK—*A Subrou-
      tine Package for Automatic Integration.* Springer-Verlag, Berlin Heidelberg New York
      Tokyo 1983.

[23] S. Pittner, J. Schneid, C. W. Ueberhuber: *Wavelet Literature Survey.* Technical Univer-
      sity Vienna, Wien 1993.

[24] W. H. Press, B. P. Flannery, S. A. Teukolsky, W. T. Vetterling: *Numerical Recipes in
      Fortran—The Art of Scientific Computing,* 2nd ed. Cambridge University Press, Cam-
      bridge 1992.

[25] W. H. Press, B. P. Flannery, S. A. Teukolsky, W. T. Vetterling: *Numerical Recipes in For-
      tran 90—The Art of Parallel Scientific Computing.* Cambridge University Press, Cam-
      bridge 1996.

[26] W. H. Press, B. P. Flannery, S. A. Teukolsky, W. T. Vetterling: *Numerical Recipes in
      C—The Art of Scientific Computing,* 2nd ed. Cambridge University Press, Cambridge
      1992.

[27] W. H. Press, B. P. Flannery, S. A. Teukolsky, W. T. Vetterling: *Numerical Recipes in
      Fortran—Example Book,* 2nd ed. Cambridge University Press, Cambridge 1992.

[28] W. H. Press, B. P. Flannery, S. A. Teukolsky, W. T. Vetterling: *Numerical Recipes in
      C—Example Book,* 2nd ed. Cambridge University Press, Cambridge 1992.

[29] J. R. Rice, R. F. Boisvert: *Solving Elliptic Problems Using* ELLPACK. Springer-Verlag,
      Berlin Heidelberg New York Tokyo 1985.

[30] B. T. Smith, J. M. Boyle, J. J. Dongarra, B. S. Garbow, Y. Ikebe, V. C. Klema, C. B.
      Moler: *Matrix Eigensystem Routines*—EISPACK *Guide.* Lecture Notes in Computer Sci-
      ence, Vol. 6, 2nd ed. Springer-Verlag, Berlin Heidelberg New York Tokyo 1990.

[31] Visual Numerics Inc.: IMSL MATH/LIBRARY—*User's Manual,* Version 3.0, Houston
      1994.

[32] Visual Numerics Inc.: IMSL STAT/LIBRARY—*User's Manual,* Version 3.0, Houston
      1994.

[33] D. Zwillinger (Ed. in Chief): *CRC Standard Mathematical Tables and Formulae,* 30th
      ed. CRC Press, Boca Raton New York London Tokyo 1996.

Textbooks

[34] D. P. Bertsekas, J. N. Tsitsiklis: *Parallel and Distributed Computation—Numerical
      Methods.* Prentice-Hall, Englewood Cliffs 1989.

[35] Å. Björck: *Numerical Methods for Least Squares Problems.* SIAM Press, Philadelphia
      1996.

[36] H. D. Brunk: *An Introduction to Mathematical Statistics,* 2nd ed. Blaisdell, New York
      1965.

[37] E. W. Cheney: *Introduction to Approximation Theory.* McGraw-Hill, New York 1966.

[38] J. B. Conway: *A Course in Functional Analysis.* Springer-Verlag, Berlin Heidelberg New
      York Tokyo 1984.

[39] J. Dagpunar: *Principles of Random Variate Generation.* Clarendon Press, Oxford 1988.

[40] P. J. Davis: *Interpolation and Approximation.* Blaisdell, New York 1963 (reprinted 1975
      by Dover Publications).

[41] P. J. Davis, P. Rabinowitz: *Methods of Numerical Integration,* 2nd ed. Academic Press,
      New York 1984.

[42] C. de Boor: *A Practical Guide to Splines.* Springer-Verlag, Berlin Heidelberg New York
      Tokyo 1978.

[43] J. E. Dennis, R. B. Schnabel: *Numerical Methods for Unconstrained Optimization and
      Nonlinear Equations.* Prentice-Hall, Englewood Cliffs 1983 (reprinted 1996, SIAM Press).

[44] P. Deuflhard, A. Hohmann: *Numerical Analysis—A First Course in Scientific Computation*. de Gruyter, Berlin New York 1995.

[45] K. Dowd: *High Performance Computing*. O'Reilly & Associates, Sebastopol 1993.

[46] H. Engels: *Numerical Quadrature and Cubature*. Academic Press, New York 1980.

[47] G. Evans: *Practical Numerical Integration*. Wiley, Chichester 1993.

[48] G. H. Golub, J. M. Ortega: *Scientific Computing and Differential Equations—An Introduction to Numerical Methods*. Academic Press, New York 1991.

[49] G. H. Golub, J. M. Ortega: *Scientific Computing—An Introduction with Parallel Computing*. Academic Press, New York 1993.

[50] G. H. Golub, C. F. Van Loan: *Matrix Computations*, 2nd ed. Johns Hopkins University Press, Baltimore 1989.

[51] W. Hackbusch: *Elliptic Differential Equations—Theory and Numerical Treatment*. Springer-Verlag, Berlin Heidelberg New York Tokyo 1993.

[52] G. Hämmerlin, K. H. Hoffmann: *Numerical Mathematics*. Springer-Verlag, Berlin Heidelberg New York Tokyo 1991.

[53] L. A. Hageman, D. M. Young: *Applied Iterative Methods*. Academic Press, New York London 1981.

[54] R. Hamming: *Numerical Methods for Scientists and Engineers*. McGraw-Hill, New York 1962 (reprinted 1987).

[55] J. L. Hennessy, D. A. Patterson: *Computer Architecture—A Quantitative Approach*, 2nd ed. Morgan Kaufmann, San Mateo 1995.

[56] P. Henrici: *Elements of Numerical Analysis*. Wiley, New York 1964.

[57] N. Higham: *Accuracy and Stability of Numerical Algorithms*. SIAM Press, Philadelphia 1996.

[58] F. B. Hildebrand: *Introduction to Numerical Analysis*. McGraw-Hill, New York 1974 (reprinted 1987).

[59] R. A. Horn, C. R. Johnson: *Matrix Analysis*. Cambridge University Press, Cambridge 1985 (reprinted 1990).

[60] R. A. Horn, C. R. Johnson: *Topics in Matrix Analysis*, paperback ed. Cambridge University Press, Cambridge 1994.

[61] E. Isaacson, H. B. Keller: *Analysis of Numerical Methods*. Wiley, New York 1966 (reprinted 1994).

[62] M. H. Kalos, P. A. Whitlock: *Monte Carlo Methods*. Wiley, New York 1986.

[63] C. T. Kelley: *Iterative Methods for Linear and Nonlinear Equations*. SIAM Press, Philadelphia 1995.

[64] A. R. Krommer, C. W. Ueberhuber: *Computational Integration*. SIAM Press, Philadelphia, 1996.

[65] C. L. Lawson, R. J. Hanson: *Solving Least Squares Problems*. Prentice-Hall, Englewood Cliffs 1974 (reprinted 1995, SIAM Press).

[66] P. Linz: *Theoretical Numerical Analysis*. Wiley, New York 1979.

[67] J. M. Ortega, W. C. Rheinboldt: *Iterative Solution of Nonlinear Equations in Several Variables*. Academic Press, New York London 1970.

[68] D. A. Patterson, J. L. Hennessy: *Computer Organization and Design—The Hardware / Software Interface*. Morgan Kaufmann, San Mateo 1994.

[69] C. S. Rees, S. M. Shah, C. V. Stanojevic: *Theory and Applications of Fourier Analysis*. Marcel Dekker, New York Basel 1981.

[70] J. R. Rice: *Matrix Computations and Mathematical Software*. McGraw-Hill, New York 1981.

[71] J. R. Rice: *Numerical Methods, Software, and Analysis*. McGraw-Hill, New York 1983.

[72] H. R. Schwarz, (J. Waldvogel): *Numerical Analysis—A Comprehensive Introduction.* Wiley, Chichester 1989.

[73] H. Schwetlick: *Numerische Lösung nichtlinearer Gleichungen.* Oldenbourg, München Wien 1979.

[74] G. W. Stewart: *Introduction to Matrix Computations.* Academic Press, New York 1974.

[75] J. Stoer, R. Bulirsch: *Introduction to Numerical Analysis,* 2nd ed. Springer-Verlag, Berlin Heidelberg New York Tokyo 1993.

[76] G. Strang: *Linear Algebra and its Applications,* 3rd ed. Academic Press, New York 1988.

[77] A. H. Stroud: *Numerical Quadrature and Solution of Ordinary Differential Equations.* Springer-Verlag, Berlin Heidelberg New York Tokyo 1974.

[78] C. W. Ueberhuber, P. Meditz: *Software-Entwicklung in Fortran 90.* Springer-Verlag, Wien New York 1993.

[79] H. Werner, R. Schaback: *Praktische Mathematik II.* Springer-Verlag, Berlin Heidelberg New York Tokyo 1979.

[80] J. H. Wilkinson: *The Algebraic Eigenvalue Problem.* Oxford University Press, London 1965 (reprinted 1988).

Technical Literature

[81] C. A. Addison, J. Allwright, N. Binsted, N. Bishop, B. Carpenter, P. Dalloz, J. D. Gee, V. Getov, T. Hey, R. W. Hockney, M. Lemke, J. Merlin, M. Pinches, C. Scott, I. Wolton: *The Genesis Distributed-Memory Benchmarks. Part 1—Methodology and General Relativity Benchmark with Results for the SUPRENUM Computer.* Concurrency—Practice and Experience 5-1 (1993), pp. 1–22.

[82] A. V. Aho, J. E. Hopcroft, J. D. Ullman: *Data Structures and Algorithms.* Addison-Wesley, Reading 1983.

[83] H. Akima: *A New Method of Interpolation and Smooth Curve Fitting Based on Local Procedures.* J. ACM 17 (1970), pp. 589–602.

[84] H. Akima: *A Method of Bivariate Interpolation and Smooth Surface Fitting for Irregularly Distributed Data Points.* ACM Trans. Math. Softw. 4 (1978), pp. 148–159.

[85] E. L. Allgower, K. Georg: *Numerical Continuation—An Introduction.* Springer-Verlag, Berlin Heidelberg New York Tokyo 1990.

[86] G. S. Almasi, A. Gottlieb: *Highly Parallel Computing.* Benjamin/Cummings, Redwood City 1989.

[87] L. Ammann, J. Van Ness: *A Routine for Converting Regression Algorithms into Corresponding Orthogonal Regression Algorithms.* ACM Trans. Math. Softw. 14 (1988), pp. 76–87.

[88] L.-E. Andersson, T. Elfving: *An Algorithm for Constrained Interpolation.* SIAM J. Sci. Stat. Comp. 8 (1987), pp. 1012–1025.

[89] M. A. Arbib, J. A. Robinson (Eds.): *Natural and Artificial Parallel Computation.* MIT Press, Cambridge 1990.

[90] M. Arioli, J. Demmel, I. S. Duff: *Solving Sparse Linear Systems with Sparse Backward Error.* SIAM J. Matrix Anal. Appl. 10 (1989), pp. 165–190.

[91] W. Arnoldi: *The Principle of Minimized Iterations in the Solution of the Matrix Eigenvalue Problem.* Quart. Appl. Math. 9 (1951), pp. 165–190.

[92] K. Atkinson: *The Numerical Solution of Laplace's Equation in Three Dimensions.* SIAM J. Num. Anal. 19 (1982), pp. 263-274.

[93] O. Axelsson: *Iterative Solution Methods.* Cambridge University Press, Cambridge 1996.

[94] O. Axelsson, V. Eijkhout: *Vectorizable Preconditioners for Elliptic Difference Equations in Three Space Dimensions.* J. Comput. Appl. Math. 27 (1989), pp. 299–321.

[95] O. Axelsson, B. Polman: *On Approximate Factorization Methods for Block Matrices Suitable for Vector and Parallel Processors.* Linear Algebra Appl. 77 (1986), pp. 3–26.

[96] L. Bacchelli-Montefusco, G. Casciola: C^1 *Surface Interpolation.* ACM Trans. Math. Softw. 15 (1989), pp. 365–374.

[97] Z. Bai, J. Demmel, A. McKenney: *On the Conditioning of the Nonsymmetric Eigenproblem.* Technical Report CS-89-86, Computer Science Dept., University of Tennessee, 1989.

[98] D. H. Bailey: *Extra High Speed Matrix Multiplication on the Cray-2.* SIAM J. Sci. Stat. Comput. 9 (1988), pp. 003–007.

[99] D. H. Bailey: MPFUN—*A Portable High Performance Multiprecision Package.* NASA Ames Tech. Report RNR-90-022, 1990.

[100] D. H. Bailey: *Automatic Translation of Fortran Programs to Multiprecision.* NASA Ames Tech. Report RNR-91-025, 1991.

[101] D. H. Bailey: *A Fortran-90 Based Multiprecision System.* NASA Ames Tech. Report RNR-94-013, 1994.

[102] D. H. Bailey, H. D. Simon, J. T. Barton, M. J. Fouts: *Floating-Point Arithmetic in Future Supercomputers.* Int. J. Supercomput. Appl. 3-3 (1989), pp. 86–90.

[103] C. T. H. Baker: *On the Nature of Certain Quadrature Formulas and their Errors.* SIAM J. Numer. Anal. 5 (1968), pp. 783–804.

[104] B. A. Barsky: *Exponential and Polynomial Methods for Applying Tension to an Interpolating Spline Curve.* Comput. Vision Graph. Image Process. 1 (1984), pp. 1–18.

[105] F. L. Bauer, C. F. Fike: *Norms and Exclusion Theorems.* Numer. Math. 2 (1960), pp. 123–144.

[106] F. L. Bauer, H. Rutishauser, E. Stiefel: *New Aspects in Numerical Quadrature.* Proceedings of Symposia in Applied Mathematics, Amer. Math. Soc. 15 (1963), pp. 199–219.

[107] R. K. Beatson: *On the Convergence of Some Cubic Spline Interpolation Schemes.* SIAM J. Numer. Anal. 23 (1986), pp. 903–912.

[108] M. Beckers, R. Cools: *A Relation between Cubature Formulae of Trigonometric Degree and Lattice Rules.* Report TW 181, Department of Computer Science, Katholieke Universiteit Leuven, 1992.

[109] M. Beckers, A. Haegemans: *Transformation of Integrands for Lattice Rules,* in "Numerical Integration—Recent Developments, Software and Applications" (T. O. Espelid, A. Genz, Eds.). Kluwer, Dordrecht 1992, pp. 329–340.

[110] J. Berntsen, T. O. Espelid: DCUTRI—*An Algorithm for Adaptive Cubature over a Collection of Triangles.* ACM Trans. Math. Softw. 18 (1992), pp. 329–342.

[111] J. Berntsen, T. O. Espelid, A. Genz: *An Adaptive Algorithm for the Approximate Calculation of Multiple Integrals.* ACM Trans. Math. Softw. 17 (1991), pp. 437–451.

[112] S. Bershader, T. Kraay, J. Holland: *The Giant Fourier Transform,* in "Scientific Applications of the Connection Machine" (H. D. Simon, Ed.). World Scientific, Singapore New Jersey London Hong Kong 1989.

[113] C. Bischof: LAPACK—*Portable lineare Algebra-Software für Supercomputer.* Informationstechnik 34 (1992), pp. 44–49.

[114] C. Bischof, P. T. P. Tang: *Generalized Incremental Condition Estimation.* Technical Report CS-91-132, Computer Science Dept., University of Tennessee, 1991.

[115] C. Bischof, P. T. P. Tang: *Robust Incremental Condition Estimation.* Technical Report CS-91-133, Computer Science Dept., University of Tennessee, 1991.

[116] G. E. Blelloch: *Vector Models for Data-Parallel Computing.* MIT Press, Cambridge London 1990.

[117] J. L. Blue: *A Portable Fortran Program to Find the Euclidean Norm.* ACM Trans. Math. Softw. 4 (1978), pp. 15–23.

[118] P. T. Boggs, R. H. Byrd and R. B. Schnabel: *A Stable and Efficient Algorithm for Non-linear Orthogonal Distance Regression.* SIAM J. Sci. Stat. Comput. 8 (1987), pp. 1052–1078.

[119] R. F. Boisvert: *A Fourth-Order-Accurate Fourier Method for the Helmholtz Equation in Three Dimensions.* ACM Trans. Math. Softw. 13 (1987), pp. 221–234.

[120] R. F. Boisvert, S. E. Howe, D. K. Kahaner: GAMS—*A Framework for the Management of Scientific Software.* ACM Trans. Math. Softw. 11 (1985), pp. 313–356.

[121] P. Bolzern, G. Fronza, E. Runca, C. W. Ueberhuber: *Statistical Analysis of Winter Sulphur Dioxide Concentration Data in Vienna.* Atmospheric Environment 16 (1982), pp. 1899–1906.

[122] M. Bourdeau, A. Pitre: *Tables of Good Lattices in Four and Five Dimensions.* Numer. Math. 47 (1985), pp. 39–43.

[123] H. Braß: *Quadraturverfahren.* Vandenhoeck und Ruprecht, Göttingen 1977.

[124] R. P. Brent: *An Algorithm with Guaranteed Convergence for Finding a Zero of a Function.* Computer J. 14 (1971), pp. 422–425.

[125] R. P. Brent: *A Fortran Multiple-Precision Arithmetic Package.* ACM Trans. Math. Softw. 4 (1978), pp. 57–70.

[126] R. P. Brent: *Algorithm 524—A Fortran Multiple-Precision Arithmetic Package.* ACM Trans. Math. Softw. 4 (1978), pp. 71–81.

[127] K. W. Brodlie: *Methods for Drawing Curves*, in "Fundamental Algorithms for Computer Graphics" (R. A. Earnshaw, Ed.). Springer-Verlag, Berlin Heidelberg New York Tokyo 1985, pp. 303–323.

[128] M. Bronstein: *Integration of Elementary Functions.* J. Symbolic Computation 9 (1990), pp. 117–173.

[129] C. G. Broyden: *A Class of Methods for Solving Nonlinear Simultaneous Equations.* Math. Comp. 19 (1965), pp. 577–593.

[130] J. C. P. Bus, T. J. Dekker: *Two Efficient Algorithms with Guaranteed Convergence for Finding a Zero of a Function.* ACM Trans. Math. Softw. 1 (1975), pp. 330–345.

[131] K. R. Butterfield: *The Computation of all Derivatives of a B-Spline Basis.* J. Inst. Math. Appl. 17 (1976), pp. 15–25.

[132] P. L. Butzer, R. L. Stens: *Sampling Theory for Not Necessarily Band-Limited Functions—A Historical Overview.* SIAM Review 34 (1992), pp. 40–53.

[133] G. D. Byrne, C. A. Hall (Eds.): *Numerical Solution of Systems of Nonlinear Algebraic Equations.* Academic Press, New York London 1973.

[134] S. Cambanis, E. Masry: *Trapezoidal Stratified Monte Carlo Integration.* SIAM J. Numer. Anal. 29 (1992), pp. 284–301.

[135] R. Carter: *Y-MP Floating-Point and Cholesky Factorization.* International Journal of High Speed Computing 3 (1991), pp. 215–222.

[136] J. Choi, J. J. Dongarra, D. W. Walker: *A Set of Parallel Block Basic Linear Algebra Subprograms.* Technical Report TM-12468, Mathematical Sciences Section, Oak Ridge National Laboratory, 1994.

[137] J. Choi, J. J. Dongarra, D. W. Walker: SCALAPACK I—*Parallel Factorization Routines (LU, QR, and Cholesky).* Technical Report TM-12470, Oak Ridge National Laboratory, Mathematical Sciences Section, 1994.

[138] W. J. Cody: *The* FUNPACK *Package of Special Function Subroutines.* ACM Trans. Math. Softw. 1 (1975), pp. 13–25.

[139] J. W. Cooley, J. W. Tukey: *An Algorithm for the Machine Calculation of Complex Fourier Series.* Math. Comp. 19 (1965), pp. 297–301.

[140] R. Cools: *A Survey of Methods for Constructing Cubature Formulae*, in "Numerical Integration—Recent Developments, Software and Applications" (T. O. Espelid, A. Genz, Eds.). Kluwer, Dordrecht 1992, pp. 1–24.

[141] R. Cools, P. Rabinowitz: *Monomial Cubature Rules Since "Stroud"—A Compilation.* Report TW 161, Department of Computer Science, Katholieke Universiteit Leuven, 1991.

[142] W. A. Coppel: *Stability and Asymptotic Behavior of Differential Equations.* Heath, Boston, 1965.

[143] P. Costantini: *Co-monotone Interpolating Splines of Arbitrary Degree—a Local Approach.* SIAM J. Sci. Stat. Comp. 8 (1987), pp. 1026–1034.

[144] W. R. Cowell (Ed.): *Portability of Mathematical Software.* Lecture Notes in Computer Science, Vol. 57, Springer-Verlag, New York 1977.

[145] M. G. Cox: *The Numerical Evaluation of B-Splines.* J. Inst. Math. Appl. 10 (1972), pp. 134–149.

[146] J. H. Davenport: *On the Integration of Algebraic Functions.* Lecture Notes in Computer Science, Vol. 102, Springer-Verlag, Berlin Heidelberg New York Tokyo 1981.

[147] J. H. Davenport: *Integration—Formal and Numeric Approaches,* in "Tools, Methods and Languages for Scientific and Engineering Computation" (B. Ford, J. C. Rault, F. Thomasset, Eds.). North-Holland, Amsterdam New York Oxford 1984, pp. 417–426.

[148] J. H. Davenport, Y. Siret, E. Tournier: *Computer Algebra—Systems and Algorithms for Algebraic Computation,* 2nd ed. Academic Press, New York 1993.

[149] T. A. Davis, I. S. Duff: *An Unsymmetric-Pattern Multifrontal Method for Sparse LU Factorization.* Technical Report TR-94-038, Computer and Information Science Dept., University of Florida, 1994.

[150] C. de Boor: CADRE—*An Algorithm for Numerical Quadrature,* in "Mathematical Software" (J. R. Rice, Ed.). Academic Press, New York 1971, pp. 417–449.

[151] C. de Boor: *On Calculating with B-Splines.* J. Approx. Theory 6 (1972), pp. 50–62.

[152] C. de Boor, A. Pinkus: *Proof of the Conjecture of Bernstein and Erdös concerning the Optimal Nodes for Polynomial Interpolation.* J. Approx. Theory 24 (1978), pp. 289–303.

[153] E. de Doncker: *Asymptotic Expansions and Their Application in Numerical Integration,* in "Numerical Integration—Recent Developments, Software and Applications" (P. Keast, G. Fairweather, Eds.). Reidel, Dordrecht 1987, pp. 141–151.

[154] T. J. Dekker: *A Floating-point Technique for Extending the Available Precision.* Numer. Math. 18 (1971), pp. 224-242.

[155] J. Demmel, B. Kågström: *Computing Stable Eigendecompositions of Matrix Pencils.* Lin. Alg. Appl. 88/89-4 (1987), pp. 139–186.

[156] J. E. Dennis Jr., J. J. Moré: *Quasi-Newton Methods, Motivation and Theory.* SIAM Review 19 (1977), pp. 46–89.

[157] R. A. De Vore, G. G. Lorentz: *Constructive Approximation.* Springer-Verlag, Berlin Heidelberg, New York Tokyo 1993.

[158] L. Devroye: *Non-Uniform Random Variate Generation.* Springer-Verlag, Berlin Heidelberg New York Tokyo 1986.

[159] D. S. Dodson, R. G. Grimes, J. G. Lewis: *Sparse Extensions to the Fortran Basic Linear Algebra Subprograms.* ACM Trans. Math. Softw. 17 (1991), pp. 253–263, 264–272.

[160] J. J. Dongarra: *The* LINPACK *Benchmark—An Explanation,* in "Evaluating Supercomputers" (A. J. Van der Steen, Ed.). Chapman and Hall, London 1990, pp. 1–21.

[161] J. J. Dongarra, J. Du Croz, I. S. Duff, S. Hammarling: *A Set of Level 3 Basic Linear Algebra Subprograms.* ACM Trans. Math. Softw. 16 (1990), pp. 1–17, 18–28.

[162] J. J. Dongarra: *Performance of Various Computers Using Standard Linear Equations Software.* Technical Report CS-89-85, Computer Science Dept., University of Tennessee, 1994.

[163] J. J. Dongarra, J. Du Croz, S. Hammarling, R. J. Hanson: *An Extended Set of Fortran Basic Linear Algebra Subprograms.* ACM Trans. Math. Softw. 14 (1988), pp. 1–17, 18–32.

[164] J. J. Dongarra, I. S. Duff, D. C. Sorensen, H. A. Van der Vorst: *Solving Linear Systems on Vector and Shared Memory Computers.* SIAM Press, Philadelphia 1991.

[165] J. J. Dongarra, E. Grosse: *Distribution of Mathematical Software via Electronic Mail.* Comm. ACM 30 (1987), pp. 403–407.

[166] J. J. Dongarra, F. G. Gustavson, A. Karp: *Implementing Linear Algebra Algorithms for Dense Matrices on a Vector Pipeline Machine.* SIAM Review 26 (1984), pp. 91–112.

[167] J. J. Dongarra, P. Mayes, G. Radicati: *The IBM RISC System/6000 and Linear Algebra Operations.* Technical Report CS-90-12, Computer Science Dept., University of Tennessee, 1990.

[168] J. J. Dongarra, R. Pozo, D. W. Walker: LAPACK++ *V. 1.0—Users' Guide.* University of Tennessee, Knoxville, 1994.

[169] J. J. Dongarra, R. Pozo, D. W. Walker: LAPACK++—*A Design Overview of Object-Oriented Extensions for High Performance Linear Algebra.* Computer Science Report, University of Tennessee, 1993.

[170] J. J. Dongarra, H. A. Van der Vorst: *Performance of Various Computers Using Standard Sparse Linear Equations Solving Techniques,* in "Computer Benchmarks" (J. J. Dongarra, W. Gentzsch, Eds.). Elsevier, New York 1993, pp. 177–188.

[171] C. C. Douglas, M. Heroux, G. Slishman, R. M. Smith: GEMMW—*A Portable Level 3 BLAS Winograd Variant of Strassen's Matrix-Matrix Multiply Algorithm.* J. Computational Physics 110 (1994), pp. 1–10.

[172] Z. Drezner: *Computation of the Multivariate Normal Integral.* ACM Trans. Math. Softw. 18 (1992), pp. 470–480.

[173] D. Dubois, A. Greenbaum, G. Rodrigue: *Approximating the Inverse of a Matrix for Use in Iterative Algorithms on Vector Processors.* Computing 22 (1979), pp. 257–268.

[174] I. S. Duff, A. Erisman, J. Reid: *Direct Methods for Sparse Matrices,* paperback ed. Oxford University Press, Oxford 1989.

[175] I. S. Duff, R. G. Grimes, J. G. Lewis: *Sparse Matrix Test Problems.* ACM Trans. Math. Softw. 15 (1989), pp. 1–14.

[176] I. S. Duff, R. G. Grimes, J. G. Lewis: *User's Guide for the Harwell-Boeing Sparse Matrix Collection* (Release I). CERFACS-Report TR/PA/92/86, Toulouse, 1992. Available via anonymous-FTP: orion.cerfacs.fr.

[177] R. A. Earnshaw (Ed.): *Fundamental Algorithms for Computer Graphics.* Springer-Verlag, Berlin Heidelberg New York Tokyo 1985 (reprinted 1991).

[178] H. Ekblom: L_p-*Methods for Robust Regression.* BIT 14 (1974), pp. 22–32.

[179] D. F. Elliot, K. R. Rao: *Fast Transforms:—Algorithms, Analyses, Applications.* Academic Press, New York 1982.

[180] T. M. R. Ellis, D. H. McLain: *Algorithm 514—A New Method of Cubic Curve Fitting Using Local Data.* ACM Trans. Math. Softw. 3 (1977), pp. 175–178.

[181] M. P. Epstein: *On the Influence of Parameterization in Parametric Interpolation.* SIAM J. Numer. Anal. 13 (1976), pp. 261–268.

[182] P. Erdös: *Problems and Results on the Theory of Interpolation.* Acta Math. Acad. Sci. Hungar., 12 (1961), pp. 235–244.

[183] P. Erdös, P. Vértesi: *On the Almost Everywhere Divergence of Lagrange Interpolatory Polynomials for Arbitrary Systems of Nodes.* Acta Math. Acad. Sci. Hungar. 36 (1980), pp. 71–89.

[184] T. O. Espelid: DQAINT—*An Algorithm for Adaptive Quadrature (of a Vector Function) over a Collection of Finite Intervals,* in "Numerical Integration—Recent Developments, Software and Applications" (T. O. Espelid, A. Genz, Eds.). Kluwer, Dordrecht 1992, pp. 341–342.

[185] G. Farin: *Splines in CAD/CAM.* Surveys on Mathematics for Industry 1 (1991), pp. 39–73.

[186] H. Faure: *Discrépances de suites associées à un système de numération (en dimension s)*. Acta Arith. 41 (1982), pp. 337–351.

[187] L. Fejér: *Mechanische Quadraturen mit positiven Cotes'schen Zahlen*. Math. Z. 37 (1933), pp. 287–310.

[188] S. I. Feldman, D. M. Gay, M. W. Maimone, N. L. Schryer: *A Fortran-to-C Converter*. Technical Report No. 149, AT&T Bell Laboratories, 1993.

[189] A. Ferscha: *Modellierung und Leistungsanalyse paralleler Systeme mit dem PRM-Netz Modell*. Oldenburg Verlag, München Wien 1995.

[190] R. Fletcher, J. A. Grant, M. D. Hebden: *The Calculation of Linear Best L_p Approximations*. Computer J. 14 (1971), pp. 276–279.

[191] R. Fletcher, C. Reeves: *Function Minimization by Conjugate Gradients*. Computer Journal 7 (1964), pp. 149–154.

[192] T. A. Foley: *Interpolation with Interval and Point Tension Controls Using Cubic Weighted ν-Splines*. ACM Trans. Math. Softw. 13 (1987), pp. 68–96.

[193] B. Ford, F. Chatelin (Eds.): *Problem Solving Environments for Scientific Computing*. North-Holland, Amsterdam 1987.

[194] L. Fox, I. B. Parker: *Chebyshev Polynomials in Numerical Analysis*. Oxford University Press, London 1968.

[195] R. Frank, J. Schneid, C. W. Ueberhuber: *The Concept of B-Convergence*. SIAM J. Numer. Anal. 18 (1981), pp. 753–780.

[196] R. Frank, J. Schneid, C. W. Ueberhuber: *Stability Properties of Implicit Runge-Kutta Methods*. SIAM J. Numer. Anal. 22 (1985), pp. 497–515.

[197] R. Frank, J. Schneid, C. W. Ueberhuber: *Order Results for Implicit Runge-Kutta Methods Applied to Stiff Systems*. SIAM J. Numer. Anal. 22 (1985), pp. 515–534.

[198] R. Franke, G. Nielson: *Smooth Interpolation of Large Sets of Scattered Data*. Int. J. Numer. Methods Eng. 15 (1980), pp. 1691–1704.

[199] R. Freund, G. H. Golub, N. Nachtigal: *Iterative Solution of Linear Systems*. Acta Numerica 1, 1992, pp. 57–100.

[200] R. Freund, N. Nachtigal: *QMR—A Quasi-Minimal Residual Method for Non-Hermitian Linear Systems*. Numer. Math. 60 (1991), pp. 315–339.

[201] R. Freund, N. Nachtigal: *An Implementation of the QMR Method Based on Two Coupled Two-Term Recurrences*. Tech. Report 92.15, RIACS, NASA Ames, 1992.

[202] F. N. Fritsch, J. Butland: *A Method for Constructing Local Monotone Piecewise Cubic Interpolants*. SIAM J. Sci. Stat. Comp. 5 (1984), pp. 300–304.

[203] F. N. Fritsch, R. E. Carlson: *Monotone Piecewise Cubic Interpolation*. SIAM J. Numer. Anal. 17 (1980), pp. 238–246.

[204] F. N. Fritsch, D. K. Kahaner, J. N. Lyness: *Double Integration Using One-Dimensional Adaptive Quadrature Routines—a Software Interface Problem*. ACM Trans. Math. Softw. 7 (1981), pp. 46–75.

[205] P. W. Gaffney, J. W. Wooten, K. A. Kessel, W. R. McKinney: NITPACK—*An Interactive Tree Package*. ACM Trans. Math. Softw. 9 (1983), pp. 395–417.

[206] E. Gallopoulos, E. N. Houstis, J. R. Rice: *Problem Solving Environments for Computational Science*. Computational Science and Engineering Nr. 2 Vol. 1 (1994), pp. 11–23.

[207] K. O. Geddes: *Algorithms for Computer Algebra*. Kluwer, Dordrecht 1992.

[208] J. D. Gee, M. D. Hill, D. Pnevmatikatos, A. J. Smith: *Cache Performance of the SPEC92 Benchmark Suite*. IEEE Micro 13 (1993), pp. 17–27.

[209] W. M. Gentleman: *Implementing Clenshaw-Curtis Quadrature*. Comm. ACM 15 (1972), pp. 337–342, 343–346.

[210] A. Genz: *Statistics Applications of Subregion Adaptive Multiple Numerical Integration*, in "Numerical Integration—Recent Developments, Software and Applications" (T. O. Espelid, A. Genz, Eds.). Kluwer, Dordrecht 1992, pp. 267–280.

[211] D. Goldberg: *What Every Computer Scientist Should Know About Floating-Point Arithmetic.* ACM Computing Surveys 23 (1991), pp. 5–48.

[212] G. H. Golub, V. Pereyra: *Differentiation of Pseudo-Inverses and Nonlinear Least Squares Problems Whose Variables Separate.* SIAM J. Numer. Anal. 10 (1973), pp. 413–432.

[213] A. Greenbaum, J. J. Dongarra: *Experiments with QL/QR Methods for the Symmetric Tridiagonal Eigenproblem.* Technical Report CS-89-92, Computer Science Dept., University of Tennessee, 1989.

[214] E. Grosse: *A Catalogue of Algorithms for Approximation,* in "Algorithms for Approximation II" (J. C. Mason, M. G. Cox, Eds.). Chapman and Hall, London New York 1990, pp. 479–514.

[215] M. H. Gutknecht: *Variants of Bi-CGSTAB for Matrices with Complex Spectrum.* Tech. Report 91-14, IPS ETH, Zürich 1991.

[216] S. Haber: *A Modified Monte Carlo Quadrature.* Math. Comp. 20 (1966), pp. 361–368.

[217] S. Haber: *A Modified Monte Carlo Quadrature II.* Math. Comp. 21 (1967), pp. 388–397.

[218] H. Hancock: *Elliptic Integrals.* Dover Publication, New York 1917.

[219] J. Handy: *The Cache Memory Book.* Academic Press, San Diego 1993.

[220] J. G. Hayes: *The Optimal Hull Form Parameters.* Proc. NATO Seminar on Numerical Methods Applied to Ship Building, Oslo 1964.

[221] J. G. Hayes: *Numerical Approximation to Functions and Data.* Athlone Press, London 1970.

[222] N. Higham: *Efficient Algorithms for Computing the Condition Number of a Tridiagonal Matrix.* SIAM J. Sci. Stat. Comput. 7 (1986), pp. 82–109.

[223] N. Higham: *A Survey of Condition Number Estimates for Triangular Matrices.* SIAM Review 29 (1987), pp. 575–596.

[224] N. Higham: *Fortran 77 Codes for Estimating the One-Norm of a Real or Complex Matrix, with Applications to Condition Estimation.* ACM Trans. Math. Softw. 14 (1988), pp. 381–396.

[225] N. Higham: *The Accuracy of Floating-Point Summation.* SIAM J. Sci. Comput. 14 (1993), pp. 783-799.

[226] D. R. Hill, C. B. Moler: *Experiments in Computational Matrix Algebra.* Birkhäuser, Basel 1988.

[227] E. Hlawka: *Funktionen von beschränkter Variation in der Theorie der Gleichverteilung.* Ann. Math. Pur. Appl. 54 (1961), pp. 325–333.

[228] R. W. Hockney, C. R. Jesshope: *Parallel Computers 2.* Adam Hilger, Bristol 1988.

[229] A. S. Householder: *The Numerical Treatment of a Single Nonlinear Equation.* McGraw-Hill, New York 1970.

[230] E. N. Houstis, J. R. Rice, T. Papatheodorou: PARALLEL ELLPACK—*An Expert System for Parallel Processing of Partial Differential Equations.* Purdue University, Report CSD-TR-831, 1988.

[231] E. N. Houstis, J. R. Rice, R. Vichnevetsky (Eds.): *Intelligent Mathematical Software Systems.* North-Holland, Amsterdam 1990.

[232] L. K. Hua, Y. Wang: *Applications of Number Theory to Numerical Analysis.* Springer-Verlag, Berlin Heidelberg New York Tokyo 1981.

[233] P. J. Huber: *Robust Regression—Asymptotics, Conjectures and Monte Carlo.* Annals of Statistics 1 (1973), pp. 799–821.

[234] P. J. Huber: *Robust Statistics.* Wiley, New York 1981.

[235] J. M. Hyman: *Accurate Monotonicity Preserving Cubic Interpolation.* SIAM J. on Scientific and Statistical Computation 4 (1983), pp. 645–654.

[236] J. P. Imhof: *On the Method for Numerical Integration of Clenshaw and Curtis.* Numer. Math. 5 (1963), pp. 138–141.

[237] M. Iri, S. Moriguti, Y. Takasawa: *On a Certain Quadrature Formula* (japan.), Kokyuroku of the Research Institute for Mathematical Sciences, Kyoto University, 91 (1970), pp. 82–118.

[238] L. D. Irvine, S. P. Marin, P. W. Smith: *Constrained Interpolation and Smoothing*. Constructive Approximation 2 (1986) pp. 129–151.

[239] ISO/IEC DIS 10967-1:1994: *International Standard—Information Technology—Language Independent Arithmetic—Part 1—Integer and Floating Point Arithmetic.* 1994.

[240] R. Jain: *The Art of Computer Systems Performance Analysis—Techniques for Experimental Design, Measurement and Simulation.* Wiley, New York 1991.

[241] M. A. Jenkins: *Algorithm 493—Zeroes of a Real Polynomial.* ACM Trans. Math. Softw. 1 (1975), pp. 178–189.

[242] M. A. Jenkins, J. F. Traub: *A Three-Stage Algorithm for Real Polynomials Using Quadratic Iteration.* SIAM J. Numer. Anal. 7 (1970), pp. 545–566.

[243] A. J. Jerri: *The Shannon Sampling—its Various Extensions and Applications—a Tutorial Review.* Proc. IEEE 65 (1977), pp. 1565–1596.

[244] S. Joe, I. H. Sloan: *Imbedded Lattice Rules for Multidimensional Integration.* SIAM J. Numer. Anal. 29 (1992), pp. 1119–1135.

[245] D. S. Johnson, M. R. Garey: *A 71/60 Theorem for Bin Packing.* J. Complexity 1 (1985), pp. 65–106.

[246] D. W. Juedes: *A Taxonomy of Automatic Differentiation Tools*, in "Automatic Differentiation of Algorithms—Theory, Implementation and Application" (A. Griewank, F. Corliss, Eds.). SIAM Press, Philadelphia 1991, pp. 315–329.

[247] D. K. Kahaner: *Numerical Quadrature by the ε-Algorithm.* Math. Comp. 26 (1972), pp. 689–693.

[248] N. Karmarkar, R. M. Karp: *An Efficient Approximation Scheme for the One Dimensional Bin Packing Problem.* 23rd Annu. Symp. Found. Comput. Sci., IEEE Computer Society, 1982, pp. 312–320.

[249] L. Kaufmann: *A Variable Projection Method for Solving Separable Nonlinear Least Squares Problems.* BIT 15 (1975), pp. 49–57.

[250] G. Kedem, S. K. Zaremba: *A Table of Good Lattice Points in Three Dimensions.* Numer. Math. 23 (1974), pp. 175–180.

[251] H. L. Keng, W. Yuan: *Applications of Number Theory to Numerical Analysis.* Springer-Verlag, Berlin Heidelberg New York Tokyo 1981.

[252] T. King: *Dynamic Data Structures—Theory and Application.* Academic Press, San Diego 1992.

[253] M. Klerer, F. Grossman: *Error Rates in Tables of Indefinite Intergrals.* Indust. Math. 18 (1968), pp. 31–62.

[254] D. E. Knuth: *The Art of Computer Programming.* Vol. 2—*Seminumerical Algorithms*, 2nd ed. Addison-Wesley, Reading 1981.

[255] P. Kogge: *The Architecture of Pipelined Computers.* McGraw-Hill, New York 1981.

[256] A. R. Krommer, C. W. Ueberhuber: *Architecture Adaptive Algorithms.* Parallel Computing 19 (1993), pp. 409–435.

[257] A. R. Krommer, C. W. Ueberhuber: *Lattice Rules for High-Dimensional Integration.* Technical Report SciPaC/TR 93-3, Scientific Parallel Computation Group, Technical University Vienna, Wien 1993.

[258] A. R. Krommer, C. W. Ueberhuber: *Numerical Integration on Advanced Computer Systems.* Lecture Notes in Computer Science, Vol. 848, Springer-Verlag, Berlin Heidelberg New York Tokyo 1994.

[259] A. S. Kronrod: *Nodes and Weights of Quadrature Formulas.* Consultants Bureau, New York 1965.

[260] V. I. Krylov: *Approximate Calculation of Integrals.* Macmillan, New York London 1962.

[261] U. W. Kulisch, W. L. Miranker: *The Arithmetic of the Digital Computer—A New Approach.* SIAM Review 28 (1986), pp. 1–40.

[262] U. W. Kulisch, W. L. Miranker: *Computer Arithmetic in Theory and Practice.* Academic Press, New York 1981.

[263] J. Laderman, V. Pan, X.-H. Sha: *On Practical Acceleration of Matrix Multiplication.* Linear Algebra Appl. 162–164 (1992), pp. 557–588.

[264] M. S. Lam, E. E. Rothberg, M. E. Wolf: *The Cache Performance and Optimizations of Blocked Algorithms.* Computer Architecture News 21 (1993), pp. 63–74.

[265] C. Lanczos: *Discourse on Fourier Series.* Oliver and Boyd, Edinburgh London 1966.

[266] C. L. Lawson, R. J. Hanson, D. Kincaid, F. T. Krogh: *Basic Linear Algebra Subprograms for Fortran Usage.* ACM Trans. Math. Softw. 5 (1979), pp. 308–323.

[267] A. R. Lebeck, D. A. Wood: *Cache Profiling and the SPEC Benchmarks—A Case Study.* IEEE Computer, October 1994, pp. 15–26.

[268] P. Ling: *A Set of High Performance Level 3 BLAS Structured and Tuned for the IBM 3090 VF and Implemented in Fortran 77.* Journal of Supercomputing 7 (1993), pp. 323–355.

[269] P. R. Lipow, F. Stenger: *How Slowly Can Quadrature Formulas Converge.* Math. Comp. 26 (1972), pp. 917–922.

[270] D. B. Loveman: *High Performance Fortran.* IEEE Parallel and Distributed Technology 2 (1993), pp. 25–42.

[271] J. Lund, K. L. Bowers: *Sinc Methods for Quadrature and Differential Equations.* SIAM Press, Philadelphia 1992.

[272] T. Lyche: *Discrete Cubic Spline Interpolation.* BIT 16 (1976), pp. 281–290.

[273] J. N. Lyness: *An Introduction to Lattice Rules and their Generator Matrices.* IMA J. Numer. Anal. 9 (1989), pp. 405–419.

[274] J. N. Lyness, J. J. Kaganove: *Comments on the Nature of Automatic Quadrature Routines.* ACM Trans. Math. Softw. 2 (1976), pp. 65–81.

[275] J. N. Lyness, B. W. Ninham: *Numerical Quadrature and Asymptotic Expansions.* Math. Comp. 21 (1967), pp. 162–178.

[276] J. N. Lyness, I. H. Sloan: *Some Properties of Rank-2 Lattice Rules.* Math. Comp. 53 (1989), pp. 627–637.

[277] J. N. Lyness, T. Soerevik: *A Search Program for Finding Optimal Integration Lattices.* Computing 47 (1991), pp. 103–120.

[278] J. N. Lyness, T. Soerevik: *An Algorithm for Finding Optimal Integration Lattices of Composite Order.* BIT 32 (1992), pp. 665–675.

[279] T. Macdonald: *C for Numerical Computing.* J. Supercomput. 5 (1991), pp. 31–48.

[280] D. Maisonneuve: *Recherche et utilisation des "bons treillis",* in "Applications of Number Theory to Numerical Analysis" (S. K. Zaremba, Ed.). Academic Press, New York 1972, pp. 121–201.

[281] M. Malcolm, R. Simpson: *Local Versus Global Strategies for Adaptive Quadrature.* ACM Trans. Math. Softw. 1 (1975), pp. 129–146.

[282] T. Manteuffel: *The Tchebychev Iteration for Nonsymmetric Linear Systems.* Numer. Math. 28 (1977), pp. 307–327.

[283] D. W. Marquardt: *An Algorithm for Least Squares Estimation of Nonlinear Parameters.* J. SIAM 11 (1963), pp. 431–441.

[284] G. Marsaglia: *Normal (Gaussian) Random Variables for Supercomputers.* J. Supercomput. 5 (1991), pp. 49–55.

[285] J. C. Mason, M. G. Cox: *Scientific Software Systems.* Chapman and Hall, London New York 1990.

[286] E. Masry, S. Cambanis: *Trapezoidal Monte Carlo Integration.* SIAM J. Numer. Anal. 27 (1990), pp. 225–246.

[287] E. W. Mayr: *Theoretical Aspects of Parallel Computation*, in "VLSI and Parallel Computation" (R. Suaya, G. Birtwistle, Eds.). Morgan Kaufmann, San Mateo 1990, pp. 85–139.

[288] G. P. McKeown: *Iterated Interpolation Using a Systolic Array*. ACM Trans. Math. Softw. 12 (1986), pp. 162–170.

[289] J. Meijerink, H. A. Van der Vorst: *An Iterative Solution Method for Linear Systems of Which the Coefficient Matrix is a Symmetric M-matrix*. Math. Comp. 31 (1977), pp. 148–162.

[290] R. Melhem: *Toward Efficient Implementation of Preconditioned Conjugate Gradient Methods on Vector Supercomputers*. Internat. J. Supercomp. Appl. 1 (1987), pp. 77–98.

[291] J. P. Mesirov (Ed.): *Very Large Scale Computation in the 21st Century*. SIAM Press, Philadelphia 1991.

[292] W. F. Mitchell: *Optimal Multilevel Iterative Methods for Adaptive Grids*. SIAM J. Sci. Statist. Comput. 13 (1992), pp. 146–167.

[293] J. J. Moré, M. Y. Cosnard: *Numerical Solution of Nonlinear Equations*. ACM Trans. Math. Softw. 5 (1979), pp. 64–85.

[294] D. E. Müller: *A Method for Solving Algebraic Equations Using an Automatic Computer*. Math. Tables Aids Comput. 10 (1956), pp. 208–215.

[295] N. Nachtigal, S. Reddy, L. Trefethen: *How Fast are Nonsymmetric Matrix Iterations?* SIAM J. Mat. Anal. Appl. 13 (1992), pp. 778–795.

[296] P. Naur: *Machine Dependent Programming in Common Languages*. BIT 7 (1967), pp. 123–131.

[297] J. A. Nelder, R. Mead: *A Simplex Method for Function Minimization*. Computer Journal 7 (1965), pp. 308–313.

[298] H. Niederreiter: *Quasi-Monte Carlo Methods and Pseudorandom Numbers*. Bull. Amer. Math. Soc. 84 (1978), pp. 957–1041.

[299] H. Niederreiter: *Random Number Generation and Quasi-Monte Carlo Methods*. SIAM Press, Philadelphia 1992.

[300] G. Nielson: *Some Piecewise Polynomial Alternatives to Splines Under Tension*, in "Computer Aided Geometric Design" (R. E. Barnhill, R. F. Riesenfeld, Eds.). Academic Press, New York San Francisco London 1974.

[301] G. Nielson, B. D. Shriver: *Visualization in Scientific Computing*. IEEE Press, Los Alamitos 1990.

[302] H. J. Nussbaumer: *Fast Fourier Transform and Convolution Algorithms*. Springer-Verlag, Berlin Heidelberg New York Tokyo 1981 (reprinted 1990).

[303] D. P. O'Leary, O. Widlund: *Capacitance Matrix Methods for the Helmholtz Equation on General 3-Dimensional Regions*. Math. Comp. 33 (1979), pp. 849–880.

[304] T. I. Ören: *Concepts for Advanced Computer Assisted Modeling*, in "Methodology in Systems Modeling and Simulation" (B. P. Zeigler, M. S. Elzas, G. J. Klir, T. I. Ören, Eds.). North-Holland, Amsterdam New York Oxford 1979.

[305] J. M. Ortega: *Numerical Analysis—A Second Course*. SIAM Press, Philadelphia 1990.

[306] A. M. Ostrowski: *On Two Problems in Abstract Algebra Connected with Horner's Rule*. Studies in Math. and Mech. presented to Richard von Mises, Academic Press, New York 1954, pp. 40–68.

[307] C. C. Page, M. A. Saunders: *LSQR: An Algorithm for Sparse Linear Equations and Sparse Least-Squares*. ACM Trans. Math. Software 8 (1982), pp. 43–71.

[308] V. Pan: *Methods of Computing Values of Polynomials*. Russian Math. Surveys 21 (1966), pp. 105–136.

[309] V. Pan: *How Can We Speed Up Matrix Multiplication?* SIAM Rev. 26 (1984), pp. 393–415.

[310] V. Pan: *Complexity of Computations with Matrices and Polynomials.* SIAM Rev. 34 (1992), pp. 225–262.

[311] B. N. Parlett: *The Symmetric Eigenvalue Problem.* Prentice Hall, Englewood Cliffs 1980.

[312] T. N. L. Patterson: *The Optimum Addition of Points to Quadrature Formulae.* Math. Comp. 22 (1968), pp. 847–856.

[313] J. L. Peterson: *Petri Net Theory and the Modeling of Systems.* Prentice Hall, Englewood Cliffs 1981.

[314] R. Piessens: *Modified Clenshaw-Curtis Integration and Applications to Numerical Computation of Integral Transforms,* in "Numerical Integration—Recent Developments, Software and Applications" (P. Keast, G. Fairweather, Eds.). Reidel, Dordrecht 1987, pp. 35–41.

[315] R. Piessens, M. Branders: *A Note on the Optimal Addition of Abscissas to Quadrature Formulas of Gauss and Lobatto Type.* Math. Comp. 28 (1974), pp. 135–140, 344–347.

[316] D. R. Powell, J. R. Macdonald: *A Rapidly Converging Iterative Method for the Solution of the Generalized Nonlinear Least Squares Problem.* Computer J. 15 (1972), pp. 148–155.

[317] M. J. D. Powell: *A Hybrid Method for Nonlinear Equations,* in "Numerical Methods for Nonlinear Algebraic Equations" (P. Rabinowitz, Ed.). Gordon and Breach, London 1970.

[318] M. J. D. Powell, P. L. Toint: *On the Estimation of Sparse Hessian Matrices.* SIAM J. Numer. Anal. 16 (1979), pp. 1060–1074.

[319] J. G. Proakis, D. G. Manolakis: *Digital Signal Processing,* 3rd ed. Macmillan, New York 1995.

[320] J. S. Quarterman, S. Carl-Mitchell: *The Internet Connection—System Connectivity and Configuration.* Addison-Wesley, Reading 1994.

[321] R. J. Renka: *Multivariate Interpolation of Large Sets of Scattered Data.* ACM Trans. Math. Softw. 14 (1988), pp. 139–148.

[322] R. J. Renka, A. K. Cline: *A Triangle-Based C^1 Interpolation Method.* Rocky Mt. J. Math. 14 (1984), pp. 223–237.

[323] R. F. Reisenfeld: *Homogeneous Coordinates and Projective Planes in Computer Graphics.* IEEE Computer Graphics and Applications 1 (1981), pp. 50–56.

[324] W. C. Rheinboldt: *Numerical Analysis of Parametrized Nonlinear Equations.* Wiley, New York 1986.

[325] J. R. Rice: *Parallel Algorithms for Adaptive Quadrature II—Metalgorithm Correctness.* Acta Informat. 5 (1975), pp. 273–285.

[326] J. R. Rice (Ed.): *Mathematical Aspects of Scientific Software.* Springer-Verlag, Berlin Heidelberg New York Tokyo 1988.

[327] A. Riddle: *Mathematical Power Tools.* IEEE Spectrum Nov. 1994, pp. 35–47.

[328] R. Rivest: *Cryptography* in "Handbook of Theoretical Computer Science" (J. van Leeuwen, Ed.). North Holland, Amsterdam, 1990.

[329] T. J. Rivlin: *The Chebyshev Polynomials,* 2nd ed. Wiley, New York 1990.

[330] Y. Robert: *The Impact of Vector and Parallel Architectures on the Gaussian Elimination Algorithm.* Manchester University Press, New York Brisbane Toronto 1991.

[331] M. Rosenlicht: *Integration in Finite Terms.* Amer. Math. Monthly 79 (1972), pp. 963–972.

[332] A. Ruhe: *Fitting Empirical Data by Positive Sums of Exponentials.* SIAM J. Sci. Stat. Comp. 1 (1980), pp. 481–498.

[333] C. Runge: *Über empirische Funktionen und die Interpolation zwischen äquidistanten Ordinaten.* Z. Math. u. Physik 46 (1901), pp. 224–243.

[334] Y. Saad: *Preconditioning Techniques for Indefinite and Nonsymmetric Linear Systems.* J. Comput. Appl. Math. 24 (1988), pp. 89–105.

[335] Y. Saad: *Krylov Subspace Methods on Supercomputers*. SIAM J. Sci. Statist. Comput. 10 (1989), pp. 1200–1232.

[336] Y. Saad: SPARSKIT—*A Basic Tool Kit for Sparse Matrix Computation*. Tech. Report CSRD TR 1029, CSRD, University of Illinois, Urbana 1990.

[337] Y. Saad, M. Schultz: *GMRES—A Generalized Minimal Residual Algorithm for Solving Nonsymmetric Linear Systems*. SIAM J. Sci. Statist. Comput. 7 (1986), pp. 856–869.

[338] T. W. Sag, G. Szekeres: *Numerical Evaluation of High-Dimensional Integrals*. Math. Comp. 18 (1964), pp. 245–253.

[339] K. Salkauskas, C^1 *Splines for Interpolation of Rapidly Varying Data*. Rocky Mt. J. Math. 14 (1984), pp. 239–250.

[340] R. Salmon, M. Slater: *Computer Graphics—Systems and Concepts*. Addison-Wesley, Wokingham 1987.

[341] W. M. Schmidt: *Irregularities of Distribution*. Acta Arith. 21 (1972), pp. 45–50.

[342] D. G. Schweikert: *An Interpolation Curve Using a Spline in Tension*. J. Math. & Physics 45 (1966), pp. 312–317.

[343] T. I. Seidman, R. J. Korsan: *Endpoint Formulas for Interpolatory Cubic Splines*. Math. Comp. 26 (1972), pp. 897–900.

[344] Z. Sekera: *Vectorization and Parallelization on High Performance Computers*. Computer Physics Communications 73 (1992), pp. 113–138.

[345] S. Selberherr: *Analysis and Simulation of Semiconductor Devices*. Springer-Verlag, Berlin Heidelberg New York Tokyo 1984.

[346] D. Shanks: *Non-linear Transformation of Divergent and Slowly Convergent Sequences*. J. Math. Phys. 34 (1955), pp. 1–42.

[347] A. H. Sherman: *Algorithms for Sparse Gauss Elimination with Partial Pivoting*. ACM Trans. Math. Softw. 4 (1978), pp. 330–338.

[348] L. L. Shumaker: *On Shape Preserving Quadratic Spline Interpolation*. SIAM J. Numer. Anal. 20 (1983), pp. 854–864.

[349] K. Sikorski: *Bisection is Optimal*. Numer. Math. 40 (1982), pp. 111–117.

[350] I. H. Sloan: *Numerical Integration in High Dimensions—The Lattice Rule Approach*, in "Numerical Integration—Recent Developments, Software and Applications" (T. O. Espelid, A. Genz, Eds.). Kluwer, Dordrecht 1992, pp. 55–69.

[351] I. H. Sloan, S. Joe: *Lattice Methods for Multiple Integration*. Clarendon Press, Oxford 1994.

[352] I. H. Sloan, P. J. Kachoyan: *Lattice Methods for Multiple Integration—Theory, Error Analysis and Examples*. SIAM J. Numer. Anal. 24 (1987), pp. 116–128.

[353] D. M. Smith: *A Fortran Package for Floating-Point Multiple-Precision Arithmetic*. ACM Trans. Math. Softw. 17 (1991), pp. 273–283.

[354] B. T. Smith, J. M. Boyle, J. J. Dongarra, B. S. Garbow, Y. Ikebe, V. C. Klema, C. B. Moler: *Matrix Eigensystem Routines—EISPACK Guide*, 2nd ed. Springer-Verlag, Berlin Heidelberg New York Tokyo 1976 (reprinted 1988).

[355] I. M. Sobol: *The Distribution of Points in a Cube and the Approximate Evaluation of Integrals*. Zh. Vychisl. Mat. i Math. Fiz. 7 (1967), pp. 784–802.

[356] P. Sonneveld: *CGS, a Fast Lanczos-type Solver for Nonsymmetric Linear Systems*. SIAM J. Sci. Statist. Comput. 10 (1989), pp. 36–52.

[357] D. C. Sorensen: *Newton's Method with a Model Trust Region Modification*. SIAM J. Numer. Anal. 19 (1982), pp. 409–426.

[358] W. Stegmüller: *Unvollständigkeit und Unbeweisbarkeit* (2. Aufl.). Springer Verlag, Berlin Heidelberg New York, Tokyo, 1970.

[359] G. W. Stewart: *On the Sensitivity of the Eigenvalue Problem $Ax = \lambda Bx$*. SIAM J. Num. Anal. 9-4 (1972), pp. 669–686.

[360] G. W. Stewart: *Error and Perturbation Bounds for Subspaces Associated with Certain Eigenvalue Problems.* SIAM Review 15-10 (1973), pp. 727–764.

[361] V. Strassen: *Gaussian Elimination Is not Optimal.* Numer. Math. 13 (1969), pp. 354–356.

[362] A. H. Stroud: *Approximate Calculation of Multiple Integrals.* Prentice-Hall, Englewood Cliffs 1971.

[363] E. E. Swartzlander (Ed.): *Computer Arithmetic—I, II.* IEEE Computer Society Press, Los Alamitos 1990.

[364] G. Tomas, C. W. Ueberhuber: *Visualization of Scientific Parallel Programs.* Lecture Notes in Computer Science, Vol. 771, Springer-Verlag, Berlin Heidelberg New York Tokyo 1994.

[365] J. F. Traub: *Complexity of Approximately Solved Problems.* J. Complexity 1 (1985), pp. 3–10.

[366] J. F. Traub, H. Wozniakowski: *A General Theory of Optimal Algorithms.* Academic Press, New York 1980.

[367] J. F. Traub, H. Wozniakowski: *Information and Computation,* in "Advances in Computers, Vol. 23" (M. C. Yovits, Ed.). Academic Press, New York London 1984, pp. 35–92.

[368] J. F. Traub, H. Wozniakowski: *On the Optimal Solution of Large Linear Systems.* J. Assoc. Comput. Mach. 31 (1984), pp. 545–559.

[369] A. Van der Sluis, H. A. Van der Vorst: *The Rate of Convergence of Conjugate Gradients.* Numer. Math. 48 (1986) pp. 543–560.

[370] H. A. Van der Vorst: *Bi-CGSTAB—A Fast and Smoothly Converging Variant of Bi-CG for the Solution of Nonsymmetric Linear Systems.* SIAM J. Sci. Statist. Comput. 13 (1992), pp. 631–644.

[371] S. Van Huffel, J. Vandewalle: *The Total Least Square Problem—Computational Aspects and Analysis.* SIAM Press, Philadelphia 1991.

[372] C. Van Loan: *Computational Framework for the Fast Fourier Transform.* SIAM Press, Philadelphia 1992.

[373] G. W. Wasilkowski: *Average Case Optimality.* J. Complexity 1 (1985), pp. 107–117.

[374] G. W. Wasilkowski, F. Gao: *On the Power of Adaptive Information for Functions with Singularities.* Math. Comp. 58 (1992), pp. 285–304.

[375] A. B. Watson: *Image Compression Using the Discrete Cosine Transform.* Mathematica Journal 4 (1994), Issue 1, pp. 81–88.

[376] L. T. Watson, S. C. Billups, A. P. Morgan: HOMPACK—*A Suite of Codes for Globally Convergent Homotopy Algorithms.* ACM Trans. Math. Softw. 13 (1987), pp. 281–310.

[377] P.-Å. Wedin: *Perturbation Theory for Pseudo-Inverses.* BIT 13 (1973), pp. 217–232.

[378] R. P. Weicker: *Dhrystone—A Synthetic Systems Programming Benchmark.* Commun. ACM 27-10 (1984), pp. 1013–1030.

[379] S. Weiss, J. E. Smith: *POWER and PowerPC.* Morgan Kaufmann, San Francisco 1994.

[380] R. C. Whaley: *Basic Linear Algebra Communication Subprograms—Analysis and Implementation Across Multiple Parallel Architectures.* LAPACK Working Note 73, Technical Report, University of Tennessee, 1994.

[381] J. H. Wilkinson: *Rounding Errors in Algebraic Processes.* Prentice-Hall, Englewood Cliffs 1963 (reprinted 1994 by Dover Publications).

[382] J. H. Wilkinson: *Kronecker's Canonical Form and the QZ Algorithm.* Lin. Alg. Appl. 28 (1979), pp. 285–303.

[383] H. Wozniakowski: *A Survey of Information-Based Complexity.* J. Complexity 1 (1985), pp. 11–44.

[384] P. Wynn: *On a Device for Computing the $e_m(S_n)$ Transformation.* Mathematical Tables and Aids to Computing 10 (1956), pp. 91–96.

[385] P. Wynn: *On the Convergence and Stability of the Epsilon Algorithm.* SIAM J. Numer. Anal. 3 (1966), pp. 91–122.

[386] F. Ziegler: *Mechanics of Solids and Fluids*, 2nd ed. Springer-Verlag, Berlin Heidelberg New York Tokyo 1994.

[387] A. Zygmund: *Trigonometric Series—I, II*, 2nd ed. Cambridge University Press, Cambridge 1988.

Author Index

Subject Index

Springer-Verlag, P. O. Box 31 13 40, D-10643 Berlin, Germany.

Springer Series in Computational Mathematics

Volume 24

H.-G. Roos, M. Stynes, L. Tobiska

Numerical Methods for Singularly Perturbed Differential Equations

Convection-Diffusion and Flow Problems

1996. XVI, 348 pages.
Hardcover DM 148,-
ISBN 3-540-60718-8

This book collects, explains and analyses basic methods and recent
results for the successful numerical
solution of singularly perturbed differential equations. Such equations
model many physical phenomena
and their solutions are characterized
by the presence of layers. The book is
a wide-ranging introduction to the
exciting current literature in this
area. It concentrates on linear convection-diffusion equations and related nonlinear flow problems, encom-
passing both ordinary and partial
differential equations. While many
numerical methods are considered,
particular attention is paid to those
with realistic error estimates. The
book provides a solid and thorough
foundation for the numerical analysis and solution of singular perturbation problems.

Volume 23

A. Quarteroni, A. Valli

Numerical Approxima-tion of Partial Dif-ferential Equations

1994. XVI, 543 pages. 59 figures,
17 tables.
Hardcover DM 128,-
ISBN 3-540-57111-6

This book deals with the numerical
approximation of partial differential
equations. Its scope is to provide a
thorough illustration of numerical
methods, carry out their stability
and convergence analysis, derive
error bounds, and discuss the algorithmic aspects relative to their
implementation. A sound balancing
of theoretical analysis, description
of algorithms and discussion of
applications is one of its main
features. Many kinds of problems
are addressed. A comprehensive
theory of Galerkin method and its
variants, as well as that of collocation
methods, are developed for the spatial discretization. These theories are
then specified to two numeral subspace realizations of remarkable interest: the finite element method and
the spectral method.

Please order by
Fax: +49-30-827 87-301
e-mail: orders@springer.de
or through your bookseller

Springer

Springer-Verlag, P. O. Box 31 13 40, D-10643 Berlin, Germany.